# Insect Molecular Genetics

# Insect Molecular Genetics

## An Introduction to Principles and Applications

## Marjorie A. Hoy

Entomology and Nematology Department
Institute of Food and Agricultural Sciences
University of Florida
Gainesville, Florida

## Academic Press

San Diego   New York   Boston   London   Sydney   Tokyo   Toronto

Copyright © 1994 by ACADEMIC PRESS, INC.
All Rights Reserved.
No part of this publication may be reproduced or transmitted in any form or by any
means, electronic or mechanical, including photocopy, recording, or any information
storage and retrieval system, without permission in writing from the publisher.

Academic Press, Inc.
A Division of Harcourt Brace & Company
525 B Street, Suite 1900, San Diego, California 92101-4495

*United Kingdom Edition published by*
Academic Press Limited
24-28 Oval Road, London NW1 7DX

Library of Congress Cataloging-in-Publication Data

Hoy, Marjorie A.
    Insect molecular genetics / by Marjorie A. Hoy.
        p.  cm.
    Includes bibliographical references and index.
    ISBN 0-12-357490-0
    1. Insects--Molecular genetics.  I. Title.
QL493.H69   1994
  595.7'87328--dc20                         94-7475
                                    CIP

PRINTED IN THE UNITED STATES OF AMERICA
94  95  96  97  98  99  BB  9  8  7  6  5  4  3  2  1

# Contents

# Chapter 3
## Transcription, Translation, and Regulation of Eukaryotic DNA

# Chapter 4
## Chromosomal and Extrachromosomal Organization of DNA in Insects

## Chapter 5
## Genes, Genome Organization, and Development in Insects

# Part II　Molecular Genetic Techniques

## Chapter 6
## Some Basic Tools: How to Cut, Paste, Copy, Measure, and Visualize DNA

# Chapter 7
# Cloning and Expression Vectors, Libraries, and Their Screening

## Chapter 8
## DNA Sequencing and Genome Analysis

## Chapter 9
## DNA Amplification by the Polymerase Chain Reaction: Molecular Biology Made Accessible

# Chapter 10
## P Elements and P-Element Vectors for Transforming *Drosophila*

# Part III  Applications to Entomological Problems

# Chapter 11
## Sex Determination in Insects

## Chapter 12
## Molecular Genetics of Insect Behavior

# Chapter 13
## Insect Molecular Systematics and Evolution

## Chapter 14
## Insect Population Ecology and Molecular Genetics

## Chapter 15
## Transgenic Pest and Beneficial Arthropods for Pest Management Programs

# Preface

The development of recombinant DNA techniques during the past 20 years has resulted in exciting advances in the detailed study of specific genes at the molecular level as well as breakthroughs in molecular, cellular, and developmental biology. Of the molecular genetics studies conducted on insects, most have been directed to *Drosophila melanogaster*. Relatively few data have been generated by molecular biological methods from analyses of other insects. Yet, the application of molecular genetics to insects other than *Drosophila* has the potential to revolutionize insect population and organismal biology.

Why have molecular genetic techniques been used so little by entomologists? There may be a number of reasons. Recombinant DNA techniques are most readily carried out by people trained in biochemistry and relatively few entomologists are so trained. The techniques have been, until recently, relatively complex and difficult, so that strong technical skills were required. Also, most entomologists have been slow to ask whether these techniques were appropriate for studies of population or organismal biology because much of the published literature has focused on fundamental issues of *Drosophila* gene structure, regulation and function, developmental regulation, and evolution.

## Goals

My goal is to introduce entomologists to the concepts of molecular genetics without assuming that they have received previous training in molecular biology. This book is not intended to substitute for formal training in biochemistry or molecular genetics. If novice readers wish to develop molecular genetics skills, they must obtain additional training in genetics and biochemistry. However, the book will provide an introduction to terminology, as well as an overview of principles, techniques, and possible applications of molecular genetics to problems of interest to entomologists.

In preference to using examples from the *Drosophila* literature, I have used examples in which other arthropods have been studied. However, without doubt, *Drosophila* is the premier model for insect molecular genetics study. One fond hope is that this book will be a bridge for entomologists seeking to apply the exciting methods developed for *Drosophila* and that it will introduce *Drosophila* workers to some of the problems and issues of interest to entomologists seeking to solve applied problems. Perhaps this book will help to break down the barriers between entomologists and *Drosophila* workers isolated from each other by perspective and technical jargon. If this book helps to achieve these goals, it will have served its purpose.

## Organization

The book was designed for a one-semester course in insect molecular genetics for upper division undergraduates or beginning graduate students. The initial portion of the book reviews basic information about DNA, RNA, and other important molecules (Chapters 1–4). Readers with a recent course in genetics could skip this section. Chapter 5 describes the genetic systems found in insects and gives an overview of development sufficient to understand subsequent topics such as P-element-mediated transformation and sex determination. Chapters 6–10 provide introductions to useful techniques, including cloning, library construction, sequencing, the polymerase chain reaction, and P-element-mediated transformation of *Drosophila*. Most molecular biologists reading this book could skip this section as well. Chapters 6–10 are not intended as a laboratory manual but, in some cases, an outline of laboratory protocols is provided in order to furnish the novice with a sense of the complexity or simplicity of the procedures and some of the issues to consider in problem solving. Throughout the book, references are provided for the reader interested in pursuing specific topics and techniques, although they are not exhaustive. Despite the value of providing an historical overview, I have not always provided references to the first publication on a subject. Rather, review articles or recent publications that include references to earlier work are cited.

Finally, in the third section of the book (Chapters 11–16), I have attempted to demonstrate how molecular genetic techniques can solve a diverse array of basic and applied problems. Part III is intended to introduce readers to the exciting molecular research that is revolutionizing insect biology, ecology, systematics, behavior, physiology, development, sex determination, and pest management. Each chapter in this section could be read by itself, assuming that the reader understands the appropriate concepts or information presented in Parts I and II.

Each chapter begins with an overview or brief summary of the material being covered. The overview should be read both before and after reading each chapter to review the concepts covered. The overview is followed by a brief introduction covering the history or rationale for the topic. References at the end of the chapter are provided for further reading. Where possible, books or reviews are cited to provide an entry into the literature. Recent references are provided, but no attempt has been made to review all the literature on a specific topic. Simple protocols may be given to provide the flavor of specific techniques, although these are not intended to be complete. References to handbooks or technique books are also provided at the end of appropriate chapters. When a term that may be unfamiliar is first introduced, it is in bold type and a brief definition or description is given in the Glossary at the end of the book. Finally, in Appendix I, a time line of some significant advances in genetics, molecular biology, and insect molecular genetics provides a perspective of the pace with which dramatic advances have been, and continue to be, made.

Progress is rapid in molecular genetics, and this book can only provide an introduction to the principles of insect molecular genetics and some of its applications. It is impossible to provide a complete review of the insect molecular genetics literature in a book of this size. The literature cited includes references through 1993 and focuses on genetics. It is not intended to be an introduction to all aspects of 'molecular entomology,' which has been defined as ". . . a blend of insect science, molecular biology, and biochemistry." The dividing line between molecular entomology and insect molecular genetics is sometimes difficult to resolve.

Shortly before this book went to the publisher, two related books were published: "Molecular Approaches to Fundamental and Applied Entomology" edited by J. Oakeshott and M. J. Whitten, and "Insect Molecular Science" edited by J. M. Crampton and P. Eggleston. Both multiauthored books cover some of the topics included here, but assume the reader is familiar with molecular genetic techniques and terminology; they would be daunting for the novice.

## Acknowledgments

This book would not have been written without the encouragement of many people. As most authors remark, it was more difficult to write than expected, in part due to a move from the University of California at Berkeley to the University of Florida in Gainesville about half-way through the endeavor. Certainly, this book would not have been written without the support of the Rockefeller Foundation which provided me with five weeks of precious time at the Bellagio Study and Conference Center in Italy. I gained a crucial grasp on the scope of the project and gathered my courage there while on sabbatical from the University of California at Berkeley. A fellowship spent at Cold Spring Harbor Laboratories in the summer of 1985 in the 'Molecular Cloning of Eukaryotic Genes' course provided my initial training in molecular genetic techniques. I thank the instructors for their patience. Likewise, G. M. Rubin kindly allowed me to participate in his cloning course at the University of California at Berkeley.

Many people have contributed information, advice, and valuable time reviewing this book. I especially thank Mary Bownes, Michael Caprio, Gary Carvalho, Howell Daly, Owain Edwards, Marilyn Houck, James Hoy, A. Jeyaprakash, Srini Kambhampati, Carolyn Kane, James Presnail, Veronica Rodriguez, Mark Tanouye, and Tom Walker, who reviewed either drafts of individual chapters or the entire book. Richard Beeman, Owain Edwards, Glenn Hall, A. Jeyaprakash, Ed Lewis, Jim Presnail, and A. Zacharopoulou provided photographs or illustrations. Lois Caprio and Denise Johanowicz assisted with many of the logistical issues. Finally, I especially thank my husband Jim, who repeatedly motivated me to clarify and elucidate the principles and applications presented here and tolerated my preoccupation with this project over many months, and John Capinera, who patiently waited for me to emerge from my compulsion. Despite the best efforts of the reviewers, errors probably persist and are my responsibility. This is University of Florida, Institute of Food and Agricultural Sciences, series number R-03561.

Marjorie A. Hoy

# Part I

# Genes and Genome Organization in Eukaryotes

# Chapter 1

# DNA and Gene Structure

## Overview

Arthropod genes are made of DNA and are located in chromosomes consisting of proteins, RNA, and DNA. DNA is a polymer of nucleotides. Each nucleotide consists of a pentose sugar, one of four different nitrogenous bases, and a phosphoric acid component. DNA consists of two complementary nucleotide strands in a helix form. Pairing of the nitrogenous bases adenine (A) with thymine (T), and cytosine (C) with guanine (G) on the two complementary strands occurs by hydrogen bonding. A pairs with T by two hydrogen bonds, and C pairs with G by three hydrogen bonds. DNA has chemically distinct 5′ and 3′ ends. The two strands are antiparallel, with one running in the 5′ to 3′ direction and the other from the 3′ to 5′ direction. Genetic information is determined by the sequence of nitrogenous bases (A, T, G, C) in one of the strands, with a three base (triplet) codon designating an amino acid. The genetic code is degenerate, with more than one codon specifying most amino acids. The genetic information is expressed when DNA is transcribed into messenger RNA, which is translated into polypeptides. Most arthropod genes have intervening sequences of noncoding nucleotides (introns) that must be removed from the primary RNA molecule before translation can occur.

## Introduction to the Central Dogma

The **Central Dogma**, as proposed by Francis Crick in 1958, stated that biological information is carried in DNA; this information is transferred first to RNA, and then to proteins. Initially, the Central Dogma also stated that the flow of information is unidirectional, with proteins unable to direct synthesis of RNA, and RNA unable to direct the synthesis of DNA (Figure 1.1). However, the Central Dogma was amended in 1970 when certain viruses were found to

3

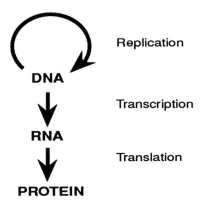

Replication

Transcription

Translation

**Figure 1.1** The Central Dogma. Biological information is transferred from DNA to RNA to proteins. Recent discoveries of viruses that transcribe information from RNA to DNA have required modification in the Dogma.

transfer information from RNA to DNA. Despite this modification, the Central Dogma remains a major tenet of modern biology.

In insects, the genes (DNA) are found in complex structures called chromosomes that consist of proteins, RNA, and DNA. This chapter reviews the structure of DNA and RNA and the basis of the genetic code.

## The Molecular Structure of DNA

**Deoxyribonucleic acid (DNA)** is a long polymeric molecule consisting of numerous individual monomers that are linked in a series and organized in a helix. Each monomer is called a **nucleotide**. Each nucleotide is itself a complex molecule made up of three components: (1) a sugar, (2) a nitrogenous base, and (3) a phosphoric acid.

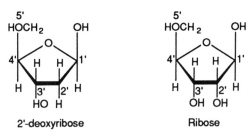

**Figure 1.2** Structure of sugars found in nucleic acids; 2'-deoxyribose is found in DNA and ribose is found in RNA.

**PURINES**

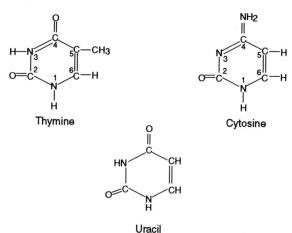

**PYRIMIDINES**

**Figure 1.3** Bases in DNA are purines (adenine and guanine) or pyrimidines (thymine and cytosine). Uracil is substituted for thymine in RNA.

In DNA, the sugar component is a pentose (with five carbon atoms) in a ring form that is called 2'-deoxyribose (Figure 1.2). The nitrogenous bases are single- or double-ring structures that are attached to the 1'-carbon of the sugar. The bases are **purines** (adenine and guanine) or **pyrimidines** (thymine and cytosine) (Figure 1.3). When a sugar is joined to a base it is called a **nucleoside**. A nucleoside is converted to a nucleotide by the attachment of a phosphoric acid group to the 5'-carbon of the sugar ring (Figure 1.4). The four different

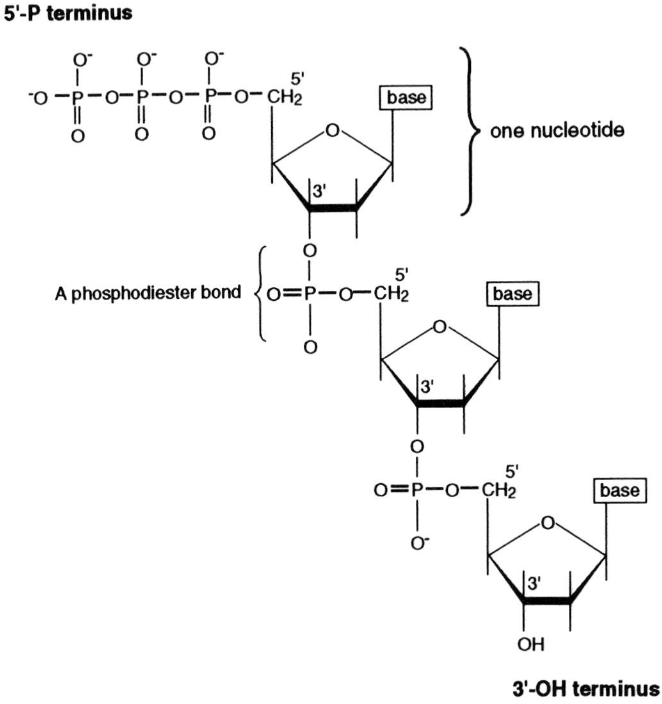

**5'-P terminus**

**3'-OH terminus**

**Figure 1.4** A nucleoside consists of a sugar joined to a base. It becomes a nucleotide when a phosphoric acid group is attached to the 5'-carbon of the sugar. Nucleotides link together by phosphodiester bonds to form polynucleotides.

nucleotides that polymerize to form DNA are 2'-deoxyadenosine 5'-triphosphate (dATP or A), 2'-deoxyguanosine 5'-triphosphate (dGPT or G), 2'-deoxycytidine 5'-triphosphate (dCTP or C), and 2'-deoxythymidine 5'-triphosphate (dTTP or T) (Figure 1.5). These names are usually abbreviated as dATP, dGTP, dCTP, and dTTP, or even as A, G, C, and T.

Individual nucleotides are linked together to form a polynucleotide by phosphodiester bonds, which are obtained by esterification of the two hydroxyl groups of the same phosphoric acid molecule (Figure 1.4). Polynucleotides have chemically distinct ends. In Figure 1.4, the top of the polynucleotide ends with a nucleotide in which the triphosphate group attached to the 5'-carbon has not participated in a phosphodiester bond. This is called the 5' or 5'-P terminus. At the other end of the molecule, the unreacted group is not the phosphate, but the 3'-hydroxyl. This is called the 3' or 3'-OH terminus. This distinction between the two ends means that polynucleotides have an orientation that is very important in many molecular genetics applications.

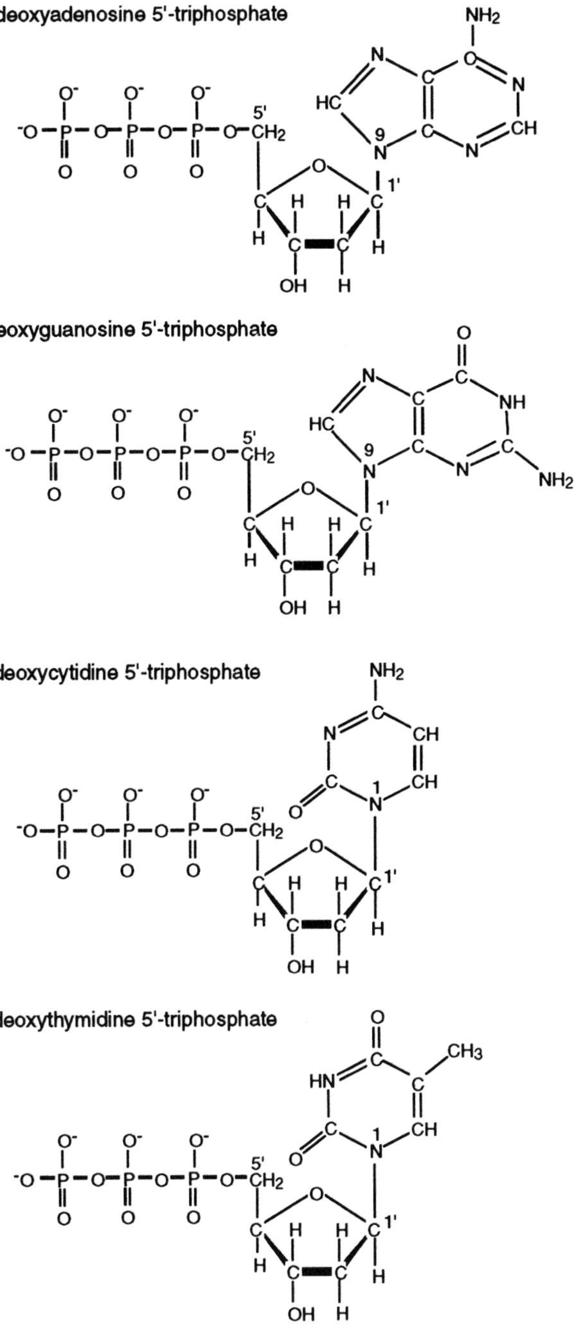

**Figure 1.5** The four trinucleotides from which DNA is synthesized are 2'-deoxyadenosine 5'-triphosphate (dATP), 2'-deoxyguanosine 5'-triphosphate (dGTP), 2'-deoxycytidine 5'-triphosphate (dCTP), and 2'-deoxythymidine 5'-triphosphate (dTTP).

Polynucleotides can be of any length and have any sequence of bases. The DNA molecules in chromosomes are probably several million nucleotides long. Because there are no restrictions on the nucleotide sequence, a polynucleotide just ten nucleotides long could have any one of $4^{10}$ (or 1,048,576) different sequences. This ability to vary the sequence is what allows DNA to contain complex genetic information.

## The Molecular Structure of RNA

RNA is also a polynucleotide, but with two important differences. First, the sugar in RNA is ribose (Figure 1.2). Second, RNA contains the nitrogenous base **uracil** (U) instead of thymine (Figure 1.3). The four nucleotides that polymerize to form RNA are adenosine 5'-triphosphate, guanosine 5'-triphosphate, cytidine 5'-triphosphate, and uridine 5'-triphosphate, which are abbreviated as ATP, GTP, CTP, and UTP, or A, G, C, or U. The individual nucleotides are linked together with 3' to 5' phosphodiester bonds. RNA is typically single-stranded, although it can form complex structures under some circumstances.

## The Double Helix

The discovery, by Watson and Crick in 1953, that DNA is a double helix of antiparallel polynucleotides ranks as one of the most important discoveries in biology. Nitrogenous bases are located inside the double helix, with the sugar and phosphate groups forming the backbone of the molecule on the outside (Figure 1.6). The nitrogenous bases of the two polynucleotides interact by **hydrogen bonding**, with an adenine pairing to a thymine and a guanine to a cytosine. Hydrogen bonds are weak bonds in which two negatively charged atoms share a hydrogen atom between them. Two hydrogen bonds form between A and T, and three between G and C. Bonding between G and C is thus stronger and more energy is required to break bonds between G and C. The hydrogen bonds, and other molecular interactions called stacking interactions, hold the double helix together.

The DNA helix turns approximately every ten base pairs (abbreviated as 10 bp), with spacing between adjacent bp of 3.4 **angstroms** (Å) so that a complete turn requires 34 Å (Figure 1.6). An angstrom is one one-hundred-millionth of a centimeter. The helix is 20 Å in diameter and right handed. This means that each chain follows a clockwise path. The strands run antiparallel to each other, with one running in the 5' to 3' direction and the other in the 3' to 5' direction. The DNA helix has two grooves, a **major** and a **minor groove** (Figure 1.6). Proteins involved in DNA replication and transcription often interact within these grooves.

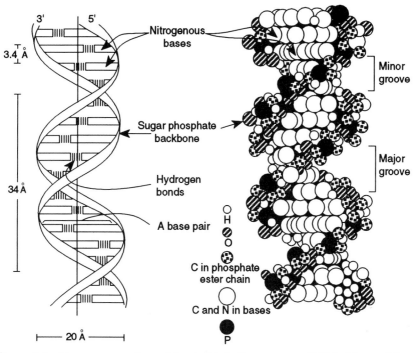

**Figure 1.6** Two representations of the double helix structure of DNA. The model on the left shows the hydrogen bonding between nitrogenous bases that holds the two antiparallel strands together. The model on the right shows the relative sizes of the atoms in the molecule.

## Complementary Base Pairing Is Fundamental

The principle of **complementary base pairing** is a fundamental element of DNA and of great practical significance in many techniques used in genetic engineering. A pairs with T, and G pairs with C, and normally no other base pairing pattern will fit in the helix or allow hydrogen bonding to occur (Figure 1.7). Complementary base pairing provides the mechanism by which the sequence of a DNA molecule is retained during replication of the molecule, which is crucial if the information contained in the gene is not to be altered or lost during cell division. Complementary base pairing is also important in the transcription and expression of genetic information in the living insect.

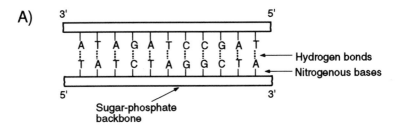

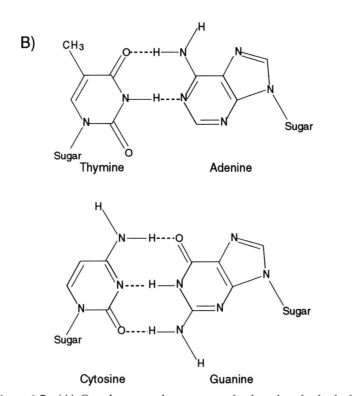

**Figure 1.7**  (A) Complementary base pairing of polynucleotides by hydrogen bonds holds the two strands of the DNA molecule together. (B) Thymine pairs with adenine with two hydrogen bonds, and guanine pairs with cytosine with three hydrogen bonds.

## DNA Exists in Several Forms

DNA actually is a dynamic molecule in living organisms and has several different variations in form. In some regions of the chromosome, the strands of the DNA molecule may separate and later come back together. DNA can form more than 20 slightly different variations of right-handed helices. In some regions, it can even form left-handed helices. If segments of nucleotides in the same strand are complementary, the DNA may even fold back upon itself in a hairpin structure.

DNA also exists in different crystalline forms, depending upon the amount of water present in the DNA solution. The B form is the structure in which DNA commonly occurs under most cellular conditions. A-DNA is more compact than B-DNA, with 11 bp per turn of the helix and a diameter of 12 Å. In addition, C-, D-, E-, and Z-DNA have been found. The Z-DNA form has a left-handed rather than a right-handed helix. A triple helical form (H) occurs, and could be important in forming the ends of chromosomes. A, H, and Z forms also are thought to occur in cells. C, D, and E forms of DNA may be produced only under laboratory conditions.

## Genes

Genes are segments of a DNA molecule which may vary in size from as few as 75 nucleotides (nt) to over 200 **kilobases** (kb) of DNA. A kilobase is 1000 nucleotides. Nucleotide sequences contain biological information by coding for the synthesis of an RNA molecule that subsequently directs the synthesis of an enzyme or other protein molecule. Thus, expression of the information contained in the genes is a two-step process of **transcription** and **translation** (Figure 1.1).

The actual genetic information is determined by just one of the two polynucleotide strands of the double helix. This is called the **coding strand** and the other strand is the noncoding complement to it. A few examples are known in which both strands in a specific region code for different genes. Typically, however, one strand of the double helix can be the coding strand over part of its length and the noncoding strand over other regions of its length (Figure 1.8).

## The Genetic Code Is a Triplet and Is Degenerate

The genetic code is based on the sequence of three bases in the DNA molecule. The triplet sequence (or **codon**) determines which amino acids are assembled in a particular sequence into proteins. It is possible to order four different bases (A, T, C, G) in combinations of three into 64 possible triplets or codons. However, there are only 20 different amino acids and the question

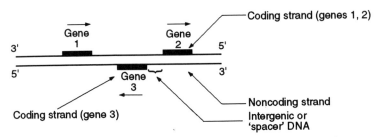

**Figure 1.8**   Genetic information is contained in genes carried on one of the two strands (coding strand). The complementary strand in that region is the noncoding strand. Genes can occur on different strands at different points of the DNA molecule. DNA between genes is called intergenic or spacer DNA.

immediately arises as to what is done with the other 44 codons. The answer is that the code is **degenerate**, with all amino acids except methionine and tryptophan determined by more than one codon (Table 1.1). The codons in Table 1.1 are represented by A, U, C, and G because the genetic information in DNA is transcribed by messenger RNA (mRNA), which uses U instead of T.

The genetic code contains punctuation codons. Three different codons (UAA, UGA, and UAG) function as 'stop' messages. One of these three **termination codons** occurs at the end of a gene to indicate where translation should stop. AUG serves as an **initiation or start codon** when it occurs at the front end of a gene. However, AUG is the sole codon for the amino acid methionine, and thus AUGs are also found in the middle of genes as well as at the beginning.

The genetic code is not universal, although it was assumed to be so initially. In 1979 it was found that mitochondrial genes use a slightly different code. For example, AGA typically codes for arginine, but in *Drosophila* mitochondria AGA codes for serine.

## Gene Organization

Genes are located on chromosomes. Each chromosome contains a single DNA molecule. These DNA molecules contain hundreds or thousands of genes. The fruit fly *Drosophila melanogaster* is estimated to have approximately 20,000 genes distributed on four chromosomes. Genes may be spaced out along the length of a DNA molecule or grouped into clusters. Genes in a cluster may be related or unrelated to each other. In addition, there are usually segments of DNA in eukaryotes in which the nucleotide sequences apparently do not code for anything; this DNA is called 'spacer' DNA. (**Eukaryotes** are organisms that are, or consist of, cells with true nuclei bounded by nuclear membranes. Cell division occurs by mitosis, reproductive cells undergo meiosis, and oxidative enzymes are packaged in mitochondria.)

**Table 1.1**

The 20 Amino Acids that Occur in Proteins and Their Codons

| Amino acid | Abbreviation | Codons | | | | |
|---|---|---|---|---|---|---|
| Alanine | ala | GCU | GCC | GCA | GCG | |
| Arginine | arg | AGA | AGG | | | |
| Asparagine | asn | AAU | AAC | | | |
| Aspartic acid | asp | GAU | GAC | | | |
| Cysteine | cys | UGU | UGC | | | |
| Glutamic acid | glu | GAA | GAG | | | |
| Glutamine | gln | CAA | CAG | | | |
| Glycine | gly | GGU | GGC | GGA | GGG | |
| Histidine | his | CAU | CAC | | | |
| Isoleucine | ile | AUU | AUC | AUA | | |
| Leucine | leu | UUA | UUG | CUU | CUC | CUA | CUG |
| Lysine | lys | AAA | AAG | | | |
| Methionine | met[a] | AUG | | | | |
| Phenylalanine | phe | UUU | UUC | | | |
| Proline | pro | CCU | CCC | CCA | CCG | |
| Serine | ser | AGU | AGC | | | |
| Threonine | thr | ACU | ACC | ACA | ACG | |
| Tryptophan | trp[a] | UGG | | | | |
| Tyrosine | try | UAU | UAC | | | |
| Valine | val | GUU | GUC | GUA | GUG | |

[a]Methionine and tryptophan are underlined because they are specified by only one codon.

**Multigene families** are clusters of related genes with identical, or nearly identical, nucleotide sequences. Multigene families may have originated from a single ancestral gene that was duplicated to produce two identical genes. These identical genes could have diverged in nucleotide sequence through time to produce two related and functional genes. In many cases, the genes of multigene families have become scattered at different positions on more than one chromosome by large-scale rearrangements that occur both within and between chromosomes. Examples of multigene families in insects include actins, tubulins, heat shock proteins, salivary glue proteins, chorion proteins, cuticle proteins, and yolk proteins.

**Pseudogenes** are DNA sequences that appear similar to those of functional genes, but the genetic information has been altered so that it is no longer functional. Once the biological information has been lost, a pseudogene can undergo rapid changes in nucleotide sequence and, given sufficient time, may evolve to the point where it is not possible to identify it as a former gene.

One of the more interesting recent discoveries in genetics was the revelation that most genes in eukaryotic organisms are discontinuous. **Discontinuous genes** contain coding and noncoding segments called **exons** and **introns**, respectively (Figure 1.9). Introns are generally absent in the genes of prokaryotes (bacteria and their viruses) and rare in some eukaryotic organisms, such as yeast. The number of introns varies from species to species and from gene to gene. Some genes in eukaryotic organisms lack introns, while other genes in the same species may have as many as 50. Introns may interrupt a coding region, or they may occur in the untranslated regions of the gene. Some eukaryotic genes contain numerous and very large introns, but they typically range from 100 to 10,000 bp in length. Oddly, a few introns contain genes themselves; how these genes got in the middle of an intron that is within a gene remains a mystery, and their function, if any, remains controversial. However, introns are now known to have a function in certain cases, perhaps in regulating gene expression. Some introns code for small nucleolar RNAs which accumulate in the nucleolus where ribosomes are formed. The small nucleolar RNAs may have a role in ribosome assembly.

The presence of introns within many eukaryotic genes requires that an additional step take place between transcription and translation in eukaryotes. Thus, when the DNA is transcribed into RNA, the initial RNA transcript is not messenger RNA (mRNA). It is a precursor to mRNA, and must undergo processing (splicing) in the nucleus to remove the introns before it appears in the cytoplasm as mRNA. This process is described in more detail in Chapter 3.

Genetic information contained in DNA molecules must be transmitted to the next generation. Chapter 2 describes how the structure of DNA helps to ensure efficient and accurate replication of DNA.

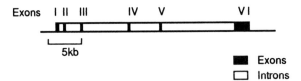

**Figure 1.9**  Genes in eukaryotic organisms are divided into introns and exons. Introns are removed from the messenger RNA before it is translated into a polypeptide. In this example, there are six exons and five introns. The genetic message is present in exons I, II, III, IV, V, and VI.

# General References

Beebee, T. and J. Burke. 1988. *Gene Structure and Transcription*. IRL Press, London.

Brown, T. A. 1989. *Genetics: A Molecular Approach*. Chapman and Hall, London.

Drlica, K. 1984. *Understanding DNA and Gene Cloning. A Guide for the Curious*. Wiley, New York.

Hartl, D. L. 1991. *Basic Genetics*. 2nd Ed. Jones and Bartlett Publ., Boston.

Watson, J. D., N. H. Hopkins, J. W. Roberts, J. A. Steitz and A. M. Weiner. 1987. *Molecular Biology of the Gene*. Vols. I and II. 4th Ed. Benjamin/Cummings, Menlo Park, CA.

# Chapter 2

# DNA Replication, Mutations, and Recombination

## Overview

Efficient and accurate replication of DNA must occur with each cell division or the cell or organism may not survive. DNA replication is semiconservative, which means that one of the nucleotide strands of each new DNA molecule is new and the other is old. The new DNA strand is complementary to the parental or template strand. DNA replication occurs in one direction, from the 5′ to the 3′ end of the strand, and thus replication takes place differently on the two parental strands. Replication on one strand, the leading strand, can occur in the 5′ to 3′ direction in a continuous manner. However, DNA replication on the other strand, the lagging strand, occurs in short segments, producing Okazaki fragments that later must be ligated together.

Replication of DNA in chromosomes begins at many specific sites called origins of replication, and involves a number of enzymes and proteins. While DNA replication is usually highly accurate, errors in DNA replication, or mutations, can result from duplications, deletions, inversions, and translocations of nucleotides, which may affect the functioning of the resultant polypeptide. New combinations of genes can occur through recombination.

## Introduction

Every living organism must make a copy of its genes in each cell every time a cell divides. Such replication ideally is rapid and accurate. If not, the organism's survival and integrity are jeopardized. Even a very small error rate of 0.001% (one mistake per 100,000 nucleotides) can lead to detrimental changes in the genetic information. However, while many changes in DNA are detrimental, not all are. Some are apparently neutral, and a few are beneficial.

Most studies on DNA replication have been conducted using prokaryotic organisms, which are more easily studied than eukaryotic organisms. DNA replication in the bacterium, *Escherichia coli*, which has a single circular DNA molecule, illustrates the basic principles. The complications added by arranging DNA molecules into linear chromosomes are described later (Chapter 4).

## DNA Replication Is Semiconservative

DNA replication is **semiconservative**. The daughter molecules each contain one polynucleotide derived from the original DNA molecule and one newly-synthesized strand (Figure 2.1). Semiconservative DNA replication requires that the base pairing of the two strands be broken so that synthesis of new strands that are complementary to each original strand can occur.

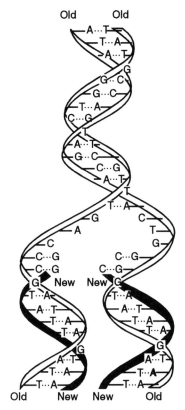

**Figure 2.1** DNA replication is semiconservative, meaning that each new DNA helix contains one old and one new complementary strand. DNA synthesis relies on complementary base pairing for accurate replication.

## Replication Begins at Replication Origins

During the replication of long DNA molecules, only a limited region of the DNA molecule is in an unpaired form at any one time. Replication occurs after the two strands separate, which involves breaking the weak hydrogen bonds holding the bases on the opposite strands together. The separation of the two strands starts at specific positions in the DNA molecule in prokaryotes and the yeast *Saccharomyces cerevisiae*. This region is called an **origin of replication** and moves along the molecule during DNA replication (Figure 2.2). Synthesis of the new complementary polynucleotides occurs as the double helix 'unzips'. The

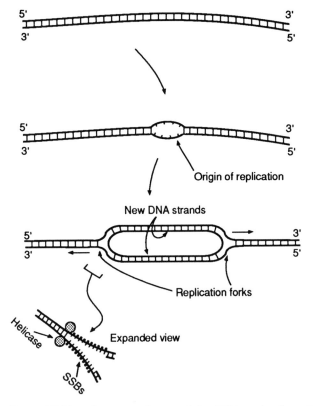

**Figure 2.2** During DNA replication only part of the DNA molecule upzips to allow synthesis of new DNA strands. In this example from a prokaryote, replication begins at the origin of replication. To keep the strands from reannealing at the replication forks, where synthesis is occurring, single-strand binding (SSB) proteins attach. Helicases break the hydrogen bonds. (Adapted from Brown 1989.)

region at which the base pairs of the parent molecule are broken and the new polynucleotides are synthesized is the **replication fork** (Figure 2.2).

The base pairing of the two strands of the parent DNA molecule is broken by enzymes called **helicases**. Once the helicase has broken the hydrogen bonding holding the two strands together, **single-strand binding proteins** (SSB) attach to the single-stranded DNA to prevent the two complementary DNA strands from immediately reannealing (Figure 2.2). This makes it possible for **DNA polymerase** to synthesize new complementary DNA strands. DNA polymerases have two properties that complicate DNA synthesis. First, DNA polymerase can synthesize only in the 5′ to 3′ direction and, second, DNA polymerase can not initiate the synthesis of new DNA strands without a primer.

Origins of replication in animals have been described only three times, but probably occur in all animals (Burhans and Huberman 1994). In some cases replication can occur without specific initiation sites or at sites throughout a region consisting of several kilobase pairs.

## Replication Occurs Only in the 5′ to 3′ Direction

Because DNA polymerase can synthesize DNA only in the 5′ to 3′ direction, the template strands must be read in the 3′ to 5′ direction. This is a straightforward process for one of the DNA template strands, called the **leading strand**, and DNA synthesis can proceed in an uninterrupted manner the entire length of the leading strand. However, DNA synthesis cannot proceed uninterrupted on the other template strand, called the **lagging strand** (Figure 2.3). DNA synthesis on the lagging strand is discontinuous, occurring in short sections, and produces short fragments of DNA called **Okazaki fragments** after their discoverer, who identified them in 1968.

## Replication of DNA Requires an RNA Primer

Another complication of DNA synthesis is that synthesis is not initiated by DNA polymerase unless there is a short double-stranded region that can act as a primer (Figure 2.4). Apparently, the first few (50–75) nucleotides attached to either the leading or lagging strands are not deoxyribonucleotides, but rather are ribonucleotides that are put in place by an RNA polymerase called **primase**. Once these ribonucleotides have been polymerized on the DNA template, the primase detaches, and polymerization of DNA is continued by DNA polymerase (Figure 2.5).

## Ligation of Replicated DNA Fragments

After the Okazaki fragments that are complementary to the lagging strand of DNA are produced, they must be joined together to produce a continuous

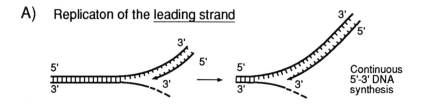

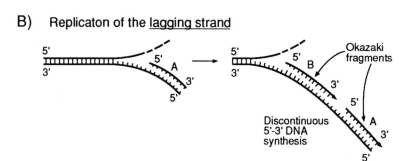

**Figure 2.3**  DNA replication occurs in a different manner on the two strands. (A) The leading strand is continuously copied, with synthesis occurring in the 5′ to 3′ direction. (B) Synthesis on the lagging strand is discontinuous, and occurs in short segments (Okazaki fragments) because DNA polymerase can only synthesize DNA in the 5′ to 3′ direction.

strand (Figure 2.5). On the lagging strand, DNA polymerase III of *E. coli* stops when it reaches the RNA primer at the 5′ end of the next Okazaki fragment. Then DNA polymerase I of *E. coli* removes the ribonucleotides from the Okazaki fragment and replaces them with deoxyribonucleotides. When all the ribonucleotides have been replaced, DNA polymerase I replaces nucleotides a short distance into the DNA region, before it dissociates from the new double helix molecule. The Okazaki fragments are then joined up by **DNA ligase**, which catalyzes the formation of a phosphodiester bond between the neighboring nucleotides (Figure 2.5).

DNA replication also requires that the double helix be unwound as well as unzipped. There are approximately 400,000 turns in 400 kb of DNA. In *E. coli*, this unwinding is accomplished with the aid of enzymes called **DNA topoisomerases**. DNA topoisomerases unwind a DNA molecule without rotating the helix by causing short term breaks in the polynucleotide backbone just in front of the replication fork. The reverse reaction is also performed by DNA topoisomerases so that DNA molecules can be coiled.

## A) DNA synthesis requires a primer

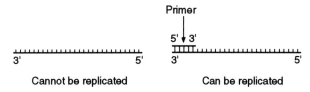

Cannot be replicated          Can be replicated

## B) Priming during DNA replication

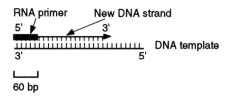

**Figure 2.4** (A) DNA must be primed or DNA polymerase cannot synthesize a complementary strand. (B) A primer of ribonucleotides is attached to a strand by RNA polymerase. DNA polymerase can then attach deoxyribonucleotides (dNTPs) to the DNA template in a sequence that is determined by the template strand. DNA synthesis occurs in the 5′ to 3′ direction. (Adapted from Brown 1989.)

## DNA Replication in Eukaryotes

DNA replication takes place during the cell cycle before the metaphase chromosomes become visible in mitosis or meiosis. The **cell cycle** consists of four distinct sections. There are two gap periods ($G_1$ and $G_2$), when the cell is carrying out its normal metabolic activities (Figure 2.6), which are separated by the **S phase**, which is when DNA replication or synthesis occurs. Mitosis (M) occurs subsequent to the $G_2$ phase.

In order to reduce the amount of time required to replicate the very long DNA molecule in eukaryotic chromosomes, DNA replication is initiated at a series of **replication origins** about 40 kb apart on the linear chromosome and proceeds bidirectionally (Figure 2.7). For example, replication in the fruit fly *Drosophila melanogaster* occurs at a rate of about 2600 nucleotide pairs per minute at 24°C. The largest chromosome in *Drosophila* is about $8 \times 10^7$ nucleotides long, so with about 8500 replication origins per chromosome, approximately 0.25 to 0.5 hour is required to replicate the chromosome. If replication occurred

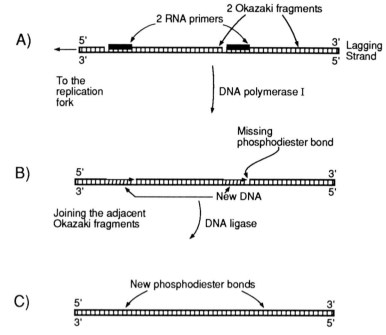

**Figure 2.5**   (A) DNA synthesis on the lagging strand requires that the RNA primers be removed, leaving the Okazaki fragments. (B) The gaps between the Okazaki fragments are filled in by DNA ligase. (C) New phosphodiester bonds are formed to produce an intact new DNA strand. (Adapted from Brown 1989.)

from a single replication fork, rather than from multiple origins, replication of a single chromosome would require about 15 days.

## Telomeres at the End

Because DNA synthesis occurs exclusively in the 5′ to 3′ direction, and initiation requires a short RNA primer, the extreme 5′ end of a linear DNA strand will consist of an RNA primer (Figure 2.4B). If this RNA primer (composed of ribonucleotides) were not replaced by deoxyribonucleotides, the DNA strands would gradually decrease in length after each mitosis, which could seriously affect gene function. However, linear chromosomes normally are stable because they have a specialized structure at their ends called a **telomere**. Telomeres contain a series of species-specific repeated nucleotide sequences that are added to the ends of eukaryotic chromosomes by an enzyme called **telomerase**. A

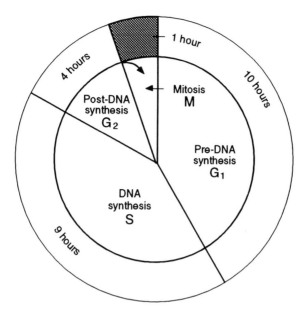

**Figure 2.6** The cell cycle of a eukaryotic cell with a generation time of 24 hours. DNA synthesis occurs during the S phase. No DNA synthesis occurs during $G_1$ and $G_2$. Mitosis (M) occurs after $G_2$.

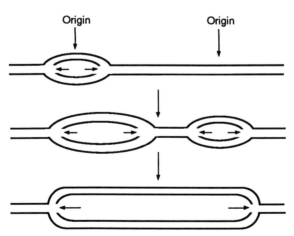

**Figure 2.7** DNA in eukaryotic organisms has multiple origins of replication. The arrows indicate the direction of movement of the replication forks when DNA is replicated.

few copies of a short repetitive sequence (called the telomere sequence) are required to prime the telomerase to add additional copies and form a telomere. There are also longer, moderately repetitive nucleotide sequences subterminal to the telomere sequences.

## Generating Variability

Changes in the genetic material (**genotype**) of an organism occur through **mutations**. A mutation is a change in the nucleotide sequence of a DNA molecule. Many kinds of mutations can occur. They may occur within a gene or in the regions located between coding segments of DNA (called the **intergenic regions**). If a mutation occurs in an intergenic region, it will probably be silent and have no detectable impact on the cell or individual. If a mutation occurs in a gene, it may alter the gene product and cause a change in the organism's **phenotype**.

An organism with the 'normal' appearance for that species is called the 'wild type' while an organism with a phenotype that has been changed is a **mutant**. A **mutagen** is a chemical or physical agent that causes changes in bases. Mutagens include ultraviolet radiation, X-irradiation, base analogues such as 5-bromouracil, acridine dyes, and nitrous acid. Mutations occur spontaneously approximately once in every $10^8$ base pairs per cell division, or they can be induced by the experimenter.

Mutations affect the DNA sequence, gene organization, gene regulation, or gene function (Table 2.1). A **point mutation** is the replacement of one nucleotide by another. An **insertion** or **deletion** is the addition or deletion of one or more base pairs of DNA. An **inversion** is the excision of a part of the DNA molecule, followed by its insertion into the same position but with a reversed orientation. Some mutations are lethal, while others have an effect on the organism that can range from phenotypically undetectable to lethal only under certain circumstances (**conditional lethal**). For example, a number of mutations are temperature sensitive and the organism can survive if reared within a specific temperature range, but not at higher temperatures.

Some mutations within a gene are silent. For example, a **silent mutation** may occur if the third base in a codon is altered but, because the genetic code is degenerate, there is no change in the amino acid specified. There is no change in protein structure or function from a silent mutation.

Some changes in codons alter the amino acid specified. **Missense mutations** are point mutations that result in changes in the amino acid. Most point mutations that occur at the first or second nucleotide positions of a codon will be missense, as will a few third position changes. A polypeptide with an amino acid change may result in a changed phenotype, depending on the precise role the altered amino acid plays in the structure or function of the polypeptide. Most

**Table 2.1**

Mutations Affect DNA Sequence, Gene Function, Gene Regulation, and the Phenotype of the Organism

| | |
|---|---|
| **Changes in DNA sequence** | |
| Point mutation | Replacement of one nucleotide by another. |
| Insertion or deletion | Addition or deletion of one or more nucleotides. |
| Inversion | Excision of a part of the DNA followed by its insertion into the same position with a reverse orientation. |
| Transition | A point mutation in which a purine is changed to a purine ($A \leftrightarrow G$) or a pyrimidine to a pyrimidine ($T \leftrightarrow C$). |
| Transversion | A point mutation in which the change is purine to pyrimidine ($A$ or $G \leftrightarrow T$ or $C$). |
| **Changes in the gene** | |
| Silent mutation | Sequence changes that occur in an intergenic region usually result in no phenotypic changes. Changes in a gene can also be silent if a point mutation occurs in the third nucleotide position of a codon which, because of the degeneracy of the code, does not alter the amino acid coded for. |
| Missense mutation | A codon is mutated to result in incorporation of a different amino acid, making an inactive or unstable enzyme. |
| Nonsense mutation | A point mutation that changes a codon specifying an amino acid into a termination codon, which will prematurely terminate the polypeptide produced, changing the activity of the protein and altering the phenotype. |
| Frameshift mutation | Insertions or delections that are not in multiples of three. These can cause changes in the amino acids downstream from the mutation, resulting in a mutant phenotype. |
| **Changes in gene regulation** | |
| | Mutations in regulatory genes alter the organism's ability to control expression of a gene normally subject to regulation. |
| **Changes in the organism** | |
| Lethal mutations | Mutations that alter the function of an essential gene product so that the organism cannot survive. |

(continues)

**Table 2.1**   (*Continued*)

| | |
|---|---|
| Conditional lethal | Mutations in which individuals can survive under a particular set of conditions, such as a specific temperature range, but die if reared outside these conditions. |
| Back mutations | Reversion to the wild-type phenotype can occur after a second mutation which restores the original nucleotide sequence of the mutated gene. |
| Second-site reversions | Restoration of the original phenotype, but not the original DNA sequence in the mutated gene, by altering a second site within the gene. |
| Suppression | A new mutation in a different gene that alters the effects of an earlier mutation. |

proteins can tolerate some changes in their sequence if the alteration does not change a segment essential for the structure or function of the protein.

**Nonsense mutations** are point mutations that change a codon specifying an amino acid into a termination codon, which will produce a truncated gene which codes for a polypeptide that is truncated. In many cases, essential amino acids will be deleted and the protein's activity will be altered, resulting in a mutant phenotype.

**Frameshift mutations** often result if an addition or deletion of base pairs occurs that is not in a multiple of three. The code is thus read with all codons shifted so that different amino acids are specified subsequent to this point. The polypeptide produced will likely have a complete new set of amino acids produced downstream of the frameshift. Frameshifts usually produce mutant phenotypes.

Occasionally, **back mutations** may occur to reverse a point mutation. **Reversions** sometimes occur when the original phenotype is restored by a new change in the nucleotide sequence. In reversions, the original mutation is not restored to its previous unmutated form; rather, the second mutation restores the code for the original amino acid because the code is degenerate. **Regulatory mutations** are mutations that affect the ability to control expression of a gene.

**Transposition** of a **transposable element** (TE) into a gene can also create mutations in genes (Figure 2.8). Transposable elements are segments of DNA that can move about within the genome, and are found in most organisms examined. There are many types of transposable elements and they are important for understanding genome evolution and for genetic engineering. TEs are discussed further in Chapters 4 and 10. Genes can be inactivated or altered in their expression if a transposable element inserts into them.

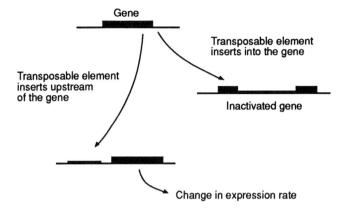

**Figure 2.8** Movement (transposition) of transposable elements into chromosomes can result in mutations that inactivate genes or alter their expression. (Adapted from Brown 1989.)

## Common Conventions in Genetic Terminology

A wild-type gene is normally found only after a mutation has disrupted the normal phenotype of an organism. Mutations are commonly given a descriptive name, such as 'white eyes'. The name of the gene is italicized (*white*), and is abbreviated using one, two, or three letters (*w*). If the mutation is dominant, the name and abbreviation have the first letter capitalized; they are in lower case if the mutation is recessive. Individuals that are homozygous for the *w* mutation are *w*/*w*, and heterozygous insects are *w*/*w*$^+$ with the wild type allele designated as *w*$^+$. The gene product is called the white product or white protein and is not italicized. The term for the gene product may also be abbreviated as the w protein.

## Independent Assortment and Recombination

In sexually reproducing organisms, the progeny from a cross between parents with different genes will have a different combination of genes than either parent. This shuffling of the genetic information is due to **independent assortment** of the homologous chromosomes into the gametes during **meiosis**. Thus, an individual of genotype AaBb, in which the genes A and B are located on different chromosomes, will produce equal numbers of four different types of gametes: AB, Ab, aB, or ab.

**Crossing over** leads to recombination between DNA molecules. Crossing over occurs between homologous chromosomes during meiosis I, and results in an exchange of genetic material. Crossing-over allows new combinations of different genes that are linked (located on the same chromosome). Thus, if a parent has two homologous chromosomes, one with allele A plus allele B, and the other with allele a plus allele b, a physical exchange of the chromatids during meiosis I can lead to gametes that have the following combinations: A plus B; A plus b; a plus B; and a plus b. The new combinations are A plus b and a plus B. Nonhomologous recombinations between DNAs lacking sequence homology may also occur, but the mechanism(s) involved is not understood.

## General References

Beebee, T. and J. Burke. 1988. *Gene Structure and Transcription*. IRL Press, London.

Burhans, W. C. and J. A. Huberman. 1994. DNA replication origins in animal cells: A question of context? Science 263: 639–640.

Brown, T. A. 1989. *Genetics: A Molecular Approach*. Chapman and Hall, London.

Drlica, K. 1984. *Understanding DNA and Gene Cloning: A Guide for the Curious*. Wiley, New York.

Hartl, D. L. 1991. *Basic Genetics*. 2nd Ed. Jones and Bartlett Publ., Boston.

Watson, J. D., N. H. Hopkins, J. W. Roberts, J. A. Steitz and A. M. Weiner. 1987. *Molecular Biology of the Gene*. Vols. I and II. 4th Ed. Benjamin/Cummings, Menlo Park, CA.

# Chapter 3

## Transcription, Translation, and Regulation of Eukaryotic DNA

### Overview

Genetic information in an organism is expressed in three classes of genes: (1) structural genes that are transcribed into messenger RNA and translated into polypeptides, (2) genes that code for ribosomal RNA or transfer RNA, in which the transcription product is used directly, and (3) regulatory sites that are not transcribed but serve as recognition sites for proteins involved in DNA replication, transcription, and repair processes. Class I genes in eukaryotic organisms are transcribed from DNA into pre-messenger RNA, which is processed into messenger RNA (mRNA). mRNA processing involves splicing to remove introns; mRNA also is capped and methylated at the 5′ end and most mRNAs are polyadenylated at the 3′ end. The information in the mRNA is then translated into proteins via protein synthesis in ribosomes.

A ribosome begins protein synthesis once the 5′ end of an mRNA is inserted into it. As the mRNA moves through the ribosome, a lengthening polypeptide chain is produced. Once the 5′ end of the mRNA emerges from a ribosome, it can attach to a second ribosome and so a second identical polypeptide can be synthesized. The assembly of amino acids into a peptide starts at the amino end and finishes at the carboxyl end. Amino acids are carried to the ribosome by transfer RNAs.

There are two binding sites for transfer RNAs (tRNAs) in the ribosome. One site holds the tRNA molecule that is attached to the growing end of the polypeptide and the other site holds the incoming tRNA molecule charged with the next amino acid. The tRNAs are held so their anticodons form base pairs with complementary codons of the mRNA. One tRNA may recognize more than one codon. According to the wobble hypothesis, the first two bases of the mRNA codon pair according to base-pairing rules, but the third base may pair with any one of a several bases.

Gene regulation in insects and other eukaryotes is complex, diverse, and the subject of intensive research. Genes may be amplified or rearranged to yield increased gene products. DNA methylation influences gene regulation and transcription is influenced by activator proteins, hormones, and enhancers. Alternative splicing, alternative promoters, and translational control are also employed in gene regulation.

## Introduction

The Central Dogma, that DNA is transcribed to RNA which is translated into proteins, describes the process by which information contained in the DNA is made available to the cell and organism (Figure 1.1). The proteins specified by the genetic code have many different functions in the cell. Structural proteins form part of the framework of the organism, such as the sclerotin in the exoskeleton of insects. Contractile proteins enable organisms to move. Catalytic enzymes regulate the diverse biochemical reactions taking place within the cell. Transport proteins carry important molecules throughout the body. Regulatory proteins control and coordinate biochemical reactions in the cell and the organism as a whole. Protective proteins (antibodies) protect against infectious agents and injury. Storage proteins store products for future use. The development of a functioning organism involves the coordinated activity of a large number of these different protein molecules, the information for which is encoded in the genes. In addition, genes carry the code for ribosomal RNA (rRNA) or transfer RNA (tRNA). Ribosomal and transfer RNA molecules are used directly (without translation into proteins).

Research to decipher how the genetic information in the DNA is utilized by the cell and organism is a complex and rapidly advancing field. Much of the early work on transcription and translation of genetic information was conducted using prokaryotes. Eukaryotes, however, differ from prokaryotes in several important ways. First, eukaryotic genes are located in more than one chromosome. Eukaryotes are genetically much more complex, with 20,000 to 100,000 genes. Furthermore, most eukaryotic genes that code for proteins are split, with one or more noncoding introns interspersed among the coding exons. Control elements, such as promoters and enhancers, are important components of gene regulation in eukaryotes. Finally, the nuclear membrane in eukaryotes separates the processes of transcription and translation in both time and space. The intricacies of transcription and translation of eukaryotic DNA are still being unraveled.

## RNA Synthesis Is Transcription

**Transcription** is the first stage of gene expression (Figure 3.1). During transcription, the coding strand of DNA serves as a template for synthesis of an ·

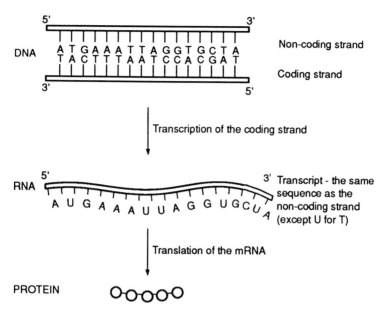

**Figure 3.1**    Gene expression involves transcription of the coding strand of DNA to pre-messenger RNA, which is then processed to messenger RNA, which is then translated into proteins.

RNA molecule. The sequence of the RNA molecule is determined by complementary base pairing so that the RNA is a complementary transcript of the coding strand of DNA.

Transcription requires four ribonucleoside 5'-triphosphates: ATP, GTP, CTP, and UTP. A sugar–phosphate bond is formed between the 3'-OH group of one nucleotide and the 5'-triphosphate of a second nucleotide with the enzyme RNA polymerase. Unlike DNA polymerase, RNA polymerase can initiate RNA synthesis without requiring a primer. The sequence of bases in the RNA molecule is determined by the sequence of bases in the DNA coding strand because base pairing occurs between the DNA bases and the newly forming RNA molecule. Nucleotides are added to the 3'-OH end of the growing end of the RNA molecule and thus synthesis occurs in the 5' to 3' direction, as it does with DNA synthesis (Figure 3.2).

Eukaryotes have three types of nuclear RNA polymerase, called RNA polymerase I, II, or III, and each is responsible for transcribing three different classes of genes (Archambault and Friesen 1993). RNA polymerase I primarily is responsible for synthesis of class I genes, those coding for the large ribosomal RNAs in the nucleolus. Class II genes include all the DNA sequences that code

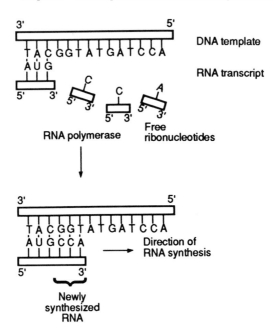

**Figure 3.2**    RNA synthesis involves polymerization of free ribonucleotides by an RNA polymerase in the 5′ to 3′ direction. Thus, the DNA template is read in the 3′ to 5′ direction.

for proteins and some small nuclear RNAs. Figure 3.3 illustrates several of the elements of a typical eukaryotic class II gene. Each gene consists of noncoding introns, coding exons, a promoter, and several (in this example, six) enhancer elements. Class III genes include the transfer RNA genes, 5S ribosomal RNA genes, and genes for some small nuclear RNAs.

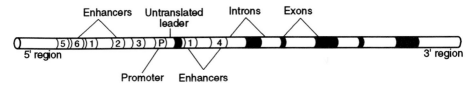

**Figure 3.3**    Components of a typical class II eukaryotic gene that codes for proteins include noncoding introns that are spliced out of the messenger RNA, coding exons (shown in black), a promoter to which RNA polymerase attaches to initiate RNA synthesis, and several enhancers (here numbered 1–6) that influence gene regulation.

# Transcription Involves Initiation, Elongation, and Termination

Transcription starts at the beginning of a gene. To initiate transcription, an RNA polymerase must bind at a specific point upstream of the gene to be transcribed. The specific attachment sites are called **promoters** and are typically 20 to 200 nucleotides long. Different eukaryotic promoter sequences are known, but certain common, or consensus, patterns occur. For example, many protein-coding genes contain the promoter sequences TATAAT and CAAT (Figure 3.4). They are often called the TATA and CAAT boxes. The location of the TATA sequence may vary, and not all genes have the TATA sequence. **Housekeeping genes** (genes that are expressed in all cells in order to maintain fundamental activities) may lack the TATA box and have a GC-rich region about 33 nucleotides upstream from the start site.

The actual sequences of the promoter vary somewhat from gene to gene, which may affect the extent to which each gene is expressed. Once RNA polymerase recognizes the specific attachment site, the base pairs are broken and the DNA double helix unwinds so that bases of the coding strand can be exposed to act as a template for transcript synthesis. Eukaryotic promoters differ from prokaryotic ones by having other types of DNA sequences called **enhancers** that influence the efficiency with which RNA polymerase II and accessory factors can

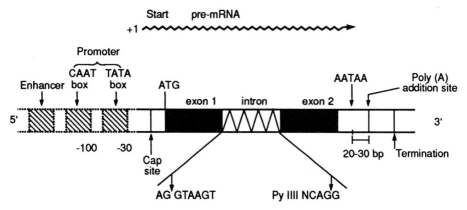

**Figure 3.4** A more detailed view of a eukaryotic class II gene. Promoters often have CAAT and TATA boxes upstream from the start site. The left junction (splice donor, AG GTAAGT) and right junction (splice acceptor, NCAGG) sequences of the intron are shown. Splice sites are indicated by the arrow between AG and GT of the splice donor and between the two Gs of the splice acceptor. mRNA synthesis is initiated at the +1 (start) site and proceeds in the 3′ direction. (Adapted from Beebee and Burke 1988, by permission of Oxford University Press.)

properly assemble at a promoter and initiate transcription of the DNA (Figures 3.3 and 3.4).

Elongation of the RNA molecule occurs as the RNA polymerase migrates along the DNA molecule, melting and unwinding the double helix as the RNA polymerase proceeds to sequentially attach ribonucleotides to the 3' end of the growing RNA molecule.

## RNA Transcripts Are Longer Than the Gene

The actual RNA transcript produced in eukaryotes is longer than the gene which it is transcribing because RNA polymerase transcribes a **leader sequence**, the length of which varies from gene to gene. Likewise, when the end of the gene has been reached, RNA polymerase continues to transcribe a **trailer segment** before terminating its activities.

The termination of class II genes remains poorly understood. Termination appears to occur hundreds or even thousands of nucleotides downstream of the 3' end of mRNA, which in turn generally lies about 35 nucleotides downstream from the site coding for a polyadenylation (poly(A)) signal, AAUAAA (Figure 3.4).

The number of noncoding introns in class II genes varies, as does their length. The boundaries between introns and exons often are determined by a consensus sequence to ensure that the introns are spliced out of the transcript in a precise manner (Figure 3.5).

## Pre-Messenger RNA Must Be Modified and Processed in Eukaryotes

In prokaryotes, the RNA transcript can be translated immediately into specific amino acid sequences. In other words, it is the messenger RNA (mRNA). In eukaryotes, however, RNA transcribed from DNA must be modified and processed before it can function as mRNA (Figure 3.5). Processing the pre-messenger RNA involves two activities: (1) modifying both ends of the RNA molecule, and (2) excising the noncoding sequences (introns) contained within the coding region.

Modifying the ends involves **capping** the 5' end and adding a **poly(A) tail** to the 3' end. Newly synthesized eukaryotic RNA molecules are capped at the 5' end by adding a terminal guanine that has been methylated on the 7 position and linked to the start site by an unusual 5'–5' triphosphate linkage (Figure 3.5). Capping appears to be necessary to enable the ribosome to bind with the mRNA before protein synthesis can begin. The methylated G nucleotide is added in a two-step process, with methylation occurring after a standard G has been added. In some eukaryotes, additional methyl groups may be added to one or both of the next two nucleotides of the mRNA molecule.

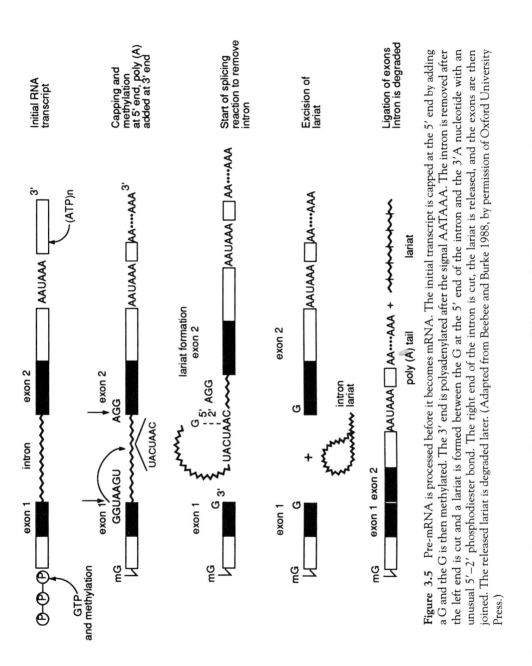

**Figure 3.5** Pre-mRNA is processed before it becomes mRNA. The initial transcript is capped at the 5′ end by adding a G and the G is then methylated. The 3′ end is polyadenylated after the signal AATAAA. The intron is removed after the left end is cut and a lariat is formed between the G at the 5′ end of the intron and the 3′A nucleotide with an unusual 5′–2′ phosphodiester bond. The right end of the intron is cut, the lariat is released, and the exons are then joined. The released lariat is degraded later. (Adapted from Beebee and Burke 1988, by permission of Oxford University Press.)

The 3' end of eukaryotic RNA is modified by adding 40–200 adenine residues to a region near the 3' end of the transcript to produce a poly(A) tail (Figure 3.5). The polyadenylation does not simply add the A residues to the end of the transcript. First, a cleavage occurs between 10 and 30 nucleotides downstream of a specific polyadenylation signal which in insects is usually AAUAAA and is found in the 3' noncoding region of the RNA. This results in an intermediate 3' end to which the poly(A) tail is added by the enzyme poly(A) polymerase. It has been speculated that the length of the poly(A) tail determines how long the mRNA survives in the cytoplasm before being degraded.

## Splicing Out the Introns

The third modification of the pre-mRNA involves splicing to remove any introns. Introns have a 5' donor and a 3' acceptor end with common consensus sequences (Figure 3.5). The 5' donor end typically has the sequence GGUAAGU. After a cut in the donor site, the G at the 5' end forms a loop by attaching to an A nucleotide a short distance upstream from the pyrimidines near the acceptor splice site. The consensus sequence of the 3' acceptor site is AGG. In the final step, a cut is made in the acceptor site and the intron is freed. The exons are then joined together. The excised loop (intron) is released as a lariat-shaped structure and is later degraded.

RNA splicing occurs in nuclear particles called **spliceosomes**. These particles are composed of protein and several types of small RNA molecules that are 100 to 200 bp long. The specificity of the splicing operation is determined by the small nuclear RNAs, which contain sequences that are complementary to the splice junctions (Lamond 1993, Wise 1993).

Introns normally have no function, and synthetic genes lacking introns can usually function quite well. However, some introns apparently are important in gene regulation and determine when, or in what tissue, the gene will be transcribed. Mutations in introns thus can be neutral or can alter gene regulation. Mutations in the splicing signals may result in two classes of mutations. If an intron is not spliced out, a mutant protein can be produced that functions abnormally. If splicing occurs at a different site than normal, an abnormal mRNA is produced and a mutant protein can be produced. Once mRNA is produced, it must be transported through the nuclear envelope to the cytoplasm, where it is translated.

## Translation Involves Protein Synthesis

Translation is the second stage of gene expression in which the information in the RNA is used to direct the synthesis of a polypeptide, the amino acid sequence of which is determined by the nucleotide sequence of the RNA. The genetic code consists of a triplet of adjacent ribonucleotides that specify an amino acid (Table 1.1). Translation requires ribosomes; transfer RNA; a set of enzymes to catalyze the attachment of each amino acid to its corresponding tRNA molecule (**aminoacyl tRNA synthetases**); and initiation, elongation, and termination factors.

Ribosomes are cellular organelles consisting of two subunits, each composed of ribosomal RNA and proteins. Ribosomes contain enzymes which create a peptide bond between amino acids, a site for binding one mRNA molecule, and two sites for holding tRNAs. Translation of the genetic information in eukaryotes begins when an mRNA molecule binds to the surface of a ribosome, and the initiation codon (AUG) is selected. Transfer RNAs carry an amino acid to the ribosome where they bind to the mRNA molecule attached to the ribosome. Transfer RNAs have a three-base sequence, called an **anticodon**, that is complementary to a specific codon in mRNA and a site to which a specific amino acid is bound (Figure 3.6). Binding between the mRNA and tRNA occurs by hydrogen bonds.

Peptide bonds are made between the successively aligned amino acids until the stop codon at the end of the mRNA is reached (UAA, UAG, or UGA) and the completed protein is released. Translation of the mRNA proceeds from the initiation codon (AUG) at the 5′ end of the molecule to the termination codon at the 3′ end. The polypeptide is thus synthesized from the amino end toward the carboxyl end by adding amino acids to the carboxyl end.

Transfer RNAs are small, single-stranded molecules ranging in size from 70 to 90 nucleotides (Figure 3.6). Internal complementary base sequences allow the molecule to form short double-stranded regions, which yields a folded molecule in which open loops are connected to each other by double-stranded stems. The three-dimensional structure of tRNA molecules is complex. One significant region is the anticodon sequence region, which consists of three bases that can base-pair with the codon in the mRNA. A second critical site is the 3′ end of the molecule where the amino acid attaches.

A specific enzyme called aminoacyl tRNA synthetase matches each amino acid with the tRNA attachment site. Transfer RNA molecules and their synthetases are designated by giving the name of the amino acid that is specific to each particular tRNA molecule. Thus, leucyl-tRNA synthetase attaches leucine to tRNA$^{Leu}$. If an amino acid is attached to a tRNA molecule, it is 'charged'. Usually, one, and only one, aminoacyl synthetase is found for each amino acid. However, there are fewer aminoacyl synthetases than there are codons for amino acids. Thus, the aminoacyl synthetases must recognize more than one codon. The **wobble hypothesis** suggests that base pairing is most critical with the first two bases but that pairing is extremely flexible in the third position.

Gene translation in eukaryotes usually involves structures more complex than a single ribosome moving along an mRNA molecule. Thus, after about 25

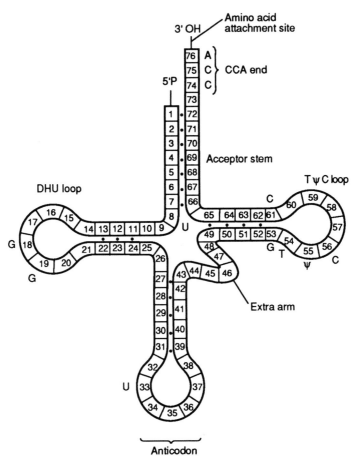

**Figure 3.6**    A tRNA molecule with its bases numbered from 1 to 76. A few bases that are present in almost all tRNA molecules are identified by letters. The Greek letter ψ is a symbol for the unusual base pseudouridine.

amino acids have been joined together in a polypeptide, the AUG initiation codon is free of the ribosome and a second polypeptide can begin to form. When the second ribosome has joined about 25 amino acids, a third ribosome can attach to the initiation site. This can result in mRNA molecules that are covered with many ribosomes, all moving in the same 5′ to 3′ direction, and this large unit is called a **polysome**. Figure 3.7 illustrates a polysome isolated from a midge larva, *Chironomus*. This electron micrograph, magnified about 140,000 times, shows the start of an mRNA molecule on the bottom right side (Kiseleva 1989). The structure at the top shows the end of the molecule, with the growing proteins attached to the ribosomes.

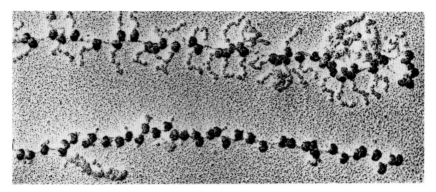

**Figure 3.7** A micrograph of polysomes from salivary gland cells of the midge *Chironomus*. Here the ribosomes are moving in order along messenger RNA, gradually extending their individual protein chains. At the bottom is the start of the polysome, at the top is the end.

## Gene Regulation in Eukaryotes

Many eukaryotic genes code for essential metabolic enzymes of cell components and are expressed constitutively at a specific level in all cells. Such genes are often called **housekeeping genes**. Many genes are not continuously expressed in complex eukaryotic organisms, however. mRNA molecules generally are short-lived, probably persisting only a few minutes or hours. Rapid turnover means that the amount of a particular mRNA in the cell can be controlled by adjusting the transcription rate of appropriate genes, or by other methods.

Genes are expressed in different cells at different times and at different levels. Thus, cells differentiate because of sequential changes in gene activity that are programmed in the genome, in response to molecular signals released by other cells, and because of cues transferred by physical contact between neighboring cells. After cells differentiate, gene regulation is influenced by environmental cues such as hormones, nutrients, or temperature. The control of gene expression is called **gene regulation.**

Gene regulation in eukaryotes is a complex and very active area of research. It is much less well understood than it is in prokaryotes. Gene expression in eukaryotes is achieved by a variety of methods, including (1) regulating the level of transcription, (2) alternative splicing of mRNA transcripts, (3) amplification of DNA, (4) programmed rearrangements of DNA, (5) methylation of cytosine bases, or (6) control of translation. There is no single regulatory mechanism. Different genes are regulated in different ways. Furthermore, regulatory mechanisms are often surprisingly complex and may employ more than one method.

Many eukaryotic genes are regulated at the level of transcription. Both negative and positive regulation can occur. Transcriptional regulation involves

**transcriptional activator proteins** that bind with an upstream DNA sequence to prepare a gene for transcription. They may help assemble a transcriptional complex that includes RNA polymerase, or they may initiate transcription by an already-assembled transcriptional complex. Some transcriptional activator proteins have a helix–turn–helix structure, which is a sequence of amino acids that forms a pair of α-helices separated by a bend. These helices fit into the grooves of a double-stranded DNA molecule and allow the proteins to bind to the DNA, although the specificity of the binding is determined by other parts of the protein (Harrison 1991). Examples of helix–turn–helix DNA-binding proteins in insects include the **homeodomain**, which is found in genes such as *engrailed* and *Antennapedia*, which are important in regulating *Drosophila* development (which is described in more detail in Chapter 5).

A second type of transcriptional activator protein is **zinc finger proteins**. These are characterized by loops (fingers) of repeating amino acid sequences, each associated with a zinc atom. Zinc finger proteins bind in the major groove of the DNA helix (Figure 3.8) of an upstream DNA sequence to prepare a gene for transcription.

A third type of DNA-binding protein is the **leucine zipper**. Leucine zippers are DNA binding proteins that contain four to five leucine residues separated from each other by six amino acids. The leucines on two protein molecules can interdigitate and dimerize in a specific interaction with a DNA recognition sequence (Abel and Maniatis 1989).

Hormones may turn on the transcription of specific sets of genes. For example, steroid hormones penetrate a target cell through diffusion because steroids

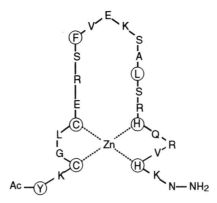

**Figure 3.8**    Amino acid sequence of a portion of a protein containing zinc fingers drawn to illustrate the finger motif. Zinc finger proteins are able to recognize specific DNA sequences. The fingers can intercalate into the DNA molecule. (Adapted from a figure provided by Peter E. Wright, Department of Molecular Biology, Research Institute of Scripps Clinic, La Jolla, California (From *Science*, 245, p. 635, 1989.))

pass freely through the cell and nuclear membranes. The nuclei of target cells contain specific receptor proteins that form complexes with the hormone, which then undergoes modification in its three-dimensional form and enables the receptor–hormone complex to bind with particular sequences in the DNA and stimulate or repress transcription. In the lepidopteran insect *Manduca*, **ecdysone**, a steroid, acts directly on the genome both to activate and repress genes. Ecdysone initiates and coordinates the molting process and thus the sequential expression of stage-specific genes (Riddiford et al. 1990).

Hormone–receptor complexes and transcriptional activator proteins bind with specific DNA sequences called enhancers. Enhancer sequences are 20 bp or less and can be found in a variety of sites in relation to the target gene (Figure 3.3). Enhancers can be long or short distances upstream (5′) from the target gene; they may be included in introns within the coding region of the target gene, or even at the 3′ end of the gene. Some enhancers respond to molecules produced inside the cell during development, and many genes are under the control of several enhancers so they can respond to a variety of internal and external molecular signals.

Figure 3.3 illustrates the components found in a typical class II eukaryotic gene. A transcriptional complex binds to the promoter to initiate RNA synthesis. The coding regions of the gene (the exons) are interrupted by introns that are eliminated in RNA processing. Transcription is regulated by enhancer elements (numbered 1–6) that respond to different molecules. Enhancers can be found both upstream and downstream of the promoter, and a specific enhancer may even be present in multiple copies, as enhancer number one is in Figure 3.3. Also, because many enhancers may exist in a eukaryote and these can respond to different signals or cell conditions, genes can be regulated by a combination of different enhancers. Combinatorial control of gene transcription makes it possible to increase the complexity of gene regulation. If transcription occurs according to the pattern of the binding state, then a small number of regulatory molecules can yield a large number of different regulatory patterns for different life stages or tissues.

Some genes have two or more promoters. Each promoter is usually active in different cell types and allows for independent regulation of transcription (Figure 3.9). The different promoters yield different primary transcripts that code for the same polypeptide. Thus, transcription in larvae could be controlled by one promoter (Figure 3.9B) and transcription in adults by another (Figure 3.9C).

Different cell types can produce different quantities of a protein or different proteins by **alternative splicing** of mRNA. Differential splicing of exons A and B can result in different rates of synthesis although the proteins may be identical. In other cases, the proteins produced are different as a result of alternative splicing.

Gene regulation can also occur by controlling translation. **Translational regulation** can occur in three different ways: (1) inability to translate an mRNA

A) Gene structure

B) Larval transcript and processing

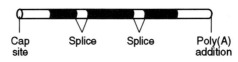

C) Adult transcript and processing

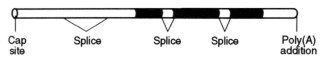

**Figure 3.9** Gene regulation can be achieved with the use of alternative promoters. (A) In this gene there are two introns within the coding region. (B) The mRNA transcript in larvae uses the promoter nearest the 5′ end of the coding region. (C) In adults, the promoter farther upstream is used and much of the leader sequence used is removed by splicing.

molecule unless a particular signal is present, (2) regulation of the longevity of a particular mRNA molecule, and (3) regulation of the rate of protein synthesis. An example of (1) is found in unfertilized eggs, which are biologically static. After fertilization, many new proteins are synthesized, including the mitotic apparatus and cell membranes. However, unfertilized eggs can store large quantities of mRNA for months in an inactive form that abruptly and rapidly becomes active within minutes after fertilization. The timing of translation is thus regulated. Eukaryotes also can coordinate the synthesis of several gene products by synthesizing a polyprotein, which is a large polypeptide that is cleaved after translation to produce several different proteins.

    Genes are contained in both chromosomal and extrachromosomal DNA in insects. The organization of DNA in chromosomes and mitochondria is described in Chapter 4.

## General References

Beebee, T. and J. Burke. 1988. *Gene Structure and Transcription*. IRL Press, London.

Brown, T. A. 1989. *Genetics: A Molecular Approach*. Chapman and Hall, London.

Drlica, K. 1984. *Understanding DNA and Gene Cloning. A Guide for the Curious.* Wiley, New York.
Hartl, D. L. 1991. *Basic Genetics.* 2nd Ed. Jones and Bartlett Publ., Boston.
Watson, J. D., N. H. Hopkins, J. W. Roberts, J. A. Steitz and A. M. Weiner. 1987. *Molecular Biology of the Gene.* Vols. I and II. 4th Ed. Benjamin/Cummings, Menlo Park, CA.

# References Cited

Abel, T. and T. Maniatis. 1989. Gene regulation: Action of leucine zippers. Nature 341: 24–25.
Archambault, J. and J. D. Friesen. 1993. Genetics of eukaryotic RNA Polymerases I, II, and III. Microbiol. Rev. 57: 703–724.
Harrison, S. C. 1991. A structural taxonomy of DNA-binding domains. Nature 353: 715–719.
Kiseleva, E. V. 1989. Secretory protein synthesis in *Chironomus* salivary gland cells is not coupled with protein translocation across endoplasmic reticulum membranes. FEBS Letters 257: 251–253.
Lamond, A. I. 1993. The spliceosome. BioEssays 15: 595–603.
Riddiford, L. M., S. R. Palli and K. Hiruma. 1990. Hormonal control of sequential gene expression in *Manduca* epidermis. Pp. 226–231. In: *Progress in Comparative Endocrinology.* Wiley-Liss, New York.
Wise, J. A. 1993. Guides to the heart of the spliceosome. Science 262: 1978–1979.

# Chapter 4

# Chromosomal and Extrachromosomal Organization of DNA in Insects

## Overview

DNA is found in chromosomes, which are very complex structures. Each chromosome contains a single linear DNA molecule combined with a variety of protein molecules, including histones. DNA and histones form structures called nucleosomes. Nucleosomes are arranged in a higher level of organization that remains only partially resolved. Chromosomes visible by light microscopy often have discrete structures, including centromeres to which spindle fibers attach so that chromosomes are distributed to the daughter cells in an orderly fashion in mitosis and meiosis. The centromeres of eukaryotic chromosomes are specific, and complex, regions along the chromatin fiber that play a fundamental role in chromosome movement. A full knowledge of centromere structure and function is not yet available.

Chromosomes always have telomeres at the ends, which are specialized structures that help to maintain the ends of chromosomes in a stable state. In polytene tissues such as salivary glands, banding patterns are visible with staining. The bands are alternating regions of euchromatin and heterochromatin. In most insects, there are two copies of each chromosome in somatic cells and hence two copies of each gene (2n = diploid complement). The reproductive cells (eggs, sperm) contain a single copy (haploid complement = n) of genes. Chromosomes contain single copy DNA (DNA present only once in the genome that codes for a polypeptide), highly-repetitive DNA, and moderately-repetitive DNA, as well as intergenic regions. Intergenic regions contain transcription and regulatory information, but a large amount of the material in this region is of unknown function. In addition, transposable elements have been identified in *Drosophila* and other arthropods. Transposable elements are DNA sequences that are capable of moving within and among chromosomes. Transposable elements, or their derivatives, may make up a significant portion of the middle-repetitive component of the genome in insects. Arthropods also contain extranuclear DNA in mitochondria.

## Introduction

DNA in eukaryotes is organized in a number of chromosomes, which are complex structures. Different insect species have a different number of haploid chromosomes, ranging from 1 to 221. There appears to be a great deal more DNA in eukaryotic organisms than is actually needed to code for the number of genes estimated for a specific species. This discrepancy is known as the **C value paradox**. For example, *Drosophila melanogaster* has approximately 165,000 kb of DNA, located in four chromosomes. If we assume that the average size of a gene is about 5 kb, and that *D. melanogaster* has approximately 20,000 genes, then *D. melanogaster* might be expected to require approximately 100,000 kb of DNA, leaving approximately 65,000 kb of 'superfluous' DNA. Even if DNA that codes for promoters, enhancers, replication origins, and other elements is added to the estimate, a large amount appears to have no function.

The genome size varies widely among insect species, with up to 75-fold differences in known C values. For example, the locust *Schistocerca gregaria* has a C value of 9,300,000 kb, 52-fold more than *Drosophila melanogaster* (Wagner et al. 1993). Among 37 species of tenebrionid beetles, nuclear DNA content varies by five-fold (Juan and Petitpierre 1991). Within the genus *Tribolium*, genome size ranges from 0.157 to 0.388 picograms (pg) of DNA (Alvarez-Fuster et al. 1991). Genome size variation seems to bear little relationship to organismal complexity or the number of genes encoded.

Nuclear DNA content also can vary within species. For example, C values in different populations of the mosquito *Aedes albopictus* vary by threefold (from 0.62 to 1.6 pg) (Kumar and Rai 1990). Intraspecific variation in nuclear DNA content has also been found in *Tribolium* species (Alvarez-Fuster et al. 1991).

The amount of DNA in insect cells is difficult to measure because many tissues are polyploid, with different tissues having different degrees of ploidy. **Polyploidy** occurs when the amount of DNA in an organism increases over the usual diploid amount, usually by duplicating the number of chromosomes. The diploid blood cells of *Bombyx mori* have 1 pg of DNA per blood cell, but polyploid silk gland cells in the same insect contain 170,000 pg of DNA per nucleus. DNA content also varies with developmental stage. At metamorphosis, the amount of DNA in *B. mori* declines by 81% after adults emerge from the pupal stage, which is probably due to histolysis of the polyploid larval silk glands and other polyploid cells.

Noncoding DNA can constitute from 30% to more than 90% of the insect genome. This noncoding DNA has been called '**junk**' or '**parasitic**' or '**selfish**'.

There are several hypotheses to explain its persistence in genomes and it is likely that several are valid. One suggests that the noncoding DNA performs essential functions, such as global regulation of gene expression. According to this hypothesis, the 'junk' DNA is wholly functional and deletions of such DNA would have a deleterious effect. Many regulatory sequences lie outside the protein-coding sequences in the 'junk' DNA. A second hypothesis is that the noncoding DNA is useless, but is maintained because it is linked physically to functional genes; the excess DNA is not eliminated because it does not affect fitness of the organism and can be maintained indefinitely in the population. A third hypothesis suggests that the noncoding DNA is a functionless parasite that accumulates and is actively maintained by selection. A fourth hypothesis is that the DNA has a structural function, perhaps for compartmentalizing genes within the nucleus, or for maintaining a structural organization (nucleoskeleton) within the nucleus (Manuelidis 1990, Manuelidis and Cher 1990).

It is impossible to discard any of these hypotheses at this time, but the origin and role of noncoding DNA is an active area of research and discussion. Eventually, analyses of the origins and roles of noncoding DNA sequences in insects may contribute to our understanding of the evolution of genomes. For example, in one study, a correlation was found between slow development and higher DNA content in the mosquito *Aedes albopictus*. This suggests that DNA content is not selectively neutral (Ferrari and Rai 1989).

## Repetitive DNA Is Common in Insects

Much of the noncoding DNA in insects is **repetitive DNA**—nucleotide sequences that are repeated several to millions of times. Repetitive DNA has been classified as **highly repetitive** (composed of sequences repeated several hundred to several million times per genome), or as **middle repetitive**. Highly-repetitive DNA is found in heterochromatin near centromeres, telomeres, and other heterochromatic regions. Middle-repetitive DNA sequences are repeated from 100 to 10,000 times and include genes that code for ribosomal RNA and transfer RNA. Middle-repetitive sequences are found in euchromatic regions, as well as in constitutive heterochromatic regions of the genome. Repetitive sequences now are known to contain DNA needed for gene regulation. Some repetitive sequences also have a crucial role in maintaining the structure of the genome.

Each species has a number of different repeated elements in its genome. For example, *Drosophila melanogaster* has about 42% of its genome as repetitive heterochromatic DNA, and about 60% of the genome of *Drosophila nasutoides* is repetitive DNA. More than 90% of the genome can be noncoding repetitive DNA. The aphid *Schizaphis graminum* has approximately 20% of its genome as middle-repetitive DNA and 78% of its genome as unique sequence DNA (Ma et al. 1992). Surprisingly, *S. graminum* has very little highly-repetitive DNA, and it is speculated that the lack of highly-repetitive DNA is associated with a faster development time. If there is a causal relationship between genome size and development time, selection for rapid development may have maintained the aphid genome relatively free of highly-repetitive DNA (Ma et al. 1992).

Repetitive and single-copy DNAs are present in different patterns in different insects. The amount and organization of repetitive DNA within the genome occurs in two general patterns. The **short-period interspersion pattern** has single-copy DNA, 1000–2000 bp long, alternating with short (200–600 bp) and moderately long (1000–4000 bp) repetitive sequences. This pattern is found in the house fly *Musca domestica*, the Australian sheep blowfly *Lucilia cuprina*, and the wild silk moth *Antheraea pernyi*.

A **long-period interspersion pattern** of genome organization involves long (>5600 bp) repeats alternating with very long (>12 kb) uninterrupted stretches of unique DNA sequences. This pattern often is found in species with small genomes (0.10–0.50 pg of DNA per haploid genome), and is found in *D. melanogaster*. The long period interspersion pattern also is found in the aphid *Schizaphis graminum* (Ma et al. 1992), *Chironomus tentans* (Wells et al. 1976), the fleshfly *Sarcophaga bullata* (Samols and Swift 1979), the honey bee *Apis mellifera* (Crain et al. 1976), and the flour beetle *Tribolium castaneum* (Brown et al. 1990).

Even within an insect family, genome organization can be variable. The genome of four mosquito species varies from 0.186 to 0.899 picograms (pg) in total DNA, and the amount of repetitive elements varies from 0.009 to 0.150 pg of foldback DNA (Black and Rai 1988). One species, *Anopheles quadrimaculatus*, has a long-period interspersion type of genome organization while *Culex pipiens*, *Aedes albopictus*, and *Aedes triseriatus* have the short-period interspersion type of genome organization. Generally, the amounts of foldback, highly-repetitive, and middle-repetitive DNA increase linearly with genome size in these mosquitoes. Intraspecific variation in the amount of highly-repetitive DNA was found in *Aedes albopictus* colonies and may be due to differences in the number or type of transposable elements. The amount of repetitive DNA in mosquitoes varies from 20% in *Anopheles quadrimaculatus* to 84% in *Aedes triseriatus* (Besansky and Collins 1992). Because the genome organization of relatively few insect species has been studied, it is difficult to determine the significance of these patterns.

**Satellite DNA** is a type of highly-repetitive DNA that differs sufficiently in its base composition from the majority of DNA in a eukaryotic species to separate out as one or more distinct bands when DNA is isolated by centrifugation with cesium chloride. Satellite DNA is rich in either A+T sequences or in G+C sequences and is found in long tandem arrays within the heterochromatic regions of chromosomes.

## Composition of Insect DNA

Insect DNA base ratios are lower than those found in vertebrates (Berry 1985). Guanine + cytosine bases comprise from 32 to 42% of the DNA, compared with 45% for vertebrates. If base composition were random, 50% of the DNA would be G+C.

DNA in eukaryotes can occur in different configurations. Most genomic DNA exists in the B-helix form, but other configurations are known, including triplex or H-DNA. In **triplex** DNA, the usual A–T and C–G base pairs of duplex DNA are present, but in addition a pyrimidine strand is bound in the major groove of the helix. DNA sequences that potentially can form triplex DNA structures appear to be common, are dispersed at multiple sites throughout the genome, and comprise up to 1% of the total genome. Triplex DNA was identified in polytene chromosomes of *Chironomus tentans* and *Drosophila melanogaster*, where it was found in the euchromatic bands (Burkholder et al. 1991). Triplex DNA is thought to play a role in chromosome organization by helping to condense chromatin.

## Chromosomes Are DNA Plus Proteins

Eukaryotes must organize and package their DNA in a form sufficiently condensed to fit into a very small space in the nucleus. This packaging must be compatible with the ability to unzip and unwind the DNA helix during DNA replication and transcription. Furthermore, this must occur rapidly. Precise and rapid replication of DNA is required in many tissues during cell division in mitosis and meiosis. How this is achieved is not yet fully resolved, although organization at the nucleosome level has been characterized in some detail.

Eukaryotic genes are found on linear DNA molecules, with each chromosome containing a single long DNA molecule. In addition, each chromosome contains an approximately equal amount of proteins with different functions. The proteins include DNA and RNA polymerases and regulatory proteins associated with the DNA. In addition, at least five **histones** are associated with the DNA in structures called **nucleosomes** (Figure 4.1). Histone proteins (called H1, H2A, H2B, H3, and H4) contain approximately 100 to 200 amino acids, of which 20 to 30% are arginine and lysine. As a result, the histones have a positive charge which helps them bind to DNA in a complementary manner by electrostatic attraction to the negatively charged phosphate groups in the sugar–phosphate DNA backbone.

The chromosomal DNA in a nucleosome (called **core DNA**) is connected by **linker DNA** to the next nucleosome (Figures 4.1 and 4.2B). The core DNA is protected from digestion by restriction enzymes, but the linker DNA is vulnerable to these enzymes. The core DNA is wound approximately 1.75 times around

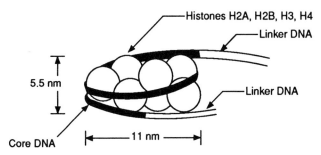

**Figure 4.1** DNA is condensed in eukaryotes by being packaged in nucleosomes, which are only the first level of condensation. A nucleosome consists of core DNA wound around two molecules each of the histones H2A, H2B, H3, and H4. These eight histone molecules are called an octamer. Nucleosomes are connected to other nucleosomes by linker DNA. In addition, a single molecule of the histone H1 (not shown) binds in the linker and helps to condense the nucleosome. Nucleosomes are organized into structures called 30-nm fibers (see text and Figure 4.2). (Adapted from Brown 1989.

a histone octamer consisting of two molecules each of H2A, H2B, H3, and H4. There is a single molecule of histone H1 that is associated with the linker DNA and it is apparently helpful in compacting the nucleosome. Linker DNA lengths vary from 20 to 100 nucleotide pairs in different species and in different cell types within the same organism. The genetic function of the linker DNA and the reasons for variation in its length are unknown.

DNA sequences that code for the major histones are very highly conserved among different eukaryotic species, which suggests that they are nearly unchanged over billions of years. Histones apparently are crucial to maintaining chromosome structure, and could be important in gene regulation by acting as a recognition signal for regulatory proteins (Tordera et al. 1993).

The origins of DNA replication in higher eukaryotes are larger and more complex than in prokaryotes (see Chapter 3). However, the underlying molecular processes are conserved (Benbow et al. 1992). Replications of origin are discussed further in Chapter 5.

## Packing Long Thin DNA Molecules into the Nucleus

Eukaryotes have to solve a serious packaging problem. If chromosomes were simply linear DNA molecules, the average length of a eukaryotic chromosome might be about 5 cm. If these long, thin chromosomes were tangled among the other chromosomes, replication would be difficult and separation of the intertwined chromosomes during mitosis could result in breakage of the chromosomes and subsequent loss of essential genetic information. Therefore, DNA is con-

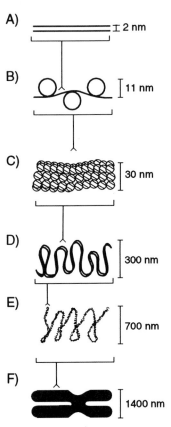

**Figure 4.2**   DNA packing in eukaryotic chromosomes must be efficient to achieve a dramatic reduction in DNA length. This figure illustrates the method by which DNA is thought to be packed, although the organization of elements in (D) through (F) is controversial. (A) Naked DNA. (B) DNA in nucleosomes. (C) The 30-nm chromatin fiber. (D) The 300-nm fiber made up of looped 30-nm fibers. (E) The 700-nm supercoiled structure that comprises the arms of a metaphase chromosome. (F) The metaphase chromosome. (Adapted from Brown 1989.)

densed, yet packaged, so that both DNA replication and transcription can occur without loss or damage. DNA packaging is achieved by a highly organized and hierarchical condensation scheme, the full details of which remain controversial (Figure 4.2).

We know that eukaryotic DNA is **supercoiled**, which means that the double helix is twisted around itself, which begins the condensation process. The next level of compaction is achieved by organizing the DNA into nucleosomes. Nucleosomes occur in a regular pattern, with linker or intervening DNA between each nucleosome (Figure 4.2B). Nucleosomes reduce the length of DNA by a factor of about six to a flexible beaded fiber.

Additional condensation of the DNA occurs when nucleosomes are condensed into a shorter, thicker fiber, called the **30-nm fiber** (Figure 4.2C). The precise structural organization of this chromatin fiber is still debated. It appears to be an irregular left-handed super helix with six nucleosomes per turn. Another level of condensation is thought to occur in metaphase chromosomes (Figures 4.2D–4.2F), although the details of this compaction remain conjectural. It is likely that the 30-nm chromatin fiber form is found in both dividing cells and cells that are not actively dividing (Figure 4.2C). Figures 4.2D–4.2F suggest one model for packing the 30-nm fiber into the highly condensed form found in metaphase chromosomes. Somehow, the length of the chromosomal DNA is reduced by a factor of approximately 10,000 in mitotic metaphase chromosomes.

Electron micrographs suggest that chromosomal DNA extends from a central core or scaffold composed of nonhistone chromosomal proteins, but the details of this organization are unknown. Additional organization within the nucleus is likely, because individual chromosomes are probably anchored in specific regions of the nuclear membrane. Chromosomes may even be located in specific regions of the nucleus, as determined by the genetic activity of the particular cell type.

## Euchromatin and Heterochromatin

**Euchromatin** is uncoiled during interphase and condensed during mitosis and meiosis, with a maximal condensation at metaphase. Euchromatin contains most of the coding DNA. In polytene salivary gland chromosomes of *Drosophila*, the banded segments are euchromatic and the intervening regions are heterochromatic.

The term **heterochromatin** was coined to define the chromosome regions that remain condensed during most of the cell cycle and have a coiling cycle out of phase with the rest of the genome. Unlike euchromatin, heterochromatin exhibits maximal condensation in nuclei during interphase. **Constitutive heterochromatin** replicates late in the cell cycle compared with euchromatin, and contains a considerable amount of middle- and highly-repetitive DNA. In many organisms, large regions of the chromosome near the centromeres and the telomeres are constitutive heterochromatin. Recombination during meiosis occurs at a lower rate in heterochromatic regions of the chromosome, and heterochromatic regions exhibit little or no transcription of the DNA. Some chromosomes are **facultatively heterochromatic**, with one homologous chromosome behaving differently than the other.

The biological significance of heterochromatin is largely unknown. At one time, heterochromatin was considered to be genetically inert, but it is now known that some heterochromatic regions in *D. melanogaster* have a genetic function. Heterochromatin can cause **position effect variegation** by inactivating euchromatic genes that have been moved to regions adjacent to hetero-chromatin by chromosomal rearrangements. Also, a number of genes are now known to be interspersed within heterochromatic regions. For example, the Y chromosome of *D. melanogaster* is heterochromatic, yet carries a set of genes, designated as fertility factors, that are required for male fertility (Gatti and Pimpinelli 1992). These elements are active only in the primary spermatocyte, where some of them form large lampbrush loops covered with RNA and pro-teins. These transcribed regions are heterochromatic and contain short, tan-demly arranged clusters of DNA interspersed with longer, more heterogeneous DNA. Some of the heterogeneous DNA appears to consist of transposable elements. The proteins may be involved in regulatory events at the post-transcriptional level later in spermiogenesis.

## Centromeres

Most chromosomes possess a special region which is important in the orga-nization of the developing spindle prior to mitosis and the separation of the daughter chromosomes at anaphase. Chromosome fragments lacking cen-tromeres, **acentric fragments,** do not get transmitted to daughter cells and are eventually lost. Some species do not seem to have localized centromeres, and the whole chromosome appears to have centromeric properties (**holocentric chro-mosomes**). If these chromosomes are fragmented, each portion can attach to the spindle and chromosome fragments are not lost at mitosis. Holocentric chromo-somes are found in the Homoptera, Heteroptera, Mallophaga, Anoplura, and Lepidoptera (White 1973).

The chromosomal region adjacent to the centromere contains very long blocks of highly-repetitive DNA in which simple sequences are repeated thou-sands of times (satellite DNA). There may be several different satellite DNA types in a given species; for example, three satellites are found near centromeric DNA of *Drosophila virilis*. One has a 5'-ACAAACT-3' repeat, the second is 5'-ATAAACT-3', and the third is 5'-ACAAATT-3'. These satellite DNAs apparently are not transcribed and they may bind proteins essential for cen-tromere function. Likewise, for the sheep blowfly *Lucilia cuprina*, several sub-families of satellite DNA are present in the centromeric regions of the chromo-somes, as well as in the sex chromosomes (Perkins et al. 1992).

While each species normally has several types of satellite DNA, two parasit-ic wasps, *Diadromus pulchellus* and *Eupelmus vuilleti*, have only one type (Bigot et al. 1990). In these two species, satellite DNA constitutes 15 and 25% of the genome, respectively. Likewise, approximately 50% of the genome of the ten-ebrionid mealworm *Tenebrio molitor* consists of only one type of satellite DNA which is distributed evenly over the centromeric regions (Plohl et al. 1992). In the tenebrionid *Palorus ratzeburgii*, approximately 31% of the genome is a single type of satellite DNA (Ugarkovic et al. 1992). The satellite sequences of the two tenebrionids are not similar.

## Telomeres

The ends of the chromosomes have distinct structures called **telomeres**. The nature of telomeres has been somewhat mysterious because they were difficult to visualize (White 1973). Telomeres have two important functions: (1) to prevent the 'sticky' ends of chromosomes from fusing with each other, and (2) to ensure that the entire chromosome is replicated during mitosis and meiosis. The telomere provides a protective cap for the end of the chromosome and may help establish the three-dimensional architecture of the interphase nucleus, because telomeres often are positioned close to the nuclear membrane. Telomeres mediate transient associations between both homologous and nonhomologous chromosomes and facilitate replication of DNA at the ends of chromosomes.

Molecular analysis indicates that telomeres consist of a series of repeated nucleotides and proteins that bind specifically to these sequences (Blackburn 1991, Wagner et al. 1993, Zakian 1989). As described in Chapter 2, DNA replication conventionally occurs only in the 5' to 3' direction and cannot be initiated without a primer, which is usually RNA. After primer removal, gaps would remain at the 5' ends of new DNA strands in eukaryotes if it were not for telomeres. Telomeres prevent the gradual loss of genetic information from the ends of chromosomes.

Neither the centromere nor the telomere of *Drosophila* has been cloned, but efforts are under way to clone these important elements (Block et al. 1990). The cloning of telomeres and centromeres offers the possibility of eventually being able to develop artificial chromosomes for insects. The ability to construct artificial yeast chromosomes has made a variety of molecular genetic studies feasible.

Telomeric DNA sequences and structure appear to be similar among widely divergent eukaryotes. The essential telomeric DNA consists of a stretch of a very simple, tandemly repeated sequence. This repeated sequence is TTGGGG (5' to 3') and TTAGGGG for the plants *Arabidopsis* and maize, respectively. A few hundred to a few thousand of these repeated sequences seem to produce a stable telomere. Partial telomeric sequences have been isolated from *Drosophila*, but their structure remains unresolved. A telomeric sequence, TTAGG, has been isolated from the silkworm *Bombyx mori* (Okazaki et al. 1993). This sequence, TTAGG, was found in telomeres from a variety of other insect species. So far, this pentanucleotide repeat has been found only in arthropods, and Okazaki et al. (1993) suggested that this telomeric sequence may have arisen during evolution of the arthropods. However, this sequence did not appear to be present in dipteran and some coleopteran insects, which suggests there are several types of telomeric structures in insects.

A **telomere terminal transferase** or **telomerase**, discovered first in the protozoan *Tetrahymena*, is essential for producing telomeres. Telomerase is a ribonucleoprotein whose RNA component is essential. Telomerase recognizes single-stranded oligonucleotides ending in -GGG-3′ in a variety of eukaryotes and adds GGGG sequences to the ends of chromosomes. Telomerase thus contains its own template, serving as a kind of reverse transcriptase, since its own RNA codes for a DNA sequence.

In addition to the terminal repetitive sequences, another group of repetitive sequences are adjacent to, and internal to, the telomeric sequences; these are called subtelomeric sequences.

Telomeres are crucial for maintaining the ends of chromosomes in an intact form. If telomeres are damaged, progressive loss of DNA occurs in the chromosome and the sticky ends of damaged chromosomes will bind to other chromosomes with sticky ends, resulting in abnormalities such as dicentric chromosomes. Genetic and physiological factors influence the length of telomeres, which appears to decline with age in humans, perhaps as a function of cell division number. The gradual loss of telomeric DNA could lead to chromosome instability and contribute to aging and senescence (Blackburn 1991).

## Chromosomes during Mitosis and Meiosis

Chromosomes are visible by light microscopy during the cell divisions called mitosis and meiosis. The following paragraphs review the basic aspects of two types of cell division. Mitosis occurs in somatic cells, while meiosis occurs only in germ-line cells in eukaryotic organisms.

### Mitosis

Somatic cells divide by **mitosis**, which produces two identical daughter cells, each containing the same chromosome complement as the parent (Figure 4.3). Prior to the onset of mitosis, the chromosomes within the nuclear membrane are not visible by light microscopy, although **nucleoli** are often visible. Nucleoli are RNA-rich spherical bodies associated with specific chromosomal segments (called the nucleolus organizer). The nucleolus organizer segment of the chromosome contains ribosomal RNA genes. The spherical nucleolus contains the primary products of the ribosomes, their associated proteins, RNA polymerase, and RNA methylases.

Cells that are not undergoing mitosis are in the **interphase** state. DNA synthesis and chromosome duplication take place during a portion of the cell cycle called the **S phase** (for synthesis). The cell cycle consists of four phases: $G_1$

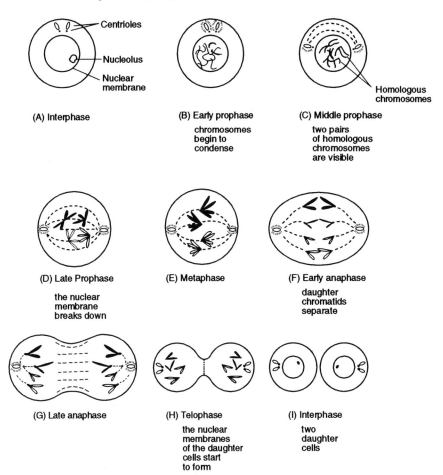

**Figure 4.3** Mitosis of a diploid cell involves duplication of each homologous chromosome and its distribution to the daughter cells. DNA replication occurs during interphase. During prophase the two daughter chromatids are attached to each other at the centromere. During metaphase, the chromosomes line up on the metaphase plate and during anaphase the daughter chromatids separate and begin moving to opposite poles. During telophase the nuclear membranes re-form and two identical daughter cells with a complete complement of chromosomes are produced.

$\rightarrow S \rightarrow G_2 \rightarrow M$. The $G_1$ and $G_2$ phases do not involve DNA synthesis. The M phase represents mitosis, in which the duplicated chromosomes and the cytoplasm are divided into two identical daughter cells. The length of a complete cell cycle varies by cell type, but typically is approximately 18–24 hours, with mitosis requiring 0.5 to 2 hours in different organisms (Figure 2.6). Cells not undergoing mitosis are in interphase, and are in either the $G_1$, S, or $G_2$ phase of the cell cycle.

Mitosis is divided into four stages: prophase, metaphase, anaphase, and telophase (Figure 4.3). In prophase, the chromosomes condense to form visible, thin threads within the nucleus. During prophase some chromosomes can be seen to be double, with two closely associated subunits called **chromatids** connected at the **centromere**. A centromere is a specialized structure that is an attachment point for the fibers that draw each of the divided chromosomes into their respective nuclei during cell division. Centromeres contain larger nucleosomes than elsewhere in the chromosome and specialized nucleotide sequences are associated with the functioning of the centromere. The complete details of the structure and function of centromeres are not yet available. Each pair of chromatids resulted from the replication of a chromosome during the S phase of the cell cycle. As prophase progresses, the chromosomes shorten and thicken. In late prophase, the nuclear membrane disappears and a mitotic spindle forms.

The spindle is a structure consisting of two bundles of fibers that extend between the opposite poles of the cell, and attach to the centromere of each chromosome. After the chromosomes are attached to spindle fibers, the chromosomes move toward the center of the cell in a plane equidistant from the spindle poles. When the chromosomes are lined up on the metaphase plate, they are at their most condensed stage, making it easiest to examine them for differences in morphology. During the next stage, anaphase, the centromeres divide and the two sister chromatids become independent chromosomes that move toward the opposite poles. At the end of anaphase, a complete set of chromosomes lies near the opposite poles of the spindle. During the last phase, telophase, a nuclear membrane forms around each group of chromosomes and the spindle disappears. The chromosomes subsequently become less condensed and invisible by light microscopy. The cytoplasm is divided into two during telophase by a gradually deepening furrow and the result is two daughter nuclei that enter interphase. The result of mitosis is, ideally, the formation of two identical cells with perfectly duplicated genetic information.

### Meiosis

The result of the cell division called **meiosis** is to form cells in the germ line tissues with a reduced number of chromosomes (the haploid or n number) so that when the germ cells (eggs and sperm) fuse, the diploid (2n) number of chromosomes is restored (Figure 4.4). Meiosis requires two cell divisions (I and

---

**Figure 4.4** Meiosis takes place in the germ line tissues and is a two-step process that results in the production of four haploid cells from a single precursor cell. Meiosis I reduces the chromosome number to the haploid state and involves a lengthy prophase, brief metaphase, anaphase, and telophase. The cells may immediately enter meiosis II. During meiosis II the cells divide to yield four haploid cells.

## MEIOSIS I

(A) Early prophase (B) Middle prophase (C) Late prophase

crossing over may
occur between
homologous
chromosomes

 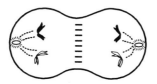 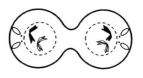

(D) Metaphase (E) Anaphase (F) Telophase

segregation of
homologous
chromosomes

## MEIOSIS II

(G) Interphase (H) Prophase

   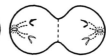

(I) Metaphase (J) Anaphase

segregation of
daughter chromatids

(K) Telophase

II) to produce daughter cells with a haploid set of chromosomes. Meiosis I is the *reductional division*, in which the number of chromosomes is reduced from 2n to n. Both divisions in meiosis have prophase, metaphase, anaphase, and telophase stages, but they are different in the two meiotic divisions (Figure 4.4). Meiosis may require days or weeks to complete. The essence of meiosis is that only *one* duplication of the chromosomes occurs, but *two* cell divisions occur, producing four haploid gametes from the original diploid cell.

Prophase of meiosis I is a long stage, and has been divided into substages (Figure 4.4). During prophase I, the chromosomes condense and become visible. Homologous chromosomes pair and become closely associated along their length. Each homologous chromosome consists of two sister chromatids joined at the centromere. The pairing of the homologous chromosomes thus produces a four-stranded structure. During prophase I, the paired chromosomes are able to exchange genetic information by **crossing over**, which results in a shuffling of the genetic information (recombination) in the gametes. The locations where genetic information was exchanged by crossing over are indicated by the formation of **chiasmata**. Chiasmata result from the physical exchange of nucleotides between chromatids of homologous chromosomes.

During metaphase I, the two homologous chromosomes are located on opposite sides of the metaphase plate (Figure 4.4). The orientation of each chromosome pair relative to the two poles is random and thus which member of each pair of chromosomes will move to a particular pole is random. This random alignment of chromosomes on the metaphase plate is the basis of Mendel's **Law of Independent Assortment**. Thus, if two different genes are not linked (located on the same chromosome), they will assort independently during meiosis.

During anaphase I, the homologous chromosomes, each composed of two chromatids joined at the centromere, separate from each other and move to opposite poles. This physical separation of homologous chromosomes during anaphase I is the physical basis of Mendel's **Law of Segregation**. Paired chromosomes segregate or separate so that each daughter cell is equally likely to contain either one. After anaphase, a haploid set of chromosomes consisting of one homolog from each pair is located near each pole of the spindle. During telophase I, the spindle breaks down (Figure 4.4). Chromosomes may pass directly from telophase I to prophase II of meiosis II. Alternatively, there may be a pause between the two meiotic divisions. Chromosome duplication does not occur between meiosis I and II, however.

Meiosis II is similar to a mitotic division, with each daughter cell from meiosis I being replicated, resulting in four haploid cells from the original diploid cell (Figure 4.4). Meiosis II is different from mitosis, because the chromatids of a chromosome are usually not identical along their entire length. This is due to crossing over, which occurred during prophase of meiosis I and resulted in an exchange of genetic information between the chromatids.

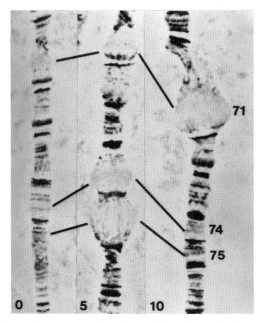

**Figure 4.5**   Polytene chromosomes from *Drosophila* salivary glands with puffing of different regions.

Each metaphase chromosome has a distinct morphology that is identifiable by staining with lactic-acetic orcein or other stains (Figure 4.5). The location of the centromere allows cytogeneticists to distinguish particular chromosomes. The arms of the chromosome take up stains in a banding pattern that is characteristic of an individual chromosome. The bands are called heterochromatic or euchromatic, depending upon the degree of staining achieved. The euchromatic bands are thought to contain single-copy DNA (genes), while the heterochromatic regions are thought to contain primarily repetitive DNA or noncoding regions. The relationship between the chromosomal bands and the nucleosomes and 30-nm fiber structures is unclear.

## Chromosome Damage

Chromosome damage probably occurs continuously in all cells, with the types of damage ranging from the mispairing of bases, which results from mistakes made by polymerases during replication, to chromosome breakage. Damage is caused by many factors, including the production of metabolic mutagens and other factors within the cells. Also damaging are ionizing radiation and ultraviolet (UV) light. Cells have active repair processes. Repairs occur by direct reversal of damage and by excision of a damaged segment of DNA followed by its replacement. Insects no doubt have many genes involved in DNA repair, with some genes encoding products that recognize DNA damage, some that can excise the damaged region, and others that can repair the damage.

Chromosome breaks can occur at any stage of the cell cycle, and generally are repaired by rejoining the broken ends, with the repaired chromosome appearing intact. Unfortunately, not all chromosomal damage is repaired, and chromosomal breaks can lead to large-scale rearrangements of chromatin within chromosomes or exchanges of chromatin between nonhomologous chromosomes. If the rearranged chromosome lacks a centromere, it is **acentric** and a cell containing the acentric fragment will be unable to transmit this fragment to its daughter cells during meiosis or mitosis, resulting in the loss of significant amounts of genetic information, which is usually lethal. Chromosomes that end up with two centromeres (**dicentric**) are also unstable, leading to breaks in the chromosome if the centromeres are distributed to opposite poles during meiosis or mitosis. This results in breakage and loss of chromosomal information.

## Polyteny

In a normal chromosome replication cycle, chromosomes contract, replicate, divide, and segregate to daughter cells. In **polytene** cells, ten or more replication cycles may occur and the daughter chromosomes remain in an extended state and do not separate. The daughter chromosomes stay synapsed, with homologous chromomeres aligned, which gives rise to a characteristic banding pattern along the length of the chromosome. In addition, the maternal and paternal homologous chromosomes may synapse in some species, which results in an apparently haploid number of giant chromosomes.

Polyteny is particularly common in larval salivary glands of Diptera, especially in flies from the families Drosophilidae, Chironomidae, Cecidomyiidae, and Sciaridae, but it also occurs in the midgut and fat body in these insects. Polyteny also occurs in Collembola. The number of rounds of replication that produce polytene chromosomes varies from tissue to tissue, with the largest chromosomes found in the salivary glands, where there may be as many as 1000 to 2000 chromatids per chromosome.

In *Drosophila* salivary gland chromosomes, the DNA packing varies by region, which produces bands and interband regions that stain differentially with DNA-specific stains. This makes it easy to identify specific sites on *Drosophila* salivary chromosomes (Figure 4.5). About 5000 bands have been identified in *D. melanogaster*, and the banding pattern provides a detailed cytological map of the chromosomes. *D. melanogaster* has four pairs of chromosomes; chromosomes 2 and 3 are large, with central centromeres, and chromosome 4 is the shortest.

Females have two X chromosomes, while males have an X and a Y chromosome. The Y chromosome is heterochromatic, and has only a few genes.

Because polytene salivary gland chromosomes of *Drosophila* are large and have a well-defined morphology, they are used to localize specific genes by *in situ* hybridization. In this process, radiolabeled DNA or RNA **probes** can be added to salivary gland cells that have been squashed on glass slides. A probe is a molecule labeled with radioactive isotopes, or another tag, that is used to identify or isolate a gene, gene product, or protein. The labeled probes will anneal to the homologous DNA after the chromosomal DNA is denatured. After any excess probe is washed off, the position of the specific gene can be located on a specific band or interband region of a specific chromosome by the presence of radioactive grains on an X-ray film. Genes, identified by a particular phenotype, also can be localized to specific sites in polytene salivary gland chromosomes if the mutation is associated with duplications, deletions, inversions, translocations, or other chromosomal abnormalities that can be detected by abnormal banding patterns.

Polytene chromosomes are thought to represent a very special case of the more general phenomenon of **endopolyploidy** (White 1973). In endopolyploidy, an increase in chromosome number occurs within the nuclei of certain tissues without a breakdown in the nuclear membrane. Thus, chromosome duplication takes place, but the chromosomes do *not* remain synapsed after replication. Many insect cells are endopolyploid, with 4n, 8n, 16n, and so on numbers of chromosomes.

## Chromosomal Puffing

At particular stages in the development of many Diptera, some of the genes in polytene chromosomes undergo swelling or **puffing**. Puffing is generally believed to be correlated with gene activity, and involves an unraveling of the DNA in a region of the chromosome approximately one to ten bands in length. The patterns of puffing differ in different instars in *D. melanogaster*, indicating that different genes are active in different instars. Puffing is controlled by the hormone ecdysone, heat shock, and other environmental conditions.

The largest puffs contain genes coding for proteins that are produced in very large amounts, such as the salivary gland secretions and silk. Puffs are associated with extensive transcription of DNA (Figure 4.5).

## B Chromosomes

**B chromosomes** are a heterogeneous class of chromosomes that are also called accessory or supernumerary chromosomes. B chromosomes are found in some plants and animals and are usually heterochromatic. They seem to have little effect on the phenotype of the organisms, vary in number from one cell type to another, occur in some individuals within a species but not in other individuals, and do not always segregate normally in mitosis or meiosis. Many species of Orthoptera have B chromosomes.

The maintenance of chromosomes without apparent function is puzzling. Perhaps B chromosomes have some genetic role, or perhaps they are parasitic remnants of chromosomes that have lost their function.

## Sex Chromosomes

In eukaryotes with identifiable sexes, there generally is a pair of chromosomes called the sex chromosomes. In most eukaryotes, the male is the **heterogametic sex**, which means that it has heteromorphic, only partially homologous sex chromosomes. These are usually called the X and Y chromosomes. These pair in the first prophase of spermatogenesis and as a result of segregation, two types of gametes, the X and the Y, are produced. The **homogametic sex**, the female, has two X chromosomes and produces only one kind of egg. Some male heterogametic species are XO, and lack a Y chromosome. The Y is usually smaller than the X, has very few of the genes that are found on the X chromosome or none, and is often composed primarily of heterochromatin. The X is usually more like an autosome (the non-X chromosomes) in function and appearance. However, because it exists in one copy in the heterogametic sex, compared with the condition of the autosomes, some form of dosage compensation is required to equalize the amount of gene product in the two sexes (see Chapter 11 for a discussion of dosage compensation). In some insect species, females are the heterogametic sex. When this occurs, the sex chromosomes are designated as W and Z, with the W analogous to the Y of the male.

## Extranuclear Inheritance in Mitochondrial Genes

Genes located in the nucleus show Mendelian inheritance because they segregate in a regular manner during meiosis. However, not all genes in eukaryotic organisms are located in the nucleus. **Mitochondria** are self-replicating organelles that occur in the cytoplasm of all eukaryotes. They are inherited cytoplasmically, and thus are primarily transmitted through the maternal gamete. Each mitochondrion is surrounded by a double membrane. The inner membrane is highly invaginated, with projections called cristae that are tubular or lamellar. These are the sites of oxidative phosphorylation which result in the formation of **adenosine triphosphate (ATP)**, the primary molecule for storing chemical energy in a cell. Mitochondrial DNA (mtDNA) is a significant component of the total DNA in insect cells. About half of the DNA in an unfertilized *D. melanogaster* egg is mtDNA.

It is assumed that paternal mtDNA either is not transmitted at fertilization or that it contributes a small fraction of the mtDNA in the developing embryo, which is lost during subsequent development. The cellular mechanisms that regulate the replication and distribution of mitochondria to daughter cells at each cell division are poorly understood. However, the coexistence of more than one type of mtDNA within a cell or individual, **heteroplasmy**, is thought to be rare in natural populations.

Mitochondria contain distinctive ribosomes, transfer RNAs, and aminoacyl-tRNA synthetases. Mitochondria are thought to be endosymbionts that were derived from aerobic bacteria associated with primitive eukaryotes (Gray 1989). Mitochondria have their own chromosomes, with a genetic code that differs slightly from the universal genetic code. The mitochondrial DNA of *Drosophila yakuba* codes for 37 genes: 2 ribosomal RNA genes, 22 transfer RNA genes, and 13 protein genes that code for subunits of enzymes functioning in electron transport or ATP synthesis (Clary and Wolstenholm 1985) (Figure 4.6). Figure 4.6 shows the **open reading frames** (ORF) that code for the subunits of the respiratory chain gene *NADH dehydrogenase* and of the genes coding for *cytochrome b*, a portion of *cytochrome c oxidase* (subunits I, II, and III), and a part of *ATPase* (subunits 5 and 6). The entire interval between the start and the stop codons is the ORF. Cytochromes are a family of heme-containing proteins that function as electron donors and acceptors during the reactions that occur in respiration. Knowledge of the organization and evolution of insect mitochondrial genomes is derived primarily from studies of *Drosophila yakuba*, the mosquitoes *Aedes albopictus* and *Anopheles quadrimaculatus*, the honey bee *Apis mellifera*, and the blowfly *Phormia regina* (Clary and Wolstenholme 1985, Cockburn et al. 1990, Crozier and Crozier 1993, Goldenthal et al. 1991).

More detailed investigation of insect mitochondrial genomes may provide some contradictions to the above generalizations, however. Mitochondrial genomes greater than 20 kb have been found in several vertebrates and recently three species of curculionid beetles (*Pissodes strobi*, *P. nemorensis*, and *P. terminalis*) were found to have mitochondrial genomes almost twice the normal size (Boyce et al. 1989). The large size (30–36 kb) in these three *Pissodes* species is due to an enlarged A+T-enriched region (9–13 kb) and a series of 0.8- to 2.0-kb tandemly repeated sequences adjacent to the A+T region. Every weevil sampled in all three species had two to five distinct size classes of mtDNA (**heteroplasmy**). The magnitude of the size differences, the number of size classes found within individual weevils, and the abundant mtDNA heteroplasmy is unprecedented to date (Boyce et al. 1989).

The perception that mtDNA is exclusively inherited in a maternal fashion has been questioned in *Drosophila* and marine mussels. Two studies have shown that incomplete maternal inheritance of mtDNA occurs in *Drosophila simulans* (Satta et al. 1988, Matsuura et al. 1991). Boyce et al. (1989) suggested that the

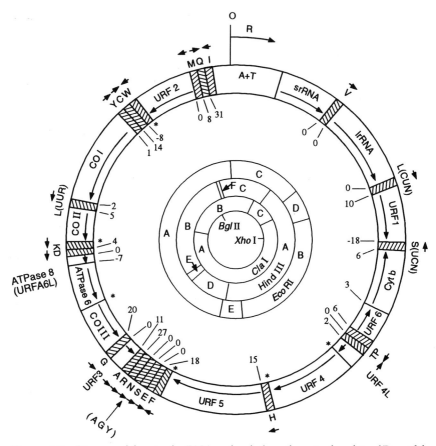

**Figure 4.6** Diagram of the circular DNA molecule from the mitochondria of *Drosophila yakuba*. The outside circle shows the open reading frames (ORF) (URF1 to URF6 and URF 4L) that code for subunits of the respiratory chain *NADH dehydrogenase* and of the genes coding for *cytochrome b; cytochrome c oxidase* (subunits I, II, and III); and *ATPase* (subunits 5 and 6). The origin and direction of replication are indicated by O and R. The variable A+T region is shaded. The arrows indicate the direction of gene transcription. The tRNA genes are crosshatched and indicated by their single letter amino acid codes. rRNA and srRNA are the large and small rRNA genes, respectively. The numbers on the inside of the outer circle are the numbers of apparently noncoding nucleotides that occur between the genes. The innermost circles indicate restriction fragments produced with the enzymes indicated. (From Clary and Wolstenholme 1984, 1985.)

high level of heteroplasmy found in the three *Pissodes* beetle species could be due to paternal transmission of mtDNA, although they did not document that paternal transmission actually occurred.

Mitochondrial chromosomes are circular, supercoiled, and double-stranded DNA molecules. In *Drosophila*, the mitochondrial chromosome contains approximately 18.5 kb of DNA. Each mitochondrion contains several copies of the chromosome. Mitochondrial genes code for all the ribosomal RNA and transfer RNA molecules involved in mitochondrial protein synthesis. In addition, an A+T-rich region, approximately 0.8 kb long, appears to exercise control over mtDNA replication and RNA transcriptions in *Drosophila*. Mitochondrial genes in insects lack introns, and intergenic regions are small or absent. The ribosomes found in the mitochondria are smaller than the ribosomes in the cytoplasm of the cell. They also have different ribosomal RNA molecules and proteins than the ribosomes found in the cytoplasm of the cell. In fact, mitochondrial ribosomes more closely resemble prokaryotic ribosomes than the versions found in eukaryotic cytoplasm, which supports the hypothesis that mitochondria are derived from microbial endosymbionts (Gray 1989). However, mitochondria do not contain sufficient genes for all the RNAs and proteins needed for their own functioning and reproduction. Thus, a number of mitochondrial components are coded for by nuclear genes; these components are imported to the mitochondrion after synthesis.

Mitochondria are **maternally inherited** (transferred primarily through the egg), with only a few exceptions. However, most eggs and somatic cells contain hundreds or thousands of mtDNA molecules, so a new mutation can either generate or add to a situation in which two or more mtDNA genotypes coexist within an individual. This heteroplasmy, however, is apparently a transitory state in germ cells. Thus, the majority of individuals tested are effectively haploid with regard to the number of types of mtDNA transmitted to the next generation.

Mitochondrial DNA evolves faster than single copy nuclear DNA in some eukaryotic organisms. This evolutionary rate is high, apparently because mitochondria are relatively inefficient in repairing errors during DNA replication or after DNA damage. Another reason is because mtDNA does not code for proteins involved directly in its own replication, transcription, or translation. As a result, mtDNA has a large number of length mutations and transitions. In Hawaiian *Drosophila*, mtDNA appears to evolve about three times faster than nuclear DNA (Moritz et al. 1987).

Mitochondrial DNA has been extensively studied because it is easier to purify than a specific segment of nuclear DNA. This is due to its buoyant density, high copy number within cells, and its location within an organelle. Isolation of mtDNA by centrifugation is thus relatively easy, making it a useful subject for systematics and evolution, and population genetics studies, as described in Chapters 13 and 14.

## Transposable Elements Are Ubiquitous Agents That Alter Genomes

**Transposable elements** are genetic elements that are able to move from one chromosomal site to another; they are usually present in multiple copies within a genome. Every genome probably contains transposable elements. For example, transposable elements have been found in humans, bacteria, frogs, mice, maize, nematodes, protozoans, and insects (Berg and Howe 1989). TEs are usually found in moderate numbers in a genome (as part of the middle-repetitive DNA) and are considered to be parasitic DNA. There are numerous 'families' of TEs found in insects and other organisms, and many are derived from retroviruses. The ubiquity of transposable elements in a diverse array of organisms has raised a number of unanswered questions about their evolutionary impact. It is believed that new TEs are still invading and spreading within insect populations. The role of TEs in insect evolution and genetic manipulation is discussed further in Chapters 5 and 10.

The diversity of arthropods and their genetic systems has only been hinted at. In Chapter 5 we explore additional details of genome organization, developmental processes, and diversity in insects.

## References Cited

Benbow, R. M., J. Zhao and D. D. Larson. 1992. On the nature of origins of DNA replication in eukaryotes. BioEssays 14:661–669.

Berg, D. E. and M. M. Howe, Eds. 1989. *Mobile DNA*. Am. Soc. Microbiol., Washington, DC.

Berry, S. J. 1985. Insect nucleic acids. Pp. 219–253. In: *Insect Physiology Biochemistry and Pharmacology*, Vol. 10. G. A. Kerkut and L. I. Gilbert, Eds. Pergamon, Oxford.

Besansky, N. J. and F. H. Collins. 1992. The mosquito genome: Organization, evolution and manipulation. Parasitol. Today 8: 186–192.

Bigot, Y., M. Hamelin and G. Periquet. 1990. Heterochromatin condensation and evolution of unique satellite-DNA families in two parasitic wasp species: *Diadromus pulchellus* and *Eupelmus vuilleti* (Hymenoptera). Mol. Biol. Evol. 7: 351–64.

Black, W. C. and K. S. Rai. 1988. Genome evolution in mosquitoes: Intraspecific and interspecific variation in repetitive DNA amounts and organization. Genet. Res. Cambridge 51: 185–196.

Blackburn, E. H. 1991. Structure and function of telomeres. Nature 350: 569–573.

Block, K., G. Ising and F. Stahl. 1990. Minichromosomes in *Drosophila melanogaster* derived from the transposing element TE1. Chromosoma 99: 336–343.

Boyce, T. M., M. E. Zwick and C. F. Aquadro. 1989. Mitochondrial DNA in the bark weevils: Size, structure and heteroplasmy. Genetics 123: 825–836.

Brown, S. J., J. K. Henry, W. C. Black and R. E. Denell. 1990. Molecular genetic manipulation of the red flour beetle: Genome organization and the cloning of a ribosomal protein gene. Insect Biochem. 20: 185–193.

Burkholder, G. D., L. J. P. Latimer and J. S. Lee. 1991. Immunofluorescent localization of triplex DNA in polytene chromosomes of *Chironomus* and *Drosophila*. Chromosoma 101: 11–18.

Clary, D. O. and D. R. Wolstenholm. 1985. The mitochondrial DNA molecule of *Drosophila yakuba*: Nucleotide sequence, gene organization, and genetic code. J. Mol. Evol. 22: 252–271.

Cockburn, A. F., S. E. Mitchell and J. A. Seawright. 1990. Cloning of the mitochondrial genome of *Anopheles quadrimaculatus*. Arch. Insect Bioch. Physiol. 14: 31–36.

Crain, W. R., E. H. Davidson and R. J. Briteen. 1976. Contrasting patterns of DNA sequence arrangement in *Apis mellifera* (honeybee) and *Musca domestica* (housefly). Chromosoma 59: 1–12.

Crozier, R. H. and Y. C. Crozier. 1993. The mitochondrial genome of the honeybee *Apis mellifera*: Complete sequence and genome organization. Genetics 133: 97–117.

Ferrari, J. A. and K. S. Rai. 1989. Phenotypic correlates of genome size variation in *Aedes albopictus*. Evolution 43: 895–899.

Gatti, M. and S. Pimpinelli. 1992. Functional elements in *Drosophila melanogaster* heterochromatin. Annu. Rev. Genet. 26: 239–275.

Goldenthal, M. J., K. A. McKenna and D. J. Joslyn. 1991. Mitochondrial DNA of the blowfly *Phormia regina*: Restriction analysis and gene localization. Biochem. Genet. 29: 1–11.

Gray, M. W. 1989. Origin and evolution of mitochondrial DNA. Annu. Rev. Cell Biol. 5: 25–50.

Juan, C. and E. Petitpierre. 1990. Evolution of genome size in darkling beetles (Tenebrionidae, Coleoptera). Genome 34: 169–173.

Kumar, A. and K. S. Rai. 1990. Intraspecific variation in nuclear DNA content among world populations of a mosquito, *Aedes albopictus* (Skuse). Theor. Appl. Genet. 79: 748–752.

Ma, R. Z., W. C. Black IV and J. C. Reese. 1992. Genome size and organization in an aphid (*Schizaphis graminum*). J. Insect Physiol. 38: 161–165.

Manuelidis, L. 1990. A view of interphase chromosomes. Science 250:1533–1540.

Manuelidis, L. and T. L. Cher. 1990. A unified model of eukaryotic chromosomes. Cytometry 11: 8–25.

Matsuura, E. T., H. Fukuda and S. I. Chigusa. 1991. Mitochondrial DNA heteroplasmy maintained in natural populations of *Drosophila simulans* in Reunion. Genet. Res., Camb. 57: 123–126.

Moritz, C., T. E. Dowling and W. M. Brown. 1987. Evolution of animal mitochondrial DNA: Relevance for population biology and systematics. Annu. Rev. Ecol. Syst. 18: 269–292.

Okazaki, S., K. Tsuchida, H. Maekawa, H. Ishikawa and H. Fujiwara. 1993. Identification of a pentanucleotide telomeric sequence, $(TTAGG)_n$, in the silkworm *Bombyx mori* and in other insects. Mol. Cell. Biol. 13:1424–1432.

Perkins, H. D., D. G. Bedo and A. J. Howells. 1992. Characterization and chromosomal distribution of a tandemly repeated DNA sequence from the Australian sheep blowfly, *Lucilia cuprina*. Chromosoma 101: 358–364.

Plohl, M., B. Borstnik, V. Lucijanic-Justic and D. Ugarkovic. 1992. Evidence for random distribution of sequence variants in *Tenebrio molitor* satellite DNA. Genet. Res., Camb. 60: 7–13.

Samols, D. and H. Swift. 1979. Genomic organization in the flesh fly *Sarcophaga bullata*. Chromosoma 75: 129–153.

Satta, Y., N. Toyohara, C. Ohtaka, Y. Tasuno, T. Watanabe, E. T. Matsuura, S. I. Chigusa and N. Takahata. 1988. Dubious maternal inheritance of mitochondrial DNA in *D. simulans* and evolution of *D. mauritiana*. Genet. Res., Camb. 52: 1–6.

Tordera, V., R. Send and J. E. Perez-Ortin. 1993. The role of histones and their modifications in the informative content of chromatin. Experientia 49: 780–788.

Ugarkovic, D., M. Plohl, V. Lucijanic-Justic and B. Borstnik. 1992. Detection of satellite DNA in *Palorus ratzeburgii*: Analysis of curvature profiles and comparison with *Tenebrio molitor* satellite DNA. Biochimie 74: 1075–1082.

Wagner, R. P., M. P. Maguire and R. L. Stallings. 1993. *Chromosomes A Synthesis*. Wiley-Liss, New York.

Wells, R. H. Royer and C. P. Hollenberger 1976. Non *Xenopus*-like DNA organization in the *Chironomus tentans* genome. Mol. Gen. Genet. 147: 45–51.

White, M. J. D. 1973. *Animal Cytology and Evolution*. 3rd Ed. Cambridge Univ. Press, Cambridge.

Zakian, V. A. 1989. Structure and function of telomeres. Annu. Rev. Genet. 23: 579–604.

# Chapter 5

# Genes, Genome Organization, and Development in Insects

## Overview

Molecular genetics has revolutionized our understanding of insect gene structure, organization, and regulation. One of the greatest surprises has been the discovery that genomes are not static, but appear to be subject to several types of changes. For example, large portions of the insect genome may consist of different families of transposable elements. Transposable elements can change gene structure and function, alter chromosomal organization, and transfer horizontally between species.

The discovery that gene amplification is involved in resistance to insecticides in aphids and mosquitoes has opened new avenues for understanding this evolutionary and economic problem. Recent research suggests that the genetic information contained in endocytobionts may play an essential role in speciation and evolution of some insects. Insects may have diploid males and females, haploid males and diploid females (arrhenotoky), and females only (thelytoky); or diploid males may undergo chromosome heterochromatinization and loss to become haploid. Insects have developed diverse gene regulation methods. One of the first analyses compared chorion gene regulation in *Drosophila* and silk moths during development. The diversity of genome organization in insects reflects their long evolutionary history and diversity.

A fundamental knowledge of the stages of development and the major genes that influence these stages in *Drosophila melanogaster* is available, although the details of their interactions are not fully understood. There are three general stages in embryonic development: first, the polarity of the embryo is determined, primarily by maternal effect genes. Next, segmentation genes influence the development of major bands or parasegments. The third determination is accomplished by the interaction of homeotic genes that provide a finer definition of the segmental structures. While a full understanding of all the genes and their interactions is still lacking, the broad outline is becoming clear. Analyses of *Drosophila* development will provide information that may have broad significance for our understanding of development in all eukaryotic organisms because homeodomains are strongly conserved among diverse higher eukaryotes.

## Introduction

Insects are extremely numerous and diverse. Estimates suggest there are at least one million species in more than 790 families organized in 32 orders. Insects live in a great variety of habitats, exhibit diverse types of life styles, have an extraordinary range of structural variations, eat an astonishing variety of food, and are among the most abundant animals on earth. Their long evolutionary history has provided sufficient time for them to develop a diversity of genetic systems (see Chapter 13 for an overview of insect evolution).

This chapter provides only a small sample of the diversity of insect genetic systems. More extensive reviews can be found in White (1973) and Wagner et al. (1993). This chapter also provides an overview of gene regulation in some insect-specific genes, as well as a brief introduction to the molecular genetics of insect development.

## Insect Genome Organization

Until recently, the eukaryote genome was considered to be relatively constant, with every cell having the same DNA sequences in the same amounts and in the same location. The genome was perceived as a relatively static entity that responded only slowly to evolutionary pressures. It is now apparent that somatic genomes are more diverse than previously imagined, with polyteny, polyploidy, and gene amplification occurring in different tissues at different developmental stages in the organism. It is also clear that DNA can move within the nuclear genome via a wide array of transposable elements (as discussed briefly in Chapter 4).

DNA has been found in interesting structures outside the nuclear chromosomes or mitochondrial genome, but the significance of this DNA is unresolved. For example, three to four covalently closed circular DNAs that appear to be derived from chromosomal DNA have been found in cell cultures of *Drosophila* (Gaubatz 1990). Much of this circular DNA is middle-repetitive chromosomal DNA and it may be associated with gene amplification during development or DNA rearrangements during aging (Gaubatz 1990). Some circular DNA molecules in *D. melanogaster* embryos apparently contain 5S ribosomal RNA genes, satellite DNA, or histone genes (DeGroote et al. 1989).

Minichromosomes have been found in *D. melanogaster* that apparently originated from the transposable element TE1 (Block et al. 1990). The minichromosome contains two structural genes, *white* and *roughest*, from the *Drosophila* X chromosome and part of chromosome 2. The minichromosome was stable and inherited by 33 to 47% of the progeny, which indicates that it contains a centromere. Centromere-like elements lacking chromosome arms have been found in the phorid fly *Megaselia scalaris* (Wolf et al. 1991). The function of these centromere-like elements is unknown, but they could be B chromosomes that have been reduced to a minimal size that can be retained during mitosis.

B chromosomes are a heterogeneous class of chromosomes that are often heterochromatic and are sometimes referred to as accessory or supernumerary chromosomes. They occur in both plants and animals. B chromosomes seem to have little effect on the phenotype, differ in ploidy from one cell type to another, and may occur in some individuals of the same species but not in others. B chromosomes usually segregate normally in mitosis and meiosis (Wagner et al. 1993). A B chromosome in the parasitic wasp *Nasonia vitripennis* causes the compaction and loss of paternally derived chromosomes in fertilized eggs. This leads to the production of all-male progeny in this haplodiploid species (Eickbush et al. 1992). Again, whether minichromosomes or B chromosomes are common elements remains unclear, but the notion that insect genomes simply consist of nuclear and mitochondrial chromosomes should be discarded (Pardue 1991).

During embryonic development of some insects, special germ line-limited chromosomes are eliminated from those cells that will become somatic cells. The loss occurs because these chromosomes lag during early cleavage divisions. Occasionally, however, these supernumerary chromosomes have been found in the somatic cells of the chironomid *Acricotopus lucidus* over many generations (Staiber 1987).

Even the recent notion that eukaryotic genes should contain introns is not always sustained. The origin of introns in eukaryotic genes is unknown. It is not clear whether eukaryotic genes acquired introns during evolution or whether introns were present in the common ancestor of prokaryotes and eukaryotes. However, hemoglobin genes sequenced from the midge *Chironomus thummi* lacked introns, even though these hemoglobin genes show sequence homology with vertebrate hemoglobin genes which contain introns (Antoine and Niessing 1984). Because the cloned *Chironomus* genes were expressed *in vivo*, the hypothesis that they are pseudogenes was rejected. (A **pseudogene** is a gene with a close resemblance to a known gene at a different locus, but nonfunctional because additions or deletions in its structure prevent normal transcription or translation.) An alternative explanation is that the genes originated by **reverse transcription** of spliced mRNA in germ line cells. Reverse transcription involves synthesis of DNA from a messenger RNA template, which lacks introns, to produce cDNA. If this intronless cDNA subsequently became integrated in the *C. thummi* genome, then the hemoglobin gene would lack introns.

Much of what we know about the genetics of insects is derived from the study of *Drosophila* species (Ashburner 1989). Extensive genetic information is available, ranging from a physical map of *D. melanogaster* (Kafatos et al. 1991, Merriam et al. 1991, Hartl et al. 1992), to the complete sequencing of more than 100 genes (Lindsley and Zimm 1992, Maroni 1993). Details of the splicing signals by which introns are removed from the mRNA precursors also have been determined for *Drosophila* (Mount et al. 1992).

By contrast, analyses of genes, development, and genetic systems in other insects lag far behind. Such analyses are important for the solution of both basic and applied problems because *D. melanogaster* may be a highly specialized insect with unique genetic characteristics. Progress is being made for at least a few species. For example, a detailed genetic map of the X chromosome of the malaria vector *Anopheles gambiae* has been produced using microsatellite DNA markers (Zheng et al. 1993). Studies of the genomes of other economically-important insects such as the Mediterranean fruit fly *Ceratitis capitata*, the silk moth *Bombyx mori*, the flour beetle *Tribolium castaneum*, the mosquito *Aedes aegypti*, and the honey bee *Apis mellifera* are under way (Crozier and Crozier 1993, Brown et al. 1990, Warren and Crampton 1991, Zheng et al. 1991, Besansky and Powell 1992, Severson et al. 1993). The following discussion focuses on insects other than *D. melanogaster*, where possible. It first describes the major categories of DNA: unique-sequence, middle-repetitive, and highly-repetitive. Within the middle-repetitive class of DNA, examples are presented of some particularly interesting genes found in insects.

### Unique-Sequence DNA

Most genetic information is contained in unique-sequence DNA. The proportion of unique sequences varies among species. For example, among four lepidopteran species, *Antheraea pernyi*, *Hyalophora cecropia*, *Bombyx mori*, and *Manduca sexta*, the proportions of unique DNA range from 55 to 80% (Berry 1985).

Some unique-sequence DNA is actually present in multiple copies in specific insect cells or tissues. This occurs by one of two different mechanisms. Multiple copies of unique sequences can occur if the cells are polyploid. They also can occur through **gene amplification**. For example, the chorion genes of *Drosophila* are amplified during specific stages of chorion production (see later discussion), although this amplification is limited to ovarian follicle cells. It now appears that some aphids and mosquitoes that are resistant to pesticides have esterase genes that have been amplified (see later discussion). Gene amplification appears to be a mechanism by which large amounts of gene product can be produced in a relatively short time.

### Middle-Repetitive DNA

Middle-repetitive elements are found in more than one copy, but still in modest numbers. They include genes that code for ribosomal RNAs (rRNA), transfer

RNAs (tRNA), histones, transposable elements (TEs), and developmentally-regulated multigene families such as actins, cuticle genes, heat shock genes, larval serum genes, silk genes, and yolk protein genes, as described later. An alternative solution to producing large amounts of gene product in a relatively short time and in a coordinated manner is to duplicate the gene.

**Heat shock genes**   The heat shock response originally was discovered in *Drosophila melanogaster* and has since been found in diverse organisms ranging from bacteria to man. Heat shock genes are activated in a small number of cytogenetic sites in response to environmental stresses such as heat or chemical shock. The products of the activated genes, the heat shock proteins, are present in small amounts in many cells in the absence of stress. However, heat shock proteins increase rapidly after stress. Heat shock genes are an evolutionarily conserved response to stress in all organisms (Morimoto et al. 1992).

If *Drosophila* are exposed to a severe heat shock (about 40°C), most flies die. If they undergo a mild shock at 33°C, this activates the synthesis of heat shock proteins without disturbing normal protein synthesis, and many flies then can survive a heat shock at 40°C. In *D. melanogaster*, nine chromosomal sites puff in response to heat shock and specific mRNAs are produced that code for seven heat shock proteins. Heat shock genes have been cloned, sequenced, and analyzed (Pauli et al. 1992). There are several types (or families), including the *hsp70*, *hsp83*, and the small heat shock gene family. The *hsp83* gene is expressed constitutively and is moderately activated by heat shock; it is also the only heat shock gene in *D. melanogaster* to have an intron. Large amounts of hsp83 protein are present at normal temperatures in maturing germ cells, salivary glands, and brains of larvae and prepupae. The small *hsp* gene family includes genes encoding hsp22, hsp23, hsp26, and hsp27 proteins. These proteins are expressed at several developmental stages.

The *hsp70* gene is present in several copies in *Drosophila* and is the most abundant and highly conserved. At the amino acid level, the *Drosophila* hsp70 protein shares 73% overall identity with human hsp70 protein and 50% identity with that of the bacterium *Escherichia coli*. In addition, seven genes that are strongly homologous to *hsp70*, called cognates, are constitutively expressed and may be important during *Drosophila* development. The hsp70 proteins are probably important in protecting vital cellular activities such as RNA processing and translation.

**Histone genes**   The histone gene family consists of genes coding for the five histones that serve as the basic proteins in eukaryotic chromosomes. The basic unit of chromosomes, the nucleosome, is composed of 146 bp of DNA coiled around the histone octamer, two molecules each of histone H2A, H2B, H3, and H4 (see Figure 4.1). Linking two nucleosomes is a small stretch of DNA to which the fifth histone, H1, is bound. The different histone genes often share regulatory sequences and are coordinately expressed. In some species, there are tissue- or stage-specific gene sets. In *Drosophila*, the genes are tandemly repeated and closely linked. The histone genes of the midge *Chironomus thummi* have been analyzed; the orientation, nucleotide sequence, and amino acid sequences, as well as chromosomal arrangement and localization, are different from those found in *D. melanogaster* (Hankeln and Schmidt 1991).

Histone genes typically lack introns. It is thought that the introns were eliminated because these genes must be expressed efficiently and rapidly during development. Histone proteins could be produced more efficiently if the primary gene product did not need to be spliced to remove introns. Having histone genes organized in a tandem repeat structure also ensures that there will be equivalent amounts of the proteins produced by the multiple gene copies.

There is a tenfold difference in copy numbers of histone genes in three species of *Drosophila* (*melanogaster, hydei*, and *hawaiiensis*) (Fitch et al. 1990). *D. melanogaster* has 100 tandemly arranged histone genes, far more than would be required for the maximal rate of transcription during development. By contrast, *D. hydei* has 5, and *D. hawaiiensis* has about 20 tandem histone repeat copies per haploid genome. In *D. melanogaster* the histone genes are located adjacent to the heterochromatic base of chromosome 2, while they are located in euchromatic regions in the other two species. This suggests that there are more histone genes in *D. melanogaster* to compensate for the fact that they are less active because they are located near heterochromatin.

**Immune response genes**   Insects defend themselves against bacteria and other parasites with both cellular and humoral immune responses. Cellular responses usually result in phagocytosis or encapsulation of the microorganism by blood cells. Humoral responses involve production of a number of antibacterial proteins and peptides, such as cecropins, attacins, lysozymes, and defensins. The genes coding for cecropins, diptericin, lysozymes, and sarcotoxins have been cloned from *Drosophila* and other insects (Hultmark 1993) (Table 5.1). These genes have been shown to have counterparts in vertebrates, including humans, and have opened up new possibilities for treatment of disease in humans (Kimbrell 1991).

**Ribosomal genes**   Ribosomes interact with mRNAs, initiation factors, and transfer RNAs during protein synthesis. Over 120 macromolecular components are needed to produce polypeptides at a high rate in ribosomes. Ribosomes consist of subunits that dissociate in the absence of magnesium ions. Intact arthropod ribosomes have a sedimentation coefficient (the rate at which a given molecule suspended in a less dense solvent sediments in a field of centrifugal force) of 80S, with two subunits that have sedimentation coefficients of 60S and 40S (Kaulenas 1985).

Different arthropod species have different numbers of ribosomal genes located in the nuclear chromosomes. For example, *Drosophila erecta* has 160 genes while *D. hydei* has over 500 (Berry 1985). The fungus fly *Sciara coprophila* contains only 65 to 70 repeat units, one of the lowest numbers reported (Kerrebrock et al. 1989). Most insects have between 200 and 500 rRNA genes for

**Table 5.1**

Selected Genes Cloned from Insects Other than *Drosophila*

| Gene coding for: | Function(s) | Insect | References |
|---|---|---|---|
| Actin | Muscle contraction | *Ceratitis capitata* | Maymer et al. 1990 |
| Attacin | Immune response | *Hyalophora cecropia* | Sun et al. 1991 |
| Cecropins | Antibacterial responses | *Bombyx mori* | Kato et al. 1993 |
| | | *Manduca sexta* | Dickenson et al. 1988 |
| | | *Ceratitis capitata* | Rosetto et al. 1993 |
| Chitinases | Digest chitin in exoskeleton | *Manduca sexta* | Kramer et al. 1993 |
| Cuticle proteins | Component of cuticle | *Manduca sexta* | Rebers et al. 1987 |
| Cytochrome P-450 mono-oxygenase, CYP6B1 | Detoxify xanthotoxin | *Papillo polyxenes* | Cohen et al. 1992 |
| Cytochrome P450, family 4 | Energy substrate mobilization | *Blaberus discoidalis* | Bradfield et al. 1991 |
| Eclosion hormone | Triggers ecdysis | *Manduca sexta* | Horodyski et al. 1989 |
| | | *Bombyx mori* | Kamito et al. 1992 |
| Histones | Chromosome structure | *Chironomus thummi* | Hankein and Schmidt 1991 |
| Insulin superfamily | Unresolved—growth and differentiation(?) | *Locusta migratoria* | Lagueux et al. 1990, Ebberink et al. 1989 |
| Luciferase | Firefly enzyme | *Luciola mingrelica* | Devine et al. 1993 |
| Lysosomal aspartic protease | Terminate oogenesis? | *Aedes aegypti* | Cho and Raikhel 1992 |
| Pesticide resistance | Amplified esterases | *Culex* mosquitoes | Mouches et al. 1990 |
| | | *Myzus persicae* | Field et al. 1988 |
| | Knockdown resistance (*kdr*) | *Musca domestica* | Williamson et al. 1993 |
| Protease | Vitellin degradation | *Bombyx mori* | Ikeda et al. 1991 |
| Pupal cuticle protein | Pupal structural protein | *Bombyx mori* | Nakato et al. 1992 |
| Sapecin | Immune response | *Sarcophaga peregrina* | Matsuyama and Natori 1988 |
| Sarcotoxin | Immune response | *Sarcophaga peregrina* | Kanai and Natori 1989 |

(*continues*)

**Table 5.1**   (*continued*)

| Gene coding for: | Function(s) | Insect | References |
|---|---|---|---|
| Serine proteases | Digestive proteases | *Haematobia irritans* | Elvin et al. 1993 |
| Serpins | Inhibitors of serine proteases | *Manduca sexta* | Jiang et al. 1994 |
| Silk genes, *Ser1*, *Ser2* | Silk | *Bombyx mori* | Michaille et al. 1986, 1990 |
| | | *Galleria mellonella* | Zurovec et al. 1992 |
| Steroid receptor superfamily, *E75* and *MHR3* | Regulation of ecdysone response | *Manduca sexta* | Segraves and Woldin 1993 |
| | | | Palli et al. 1992 |
| Triosephosphate isomerase | Glycolysis | *Culex tarsalis* | Tittiger et al. 1993 |
| Vitellogenin | Yolk proteins | *Anthonomus grandis* | Trewitt et al. 1992 |
| | | *Ceratitis capitata* | Rina and Savakis 1991 |
| | | *Bombyx mori* | Sato and Yamashita 1991 |
| Vitelline membrane proteins | Membrane ? | *Aedes aegypti* | Lin et al. 1993 |

diploid genome. The ribosomal genes of *Drosophila* are arranged in two clusters, one in the nucleolar organizer of each of the sex chromosomes (Williams and Robbins 1992). The 5.8S, 18S, and 28S rRNAs are transcribed as a single unit, which is then processed in the nucleus to provide the separate subunit RNAs. Ribosomal genes comprise 2% of the total genome and about 20% of the middle-repetitive sequences of *D. melanogaster*.

**Silk genes**   The silk gland provides a model system for cell biologists and molecular geneticists to study gene regulation and development. Silk gland cells of *Bombyx mori* are polyploid (up to 20-fold endopolyploidy), which may explain how silk moth larvae produce huge amounts of silk proteins within a short period of time prior to pupating. When the insect is secreting silk for its cocoon, it can synthesize enormous amounts of protein within the salivary gland within a period of five to six days. The cells from the posterior silk gland produce **fibroin**, a very large insoluble polypeptide that forms the fibers of silk. Those from the middle part store fibroin and secrete **sericin**, a mixture of four to six hot-water soluble polypeptides. Sericin binds strands of raw silk fibers together.

Silk proteins have an unusual amino acid composition, with a predominance of glycine and alanine in fibroin and serine in sericin (Prudhomme et al.

1985). A sericin gene (*Ser1*) has been cloned and partially sequenced. This gene produces a single mRNA transcript which is differentially spliced to produce the four major cocoon sericins. A second sericin gene (*Ser2*) has been cloned that produces two proteins from a single mRNA by differential splicing. Three middle silk gland specific genes (MSGS 3, 4, and 5) are thought to code for minor cocoon proteins.

Although *Drosophila melanogaster* does not produce silk, a silk-encoding gene (*P25*) of *Bombyx mori* was expressed in the anterior salivary gland of *D. melanogaster* after flies received the moth silk gene by P-element-mediated transformation (Bello and Couble 1990) (see Chapter 10 for a discussion of P-element-mediated transformation). The *P25* gene was appropriately expressed in the fly larval salivary glands and no activity was seen in other tissues. These results indicate that salivary glands of *Drosophila* contain regulatory factors capable of recognizing a *Bombyx* silk protein gene and controlling its expression, despite the evolutionary divergence of flies and moths over 250 million years ago.

**Transfer RNA genes** Over 90 tRNAs have been identified during *Drosophila* development that are encoded by at least 670 genes, which can be divided into 60 separate groups. One to 18 tRNA genes are contained in 30 chromosomal sites, but there are no tandem repeats.

**Vitellogenin genes** Yolk proteins provide embryos with nutrients essential for growth within the egg. Most are phosphoglycoproteins and provide a source of amino acids, phosphate, lipids, and carbohydrates. The major yolk proteins are derived from **vitellogenins**, which are produced by the fat body and secreted for uptake by maturing oocytes. Vitellogenin gene structure and regulation is being studied in a number of insects, including the migratory locust *Locusta migratoria*, the tobacco hornworm *Manduca sexta*, the silk moth *Bombyx mori*, the boll weevil *Anthonomus grandis*, the Mediterranean fruit fly *Ceratitis capitata*, and *Drosophila* (Bownes 1986, Rina and Savakis 1991, Trewitt et al. 1992). The fat body of the mother is the primary producer of yolk proteins, but some vitellogenin is synthesized by the follicular epithelium of the ovary in *D. melanogaster*.

Vitellogenins produced by the higher Diptera are significantly smaller than those produced by the boll weevil and vertebrates. Higher dipteran yolk protein genes (*Drosophila*, *Aedes*, and *Ceratitis*) are closely related to the triacylglycerol lipase gene family (Rina and Savakis 1991), while boll weevil and other insect vitellogenins are more closely related to vitellogenins in nematodes and vertebrates (Bownes 1986, Trewitt et al. 1992). There is no significant degree of similarity in DNA sequences in the two types of vitellogenins.

Yolk proteins in *Drosophila* consist of three polypeptides: YP1, YP2, and YP3. YP1 is expressed by the fat body and, after post-translational processing and glycosylation, the YP1 proteins are secreted into the hemolymph and delivered to the oocyte. YP2 is expressed in ovaries. The production and delivery of the three proteins is coordinately regulated and under the control of two hormones, 20-hydroxyecdysone and juvenile hormone (Bownes 1986). (These two hormones regulate molting and metamorphosis during development as well.)

Production of yolk proteins begins during the first day of adult life. The production rate is high, with yolk protein concentrations representing about one third of the total proteins found in the hemolymph. The genes for YP1 and YP2 are closely linked on the X chromosome, while the gene for YP3 also is sex-linked but more distant. *YP1* and *YP2* show much sequence homology and probably resulted from a fairly recent gene duplication event. Only one small intron is found in *YP1* and *YP2*, and two in *YP3*. Amplification of the yolk protein genes does not occur. Rather, extensive yolk protein synthesis in *Drosophila* is achieved by production in tissues that are polytene and polyploid.

**Transposable elements**   Transposable elements (TEs) are middle-repetitive DNA sequences that can move (transpose) to new sites, invert, and undergo deletion or amplification (Table 5.2). Transposable elements have been divided into two classes according to their structure and mechanism of transposition.

**Table 5.2**

A Classification of Transposable Elements by their Method of Transposition[a]

---

Class I Transposable Elements Transpose by Means of RNA Intermediates
    A. Viral superfamily (retrovirus-like retrotransposons)
        Have long direct repeats (LTRs), encode reverse transcriptase from open reading frames (ORFs) in DNA between LTRs, able to generate 4-bp to 6-bp target site duplications, have no 3' terminal poly(A) tract, are dispersed in genome.
        Examples: *Copia*-like elements in *Drosophila melanogaster*
                *Gypsy*-like elements in *D. melanogaster*
                *Pao* in *Bombyx mori*
    B. Nonviral superfamily (nonviral retroposons)
        Have no terminal repeats, have ORFs, do not encode enzymes responsible for their transposition, have 3' terminal poly(A) tract, are dispersed in genome.
        Examples: F family in *Drosophila melanogaster*
                R2 retroposons in many insects
                HeT-A retroposons in telomeres of *D. melanogaster*

Class II Transposable Elements Transpose Directly From DNA to DNA (All have a transposase enzyme and terminal inverted repeats.)
    A. With short terminal inverted repeats (SIRs)
        Examples: *P* and *hobo* in *D. melanogaster*
                *mariner* in many insect species
    B. With long terminal inverted repeated (LIRs)
        Example: FB (*foldback*) in *D. melanogaster*

---

[a]From Finnegan 1990, Robertson 1993, Xiong et al. 1993.

**Table 5.3**

Examples of Transposable Elements Identified from Insects Other than *Drosophila*

| Element | Type | Insect/mite | References |
|---|---|---|---|
| G | LINE-like | *Apis cerana* | Kimura et al. 1993 |
| *jockey* | LINE-like | *Apis mellifera* *Bombyx mori* | Kimura et al. 1993 |
| Juan | LINE-like, retrotransposon-like | *Culex* and *Aedes* mosquitoes | Mouches et al. 1991, 1992; Agarwal et al. 1993 |
| Lu-P1 and Lu-P2 | Homologous to P elements | *Lucilia cuprina* | Perkins and Howells 1992 |
| *mariner* | Short inverted terminal repeat with a DNA intermediate | Many arthropods | Robertson 1993, Robertson et al. 1992 |
| *Pao* | Retrotransposable element with long terminal repeats and unusual tandem repeats | *Bombyx mori* | Xiong et al. 1993 |
| R1 and R2 | Retrotransposable elements without long terminal repeats | Found in rRNA genes of many insects | Jakubczak et al. 1991, Bigot et al. 1992, Burke et al. 1993, Luan et al. 1993 |
| TED | Retrotransposon | *Trichoplusia ni* | Friesen and Nissen 1990 |
| TFB1 | Foldback TE | *Chironomus thummi* | Hankeln and Schmidt 1990 |
| T1 family | Retrotransposable elements without long terminal repeats | *Anopheles gambiae* | Besansky 1990a,b |

Class I TEs transpose by reverse transcription of an RNA intermediate. They include elements related to retroviruses that have long terminal repeats (LTRs). They also include elements that have no terminal repeats. Class II TEs transpose directly from DNA to DNA. They include elements with short inverted terminal repeats and have a coding region for a transposase. They also include elements with long terminal inverted repeats. Many transposable elements have been discovered in *D. melanogaster*, and a number of types of TEs are known from other insects as well (Table 5.3).

At least half of all spontaneous mutations in *D. melanogaster* are due to insertions of TEs. For example, P elements in *D. melanogaster* cause excisions, chromosome rearrangements, and insertions. Another type of transposable element, the **foldback** (FB) transposon group, is associated with deletions, inversions, reciprocal translocations, and insertional translocations in which normally unique *Drosophila* DNA is flanked by two FB elements. All well-characterized, *highly* unstable genes in *D. melanogaster* are unstable because they contain either the P element or FB elements (Berg and Howe 1989). Different TEs are found in *D. melanogaster* with different characteristics. For example, members of the HeT-A family of TEs are found only at telomeres and in centromeric heterochromatin and never in the euchromatin regions of chromosomes in *D. melanogaster*. The HeT-A elements may be involved in maintaining the length of telomeres in *D. melanogaster* chromosomes (Biessmann et al. 1992, 1993).

It has been proposed that TEs could carry genetic information, regulate genes, or initiate genetic changes. Wilson (1993) suggested that TEs could lead to resistance to pesticides, although he did not provide any direct evidence for this. Agarwal et al. (1993) found a TE named Juan associated with amplification of the esterase gene in pesticide-resistant *Culex* mosquitoes, but a direct involvement in inducing gene amplification has not been demonstrated. Waters et al. (1992) suggested that the TE called *17.6* is involved in susceptibility to pesticides in *Drosophila* associated with a P450 gene. However, Delpuech et al. (1993) screened colonies of *D. melanogaster* and *D. simulans* from around the world and found no relationship between the presence or absence of *17.6* and resistance or susceptibility. One example of TEs containing genetic information may be found in *Drosophila hydei*. TEs and repetitive DNA sequences comprise the majority of the Y chromosome of *D. hydei*. Apparently the lampbrush-loop-forming fertility genes on the Y chromosome consist, at least in part, of retrotransposons of the *micropia* family (Huijser et al. 1988). However, the precise function of these TEs remains unclear.

*R1* and *R2* are transposable elements that lack long terminal repeats and were originally found in some of the 28S rRNA genes of the silk moth *Bombyx mori* and several Diptera. A recent survey suggests that *R1* and *R2* elements occur within the rRNA genes of most insects (Jakubczak et al. 1991). Of 47 species surveyed, 43, including species in the orders Odonata, Orthoptera, Dermaptera, Hemiptera, Homoptera, Coleoptera, Hymenoptera, Lepidoptera, and Diptera, contained the insertions in 5 to 50% of their 28S genes. Four species from four orders (the house cricket *Acheta domestica*, the mealybug *Pseudococcus affinis*, a carabid ground beetle, and the Eastern tent caterpillar *Malacosoma americanum*) appeared to lack the insertions. The broad distribution of these elements raises the question of whether they could have been present in insects before their radiation more than 300 million years ago.

Very little is known about the origin and evolutionary history of transposable elements. A TE family might originate in a species, or TEs might be acquired by **horizontal transmission** from another species. Normally, DNA or RNA sequences are transmitted *vertically* from parent to progeny, but in *horizontal* transfer, DNA sequences are transferred laterally across taxonomic borders that were once thought to be inviolable (Daniels et al. 1990, Kidwell 1992). For example, the *hobo* transposable element of *D. melanogaster* was shown to have a sequence similar to TEs from plants (*Activator* from corn and *Tam3* from snapdragon), which suggests that horizontal transmission of genetic information between plants and animals might have occurred (Calvi et al. 1991). Another transposable element, *jockey*, identified from *D. melanogaster*, has been found in the distantly-related *D. funebris*, but not in species closely related to *funebris*. This suggests that the horizontal transfer of *jockey* from *D. melanogaster* into the genome of *D. funebris* occurred after *D. funebris* diverged from closely related species (Mizrokhi and Mazo 1990).

A possible superstar at horizontal transfer may be *mariner*. The *mariner* transposable element was originally found in *Drosophila mauritiana* and shown to be present in several species of *Drosophila*, as well as the moth *Hyalophora cecropia*. Subsequently, Robertson (1993) found that several types (subfamilies) of *mariner* are present in six additional orders of insects, including the mosquito *Anopheles gambiae*. The *mariner* element was also found in the predatory mite, *Metaseiulus occidentalis* (Acari: Phytoseiidae). It is likely that *mariner* has moved horizontally among these diverse insect and mite species, although the frequency of horizontal transfer is rare on a human time scale. It is less likely that the *mariner* element became inserted into the genome of an early arthropod lineage giving rise to insects and mites and has since evolved into different *mariner* families in different insects. A low frequency of horizontal transfer has occurred, but many *mariner* elements have since degenerated and become inactive.

*If* horizontal transmission occurs with some regularity, the implications will have a dramatic effect on many aspects of evolutionary theory and could also influence regulations regarding the risks associated with releases of transgenic arthropods into the environment (Brosius 1991, Kidwell 1993, Plasterk 1993). As discussed in Chapter 10, studies of the transfer of P elements by a mite vector provide an intriguing glimpse of one possible mechanism by which TEs are able to move between species (Houck et al. 1991). Other possible vectors of TEs are insect symbionts, which include viruses, rickettsia, fungi, and bacteria. Insect viruses may carry DNA from their hosts, and a proportion of the foreign DNA within insect viruses consists of TEs from their insect hosts (Fraser 1985). The clearest cases of horizontal transfer of TEs are seen with P and *mariner*, but the frequency of horizontal transfer is unknown and other interpretations are possible (Capy et al. 1994).

## Highly-Repetitive DNA

Highly-repeated DNA sequences with a uniform nucleotide composition can, upon fractionalization of the genomic DNA and separation by density gradient centrifugation, form one or more bands that are clearly different from the main band of DNA and from the smear created by other fragments of a more heterogeneous composition. These sequences are called 'satellite DNA'. Satellite DNA sometimes is described as minisatellite or microsatellite DNA, depending on the length of the repeated sequences. Satellite DNA is found in many insects and can constitute a substantial portion of the genome. For example, the darkling beetle *Misolampus goudoti* contains two satellite DNA families, one with about 120,000 copies and the other with about 70,000 copies per haploid genome (Pons et al. 1993).

The role of highly-repetitive sequences is not well understood (Ohno and Yomo 1991, Pardue and Hennig 1990), but such sequences are associated with heterochromatin in the centromeres and telomeres and could be important in chromosome pairing. Relatively little is known about the structure of either centromeres or telomeres in insects. Because they are very large and complex, centromeres have not been well characterized in insects. However, at least four different satellite DNA classes are found in the centromeric region of *Drosophila* autosomes and along most of the Y chromosome. Likewise, in the midge *Chironomus pallidivittatus* the centromere contains 200 kb of repetitive DNA. This DNA consists of 155 bp tandem repeats and is surrounded with AT base pairs, a feature with some similarity to centromeres in the yeast *Saccharomyces cerevisiae* (Rovira et al. 1993).

Telomeric DNA fragments have been isolated from *D. melanogaster*, *Chironomus*, and *Bombyx mori*. The sequence $(TTAGG)_n$ is found at the extreme terminal region of all *B. mori* chromosomes, and is associated with the ends of chromosomes in the Isoptera, Orthoptera, Hymenoptera, Trichoptera, Mecoptera, some Coleoptera, Hemiptera, and Lepidoptera (Okazaki et al. 1993). The sequence $(TTAGG)_n$ appears to be found only in arthropod telomeres, although not all arthropods have this telomeric sequence. *D. melanogaster* and some Coleoptera appear to have different telomeric structures. The telomeres of *Drosophila* apparently lack the simple telomeric repeats found in other eukaryotes and have transposable HeT-A elements in the subtelomeric region (Biessmann et al. 1993).

## Gene Amplification

**Gene Amplification** is a term originally coined to describe the presence in mammalian cell cultures of multiple copies of a specific gene and some of its adjacent DNA (5′ and 3′), which results in the production of more protein. Gene amplification can develop in response to a toxic substance, such as an anticancer drug, especially if the initial drug concentration is low and the surviving cells are subjected to multiple rounds of selection with increasing concentrations of toxin. When amplified DNA sequences are treated with restriction enzymes, unique restriction fragments are found that do not match those in normal cells. These unique restriction fragments are probably caused by flanking sequences that are variable in length.

Ribosomal and tRNA genes probably are amplified in all eukaryotes and amplification of rDNA occurs in the eggs of many insects. Both gene amplification and gene duplication occur in chorion genes in *Drosophila* and the silk moths *Bombyx mori* and *Antheraea polyphemus*. Gene amplification may be important to economic entomologists because at least one species of aphid and one species of mosquito which is resistant to insecticides have amplified esterase genes, as described later.

**Gene duplication in chorion genes** The genes that code for the egg shell of insects, the **chorion**, have been studied extensively, particularly in *Drosophila* and *Bombyx mori* and *Antheraea polyphemus* (Kafatos 1981, Eickbush and Burke 1985, Kafatos et al. 1986, Orr-Weaver 1991, Carminati et al. 1992). Analyses of the chorion genes have resulted in significant advances in our knowledge of the mechanisms of gene regulation and development.

In both *Drosophila* and silk moths, the egg is produced in the ovary, which consists of follicles composed of three cell types: (1) the oocyte, (2) a small number of nutritive nurse cells connected to the oocyte, and (3) follicular epithelial cells that surround the oocyte and nurse cells. There are approximately 1000 epithelial cells per follicle in *Drosophila* and up to 10,000 per oocyte in silk moths. These cells synthesize a complex mixture of proteins and secrete them onto the surface of the oocyte to form the outer covering, or chorion. The chorion protects the embryo after fertilization and oviposition, preventing desiccation yet enabling respiration to occur.

*Drosophila* and silk moth chorions are quite different. The *Drosophila* chorion is comparatively simple, with an endochorion and exochorion composed of six major and fourteen minor proteins that are produced over approximately five hours. In silk moths, the number of genes and the time devoted to producing the chorion is much greater. There are three gene families in the wild silk moth *A. polyphemus*, and the same three families, plus two others, in the domesticated silk moth *B. mori*. Approximately 100 chorion proteins are produced during a period of approximately 51 hours (Kafatos 1981).

Moth and fly chorion genes are organized differently. In *Drosophila*, large amounts of the chorion proteins are produced in a relatively short time. This is facilitated by *amplification* of the chorion genes. In *Drosophila*, there are two chorion gene clusters, 5 to 10 kb in size, each encoding tandemly oriented chorion genes. One gene cluster is found on the X and one on the third chromosome. Because each chorion gene cluster is represented only once in the haploid genome, the chorion proteins cannot be synthesized quickly and in sufficiently

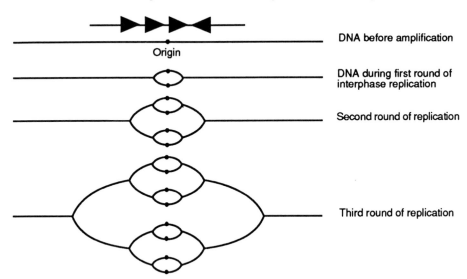

**Figure 5.1**   Amplification of the *Drosophila* chorion genes in follicle cells. The first three rounds of DNA replication at the 66D locus on chromosome 3. The three small arrows represent three well-characterized chorion genes in this cluster. The polarity of a fourth chorion gene and the precise location of the origin is unknown. The boundaries of the amplified DNA are much larger than the chorion protein transcription units within it.

large quantities unless gene amplification occurs. A 20-fold amplification of the chorion genes on the X chromosome and an 80-fold amplification of the genes on chromosome 3 is found in follicle cells. Amplification is achieved by replicating the DNA segments at multiple replication origins (Heck and Spradling 1990). DNA amplification extends bidirectionally for a distance of up to 40–50 kb to produce a multiforked 'onionskin' structure that contains multiple copies of DNA containing the chorion genes (Figure 5.1).

Gene amplification also occurs in the chorion genes of the Mediterranean fruit fly *Ceratitis capitata*. The overall organization of the cluster is similar to that of *Drosophila*, with the same four genes maintained in tandem, in the same order, and with similar spacing (Konsolaki et al. 1990). Despite the divergence of *Drosophila* and *Ceratitis* family lineages approximately 120 million years ago, there is a high conservation in the coding sequences and regulatory properties of their chorion genes.

Silk moth chorion proteins are produced over a longer time interval and involve larger numbers of genes that have probably arisen by **gene duplication**. More than 100 structural proteins have been found in the chorion of the silk moth *A. polyphemus* and the chorion has an extremely elaborate organization. In

*B. mori*, the chorion genes are on chromosome 2 and consist of two segments that total more than 1000 kb of DNA. Subsets of the genes are expressed at different periods of choriogenesis (early, middle, late, very late), with the early proteins associated with framework formation, middle proteins with framework expansion, late proteins with densification, and very late proteins with surface sculpturing of the chorion.

Silk moths have solved the problem of producing large amounts of protein quickly by gene duplication. Silk moth chorion genes are found in multiple copies of divergently transcribed, coordinately expressed pairs (Kafatos et al. 1986). For example, all members from each of two late gene families are arranged in 15 pairs on a 140-kb segment. The members of each family have a high degree of sequence homology, although they are not identical. This homology could be maintained by a process called **concerted evolution** (Eickbush and Burke 1985). Concerted evolution often occurs in multigene families and could be maintained by two different mechanisms: either unequal crossing over or gene conversion. Unequal crossing over may occur between the two sister chromatids of a chromosome during mitosis of a germ line cell or between two homologous chromosomes at meiosis. It is a reciprocal recombination that results in a sequence duplication in one chromatid or chromosome and a corresponding deletion in the other. As a result of unequal exchange, daughter chromosomes become more homogeneous than the parental chromosomes. If the process is repeated, the numbers of each variant repeat on a chromosome will fluctuate with time and eventually one will become dominant in the family. **Gene conversion** is a nonreciprocal recombination process in which two sequences interact so that one is converted by the other.

Despite the very different organization of chorion genes in *Drosophila* and silk moths, silk moth chorion genes can function in *D. melanogaster*. Moth chorion genes were cloned into P-element vectors and inserted into the *D. melanogaster* germ line (Mitsialis and Kafatos, 1985). (Chapter 10 contains a description of the methods employed in inserting moth genes into *Drosophila* using a P transposable element that has been modified to carry exogenous genes.) Analysis of RNAs from transformed flies indicated that moth genes are expressed in an appropriate manner in the correct sex, tissue, and time in *D. melanogaster*. Fly and moth lineages are thought to have diverged over 250 million years ago, yet regulatory elements conferring sex, tissue, and temporal specificity of gene expression must have been conserved. Subsequent analysis showed chorion gene promoter sequences from the silk moth species *Antheraea pernyi* and *A. polyphemus* also functioned in *D. melanogaster* after P-element-mediated transformation, although some regulatory interactions had diversified (Mitsialis et al. 1989).

**Insecticide resistance**   With the availability of molecular genetic techniques, geneticists have identified a new mechanism by which insects become resistant to pesticides (Mouches et al. 1990, Devonshire and Field 1991, Devonshire et al. 1992). Amplification of esterase genes in the aphid *Myzus persicae* and the mosquito *Culex pipiens quinquefasciatus* has been found, although the precise method of DNA duplication is not yet understood. Apparently, many identical gene copies are present in tandem arrays in each cell and large amounts of a specific esterase are produced. It is not known if gene amplification is responsible for pesticide resistances in other insect species, but gene amplification could occur if increased esterase activity is associated with the insecticide resistance.

Whether exposure to pesticides can induce resistance in insects by gene amplification is an interesting, but unresolved, question. It has long been assumed that pesticide resistance in insects is due to the presence of rare alleles in populations that are selected for by pesticide applications (preadaptive mutations). However, amplification of genes in mammalian cells, plants, yeast, and microorganisms has been shown to occur in response to exposure to toxins. For example, amplification of the dihydrofolate reductase gene in mammalian cells in tissue culture occurs in response to exposure to methotrexate. A 100-fold amplification in a cholinesterase gene in two generations of a human family subjected to prolonged exposure to organophosphorus insecticides has been demonstrated and could be due to genetic changes induced by prolonged exposure to parathion (Prody et al. 1989). Cultures of mosquito cells selected with methotrexate became amplified (Fallon 1984). Thus, insecticide resistances due to gene amplification in insects could, at least in some cases, be induced by exposure to insecticides.

There are several models for how amplified genes are generated. These include gene duplication by random unequal crossing over between sister chromatids, with a subsequent misalignment resulting in increased numbers of tandem repeats with intervening spacer DNA (Stark and Wahl 1984). Another model involves replication of DNA more than once at the same origin of replication within a cell cycle (repeated replication model), which generates multiple unattached DNA molecules that are either released or are integrated into the chromosome by end-to-end ligation and recombination. The repeated replication model is consistent with the sudden appearance of many gene copies and the initiation of chromosome breaks or translocations, which are often found associated with gene amplification. The mechanism by which resistance genes are amplified in insects remains unknown. In both aphids and mosquitoes the increase in gene copy number appears to occur in a stepwise manner.

In the aphid *Myzus persicae*, only one of several esterases responsible for the increase in total esterase activity is associated with resistance to insecticides. Some of the resistant aphids also have a chromosomal translocation, but it is not clear that the translocation is associated with the underlying mechanism(s) of resistance. To study resistance, different aphid colonies initiated by individual parthenogenic females (which produce clonal colonies) were established in the laboratory. These clonal colonies produced different amounts of the detoxifying esterase, with a 64-fold difference between the most susceptible and the most tolerant aphid clones. The most resistant aphid clones tested produced approximately 10 picomoles (pmoles) of esterase per insect, which is equivalent to 1% of its total proteins. The esterase is found in two closely related forms, E4 and FE4, in different resistant aphid clones. Any one clonal colony produces only one form, which can be distinguished electrophoretically. Aphids with a normal karyotype overproduce the FE4 esterase and aphids with the translocation (A1,3) overproduce the E4 esterase.

In several species of mosquitoes (*Culex pipiens*, *C. tarsalis*, *C. tritaeniorhynchus*), resistance to organophosphorus insecticides is associated with increased activity of two different esterases, A and B. The A and B esterases seem to comprise two gene families (A1, A2 and B1, B2, B3) with little homology between them. Amplification of B1, B2, and B3 esterases has occurred, probably in independent events. In a California population of *Culex pipiens quinquefasciatus*, the B1 gene is amplified at least 250-fold and confers about 800-fold levels of resistance. The amplified esterase gene and surrounding sequences have been cloned (Mouches et al. 1990). The esterase gene is surrounded by two 25-kb conserved DNA sequences. These DNA sequences also are found in other parts of the genome of both susceptible and resistant mosquito colonies, and are called 'Juan'. The Juan elements appear to consist of about 3% of the total DNA and are scattered throughout the genome of all *C. pipiens* and *C. tarsalis* strains analyzed to date. Agarwal et al. (1993) showed that Juan elements from *C. pipiens* are similar to LINE retroposons and can probably be transposed by reverse transcription of an RNA intermediate.

In both aphids and mosquitoes, traditional analyses of the mode of inheritance of resistance indicate that high esterase activity is inherited as a single factor, which is expected if the amplified genes are located on the same chromosome and inherited as a unit. However, in both aphids and mosquitoes, resistance can be unstable in the absence of selection. In the parthenogenic aphid clones, resistance is usually stable for long periods but can be lost in some progeny of some clones. This reversion is associated with clones that carry the translocation, and the revertant lines can often be reselected for resistance. Because reversion involves loss of elevated esterase production and mRNA, but not of the amplified genes, it is likely that control over gene transcription is responsible for the loss of resistance. In *Culex pipiens*, resistance and amplification are both lost when strains are not homozygous, but both are stable when strains are homozygous, suggesting that mutations that would reverse amplification are rare (Raymond et al. 1993).

Information on the molecular basis of resistance to pesticides is increasing rapidly and promises to provide fundamental advances in the knowledge of how insects respond to environmental stresses, as well as information useful for practical pest management programs. Whether resistance to other pesticides is due to amplification remains to be determined. Taylor et al. (1933) found that pyrethroid-resistant strains of *Heliothis virescens* exhibited endopolyploidy, but found no evidence for large-scale DNA amplification associated with the pyrethroid resistance. See ffrench-Constant et al. (1993) for a review and Chapter 15 for information on cloned resistance genes.

## Endocytobiosis in Insects

Mitochondria are now generally accepted as microbial **symbionts** that were modified after a long process of evolution within eukaryotic cells (Gray 1989). Mitochondria retain a distinctive genome that is replicated and expressed, but mitochondria are now incapable of independent existence. Most of the genes in mtDNA are retained from the original endosymbiont, but in the course of evolution some genes were transferred from the endosymbiont to the nuclear genome of its host.

In addition to mitochondria, many insects have intimate intracellular relationships with a diverse array of microorganisms, including viruses, bacteria, yeasts, and rickettsia (Schwemmler and Gassner 1989, Douglas 1992). The details of the relationship between the host and the microorganism usually are unknown, and the term **endocytobiosis** was coined to reflect this lack of knowledge. Thus, endocytobiosis includes both the intracellular symbionts (such as viruses and bacteria) as well as the intracellular parasites. The term also includes DNA-containing organelles such as mitochondria.

In many cases, the endocytobionts possess metabolic capabilities that the insect host lacks and the insect hosts use these capabilities to survive on poor or unbalanced diets. The insect and its microbes often require their association. Many insects freed of their symbionts grow slowly and produce few or no progeny, and many microorganisms cannot grow outside their insect host. In other cases, the microorganisms influence the behavior of other insects. For example, gut microorganisms in the leek moth *Acrolepiopsis assectella* produce chemicals that are found in the larval frass. These volatile odors are used by the specialist parasitoid *Diadromus pulchellus* to find its host (Thibout et al. 1993).

Intracellular microorganisms, or endocytobionts, are found in the Anoplura, Mallophaga, Isoptera, Orthoptera, Homoptera, Coleoptera, Diptera, and Hymenoptera. The amazing diversity of relationships and microorganisms involved has raised many questions, but provided few clear-cut answers, in large part because most symbionts are very difficult to culture outside their hosts. Many microorganisms are contained in special structures and transmitted to progeny by a highly specific method, including transovarial transmission. Transmission also can occur when larvae feed on contaminated egg shells or feces.

Some insect species contain several different types of endocytobionts in different tissues, including the gut, Malpighian tubules, fat body, or gonads. Usually these microorganisms are neutral or beneficial to their hosts. Endocytobionts of some aphid species (Homoptera) synthesize essential amino acids, sterols, and vitamins and may function as 'genetic elements' distinct from the host nuclear genome. Endocytobiosis in scale insects (Homoptera) is particularly diverse, with almost 20 different types of associations described so far.

In the leafhopper species *Euscelidium variegatus*, specific bacteria are thought to be essential for normal growth and development, breaking down uric acid in the host cells and synthesizing amino acids and vitamins. It appears that symbionts are involved in normal egg development of *E. variegatus*; embryos in eggs artificially lacking symbionts fail to develop normal abdomens. It is hypothesized that some genes from this microorganism have been transferred to the nuclear genome of *E. variegatus* in a manner parallel to that of mitochondria.

Bacteroids, spiroplasmas, rickettsia, mycoplasmas, or virus-like endocytobionts are found in dipteran testes, ovaries, pole cells, nurse cells, and gut wall cells. The relationships can be intimate and even essential. In the mosquito *Culex pipiens*, the rickettsia *Wolbachia pipientis* is transovarially transmitted and eggs lacking *Wolbachia* fail to hatch. Both *Culex pipiens* and *Aedes scutelaris* exhibit intraspecific reproductive incompatibility caused by *Wolbachia*. When *Wolbachia* are present in a population, either no hybrids are produced when different populations of the same species are crossed or no (or fewer) progeny are produced by females from one reciprocal cross but not the other.

Testes of 14 species and strains of *Heliothis* and *Helicoverpa* contain virus-like particles in spermatocyst cells that are similar to virus-like particles found in *D. melanogaster* and could be transposable elements similar to the *copia*-like family of retrotransposons (Degrugillier and Newman 1993). Many insects contain several different types of microorganisms. For example, tsetse flies in the genus *Glossina* contain intracellular bacterial symbionts in the midgut from the gamma subdivision of the Proteobacteria as well as bacteria from the alpha subdivision (probably genus *Wolbachia*) in the ovaries (O'Neill et al. 1993).

In the case of the parasitic wasps *Nasonia giraulti* and *Nasonia vitripennis*, microorganisms cause *interspecific* incompatibility by destroying the paternal chromosome set (Breeuwer and Werren 1990). These two closely related sympatric species can produce hybrid progeny if microorganisms have been eliminated by treatment with antibiotics. If the microorganisms are present, the paternally derived chromosomes are destroyed in the fertilized eggs and no female progeny are produced in these haplodiploid species. Thus, it is possible that the microorganisms have been involved in reproductive isolation and a rapid speciation event. (See Chapter 11 for additional examples of the impact of microorganisms on sex ratio.)

In other cases, insects lacking their endocytobionts are apparently com-

pletely normal. For example, in the beetle family Cerambycidae, all of which
live in wood, some species have endocytobionts while others lack them. The
hypothesis that endocytobionts supply a deficiency in the insect's diet appears to
be too simplistic: some insects feeding on a well-balanced diet have symbionts.

In some cases, rickettsia-like organisms are important in determining
whether an insect vector is able to transmit (vector) a disease. For example,
rickettsia-like organisms from the tsetse fly *Glossina morsitans morsitans* are associated with the maternally inherited condition of susceptibility to trypanosome
infection (Welburn et al. 1993). The rickettsia-like organisms from tsetse produce endochitinases in the gut that inhibit midgut lectins in newly emerged
adult flies. Tsetse flies lacking the rickettsia-like organisms are less susceptible to
trypanosomes (refractory).

Endocytobionts are frequently found in insects, but a full understanding of
their genetic and evolutionary role remains to be determined. In the few cases
that have been well studied, a genetic interplay between insect host and symbiont occurs; factors are supplied to each from the other; and the endocytobiont
has specific means of movement and relocation within the insect. Insects commonly control movement and multiplication of the endocytobiont and the
endocytobiont often influences growth and reproduction of the insect. An endocytobiont must be recognized by the insect as 'self' rather than as foreign or it
would be subject to the insect's immune system. Our understanding of how
microorganisms were incorporated into insect tissues and cells and have evolved
remains fragmentary (Schwemmler and Gassner 1989). A particularly interesting example of an intimate relationship between insects and endocytobionts is
illustrated by the relationship between polydnaviruses and parasitoid wasps.

### Polydnaviruses in Parasitic Wasps

The **polydnaviruses** are relatively newly recognized viruses that are found
only in the Braconidae and Ichneumonidae among the parasitic Hymenoptera
(Krell 1991, Fleming 1992, Stoltz and Whitfield 1992). Polydnaviruses have
double-stranded circular DNA genomes; they are literally 'poly-DNA-viruses',
having segmented genomes composed of several circular DNA molecules. For
example, the genome of the virus within the parasitic wasp *Campoletis sonorensis*
consists of 28 DNA molecules ranging in size from approximately 5.5 to 21 kb,
with the total genome size approximately 150 kb.

Polydnaviruses appear to be important in ensuring that many species of
braconids and ichneumonids are able to successfully parasitize their insect hosts.
At least 50 species of parasitic wasps have been shown to contain polydnaviruses, but not all species carry them (Stoltz and Whitfield 1992). Polydnaviruses alter the host insect's cellular immune response, prevent encapsulation
of wasp eggs and larvae by host hemocytes, and influence development of the
host to benefit the wasp.

Polydnaviruses replicate only in braconid or ichneumonid wasp ovaries and are secreted into the oviducts from where, during oviposition, they are injected into host lepidopteran larvae. The viruses appear to be vertically transmitted and integrated into the chromosome of the wasp (Fleming and Summers 1991). Each wasp species appears to carry a polydnavirus characteristic of that species. If one species within a particular genus carries a polydnavirus, they all are likely to do so (Stoltz and Whitfield 1992).

Insects possess immune mechanisms that protect them from microorganisms, other invertebrates, and abiotic materials (Hultmark 1993). Protection occurs through constitutive factors or by inducible humoral and cellular responses. Many parasitic wasps are internal parasites and spend part of their lives in the bodies of other insects. Many behavioral, morphological, nutritional, and endocrine factors determine whether the interactions between a host and a parasitic wasp will lead to development of the wasp or to its destruction (Fleming 1992).

It appears that one or more gene in the polydnavirus influences the immune system of the insect host, which allows the wasp eggs and larvae to survive. How the polydnaviruses cause such changes in the host insect has not been fully resolved (Schmidt and Theopold 1991, Fleming 1992). The virus alone can induce altered immune responses in some hosts, but in other hosts the venom injected by the wasp also must be present for the full effect of the virus to occur. Parasitic wasps thus appear to benefit significantly from the polydnaviruses that replicate in their reproductive tracts. The virus also clearly benefits if the wasp is able to reproduce, because polydnaviruses are known to replicate only within the wasps.

The origin of the polydnaviruses is unknown, but there are two possibilities: (1) They could have arisen from the insect genome itself, or (2) they originally existed as typical viruses, such as baculoviruses. The polydnavirus–parasite relationship is ancient, and it is likely that genetic transfer between viral and host genomes has occurred as the polydnavirus and parasitic wasp genomes evolved to become a single genome (Stoltz and Whitfield 1992).

A possible second, unnamed, virus family recently was found in the parasitic Hymenoptera. These nonpathogenic viruses (to the wasps) have a long, filamentous, enveloped nucleocapsid, and have been found in both braconids and ichneumonids. They resemble filamentous viruses found in the tsetse fly *Glossina pallidipes* and in honey bees, but their relationships to their hosts remain relatively obscure (Stoltz and Whitfield 1992). Our understanding of the role that microorganisms play in the genetics and evolution of insects remains fragmentary, but each new discovery provides intriguing glimpses into complex interrelationships.

## Genetic Systems in Insects

Most insects are diploid (2n) in their somatic cells and haploid (n) in their gametes. However, some insect groups are parthenogenetic and may be polyploid, including species in the Orthoptera, Homoptera, Lepidoptera, Diptera, Coleoptera, and Hymenoptera. Polyploid insects usually are 3n or 4n, but exceptions include five species of curculionid weevils that are 5n and two that are 6n (Retnakaran and Percy 1985, Saura et al. 1993). Parthenogenesis is found in all orders of insects except the Diplura, Protura, Odonata, Plecoptera, Dermaptera, Grylloblattodea, Zoraptera, Megaloptera, Mecoptera, and Siphonaptera.

Parthenogenesis can be divided into three major types: arrhenotoky, thelytoky, and deuterotoky. **Deuterotoky** involves the development of unfertilized eggs into either males or females, and at least one insect, a mayfly, is reported to exhibit facultative deuterotoky (White 1973). In **arrhenotoky**, insects are haplodiploid, with males developing from unfertilized eggs while females develop from fertilized eggs. The entire order Hymenoptera and many species in the Homoptera, Thysanoptera, and Coleoptera are arrhenotokous (Hartl and Brown 1970, White 1973). When the male of a species is haploid, its germ line nuclei contain half the number of chromosomes present in the corresponding diploid nuclei of the female. However, in most insects, some of the somatic tissues exhibit high levels of endopolyploidy. For example, in honey bees, haploid males have about the same amount of DNA as females in some of their somatic tissues because nuclei of the male undergo compensatory endomitosis so that equal amounts of DNA are present. In some cases, haploid males are known to exhibit higher levels of endopolyploidy in some tissues than the females of the same species.

Thelytokous insect species have females only. **Thelytoky** has arisen repeatedly in evolution, consists of several types, and can be induced experimentally in a number of ways (White 1973) (see Chapter 11 for examples). For example, in some cases of thelytoky, eggs develop only after penetration by a sperm (pseudogamy or gynogenesis), but the sperm nucleus degenerates without fusing with the egg nucleus so that it makes no genetic contribution to the embryo. The sperm may be derived from the testis or ovotestis of a hermaphrodite or from a male of a different but closely related species.

Thelytoky may be the sole mode of reproduction in a species or it may alternate with sexual reproduction in a regular manner (cyclical thelytoky), as happens in aphids, gall wasps, and some cecidomyiids. In species that reproduce by cyclical thelytoky, genetic recombination is possible, but in species with complete thelytoky there is no way in which mutations that have occurred in two unrelated individuals can be combined in a third. Thelytokous reproduction can be induced in the eggs of many species by a variety of treatments such as pricking the egg or on exposing it to chemical agents or heat. In a number of normally bisexual insects, eggs deposited by virgin females can hatch spontaneously and the incidence of such egg hatches can be increased by artificial selection. White (1973) suggests that the capacity for artificial parthenogenesis, induced thelytoky, or facultative thelytoky indicates that some capacity for parthenogenetic development is probably present in all eggs. Thelytokous species or thelytokous populations of bisexual species have been found in the Diptera, Hymenoptera, Lepidoptera, Orthoptera, and Coleoptera. Chapter 11 describes how some microorganisms can alter sex ratios in some species of parasitic Hymenoptera.

Parthenogenetic insects exhibit different modes of reproduction and sex determination, indicating that parthenogenesis has developed independently in different groups. In the Homoptera, both arrhenotoky and thelytoky occur (Retnakaran and Percy 1985). However, more complex genetic systems are also found (White 1973, Haig 1993). In mealybugs (Pseudococcidae), both males and females develop from fertilized eggs, but in the embryos that develop into males, the paternally derived chromosomes become heterochromatic, genetically inactive, and are not transmitted to the male progeny (Brown and Nur 1964, Nur 1990).

Some method of **chromosome imprinting** is probably involved in ensuring that the paternally derived chromosomes and not the maternally derived ones are eliminated in many insects. The mechanisms involved in chromosomal imprinting are not understood in insects, but in other organisms imprinting is associated with methylation of DNA (Sapienza et al. 1987, Solter 1988, Peterson and Sapienza 1993). It is possible that imprinting makes the paternal chromosomes susceptible to endonucleases, which leads to the destruction of the chromosomes (Wagner et al. 1993).

Heckel (1993) pointed out that relatively little genetic information is available for the vast majority of insect species. For example, sufficient genetic information is available to develop genetic linkage maps for only 27 species. A linkage map involves identifying specific chromosomes by one or more genetic markers, usually by phenotypic mutants. Mutations on the sex (X) chromosomes are most easily identified, because they exhibit a characteristic mode of inheritance. Most of the well-studied species are dipterans—21 of the 27 species. Linkage maps are available for six species of *Drosophila*, nine mosquitoes, the screwworm *Cochliomyia hominivorax*, the sheep blowfly *Lucilia cuprina*, the Mediterranean fruit fly *Ceratitis capitata*, the Oriental fruit fly *Bactrocera dorsalis*, and the house fly *Musca domestica*. Two coleopterans (*Tribolium castaneum* and *T. confusum*), one orthopteran (the cockroach *Blattela germanica*), two hymenopterans (*Habrobracon juglandis* and *Nasonia vitripennis*) and one lepidopteran (*Bombyx mori*) make up the rest (Wright and Pal 1967, King 1975, Robinson 1971, Sokoloff 1966, 1977, Steiner et al. 1982, Heckel 1993, Severson et al. 1993).

Other insects have been studied, but insufficient markers are available to develop linkage maps. References are provided by Heckel (1993) to the genetics of the honey bee *Apis mellifera*, the Mediterranean flour moth *Ephestia kuehniella*, *Heliconius* butterflies, the butterflies *Papilio glaucus* and *Colias eurhytheme*, and the tobacco budworm *Heliothis virescens*. Some genetic linkage studies have been conducted on the tenebrionid beetle *Latheticus oryzae*, the fruit fly *Rhagoletis pomonella*, and the grasshoppers *Melanoplus sanguinipes* and *Locusta migratoria* (Chapco 1983, Sokoloff 1966). Genetic studies of honey bees and silk moths have focused on practical management of these useful insects (Rinderer 1986, Robinson 1971, Tazima 1964, Tazima et al. 1975). Genetic analysis of the haplodiploid sawfly *Athalia rosae* has focused on development in sex determination (Oishi et al. 1993).

## Insect Development

Most of what we know about the genetics of development in insects has been learned by studying *Drosophila melanogaster* (Lawrence 1992, Wilkins 1993). Extensive analyses of insect development have become feasible with the tools of molecular genetics, and thousands of papers have been published on the molecular genetics of development in *D. melanogaster*. Numerous review articles have been published on this rapidly advancing field, as reviewed by Lawrence (1992) and Wilkins (1993). A complete discussion of development is beyond the scope of this book. However, the following paragraphs provide a brief outline of insect development that will be useful in understanding sex determination, behavior, and P-element-mediated transformation (Chapters 10, 11, and 12).

### Oocyte Formation in Drosophila melanogaster

A substantial amount of development of the insect embryo is determined in the oocyte, before oocyte and sperm pronuclei fuse to form an embryo. An understanding of the development of the oocyte is particularly critical to understanding development during embryogenesis. Oocyte formation in *D. melanogaster* is complex, involving both somatic and germ line cells. The ovaries contain oocytes, which are formed from the pole cells, but the cells that surround each egg chamber and make up the walls of the egg chambers are derived from mesoderm (somatic tissues). The pro-oocyte arises in a set of cell divisions within the ovary from an oogonial stem cell. Each oogonial stem cell divides to give a daughter stem cell and a cystoblast cell. The cystoblast cell gives rise to a set of 16 sister cells in four mitotic divisions, which provides a cyst. One of these 16 cells becomes the pro-oocyte, and eventually the oocyte, while its 15 sister cells become nurse cells whose function is to synthesize materials to supply the growing oocyte. The 16-cell cyst, surrounded by a layer of somatic cells, is termed the egg chamber. The final stages of egg chamber development involve covering the cyst with a monolayer of prefollicle cells, which are somatic in origin. These 80 somatic cells divide an additional four times to give 1200 follicle cells which cover each cyst.

Initially *Drosophila* oocytes and nurse cells are roughly the same size, but increase in volume by approximately 40-fold when vitellogenin begins to accumulate about halfway through development of the oocyte. Some vitellogenin is derived from the follicle cells, but most is produced in the fat body and transported to the ovary (Raikhel and Dhadialla 1992). The later stages of oocyte development involve very rapid growth, with the oocyte increasing in volume 1500-fold. While the oocyte is increasing in size, the nurse cells are decreasing because their contents are being deposited in the oocyte. Nurse cells, derived from the germ line, are polyploid, containing 512 and 1024 times the haploid DNA content. The polyploid nurse cells synthesize proteins, ribosomes, and mRNAs. These products, and mitochondria, are transferred to the oocyte by intercellular channels. Thus, the oocyte contains products produced by the mother, which means that initial development in the oocyte is highly dependent upon the genome of the mother. Finally, the vitelline membrane and the chorion are secreted around the oocyte by follicle cells and the oocyte enters metaphase of meiosis I. Follicle cells are polyploid, secreting the vitelline membrane of the oocyte and the chorion. The oocyte remains arrested at metaphase of meiosis I until after fertilization.

The oocyte increases in total volume during its development by approximately 90,000-fold. Oogenesis is a complex developmental pathway that is estimated to require 70 to 80% of all genes in the *Drosophila* genome to function. While these genes are active during oogenesis, the great majority are expressed during other stages of development as well. Only about 75 genes are expressed exclusively during oogenesis (Perrimon et al. 1986). The egg of *D. melanogaster* is rich in stored RNA molecules, including ribosomal RNA and mRNA. The bulk of the maternally produced, stored mRNA is derived from transcription of nurse cell nuclei during egg chamber growth, but some mRNA may be derived from the oocyte nucleus itself, which is active briefly about halfway through development. The total amount of mRNA in the oocyte is equal to about 10% of the single-copy DNA of the *Drosophila* genome and corresponds to approximately 8000 distinct protein coding sequences. Most of the mRNA is used to code for proteins that are required early in embryogenesis, including proteins such as tubulins and histones. There is some evidence that products from a few maternal genes continue to affect development in *D. melanogaster* during the larval stage.

### Embryogenesis in Drosophila melanogaster

Fertilization occurs when the mature oocytes are released into the oviducts. A single sperm enters the egg cytoplasm through a special channel in the anterior region of the oocyte called the micropyle. Fertilization initiates the completion of meiosis I and II, producing two polar body nuclei and the female pronucleus. After the haploid male and female pronuclei unite (syngamy), early embryogenesis takes place so rapidly there is no time for cell growth (Figure 5.2).

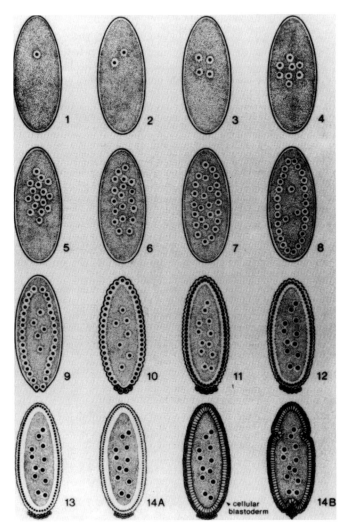

**Figure 5.2**   Early embryonic stages of *Drosophila melanogaster* from fertilization to just before gastrulation, showing the appearance of pole and somatic buds and cessation of division of yolk nuclei. Numbers indicate division cycles; each cycle begins with the start of interphase and ends at the conclusion of mitosis. Embryos are in longitudinal section without the vitelline membrane. All nuclei (black circles) are shown for cycles 1–5 and afterward only some are shown. Stippled areas represent yolk and open areas represent yolk-free cytoplasm. Yolk-free cytoplasm is found both at the periphery (periplasm) and in islands around the nuclei.

During cycles 1–7, nuclei multiply exponentially in the central region of the fertilized egg. Cycle 8 illustrates migration of the majority of the nuclei to the periphery, leaving the future yolk nuclei behind in the center. Yolk nuclei continue to divide in

Initial mitoses are atypical because the first nine divisions result in a syncytium containing approximately 512 nuclei that lack cellular membranes. Thus, during embryogenesis the relatively large egg is divided into many smaller cells.

After seven nuclear divisions, and when there are 128 nuclei in the central region of the egg, most of the nuclei and their surrounding cytoplasm migrate outward as they continue to divide. A few nuclei are left behind which divide once to become yolk nuclei that do not become incorporated in the embryo (Figure 5.2). After nine divisions, most of the nuclei have migrated to the egg surface. At this time, the soma and germ line nuclei segregate when about 15 nuclei move to the posterior region of the egg, bud off, and eventually become the nuclei in the pole cells. These nuclei divide about twice more and become pole cells that will give rise to the germ line tissues of the fly. Meanwhile, the other nuclei migrate to the surface of the egg and divide four times more in synchrony to produce a syncytial blastoderm.

Finally, the membrane covering the egg invaginates to enclose each nucleus in a separate membrane, to form a cellularized **blastoderm** (Figure 5.2). The blastoderm is the layer of cells in an insect embryo that completely surrounds an internal yolk mass. The cellular blastoderm develops from a syncytial blastoderm by the partitioning of the cleavage nuclei, with membranes derived from infolding of the oolemma. During the cellular blastoderm stage, D. *melanogaster* exhibits the **long germ band** type of development in which the pattern of segmentation is established by the end of the blastoderm stage. Some other insects exhibit the **short germ band** type of development in which all or most of the metameric pattern is completed by the sequential addition of segments during elongation of the caudal region of the embryo.

Prior to the cellularized blastoderm stage, the dividing nuclei are equivalent and totipotent, but after the cellularized blastoderm stage is reached, specific body segments have been determined. The cellularized blastoderm stage is a key transition point in embryogenesis in D. *melanogaster* because this is the period during which the products of maternal genes become less important. It is thought that only a few zygotic genes are active prior to cellularization. After the

---

synchrony with other nuclei in cycles 8–10; they then cease dividing and become polyploid. Early in cycle 9, a few nuclei appear in the posterior periplasm and cause protrusions of the cytoplasm, called pole buds. During cycle 10, the remaining migrating nuclei enter the periplasmic region, forming somatic buds over the entire embryonic surface. During the tenth cycle, pole buds are pinched off to form pole cells. After this, synchrony between the pole cells and the syncytium is lost. The syncytial nuclei continue to divide synchronously. The periplasm begins to thicken in cycle 13. During cycle 14, the formation of a plasma membrane begins to separate cells over the entire surface of the embryo, with nuclei elongating to match elongated cells formed by late cycle 14A. During 14B, gastrulation movements begin with the infolding of the cephalic furrow (anterior) and posterior midgut furrow, and subsequently the cells no longer divide synchronously.

cellularized blastoderm stage, the genes in the zygote begin to dominate development of the embryo, although maternal genes continue to have some effect. After additional development, the embryo gives rise to a segmented larva with three major tagmata: head, thorax, and abdomen.

### Postembryonic Development

D. melanogaster is a holometabolous insect with sequential life stages: egg → larval stage 1 → molt → larval stage 2 → molt → larval stage 3 → molt → pupa → molt → adult. The larva emerges (**ecloses**) from the egg, grows, and molts after each larval stadium. After the third larval stadium, the insect molts, pupates, and undergoes metamorphosis to the adult form. During metamorphosis, most of the larval tissues are digested.

Adult structures develop from cells in structures called **imaginal discs** and abdominal histoblast nests that will give rise to the abdominal epithelium. The cells that give rise to the 19 imaginal discs become segregated from surrounding cells during the first half of embryogenesis. By the time the larva ecloses, the imaginal discs and histoblast cells are visibly distinct from the surrounding larval cells because they have smaller nuclei and an undifferentiated appearance. The labial, clypeolabral, antennal + eye, thoracic, three leg, wing, and haltere (wings on metathorax of other insects) imaginal discs are paired. In addition, there is a single fused genital disc. Imaginal cells are diploid and able to divide. By contrast, most of the 6000 somatic cells of D. melanogaster grow in size but do not undergo mitosis or cell division, although the chromosomes of the larval cells continue to undergo replication and become polytene.

At the end of the third larval stage, the larva is transformed into a pupa. During the pupal stage, the imaginal discs, each consisting of about 40 cells, develop into adult structures such as legs, wings, eyes, ovaries or testes, and antennae (Larsen-Rapport 1986). Because the imaginal discs were determined during embryonic development, the basic body plan of the adult fly was laid down long before the first instar larva eclosed from the egg. The wings, halteres, and legs of the adult, with as many as 50,000 cells each, are formed from the imaginal discs.

Because Drosophila undergoes a complete metamorphosis, its development traditionally has been considered to occur as a two-tiered process in which the imaginal precursors are segregated early in embryogenesis so that the development of larval and imaginal tissues is parallel and independent from the beginning. However, Couso and Gonzalez-Gaitan (1993) suggest that development of imaginal and larval precursor cells still is not segregated at blastoderm or early gastrula. Separation of imaginal and larval cells does not occur before embryonic stage 12. The expression and function of some genes suggest that the imaginal discs inherit some basic positional cues from the embryo, and these cues must be maintained throughout imaginal development to ensure correct growth and patterning. Not only is segmentation in *Drosophila* based on cues obtained from the mother, but the position and organization of adult structures may also be determined by coordinates provided by the mother (Couso and Gonzalez-Gaitan 1993). After emergence as an adult, the insect mates and her progeny begin this developmental cycle again.

## Dissecting Development with Mutants

The study of development in *Drosophila* is dependent upon the availability of mutants so that the process can be dissected. In fact, in discussing development in *Drosophila*, the genes influencing development are called by names that reflect their mutant form. Nusslein-Volhard and Wieschaus (1980) began a systematic program of mutagenizing *Drosophila* females in order to obtain many developmental mutants in insect embryos. However, many useful developmental mutants were discovered by E. B. Lewis in his pioneering work on *Drosophila* development involving the bithorax homeotic gene complex.

The term **homeotic** was coined to describe the replacement of one part of the body by a serially homologous part. Lewis (1978) developed the hypothesis that families of structurally related genes control the specification of the insect body plan because insect bodies are metameric (composed of serially repeating units or body segments) which differentiate into specific structures according to their position. Likewise, the appendages in each major body segment develop into appropriate structures, with antennae located on the head, legs on the thoracic segments, and wings on the second thoracic segment. However, this normal pattern is disrupted in a number of homeotic mutants.

*Drosophila* embryos appear to go through two different phases in their development. During the first phase, many genes appear to encode transcription factors or nuclear proteins. This suggests that development is regulated by a transcriptional regulation cascade. There is a successive subdivision of the embryo into smaller and smaller domains that is accomplished by the differential and combinatorial action of transcription factors and is initiated by the activities of a limited number of localized determinants. The first phase is completed by the time cells are fully formed at the end of the blastoderm stage.

The second phase begins after the formation of the cellular blastoderm and consists of elaborating the information provided from reference points that have been deposited along the dorsal–ventral and anterior–posterior axes. This requires the communication of information between cells by intercellular signal molecules.

Analysis of the genes that control *Drosophila* embryonic development provides details about the molecular processes during development. These genes can be divided into three classes: (1) the **maternal effect genes**, which specify egg polarity and the spatial coordinates of the egg and future embryo; (2) the

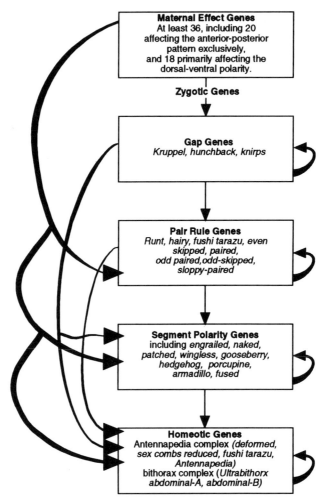

**Figure 5.3** Development of segments in the embryo of *Drosophila melanogaster* involves a hiearachy of regulatory genes. The earliest-acting genes are maternal effect genes that regulate the spatial expression of later-acting genes. Some maternal gene products are present in concentration gradients and regulate downstream genes in a concentration-dependent manner. In many cases, genes within a given class regulate other members of the class (depicted as arrows beginning and ending at the same class). Many other genes not shown may also play a role in segmentation.

**segmentation genes,** including the **gap, pair-rule,** and **segment polarity** classes of genes, which determine the number and polarity of the body segments; and (3) the **homeotic genes,** which determine the identification and sequence of the segments. Although most genes with a homeodomain are in the homeotic class, a few are found among the segmentation genes (Figure 5.3).

## Maternal Effect Genes

Maternal effect genes are genes that function in the mother, and their products (mRNA or stored materials) influence development of oocytes and embryos. Maternal effect genes are discovered by determining if the mother carries a mutant gene and her embryo cannot be rescued by a wild-type gene contributed by the father at the time the egg is fertilized. Maternal effect genes are important because, before the *Drosophila* egg is fertilized, they determine the basic organization of the oocyte. There at least 38 maternal effect genes, with about 20 known to affect the anterior–posterior pattern and about 18 known to affect the dorsal–ventral polarity of the embryo (Figure 5.3).

The 20 genes influencing differentiation into proper anterior–posterior polarity can be subdivided into a group affecting the anterior half of the embryo (including *bicaudal, Bicaudal C, Bicaudal D, bicoid, exuperantia, swallow*), a group affecting the posterior half of the embryo [including *torso, trunk, fs (1) polehole, fs(1) nasrat, lethal (1) polehole*], and a group affecting both the anterior and posterior ends of the embryo (including *oskar, staufen, tudor, valois, vasa, cappuccino, spire, nanos,* and *pumilio*).

Of the 18 genes affecting dorsal–ventral polarity, several are required for the polarity of both the eggshell and embryo (including *fs(1) K10, cappuccino, spire, torpedo, gurken, cornichon*], and several are required for the polarity of the embryo only (including *nudel, pipe, windbeutel, gastrulation, snake, easter, spatzle, Toll, pelle, tube, dorsal, cactus*) (Wilkins 1993).

Determining the dorsal–ventral and anterior–posterior polarity in the embryo is a highly significant step. For example, the $Toll^+$ gene is a dorsalizing gene. Mutations in $Toll^+$ produce embryos lacking both ventral and lateral structures. The $Toll^+$ gene product appears be a **morphogen** (molecules whose local concentration directly *determines* the local pattern of differentiation) because when the wild-type gene product is injected into mutant embryos, it can make the affected region become the dorsal region of the fly. Thus, the $Toll^+$ gene product itself determines the polarity of the embryo.

A major determinant of anterior-posterior polarity is the product of the $bicoid^+$ gene. It is transcribed in the nurse cells of the ovary and the mRNA passes into the oocyte where it becomes localized in the anterior of the egg, apparently aided by components of the cytoskeleton that are encoded by products of the genes $swallow^+$ and $exuperantia^+$.

Several of the maternal-effect genes (*nanos, cappuccino, spire, staufen, oskar, vasa, valois,* and *tudor*) are required for the localization of factors that determine the germ line. In addition, mitochondrial large ribosomal RNA (mtlrRNA) may be important for pole cell formation (Kobayashi et al. 1993). Thus, both the mitochondrial and nuclear genomes are involved in determination of the germ line during embryonic development.

Maternal effect genes are most important during development of the egg up to the blastoderm stage. After that, genes inherited by the zygote from both parents become dominant factors determining development. However, because development is an elaborative process and the adult phenotype is a summation of the developmental effects accrued over the lifespan of the individual, developmental events occurring early in the life cycle can significantly influence the phenotype at later stages. For example, maternal effects have significant and diverse effects on insect life histories, including incidence and intensity of diapause, production of sexual forms, wing polyphenism, dispersal behavior, development time, growth rate, resistance to chemicals and microbial infection, and survival. Some of these influences are caused by maternal age and diet, but some are genetically determined (Mousseau and Dingle 1991).

### Zygotic Segmentation Genes

During blastoderm formation, the embryo begins to develop a pattern of repeating body segments. The genetic control of segmentation is determined by *zygotic genes*, which have been divided into three categories: (1) pair-rule genes, (2) gap genes, and (3) segment polarity genes.

The segmentation mutants found in *Drosophila* embryos were initially difficult to interpret because they did not affect what appeared to be a single segment but, rather, half of one 'segment' and the adjacent half of the next 'segment'. Eventually, it was determined that true segments are not reflected by the visible cuticular patterns of sclerites and sutures, but the visible 'segments' are, in fact, **parasegments**. There are 14 complete parasegments in *D. melanogaster* that are defined early in development; each is a precise set of cells.

The gap gene mutants cause deletions in groups of adjacent segments, the pair-rule mutants cause pattern deletions in alternating segments, and the segment-polarity mutants cause pattern defects in every segment (Figure 5.4). Most segmentation mutants are lethals in the zygote, but some gap genes have a maternal effect and are also expressed during oogenesis. Four of the cloned segmentation genes, *fushi tarazu, even-skipped, paired,* and *engrailed,* have been found to contain a homeobox (see following discussion). Thus these genes encode DNA regulatory proteins or transcription factors that bind to specific DNA or RNA sequences.

**Gap genes   Gap genes** were so named because large areas of the normal cuticular pattern are deleted in individuals with mutant phenotypes (Figure 5.4).

Segment polarity       Pair-rule       Gap

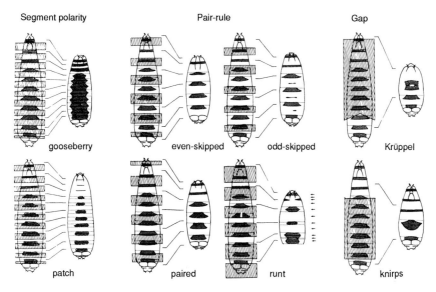

**Figure 5.4** Embryonic segment pattern defects are illustrated with selected mutants: *gooseberry* and *patch* are segment polarity mutants; *even-skipped, odd-skipped, paired,* and *runt* are pair-rule mutants; and *Krüppel* and *knirps* are gap genes. Dotted regions represent denticle bands on the developing embryo and dotted lines show the boundaries of segments. Hatched regions indicate the parts of the pattern that are missing in the mutants. Transverse lines link the corresponding regions in mutant and wild-type embryos. Arrows indicate where lines of polarity have been reversed. (From Nature 287: 796, 1980.)

The three wild-type versions of the gap genes—*Krüppel, hunchback,* and *knirps*—regionalize the embryo by delimiting domains of homeotic gene expression and effect position-specific regulation of the pair-rule genes (*runt, fushi tarazu, even skipped, paired,* and *odd-paired*). All three gap gene products contain DNA-binding domains. All three are expressed first in the syncytium, two division cycles before cellularization (cellular blastoderm). *Krüppel* embryos lack all of the thoracic and most of the abdominal segments, while *knirps* embryos have a normal thoracic region but lack nearly all the abdominal segments. Embryos homozygous for *hunchback* lack head segments, mesothorax, and metathorax while showing a normal abdominal segmentation.

The gap genes interact to produce sharp boundaries, with wild type alleles of *hunchback* and *Krüppel* repressing one another and *knirps* acting as a negative regulator of *Krüppel*. Thus, the establishment of stable domains by the gap genes is a two-step process: (1) a differential response to graded levels of maternal determinants, (2) followed by mutual repression leading to the generation of stable boundaries between adjacent domains. Gap genes regionalize the embryo by delimiting the domains of later homeotic gene expression and this results in position-specific regulation of the pair-rule class of genes, which give rise to the metamerization of the embryo.

**Pair-rule genes**   The **pair-rule genes** were so named because mutant flies have a repetitive aberration throughout the germ band, with the removal of integral, alternate segment-width areas (Figure 5.4). The pair-rule genes (including *runt, hairy, fushi tarazu, even-skipped, paired, odd-paired, odd-skipped, sloppy-paired*) are transiently expressed in seven or eight stripes during cellularization of the blastoderm. However, each is otherwise unique in its expression. Both *runt* and *hairy* are initially expressed rather uniformly throughout the embryo but begin restricted expression earlier than the others in the class. Because *runt* and *hairy* appear to have a major role in generating the striped pattern of the other pair-rule genes, they are called the primary pair-rule genes. *Fushi tarazu (ftz)* is a Japanese name meaning 'not enough segments' and mutants of this pair-rule gene have half the normal number of segments. The pair-rule genes are essential, directly or indirectly, for the initial establishment of segmentation.

**Segment polarity genes**   The **segment polarity genes** appear to determine a linear sequence of repeated positional values within each segment. The segment polarity mutants have a repetitive deletion of pattern, but the deletions occur within each segment and are followed, for many mutants in this group, by a partial mirror-image duplication of the part that remains (Figure 5.4). Segment polarity genes (including *engrailed, naked, patched, wingless, gooseberry, patched hedgehog, porcupine, armadillo, fused*) are required either continuously or over extensive periods to maintain the segmental pattern. Most or all are required to maintain the patterns in the imaginal tissues.

**Homeotic genes**   The periodic pattern of body segments generated by segmentation genes (e.g., gap genes, pair-rule genes, and segment polarity genes) has to be converted into segments with different characteristics (Figure 5.3). Thus, in insects, thoracic segment two is different from thoracic segment three and abdominal segment two will be different from the terminal abdominal segments, which typically have genital structures. This fine tuning is determined by **homeotic genes**. Both homeotic and some segmentation genes contain a special sequence called the **homeobox**. The homeobox consists of about 180 bp, which is found in the 3′ exon of the *Ultrabithorax, abdominal-A, Abdominal-B, Deformed, sex combs reduced, fushi tarazu,* and *Antennapedia* genes. The sequences of the different homeoboxes are nearly identical and they mediate the binding of homeotic proteins to specific DNA sequences and thus regulate gene function.

Since the first homeobox sequence was isolated from the *Antennapedia* gene in late 1983, it has been used as a probe to identify and isolate previously unknown homeotic genes from *Drosophila*. Furthermore, because the homeobox is evolutionarily conserved, it has been used as a probe to identify homeotic genes from other species, including humans (Gehring 1985, Kappan et al. 1993). The homeobox sequence codes for a sequence of 60 amino acids and its presence labels a protein as a DNA binding protein and the gene containing it as one that can control other genes.

One of the homeobox sequences of *Xenopus* is so similar to the *Antennapedia* homeobox in *Drosophila* that only one amino acid out of 60 is different. The reason for this extraordinary conservation during evolution is not fully understood (Gehring 1987). However, not only are the sequences of the different elements in the vertebrate and insect gene clusters comparable, but the order of those elements on the chromosome is conserved. Thus, the position of the anterior margin of expression in the body axis (*Ultrabithorax* affects parasegment 5; *abdominal-A* affects parasegment 7; *Abdominal-B* affects parasegment 10) and their order on the chromosome correlate. The order of the homologous homeobox genes in vertebrates also is conserved. This remarkable conservation suggests that there may have been an ancestral sequence common to flies and humans that survives in both the hindbrain of humans and the parasegments of insects (Lawrence 1992).

There are probably more than 100 different homeodomain proteins in *D. melanogaster* and these are classified into families according to their amino acid sequences. Homeotic mutants may have segments which are transformed dramatically. For example, antennal segments may be transformed into leglike segments, and metathoracic segments with halteres may be transformed into mesothoracic segments with a set of wings (Figure 5.5).

E. B. Lewis (1978) proposed a combinatorial model that assumes each insect segment is specified by a unique combination of homeotic genes that are expressed in that particular segment. Thus, the fewest number of homeotic genes would be required in thoracic segment two, which would be the prototypical segment, and progressively more genes would be active in the more posterior segments. While this model has been modified, it provided a useful conceptual framework for investigating *Drosophila* development.

Homeotic genes are classified in two complexes: the Antennapedia complex (including *Deformed*, *sex combs reduced*, *fushi tarazu*, and *Antennapedia*) and the bithorax complex (including *Ultrabithorax*, *abdominal-A*, and *Abdominal-B*). The genes of the Antennapedia complex control the anterior thorax and head, while the genes of the bithorax complex control the posterior thoracic and abdominal segments. Homeotic genes appear to be activated by a gradient of positional information laid down in the egg by the maternal effect genes and by interactions among the homeotic genes themselves. A number of the homeotic genes have been cloned and demonstrated to be expressed in a spatially restricted manner, in part supporting the Lewis hypothesis. However, all the genes are

**Figure 5.5**   A four-winged *Drosophila melanogaster* fly showing a complete transformation of the third thoracic segment into a second thoracic segment. The fly carries one chromosome with a deletion of the homeotic bithorax complex, while the other chromosome carries mutations of the bithorax complex (*Abx*, *bx3*) and *postbithorax* (*pbx*) locus. (Photograph provided by Edward Lewis, California Institute of Technology.)

expressed in more than one segment. Thus, the regulatory process is more complex than that proposed by Lewis.

Homeotic genes have some unusual characteristics. First, several homeotic genes appear to be very large relative to most other genes in *Drosophila*. For example, the *Antennapedia* primary gene transcript is approximately 100 kb long and the *Ultrabithorax* transcript is about 75 kilobases long. However, after the introns are spliced out, the remaining sequences are only a few kilobases long. Many of the exons in homeotic genes appear to encode protein domains with distinct structural or enzymatic functions. As a result, alternative splicing patterns in large genes such as is found in the Antennapedia and bithorax complexes may allow organisms to adapt one basic protein structure to different but related developmental uses. By adding or subtracting functional protein domains encoded by optional exons, the structural and enzymatic properties of the homeotic gene product can be modified and the ability of the protein to interact with other cellular components can be altered as development proceeds.

One example of a fly influenced by homeotic mutations in the Ultrabithorax complementation group is shown in Figure 5.5. Three separate mutated genes were combined in one fly to produce a *D. melanogaster* with four wings! Normally, of course, a pair of wings is found on the second thoracic segment and a small pair of balancing organs, called halteres, is found on the third thoracic segment. However, this fly has two essentially normal second thoraces and no third thoracic segment because the combined effect of the three mutations is to transform the third thoracic segment into the second without affecting any other parts of the fly.

## Interactions during Development

Normal development requires the coordinated expression of thousands of structural genes in a controlled manner. Since independent control of individual structural genes would result in chaotic development, controlling genes regulate the activity of groups of structural genes in a coordinate manner. Such genes are presumably arranged hierarchically or form a controlling network that ensures the proper timing of development so that the proper pattern develops. Perhaps the most important conclusion reached from analyses of the molecular genetics of development in *Drosophila* is that the problem of understanding development appears to be resolvable (Wilkins 1993). While development in *Drosophila* is not fully understood, many of the genes and their products are known. Progress toward understanding the full process is rapid. Furthermore, it is likely that an understanding of development in *D. melanogaster* will elucidate many of the principles by which other higher eukaryotes develop.

Exciting advances are being made in understanding other aspects of the development of *D. melanogaster*. For example, the molecular genetics of the development of the compound eye is providing an understanding of small-scale patterns of development (for reviews see Lawrence 1992, Rubin 1991, Tearle et al. 1993). Studies of how individual cells in imaginal discs know their physical position in the embryo and behave accordingly suggest that a polar coordinate system of positional information may operate during metamorphosis of imaginal discs (Bryant 1993, Cohen and Di Nardo 1993). The genetic basis by which ecdysone exerts its effects during development and metamorphosis is being resolved (Thummel 1990, Andres and Thummel 1992). Analysis of the development of *Drosophila* sensory organs, appendages, viscera, and neurogenesis also is advancing (Campuzano and Modolell 1992, Woods and Bryant 1992, Bienz 1993, Williams and Carrol 1993, Couso and Gonzalez-Gaitan 1993). Details of the molecular genetics of sex determination are discussed in Chapter 11.

## Similarities and Differences in Development in Other Insects

While developmental studies are most advanced in *D. melanogaster*, it is important to carry out comparable research with other insects to determine whether our knowledge of development in *D. melanogaster* can be extrapolated to other insects. The bodies of insects are composed of reiterated segments, which are modified to varying degrees in different insect orders. *Drosophila*

represents a relatively specialized type of development, the 'long germ' development pattern, in which segmentation occurs essentially simultaneously along the
anterior–posterior axis with the process of segmental specification under the
control of homeotic genes in the Antennapedia and bithorax complexes.

The locust (Orthoptera) *Schistocerca* is a 'short-germ' insect, in which most
segments are formed sequentially after gastrulation, in an anterior-to-posterior
sequence. In *Drosophila*, the segmental pattern is generated before cell membranes have formed, with patterning depending on the diffusion of gap and pair-
rule gene products. In the desert locust *Schistocerca gregaria*, segment patterning
appears after cellularization, and some of the genes that act early to divide the
*Drosophila* embryo into segments appear to play no corresponding role in *Schistocerca* (Kelsh et al. 1993). For example, the homolog of the *Drosophila* pair-rule
gene *even-skipped* was cloned from *S. americana* (Patel et al. 1992). The *even-
skipped* gene serves as a pair-rule gene in early development of *D. melanogaster*,
but does not play such a role in the early development of *S. americana*. The gene
does have a similar function during neurogenesis in both insects later in development, however, and Patel et al. (1992) have suggested that *even-skipped* had a
role in neurogenesis in the common ancestor to vertebrates and arthropods.
During the course of insect evolution, they suggest that the gene acquired an
additional function (pair-rule patterning) in insects such as *D. melanogaster* that
use a long-band mode of development.

Homeotic genes have been found in a number of other insects, either by
mutations or by sequence homology. Developmental research has been conducted on endopterygote insects, including the silk moth *Bombyx mori* (Lepidoptera), the red flour beetle *Tribolium castaneum* (Coleoptera), the honey bee *Apis
mellifera* (Hymenoptera), the lepidopteran *Manduca sexta*, the mosquito *Aedes
aegypti* (Diptera), and the house fly *Musca domestica* (Diptera). Development in
the exopterygote grasshopper *Schistocerca americana* (Orthoptera) also has been
studied.

The segmentation gene expression in *Musca domestica* is very similar to that
of *Drosophila* despite their divergence over 100 million years ago (Sommer and
Tautz 1991). Analysis of a number of genes, including *bicoid, hunchback, Krüppel,
knirps, tailless, hairy, engrailed,* and *Ultrabithorax*, indicated that the major steps in
segmentation during early stages of pattern formation in dipteran flies have been
conserved. *Ubx* also has been isolated from *B. mori* (Ueno et al. 1992).

Homologues of the *Drosophila abdominal-A* (*abd-A*) gene were found in *Aedes
aegypti, Schistocerca gregaria, Manduca sexta, Bombyx mori, Tribolium castaneum,*
and *Apis mellifera* (Eggleston 1992, Tear et al. 1990, Nagy et al. 1991, Ueno et al.
1992, Stuart et al. 1993, Walldorf et al. 1989). All the predicted proteins are
identical throughout the homeodomain in these insect orders, except for the
sequence from *Apis*, which differed in one amino acid from the other sequences.
The *Abdominal-B* gene (*Abd-B*) likewise was cloned from *Schistocerca* and *B.*

*mori*, and found to be homologous to that of *Drosophila* (Kelsh et al. 1993, Ueno et al. 1992). *Abd-B*, along with *abd-A*, controls segment specification in the most posterior region of the abdomen, including the genital segments.

Honey bee embryos were stained with a polyclonal antibody raised against the *Drosophila engrailed* protein. Expression of the *engrailed* pattern followed rules shared by all arthropods studied (Fleig 1990). A homolog of the *Drosophila* gene *Deformed* was also found in the honey bee (Fleig et al. 1988). In both the honey bee and *Drosophila*, the *Deformed* gene is involved in establishing head structures. Seven homeobox-containing genes (*Deformed, Sex combs reduced, Antennapedia, abdominal-A*, two *engrailed* genes, and the muscle segment homeobox (*msh*) were isolated and sequenced from honey bees by Walldorf et al. (1989). Six of the seven genes exhibited more than 90% sequence similarity to *Drosophila* homologues in the homeobox domain. Walldorf et al. (1989) concluded that there is a relatively high degree of conservation of genes involved in embryonic development of holometabolus insects.

The E locus in the silkworm *Bombyx mori* contains a homeotic gene complex that specifies larval abdominal segments. These genes are homologues of the *Drosophila* bithorax complex, which consists of *Ultrabithorax, abdominal-A*, and *Abdominal-B* (Ueno et al. 1992). Likewise, a homologue of the binding protein that binds the *fushi tarazu* gene of *Drosophila* has been found in B. *mori* (Ueda and Hirose 1991). Stuart et al. (1991) found no homologue of the segmentation gene *fushi tarazu* in *Tribolium*. Sommer and Tautz (1993) found that genes similar to *hairy* and *Krüppel* are present in *T. castaneum* and suggested that the segment patterning mechanism that acts in the long-banded *Drosophila* works in a similar way in the short-banded *Tribolium* embryo.

Beeman (1987) showed that six loci of homeotic genes in a single cluster (HOM-C) of the flour beetle, *Tribolium castaneum*, contain elements homologous to the homeotic genes in the Antennapedia and bithorax complexes of *Drosophila*. These genes map along the chromosome in the same order in *Tribolium* from anterior to posterior as their effects occur, but they occur in a single cluster rather than in two as in *Drosophila* (Beeman et al. 1989, 1993a). In *Drosophila*, these genes occur in the order: *labial, proboscipedia, Deformed, Sex combs reduced* and *Antennapedia*, while the order in *Tribolium* is *maxillopedia, Cephalothorax, prothoraxless, Ultrathorax, Abdominal*, and *extra urogomphi* (Beeman et al. 1993b).

Despite unique components and differences between *D. melanogaster* and other arthropods, the results described here suggest that *D. melanogaster* generally serves as a useful model for understanding development in other arthropods. For example, Patel et al. (1994) examined the functioning of the *even-skipped* gene in three beetle genera, *Callosobruchus, Dermestes*, and *Tribolium*, which have been classed as long-germ, short-germ, or intermediate. The expression of *even-skipped* presents itself rather differently in these beetles compared to *Dro-*

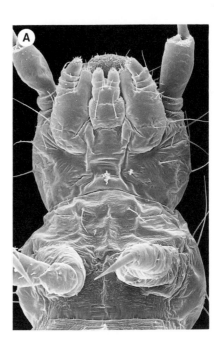

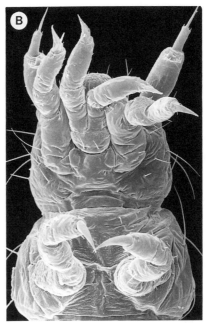

**Figure 5.6** (A) A ventral view of a wild-type embryo of *Tribolium castaneum*. (B) A ventral view showing a homeotic mutation, *maxillopedia*, in which the maxillary and labial palps are modified into leg-like appendages. (C) A side view of a mature wild-type

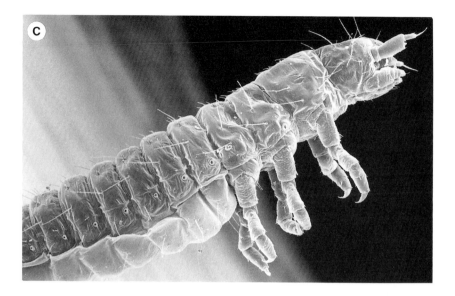

embryo. (D) A side view showing the homeotic mutant *cephalothorax*, which results in incorporation of the prothorax into the head and transformation of the labial palps into antennae. *maxillopedia* and *cephalothorax* are in the HOM-C complex and correspond to *proboscipedia* and *sex combs reduced* in *Drosophila*. (Photographs provided by Richard W. Beeman.)

*sophila*. The segmentation mechanisms seem to be different because the relative timing of the processes is shifted. However, more radical innovations also could have occurred with the transition from 'lower' to 'higher' insects, because the *even-skipped* gene is expressed differently in the grasshopper *Schistocerca* (Patel et al. 1994). Thus, comparative studies on the molecular evolution of genes involved in development in insects and other organisms may provide information about the evolution of gene families that regulate development, as well as contribute to our understanding of the basic mechanisms underlying the genetic control of development.

*Tribolium* with the *maxillopedia* mutation have transformations of labial and maxillary palps into leg-like structures (Figure 5.6). *Tribolium* with the mutations *Cephalothorax* and *prothoraxless* also exhibit significant homeotic transformations, with beetles carrying the *Cephalothorax* mutation having the head and first thoracic segment fused and the labial palps transformed into antennae (Figure 5.6). The *prothoraxless* mutants exhibit fusion of head segments with the entire thorax, and transformation of all three pairs of thoracic legs into antennae.

The developmental control genes that contain an Antennapedia-type homeobox are clustered in insects. In *Tribolium*, the homeotic genes are arranged in one complex while in *Drosophila* they are split into two (the bithorax complex and the Antennapedia complex) (Beeman 1987, Stuart et al. 1991, Beeman et al. 1993b, Schubert et al. 1993). In mammals, there are four clusters of homeotic genes that are homologous to these genes. Furthermore, the homeotic gene order in both insects and vertebrates is similar. This suggests there is a common ancestral gene cluster for both insects and vertebrates (Beeman et al. 1993b).

Schubert et al. (1993) suggest that the Antennapedia-type of homeobox genes were subdivided into three classes early in metazoan evolution and propose a model for their evolution based on homeobox sequence analysis. The model suggests there was a single ancestral gene (*Abdominal-B*), which produced the three precursors of the three classes of Antennapedia-type genes by gene duplication (*abdominal-A, Ultrabithorax, Antennapedia*). Before the divergence of vertebrates and arthropods in the Precambrian, another duplication occurred, so that the common ancestor of higher insects and vertebrates contained a cluster of at least five genes (*Abdominal-B, Ultrabithorax, Deformed, proboscipedia,* and *labial*). During the evolution of insects, further duplications generated the eleven Antennapedia-type homeobox genes found in *Drosophila* (*Abdominal B, Ultrabithorax, abdominal-A, Antennapedia, sex combs reduced, fushi tarazu, Deformed, zen1, zen2, probosipedia,* and *labial*). Likewise, gene duplications led to 13 Antennapedia-type genes in vertebrates. In *Drosophila*, multiple promoters and alternative splicing processes may have been developed to increase the complexity of this anterior–posterior differentiation system.

# References Cited

Agarwal, M., N. Bensaadi, J. C. Salvado, K. Campbell and C. Mouches. 1993. Characterization and genetic organization of full-length copies of a LINE retroposon family dispersed in the genome of *Culex pipiens* mosquitoes. Insect Biochem. Mol. Biol. 23: 621–629.

Alvarez-Fuster, A., C. Juan and E. Petitpierre. 1991. Genome size in *Tribolium* flour-beetles: Inter- and intraspecific variation. Genet. Res., Camb. 58: 1–5.

Andres, A. J. and C. S. Thummel. 1992. Hormones, puffs and flies: The molecular control of metamorphosis by ecdysone. Trends Genet. 8: 132–138.

Antoine, M. and J. Niessing. 1984. Intron-less globin genes in the insect *Chironomus thummi thummi*. Nature 310: 795–798.

Ashburner, M. 1989. Drosophila: *A Laboratory Manual*. Cold Spring Harbor Laboratory Press, Cold Spring Harbor, NY.

Beeman, R. W. 1987. A homeotic gene cluster in the red flour beetle. Nature 327: 247–249.

Beeman, R. W., J. J. Stuart, M. S. Haas, and R. E. Denell. 1989. Genetic analysis of the homeotic gene complex (Hom-C) in the beetle *Tribolium castaneum*. Dev. Biol. 133: 196–209.

Beeman, R. W., J. J. Stuart, S. J. Brown and R. E. Denell. 1993a. Structure and function of the homeotic gene complex (HOM-C) in the beetle, *Tribolium castaneum*. BioEssays 15: 439–444.

Beeman, R. W., S. J. Brown, J. J. Stuart and R. Denell. 1993b. The evolution of genes regulating developmental commitments in insects and other animals. Pp. 71–84. In: *Evolutionary Conservation of Developmental Mechanisms*. Wiley-Liss, New York.

Bello, B. and P. Couble. 1990. Specific expression of a silk-encoding gene of *Bombyx* in the anterior salivary gland of *Drosophila*. Nature 346: 480–482.

Berg, D. E. and M. M. Howe, Eds. 1989. *Mobile DNA*. Am. Soc. Microbiol., Washington, DC.

Berry, S. J. 1985. Insect nucleic acids. Pp. 219–253. In: *Comprehensive Insect Physiology Biochemistry and Pharmacology*. Vol. 10. G. A. Kerkut and L. I. Gilbert, Eds. Pergamon Press, Oxford.

Besansky, N. J. 1990a. A retrotransposable element from the mosquito *Anopheles gambiae*. Mol. Cell. Biol. 10: 863–871.

Besansky, N. J. 1990b. Evolution of the T1 retroposon family in the *Anopheles gambiae* complex. Mol. Biol. Evol. 7: 229–246.

Besansky, N. J. and J. R. Powell. 1992. Reassociation kinetics of *Anopheles gambiae* (Diptera: Culicidae) DNA. J. Med. Entomol. 29: 125–128.

Beinz, M. 1993. Homeotic genes and positional signalling in the *Drosophila* viscera. Trends Genet. 10: 22–26.

Biessmann, H., K. Valgeirsdottir, A. Lofsky, C. Chin, B. Ginther, R. W. Levis, and M. L. Pardue. 1992. HeT-A, a transposable element specifically involved in 'healing' broken chromosome ends in *Drosophila*. Mol. Cell Biol. 12: 3910–3918.

Biessmann, H., B. Kasravi, K. Jakes, T. Bui, K. Ikenaga and J. M. Mason. 1993. The genomic organization of HeT-A retroposons in *Drosophila melanogaster*. Chromosoma 102: 297–305.

Bigot, Y., F. Lutcher, M. H. Hamelin and G. Periquet. 1992. The 28S ribosomal RNA-encoding gene of Hymenoptera: Inserted sequences in the retrotransposon-rich regions. Gene 121: 347–352.

Block, K., G. Ising and F. Stahl. 1990. Minichromosomes in *Drosophila melanogaster* derived from the transposing element TE1. Chromosoma 99: 336–343.

Bownes, M. 1986. Expression of the genes coding for vitellogenin (yolk protein). Annu. Rev. Entomol. 31: 507–531.

Bradfield, J. Y., Y. H. Lee and L. L. Keeley. 1991. Cytochrome P450 family 4 in a cockroach: Molecular cloning and regulation by hypertrehalosemic hormone. Proc. Natl. Acad. Sci. USA 88: 4558–4562.

Breeuwer, J. A. J. and J. H. Werren. 1990. Microorganisms associated with chromosome destruction and reproductive isolation between two insect species. Nature 346: 558–560.

Brosius, J. 1991. Retroposons—Seeds of evolution. Science. 251: 753.

Brown, S. W. and U. Nur. 1964. Heterochromatic chromosomes in coccids. Science 145: 130–136.

Brown, S. J., J. K. Henry, W. C. Black IV and R. E. Denell. 1990. Molecular genetic manipulation of the red flour beetle: Genome organization and cloning of a ribosomal protein gene. Insect Biochem. 20: 185–193.

Bryant, P. J. 1993. The polar coordinate model goes molecular. Science 259: 471–472.

Burke, W. D., D. G. Eickbush, Y. Xiong, J. Jakubczak and T. H. Eickbush. 1993. Sequence relationship of retrotransposable elements *R1* and *R2* within and between divergent insect species. Mol. Biol. Evol. 10: 163–185.

Calvi, B. R., T. J. Hong, S. D. Findley and W. M. Gelbart. 1991. Evidence for a common evolutionary origin of inverted repeat transposons in *Drosophila* and plants: *hobo, activator,* and *Tam3.* Cell 66: 465–471.

Campuzano, S. and J. Modolell. 1992. Patterning of the *Drosophila* nervous system: The *achaete–scute* gene complex. Trends Genet. 8: 202–208.

Capy, P., D. Anxolabehere and T. Langin. 1994. The strange phylogenies of transposable elements: Are horizontal transfers the only explanation? Trends Genet. 10: 7–12.

Carminati, J. L., C. G. Johnston and T. L. Orr-Weaver. 1992. The *Drosophila* ACE3 chorion element autonomously induces amplification. Mol. Cell. Biol. 12: 2444–2453.

Chapco, W. 1983. Genetics and population genetics of grasshoppers and locusts—A bibliography. Bibliog. Entomol. Soc. Am. 2: 11–16.

Cho, W. L. and A. S. Raikhel. 1992. Cloning of cDNA for mosquito lysosomal aspartic protease. J. Biol. Chem. 267: 21823–21829.

Cohen, S. M. and S. Di Nardo. 1993. wingless: From embryo to adult. Trends Genet. 9: 189–192.

Cohen, M. B., M. A. Schuler and M. R. Berenbaum. 1992. A host-inducible cytochrome P-450 from a host-specific caterpillar: Molecular cloning and evolution. Proc. Natl. Acad. Sci. USA 89: 10920–10924.

Couso, J. P. and M. Gonzalez-Gaitan. 1993. Embryonic limb development in *Drosophila.* Trends Genet. 9: 371–373.

Crozier, R. H. and Y. C. Crozier. 1993. The mitochondrial genome of the honeybee *Apis mellifera*: Complete sequence and genome organization. Genetics 133: 97–117.

Daniels, S. B., K. R. Peterson, L. D. Strausbaugh, M. G. Kidwell and A. Chovnick. 1990. Evidence for horizontal transmission of the P transposable element between *Drosophila* species. Genetics 124: 339–355.

DeGroote, F., G. Pont, D. Micard and G. Picard. 1989. Extrachromosomal circular DNAs in *Drosophila melanogaster*: Comparison between embryos and KcO% cells. Chromosoma 98: 201–206.

Degrugillier, M. E. and S. M. Newman, Jr. 1993. Hereditary viruses of *Heliothis*? Chromatin-associated virus-like particles in testes of six species of *Heliothis* and *Helicoverpa*, F$_1$, and backcross males. J. Invertebr. Pathol. 61: 147–155.

Delpuech, J. M., C. F. Aquadro and R. T. Roush. 1993. Noninvolvement of the long terminal repeat of transposable element *17.6* in insecticide resistance in *Drosophila.* Proc. Natl. Acad. Sci. USA 90: 5643–5647.

Devine, J. H., G. D. Kutuzova, V. A. Green, N. N. Ugarova and T. O. Baldwin. 1993. Luciferase from the East European firefly *Luciola mingrelica*: Cloning and nucleotide sequence of the cDNA, overexpression in *Escherichia coli* and purification of the enzyme. Biochim. Biophy. Acta 1173: 121–132.

Devonshire, A. L. and L. M. Field. 1991. Gene amplification and insecticide resistance. Annu. Rev. Entomol. 36: 1–23.

Devonshire, A. L., L. M. Field and M. S. Williamson. 1992. Molecular biology of insecticide resistance. Pp. 173–183. In: *Insect Molecular Science*. J. M. Crampton and P. Eggleston, Eds. Academic Press, London.

Dickenson, L., V. Russell and P. E. Dunn. 1988. A family of bacteria-regulated, cecropin D-like peptides from *Manduca sexta*. J. Biol. Chem. 263: 19424–19429.

Douglas, A. E. 1992. Symbiotic microorganisms in insects. Encyc. Microbiol. 4: 165–178.

Ebberink, H. M., A. B. Smit and J. van Minnen. 1989. The insulin family: Evolution of structure and function in vertebrates and invertebrates. Biol. Bull. 177: 176–182.

Eggleston, P. 1992. Identification of the *abdominal-A* homologue from *Aedes aegypti* and structural comparisons among related genes. Nucleic Acids Res. 20: 4095.

Eickbush, T. H. and W. D. Burke. 1985. Silkmoth chorion gene families contain patchwork patterns of sequence homology. Proc. Natl. Acad. Sci. USA 82: 2814–2818.

Eickbush, D. G., T. H. Eickbush and J. H. Werren. 1992. Molecular characterization of repetitive DNA sequences from a B chromosome. Chromosoma 101: 575–583.

Elvin, C. M., V. Whan and P. W. Riddles. 1990. A family of serine protease genes expressed in adult buffalo fly (*Haematobia irritans exigua*). Mol. Gen. Genet. 240: 132–139.

Fallon, A. M. 1984. Methotrexate resistance in cultured mosquito cells. Insect Biochem. 14: 697–704.

ffrench-Constant, R. H., R. T. Roush and F. A. Carino. 1993. *Drosophila* as a tool for investigating the molecular genetics of insecticide resistance. Pp. 1–37. In: *Molecular Approaches to Fundamental and Applied Entomology*. J. Oakeshott and M. J. Whitten, Eds. Springer-Verlag, New York.

Field, L. M., A. L. Devonshire and B. G. Forde. 1988. Molecular evidence that insecticide resistance in peach-potato aphids (*Myzus persicae* Sulz.) results from amplification of an esterase gene. Biochem. J. 251: 309–312.

Finnegan, D. J. 1990. Transposable elements. Pp. 371–382. In: *The Genome of Drosophila melanogaster. Part 4. Genes L-Z, Balancers, Transposable Elements*. Drosophila Inform. Service. Vol. 68.

Fitch, D. H. A., L. D. Strausbaugh and V. Barrett. 1990. On the origins of tandemly repeated genes: Does histone gene copy number in *Drosophila* reflect chromosomal location? Chromosoma 99: 118–124.

Fleig, R. 1990. *engrailed* expression and body segmentation in the honeybee *Apis mellifera*. Roux's Arch. Dev. Biol. 198: 467–473.

Fleig, R., U. Walldorf, W. J. Gehring and K. Sander 1988. In situ localization of the transcripts of a homeobox gene in the honeybee *Apis mellifera* L. (Hymenoptera). Roux's Arch. Dev. Biol. 197: 269–274.

Fleming, J. G. W. 1992. Polydnaviruses: Mutualists and pathogens. Annu. Rev. Entomol. 37: 401–425.

Fleming, J. G. W. and M. D. Summers. 1991. Polydnavirus DNA is integrated in the DNA of its parasitoid wasp host. Proc. Natl. Acad. Sci. USA 88: 9770–9774.

Fraser, M. J. 1985. Transposon-mediated mutagenesis of baculoviruses: Transposon shuttling and implications for speciation. Ann. Entomol. Soc. Am. 79: 773–783.

Friesen, P. D. and M. S. Nissen. 1990. Gene organization and transcription of TED, a lepidopteran retrotransposon integrated within the baculovirus genome. Mol. Cell. Biol. 10: 3067–3077.

Gaubatz, J. W. 1990. Extrachromosomal circular DNAs and genomic sequence plasticity in eukaryotic cells. Mutat. Res. 237: 271–292.

Gehring, W. J. 1985. The homeo box: A key to the understanding of development? Cell 40: 3–5.

Gehring, W. J. 1987. Homeo boxes in the study of development. Science 236: 1245–1252.

Gray, M. W. 1989. Origin and evolution of mitochondrial DNA. Annu. Rev. Cell Biol. 5: 25–50.

Haig, D. 1993. The evolution of unusual chromosomal systems in sciarid flies: Intragenomic conflict and the sex ratio. J. Evol. Biol. 6: 249–261.

Hankeln, T. and E. R. Schmidt. 1990. New foldback transposable element TFB1 found in histone genes of the midge *Chironomus thummi*. J. Mol. Biol. 215: 477–482.

Hankeln, T. and E. R. Schmidt. 1991. The organization, localization and nucleotide sequence of the histone genes of the midge *Chironomus thummi*. Chromosoma 101: 25–31.

Hartl, D. L. and S. W. Brown. 1970. The origin of male haploid genetic systems and their expected sex ratio. Theoret. Popul. Biol. 1: 165–190.

Hartl, D. L., J. W. Ajioka, H. Cai, A. R. Lohe, E. R. Lozovskaya, D. A. Smoller and I. W. Duncan. 1992. Towards a *Drosophila* genome map. Trends Genet. 8: 70–75.

Haymer, D. S., J. E. Anleitner, M. He, S. Thanaphum, S. H. Saul, J. Ivy, K. Houtchens and L. Arcangeli. 1990. Actin genes in the Mediterranean fruit fly, *Ceratitis capitata*. Genetics: 125: 155–160.

Heck, M. M. S. and A. C. Spradling. 1990. Multiple replication origins are used during *Drosophila* chorion gene amplification. J. Cell Biol. 110: 903–914.

Heckel, D. G. 1993. Comparative genetic linkage mapping in insects. Annu. Rev. Entomol. 38: 381–408.

Horodyski, F. M., L. M. Riddiford and J. W. Truman. 1989. Isolation and expression of the eclosion hormone gene from the tobacco hornworm, *Manduca sexta*. Proc. Natl. Acad. Sci. USA 86: 8123–8127.

Houck, M. A., J. B. Clark, K. R. Peterson and M. G. Kidwell. 1991. Possible horizontal transfer of *Drosophila* genes by the mite *Proctolaelaps regalis*. Science 253: 1125–1129.

Huijser, P., C. J. Kirchhoff, D. H. Lankenau, and W. Hennig. 1988. Retrotransposon-like sequences are expressed in the Y chromosomal lampbrush loops of *Drosophila hydei*. J. Mol. Biol. 203: 689–697.

Hultmark, D. 1993. Immune reactions in *Drosophila* and other insects: A model for innate immunity. Trends Genet. 9: 178–183.

Ikeda, M., T. Yaginuma, M. Kobayashi and O. Yamashita. 1991. cDNA cloning, sequencing and temporal expression of the protease responsible for vitellin degradation in the silkworm, *Bombyx mori*. Comp. Biochem. Physiol. B99: 405–411.

Jakubczak, J. L., W. D. Burke, and T. H. Eickbush. 1991. Retrotransposable elements R1 and R2 interrupt the rRNA genes of most insects. Proc. Natl. Acad. Sci. USA 88: 3295–3299.

Jiang, H., Y. Wang and M. R. Kanost. 1994. Mutually exclusive exon use and reactive center diversity in insect serpins. J. Biol. Chem. 269: 55–58.

Kafatos, F. C. 1981. Structure, evolution and developmental expression of the silkmoth chorion multigene families. Am. Zool. 21: 707–714.

Kafatos, F. C., C. Delidakis, W. Orr, G. Thireos, K. Komitopoulou and Y. Wong. 1986. Studies on the developmentally regulated amplification and expression of *Drosophila* chorion genes. Pp. 85–101. In: *Molecular Developmental Biology*. A. R. Liss, New York.

Kafatos, F. C., C. Louis, C. Savakis, D. M. Glover, M. Ashburner, A. J. Link, I. Siden-Kiamos and R. D. C. Saunders. 1991. Integrated maps of the *Drosophila* genome: Progress and prospects. Trends Genet. 7: 155–161.

Kamito, T. H. Tanaka, B. Sato, H. Nagasawa and A. Suzuki. 1992. Nucleotide sequence of cDNA for the eclosion hormone of the silkworm, *Bombyx mori* and the expression in a brain. Biochem. Biophys. Res. Commun. 182: 514–519.

Kanai, A. and S. Natori. 1989. Cloning of gene cluster for Sarcotoxin I, antibacterial proteins of *Sarcophaga peregrina*. FEBS Lett. 258: 199–202.

Kappen, C., K. Schughart and F. H. Ruddle. 1993. Early evolutionary origin of major homeodomain sequence classes. Genomics 18: 54–70.

Kato, Y., K. Taniai, H. Hirochika and M. Yamakawa. 1993. Expression and characterization of cDNAs for cecropin B, an antibacterial protein of the silkworm, *Bombyx mori*. Insect Biochem. Mol. Biol. 23: 185–190.

Kaulenas, M. S. 1985. Molecular biology: Protein synthesis. Pp. 255–305. In: *Comprehensive Insect*

*Physiology Biochemistry and Pharmacology.* Vol. 10. G. A. Kerkut and L. I. Gilbert, Eds. Pergamon, Oxford.

Kelsh, R., I. Dawson and M. Akam. 1993. An analysis of Abdominal-B expression in the locust *Schistocerca gregaria.* Development 117: 293–305.

Kerrebrock, A. W., R. Srivastava and S. A. Gerbi. 1989. Isolation and characterization of ribosomal DNA variants from *Sciara coprophila.* J. Mol. Biol. 210: 1–13.

Kidwell, M. G. 1992. Horizontal transfer. Curr. Opin. Genet. Dev. 2: 868–873.

Kidwell, M. G. 1993. Lateral transfer in natural populations of eukaryotes. 1993. Annu. Rev. Genet. 27: 235–256.

Kimbrell, D. A. 1991. Insect antibacterial proteins: Not just for insects and against bacteria. Bio-Essays 13: 657–663.

Kimura, K., T. Okumura, O. Ninaki, M. G. Kidwell and K. Suzuki. 1993. Transposable elements in commercially useful insects. I. Southern hybridization study of silkworms and honeybees using *Drosophila* probes. Jpn. J. Genet. 68: 63–71.

King, R. E., Ed. 1975. *Handbook of Genetics.* Vol. 3. *Invertebrates of Genetic Interest.* Plenum, New York.

Kobayashi, S., R. Amikura and M. Okada. 1993. Presence of mitochondrial large ribosomal RNA outside mitochondria in germ plasm of *Drosophila melanogaster.* Science 260: 1521–1524.

Konsolaki, M., K. Komitopoulou, P. P. Tolias, D. L. King, C. Swimmer and F. C. Kafatos. 1990. The chorion genes of the medfly, *Ceratitis capitata.* I. Structural and regulatory conservation of the s36 gene relative to two *Drosophila* species. Nucleic Acids Res. 18: 1731–1737.

Kramer, K. J., L. Corpuz, H. K. Choi and S. Muthukrishnan. 1993. Sequence of cDNA and expression of the gene encoding epidermal and gut chitinases of *Manduca sexta.* Insect Biochem. Mol. Biol. 23: 691–701.

Krell, P. J. 1991. The polydnaviruses: Multipartite DNA viruses from parasitic Hymenoptera. Pp. 141–177. In: *Viruses of Invertebrates.* E. Kurstak, Ed. Dekker, New York.

Lagueux, M., L. Lwoff, M. Meister, F. Goltzene and J. A. Hoffmann. 1990. cDNAs from neurosecretory cells of brains of *Locusta migratoria* (Insecta, Orthoptera) encoding a novel member of the superfamily of insulins. Eur. J. Biochem. 187: 249–254.

Larsen-Rapport, E. W. 1986. Imaginal disc determination: Molecular and cellular correlates. Annu. Rev. Entomol. 31: 145–175.

Lawrence, P. A. 1992. *The Making of a Fly: The Genetics of Animal Design.* Blackwell Scientific, Oxford.

Lewis, E. 1978. A gene complex controlling segmentation in *Drosophila.* Nature 176: 141–152.

Lin, Y., M. T. Hamblin, M. J. Edwards, C. Barillas-Mury, M. R. Kanost, D. C. Knipple, M. F. Wolfner and H. H. Hagedorn. 1993. Structure, expression, and hormonal control of genes from the mosquito, *Aedes aegypti,* which encode proteins similar to the vitelline membrane proteins of *Drosophila melanogaster.* Dev. Biol. 155: 558–568.

Lindsley, D. L. and G. G. Zimm. 1992. *The Genome of* Drosophila melanogaster. Academic Press, San Diego.

Luan, D. D., M. H. Korman, J. L. Jakubczak and T. H. Eickbush. 1993. Reverse transcription of R2Bm RNA is primed by a nick at the chromosomal target site: A mechanism for non-LTR retrotransposition. Cell 72: 595–605.

Maroni, G. 1993. *An Atlas of* Drosophila Genes. IRL Press, Oxford.

Matsuyama, K. and S. Natori. 1988. Molecular cloning of cDNA for sapecin and unique expression of the sapecin gene during the development of *Sarcophaga peregrina.* J. Biol. Chem. 262: 17117–17121.

Merriam, J., M. Ashburner, D. L. Hartl and F. C. Kafatos. 1991. Toward cloning and mapping the genome of *Drosophila.* Science 254: 221–225.

Michaille, J. J., P. Couble, J. C. Prudhomme and A. Garel. 1986. A single gene produces multiple sericin messenger RNAs in the silk gland of Bombyx mori. Biochimie 68: 1165–1173.

Michaille, J. J., A. Garel and J. C. Prudhomme. 1990. Cloning and characterization of the highly polymorphic Ser2 gene of Bombyx mori. Gene 86: 177–184.

Mitsialis, S. A. and F. C. Kafatos. 1985. Regulatory elements controlling chorion gene expression are conserved between flies and moths. Nature 317: 453–456.

Mitsialis, S. A., S. Veletza and F. C. Kafatos. 1989. Transgenic regulation of moth chorion gene promoters in Drosophila: Tissue, temporal, and quantitative control of four bidirectional promoters. J. Mol. Evol. 29: 486–495.

Mizrokhi, L. J. and Mazo, A. M. 1990. Evidence for horizontal transmission of the mobile element jockey between distant Drosophila species. Proc. Natl. Acad. Sci. USA 87: 9216–9220.

Morimoto, R. I., K. D. Sarge and K. Abravaya. 1992. Transcriptional regulation of heat shock genes: A paradigm for inducible genomic responses. J. Biol. Chem. 31: 21987–21990.

Mouches, C., Y. Pauplin, M. Agarwal, L. Lemieux, M. Herzog, M. Abadon, V. Beyssat-Arnaouty, O. Hyrien, B. R. deSaint Vincent, G. P. Georghiou and N. Pasteur. 1990. Characterization of amplification core and esterase B1 gene responsible for insecticide resistance in Culex. Proc. Natl. Acad. Sci. USA 87: 2574–2578.

Mouches, C., M. Agarwal, K. Campbell, L. Lemieux and M. Abadon. 1991. Sequence of a truncated LINE-like retroposon dispersed in the genome of Culex mosquitoes. Gene 106: 279–280.

Mouches, C., N. Bensaadi and J. C. Salvado. 1992. Characterization of a LINE retroposon dispersed in the genome of three non-sibling Aedes mosquito species. Gene 120: 183–190.

Mount, S. M., C. Burks, G. Hertz, G. D. Stormo, O. White and C. Fields. 1992. Splicing signals in Drosophila: Intron size, information content, and consensus sequences. Nucleic Acids Res. 20: 4255–5262.

Mousseau, T. A. and H. Dingle. 1991. Maternal effects in insect life histories. Annu. Rev. Entomol. 36: 511–534.

Nagy, L. M. and S. Carroll. 1994. Conservation of wingless patterning functions in the short-germ embryos of Tribolium castaneum. Nature 367: 460–463.

Nagy, L. M., R. Booker and L. M. Riddiford. 1991. Isolation and embryonic expression of an abdominal-A-like gene from the lepidopteran, Manduca sexta. Development 112: 119–131.

Nakato, H., S. Izumi and S. Tomino. 1992. Structure and expression of gene coding for a pupal cuticle protein of Bombyx mori. Biochim. Biophys. Acta 1132: 161–167.

Nusslein-Volhard, C. and E. Wieschaus. 1980. Mutations affecting segment number and polarity in Drosophila. Nature 187: 795–801.

Nur, U. 1990. Heterochromatization and euchromatization of whole genomes in scale insects (Coccoidea: Homoptera). Development 1990 Supplement: 29–34.

Ohno, S. and T. Yomo. 1991. The grammatical rule for all DNA: Junk and coding sequences. Electrophoresis 12: 103–108.

Oishi, K., M. Sawa, M. Hatakeyama and Y. Kageyama. 1993. Genetics and the biology of the sawfly, Athalia rosae (Hymenoptera). Genetica 88: 119–127.

Okazaki, S., K. Tsuchida, H. Maekawa, H. Ishikawa and H. Fujiwara. 1993. Identification of a pentanucleotide telomeric sequence, (TTAGG)n, in the silkworm Bombyx mori and in other insects. Mol. Cell. Biol. 13: 1424–1432.

O'Neill, S. L., R. H. Gooding and S. Aksoy. 1993. Phylogenetically distant symbiotic microorganisms reside in Glossina midgut and ovary tissues. Med. Vet. Entomol. 7: 377–383.

Orr-Weaver, T. L. 1991. Drosophila chorion genes: Cracking the eggshell's secrets. BioEssays 113: 97–105.

Palli, S. R., K. Hiruma and L. M. Riddiford. 1992. An ecdysteroid-inducible Manduca gene similar to the Drosophila DHR3 gene, a member of the steroid hormone receptor superfamily. Dev. Biol. 150: 306–318.

Pardue, M. L. 1991. Dynamic instability of chromosomes and genomes. Cell 66: 427–431.

Pardue, M. L. and W. Hennig. 1990. Heterochromatin: Junk or collector's item? Chromosoma 100: 3–7.

Patel, N. H., E. E. Ball and C. S. Goodman. 1992. Changing role of even-skipped during the evolution of insect pattern formation. Nature 357: 339–341.

Patel, N. H., B. G. Condron and K. Zinn. 1994. Pair-rule expression patterns of even-skipped are found in both short- and long-germ beetles. Nature 367: 429–434.

Pauli, D., A. P. Arrigo and A. Tissieres. 1992. Heat shock response in Drosophila. Experientia 48: 623–629.

Perkins, H. D. and A. J. Howells. 1992. Genomic sequences with homology to the P element of Drosophila melanogaster occur in the blowfly Lucilia cuprina. Proc. Natl. Acad. Sci. USA 89: 10753–10757.

Perrimon, N., D. Mohler, L. Engstrom and A. Mahowald. 1986. X-linked female-sterile loci in Drosophila melanogaster. Genetics 113: 695–712.

Peterson, K. and C. Sapienza. 1993. Imprinting the genome: Imprinted genes, imprinting genes, and a hypothesis for their interaction. Annu. Rev. Genet. 27: 7–31.

Plasterk, R. H. A. 1993. Molecular mechanisms of transposition and its control. Cell 74: 781–786.

Pons, J., E. Petitpierre and C. Juan. 1993. Characterization of the heterochromatin of the darkling beetle Misolampus goudoti: Cloning of two satellite DNA families and digestion of chromosomes with restriction enzymes. Hereditas 119: 179–185.

Prody, C. A., P. Dreyfus, R. Zamir, H. Zakut and H. Soreq. 1989. De novo amplification within a 'silent' human cholinesterase gene in a family subjected to prolonged exposure to organophosphorous insecticides. Proc. Natl. Acad. Sci. USA 86: 690–694.

Prudhomme, J. C., P. Couble, J. P. Garel and J. Daillie. 1985. Silk synthesis. Pp. 571–94. In: Comprehensive Insect Physiology Biochemistry and Pharmacology. Vol. 10. G. A. Kerkut and L. I. Gilbert, Eds. Pergamon, Oxford.

Raikhel, A. S. and T. S. Dhadialla. 1992. Accumulation of yolk proteins in insect oocytes. Annu. Rev. Entomol. 37: 217–251.

Raymond, M., E. Poulin, V. Boiroux, E. Dupont and N. Pasteur. 1993. Stability of insecticide resistance due to amplification of esterase genes in Culex pipiens. Heredity 70: 301–307.

Rebers, J., F. Horodyski, R. Hice and L. M. Riddiford. 1987. Genomic cloning of developmentally regulated larval cuticle genes from the tobacco hornworm, Manduca sexta. Pp. 201–210. In: Molecular Entomology. J. H. Law, Ed. A. R. Liss, New York.

Retnakaran, A. and J. Percy. 1985. Fertilization and special modes of reproduction. Pp. 231–293. In: Comprehensive Insect Physiology Biochemistry and Pharmacology. Vol. 1. G. A. Kerkut and L. I. Gilbert, Eds. Pergamon, Oxford.

Rina, M. and C. Savakis. 1991. A cluster of vitellogenin genes in the Mediterranean fruit fly Ceratitis capitata: Sequence and structural conservation in dipteran yolk proteins and their genes. Genetics 127: 769–780.

Rinderer, T. E., Ed. 1986. Bee Genetics and Breeding. Academic Press, Orlando, FL.

Robertson, H. M. 1993. The mariner transposable element is widespread in insects. Nature 362: 241–245.

Robertson, H. M., D. L. Lampe and E. G. MacLeod. 1992. A mariner transposable element from a lacewing. Nucleic Acids Res. 20: 6409.

Robinson, R. 1971. Lepidoptera Genetics. Pergamon, Oxford.

Rosetto, M., A. G. O. Manetti, D. Marchini, R. Dallai, J. L. Telford and C. T. Baldari. 1993. Sequences of two cDNA clones from the medfly Ceratitis capitata encoding antibacterial peptides of the cecropin family. Gene 134: 241–243.

Rovira, C., W. Beermann and J. E. Edstrom. 1993. A repetitive DNA sequence associated with the centromeres of Chironomus pallidivittatus. Nucleic Acids Res. 21: 1775–1781.

Rubin, G. M. 1991. Signal transduction and the fate of the R7 photoreceptor in *Drosophila*. Trends Genet. 7: 372–377.

Sapienza, C., A. C. Peterson, J. Rossant and R. Balling. 1987. Degree of methylation of transgenes is dependent on gamete of origin. Nature 328: 251–254.

Sato, Y. and O. Yamashita. 1991. Structure and expression of a gene coding for egg-specific protein in the silkworm, *Bombyx mori*. Insect Biochem. 21: 495–505.

Saura, A., J. Lokki and E. Suomalainen. 1993. Origin of polyploidy in parthenogenetic weevils. J. Theor. Biol. 163: 449–456.

Schmidt, O. and U. Theopold. 1991. Immune defense and suppression in insects. BioEssays 13: 343–346.

Schubert, F. R., K. Nieselt-Struwe and P. Gruss. 1993. The antennapedia-type homeobox genes have evolved from three precursors separated early in metazoan evolution. Proc. Natl. Acad. Sci. USA 90: 143–147.

Schwemmler, W. and G. Gassner, Eds. 1989. *Insect Endocytobiosis: Morphology, Physiology, Genetics, Evolution*. CRC Press, Boca Raton, FL.

Segraves, W. A. and C. Wolfin. 1993. The *E75* gene of *Manduca sexta* and comparison with its *Drosophila* homolog. Insect Biochem. Mol. Biol. 23: 91–97.

Severson, D. W., A. Mori, Y. Zhang and B. M. Christensen. 1993. Linkage map for *Aedes aegypti* using restriction fragment length polymorphisms. J. Hered. 84: 241–247.

Sokoloff, A. 1966. *The Genetics of* Tribolium *and Related Species*. Academic Press, New York.

Sokoloff, A. 1977. *The Biology of* Tribolium, *with Special Emphasis on Genetic Aspects*. Vol. 3. Oxford, Clarendon.

Solter, D. 1988. Differential imprinting and expression of maternal and paternal genomes. Annu. Rev. Genet. 22: 127–146.

Sommer, R. and D. Tautz. 1991. Segmentation gene expression in the housefly *Musca domestica*. Development 113: 419–430.

Sommer, R. and D. Tautz. 1993. Involvement of an orthologue of the *Drosophila* pair-rule gene *hairy* in segment formation of the short germ-band embryo of *Tribolium* (Coleoptera). Nature 361: 448–450.

Staiber, W. 1987. Unusual germ line limited chromosomes in *Acricotopus lucidus* (Diptera, Chironomidae). Genome 29: 702–705.

Stark, G. R. and G. M. Wahl. 1984. Gene amplification. Annu. Rev. Biochem. 53: 447–91.

Steiner, W. W. M., W. J. Tabachnick, K. S. Rai and S. Narang, Eds. 1982. *Recent Developments in the Genetics of Insect Disease Vectors*. Stipes Publ., Champaign, IL.

Stoltz, D. and J. B. Whitfield. 1992. Viruses and virus-like entities in the parasitic Hymenoptera. J. Hym. Res. 1: 125–139.

Stuart, J. J., S. J. Brown, R. W. Beeman and R. E. Denell. 1991. A deficiency of the homeotic complex of the beetle *Tribolium*. Nature 350: 72–74.

Stuart, J. J., S. J. Brown, R. W. Beeman and R. E. Denell. 1993. The *Tribolium* homeotic gene *Abdominal* is homologous to *abdominal-A* of the *Drosophila* bithorax complex. Development 117: 233–243.

Sun, S. C., I. Lindstrom, J. Y. Lee and I. Faye. 1991. Structure and expression of the attacin genes in *Hyalophora cecropia*. Eur. J. Biochem. 196: 247–254.

Taylor, M., J. Zawdzki, B. Black and M. Kreitman. 1993. Genome size and endopolyploidy in pyrethroid-resistant and susceptible strains of *Heliothis virescens* (Lepidoptera: Noctuidae). J. Econ. Entomol. 86: 1030–1034.

Tazima, Y. 1964. *Genetics of the Silkworm*. Logos, London.

Tazima, Y., H. Doira and H. Akai. 1975. The domesticated silkmoth, *Bombyx mori*. Pp. 63–124. In: *Handbook of Genetics*. Vol. 3. *Invertebrates of Genetic Interest*. R. C. King, Ed. Plenum, New York.

Tear, G., M. Akam and A. Martinez-Arias. 1990. Isolation of an *abdominal-A* gene from the locust *Schistocerca gregaria* and its expression during early embryogenesis. Development 110: 915–926.

Tearle, R. G., T. J. Lockett, W. R. Knibb, J. Garwood and R. B. Saint. 1993. The regulation of cellular pattern formation in the compound eye of *Drosophila melanogaster*. Pp. 267–291. In: *Molecular Approaches to Fundamental and Applied Entomology*. J. Oakeshott and M. J. Whitten, Eds. Springer-Verlag, New York.

Thibout, E., J. F. Guillot and J. Auger. 1993. Microorganisms are involved in the production of volatile kairomones affecting the host seeking behaviour of *Diadromus pulchellus*, a parasitoid of *Acrolepiopsis assectella*. Physiol. Entomol. 18: 176–182.

Thummel, C. S. 1990. Puffs and gene regulation—Molecular insights into the *Drosophila* ecdysone regulatory hierarchy. BioEssays 12: 561–568.

Tittiger, C., S. Whyard and V. K. Walker. 1993. A novel intron site in the triosephosphate isomerase gene from the mosquito *Culex tarsalis*. Nature 361: 470–472.

Trewitt, P. M., L. J. Heilmann, S. S. Degrugillier and A. K. Kumaran. 1992. The boll weevil vitellogenin gene: Nucleotide sequence, structure, and evolutionary relationship to nematode and vertebrate vitellogenin genes. J. Mol. Evol. 34: 478–492.

Ueda, H. and S. Hirose. 1991. Identification and purification of a *Bombyx mori* homologue of FTZ-F1. Nucleic Acids Res. 18: 7229–7234.

Ueno, K., C. C. Hui, M. Fukuta and Y. Suzuki. 1992. Molecular analysis of the deletion mutants in the E homeotic complex of the silkworm *Bombyx mori*. Development 114: 555–563.

Wagner, R. P., M. P. Maguire and R. L. Stallings. 1993. *Chromosomes A Synthesis*. Wiley-Liss, New York.

Walldorf, U. R. Fleig and W. J. Gehring. 1989. Comparison of homeobox-containing genes of the honeybee and *Drosophila*. Proc. Natl. Acad. Sci. USA 86: 9971–9975.

Warren, A. M. and J. M. Crampton. 1991. The *Aedes aegypti* genome: Complexity and organization. Genet. Res., Camb. 58: 225–232.

Waters, L. C., A. C. Zelhof, B. J. Shaw and L. Y. Chang. 1992. Possible involvement of the long terminal repeat of transposable element *17.6* in regulating expression of an insecticide resistance-associated P450 gene in *Drosophila*. Proc. Natl. Acad. Sci. USA 89: 4855–4859.

Welburn, S. C., K. Arnold, I. Maudlin and G. W. Gooday. 1993. Rickettsia-like organisms and chitinase production in relation to transmission of trypanosomes by tsetse flies. Parasitology 107: 141–145.

White, M. J. D. 1973. *Animal Cytology and Evolution*. 3rd Ed. Cambridge Univ. Press, Cambridge.

Wilkins, A. S. 1993. *Genetic Analysis of Animal Development*. 2nd Ed., Wiley-Liss, New York.

Williams, J. A. and S. B. Carroll. 1993. The origin, patterning and evolution of insect appendages. BioEssays 15: 567–577.

Williams, S. M. and L. G. Robbins. 1992. Molecular genetic analysis of *Drosophila* rDNA arrays. Trends Genet. 8: 335–340.

Williamson, M. S., I. Denholm, C. A. Bell and A. L. Devonshire. 1993. Knockdown resistance (*kdr*) to DDT and pyrethroid insecticides maps to a sodium channel gene locus in the housefly (*Musca domestica*). Mol. Gen. Genet. 240: 17–22.

Wilson, T. G. 1993. Transposable elements as initiators of insecticide resistance. J. Econ. Entomol. 86: 645–651.

Woods, D. F. and P. J. Bryant. 1992. Genetic control of cell interactions in developing *Drosophila* epithelia. Annu. Rev. Genet. 26: 305–350.

Wolf, K. W., H. G. Mertl and W. Traut. 1991. Structure, mitotic and meiotic behaviour, and stability of centromere-like elements devoid of chromosome arms in the fly *Megaselia scalaris* (Phoridae). Chromosoma 101: 99–108.

Wright, J. W. and R. Pal, Eds. 1967. *Genetics of Insect Vectors of Disease*. Elsevier, Amsterdam.

Xiong, Y., W. D. Burke and T. H. Eickbush. 1993. *Pao*, a highly divergent retrotransposable element from *Bombyx mori* containing long terminal repeats with tandem copies of the putative R region. Nucleic Acids Res. 21: 2117–2123.

Zheng, L., R. D. C. Saunders, D. Fortini, A. della Torre, M. Coluzzi, D. M. Glover and F. C. Kafatos. 1991. Low-resolution genome map of the malaria mosquito *Anopheles gambiae*. Proc. Natl. Acad. Sci. USA 88: 11187–11191.

Zheng, L., F. H. Collins, V. Kumar and F. C. Kafatos. 1993. A detailed genetic map for the X chromosome of the malaria vector, *Anopheles gambiae*. Science 261: 605–608.

Zurovec, M., F. Sehnal, K. Scheller and A. K. Kumaran. 1992. Silk gland specific cDNAs from *Galleria mellonella*. Insect Biochem. Mol. Biol. 22: 55–67.

# Part II

# Molecular Genetic Techniques

# Chapter 6

## Some Basic Tools: How To Cut, Paste, Copy, Measure, and Visualize DNA

### Overview

Genetic engineers use a number of techniques to isolate DNA, cut and join molecules, and monitor the results. In order to clone a gene, determine its sequence, or alter the genetic makeup of an arthropod, various microbiological techniques are employed, particularly with the bacterium *Escherichia coli*. *E. coli* has become a molecular biology workhorse because it can be induced to produce large amounts of recombinant DNA molecules by inserting plasmids, the bacteriophage λ, or genetically engineered variants of these agents into it. A number of enzymes from different organisms are used to modify, ligate, or splice DNA. Purifying plasmids from *E. coli*, visualizing DNA by electrophoresis through agarose or polyacrylamide gels, Southern blot analyses, and producing labeled probes are techniques that are basic to the molecular geneticist. As a mechanism to introduce these techniques, this chapter describes the steps involved in inserting a foreign gene into a plasmid, inserting the plasmid into *E. coli*, and isolating and analyzing the amplified DNA by Southern blot analysis and restriction site mapping.

### Introduction to a Simple Experiment

A diverse array of techniques, some arising from research on apparently nonapplied topics, became crucial tools that allow scientists to manipulate DNA from living organisms. This molecular genetic revolution only began about 20 years ago. Prior to 1970 there was no way to cut a DNA molecule into discrete and predictable fragments, nor could specific DNA fragments be joined together. The discovery of enzymes called **restriction endonucleases** and **ligases** solved this problem. Much of genetic engineering technology is dependent upon our ability to cut DNA molecules at specific sites and combine them into new molecules.

125

Another significant development was the harnessing of **plasmids** and **bacteriophages** as vehicles to replicate foreign DNA within the bacterium *Escherichia coli*. This allowed nearly unlimited amounts of specific DNA to be produced for study and manipulation. Various techniques for monitoring the results of such manipulations were developed so that researchers could identify changes in DNA molecules as small as a single base modification.

A simple cloning project is outlined in Figure 6.1. The project involves inserting a piece of **exogenous** (or foreign) DNA extracted from one organism into a plasmid that has been engineered to serve as a **vector** to carry the exogenous DNA into *E. coli*. The *E. coli* cells with the exogenous DNA (in the plasmid vector) can be mass produced to yield large numbers of the desired DNA molecules. Once the DNA has been mass produced (**cloned**), it can be studied in detail.

While the experiment in Figure 6.1 is conceptually very simple, cloning a fragment of foreign DNA in a vector demands that several steps be achieved: (1) The vector DNA must be purified and cut open. (2) The foreign DNA must be extracted, purified, and cut. (3) The vector DNA and foreign DNA must be joined together. (4) The reactions should be monitored. (5) The recombinant plasmid or vector containing the exogenous DNA must be put back into *E. coli* to be amplified. (6) The recombinant plasmid must then removed from *E. coli* and purified for analysis or use of the exogenous DNA.

This chapter provides an introduction to the procedures that could be employed to carry out the experiment in Figure 6.1. Simplified protocols of some procedures are provided for those interested in knowing something about the steps involved, although the methods described are illustrative rather than complete laboratory protocols. There are many excellent laboratory manuals available that provide detailed techniques. Furthermore, many of the techniques have been simplified and are available in kits provided by commercial sources.

## Extracting DNA

The DNA to be manipulated must be extracted from its source, either from an intact organism or from cells. This DNA also must be purified before it can be used. The degree of purity needed is determined by the goals of the experiment. One of the most common methods for extracting and purifying nucleic acids uses phenol to extract DNA (or RNA) in large- or small-scale procedures, from complex extracts or from simpler *in vitro* manipulations. A plethora of different phenol extraction methods have been published, but the primary function of phenol is to remove proteins from an aqueous solution containing nucleic acids. Some of the proteins may be nucleases that could damage the DNA, while others simply could interfere with later manipulations. EDTA (ethylenediaminetetraacetic acid) is often added; it is a chelating agent that binds $Mg^{++}$ which is required for nucleases to act on the DNA.

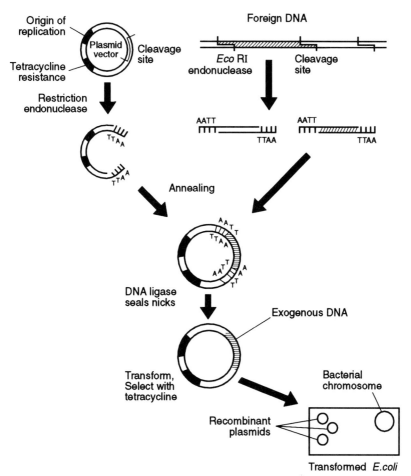

**Figure 6.1** Outline of a simple cloning project involving insertion of foreign DNA into a plasmid vector, and its subsequent insertion into a bacterial host, *E. coli*, to produce multiple copies of the foreign DNA in recombinant plasmids. Both foreign and plasmid DNA are cut with a restriction endonuclease to produce ends that will allow annealing of the plasmid and foreign DNA fragments. The addition of DNA ligase combines the two DNA molecules and the plasmid is inserted into *E. coli* where it will be selected with tetracycline to produce multiple copies of the new, recombinant DNA molecule. (Adapted from Old and Primrose 1980.)

Highly purified phenol is mixed with the sample under conditions that favor the dissociation of proteins from the nucleic acids and the sample is then centrifuged (Table 6.1). Centrifugation yields two phases: (1) a lower organic phenol phase carrying the protein, and (2) the less dense aqueous phase containing the nucleic acids. Highly purified liquefied phenol ready for use in DNA

**Table 6.1**

Rapid Phenol Extraction of Genomic DNA from *Drosophila melanogaster*[a]

---

1. Homogenize 50–200 flies (frozen in liquid nitrogen) in a 15-ml polypropylene tube with a teflon pestle in 2 ml of lysis buffer. (Lysis buffer contains 100 mM Tris–HCl (pH 8.0), 50 mM NaCl, 50 mM EDTA, 1% SDS, 0.15 mM spermine, 0.5 mM spermidine.)

2. Add 20 µl of proteinase K solution (10 mg/ml).

3. Leave at 37°C for 1–2 hours, but occasionally swirl and invert the tube to mix.

4. Extract once with an equal volume of phenol + chloroform + isoamyl alcohol. (The phenol, chloroform, and isoamyl alcohol should be in a 24:24:1 ratio. The isoamyl alcohol serves as an antifoaming agent.)

5. Spin in a bench centrifuge for 5 minutes at room temperature.

6. Decant the aqueous layer with a Pasteur pipette into a new tube.

7. Extract twice more with phenol + chloroform + isoamyl alcohol. Respin and decant the aqueous layer each time.

8. Extract the aqueous layer with chloroform and isoamyl alcohol (24:1). The interface between the organic and aqueous layer should be clean.

---

[a]Modified from Jowett 1986.

extraction and purification can be purchased from commercial sources. Some phenol extraction protocols include chloroform, which denatures proteins, removes lipids, and improves the efficiency of the extractions. To reduce foaming caused by chloroform, isoamyl alcohol also is usually added.

Extraction of DNA from cells or organisms should be carried out as quickly as possible in ice with refrigerated buffers to minimize the activity of any nucleases present in the cells that can degrade the DNA. Extraction of DNA from *in vitro* experiments that lack nucleases can be carried out at room temperature. Table 6.1 outlines one protocol for extracting DNA from *Drosophila melanogaster*. Other DNA extraction methods are available (e.g., Henry et al. 1990, Pinol et al. 1988), and various commercial kits offer methods for rapid extraction of DNA without using toxic chemicals. Modifications of these DNA extraction methods may be necessary for DNA with unusual characteristics. For example, DNA with different ratios of G+C can be extracted from the honey bee *Apis mellifera* (Beye and Raeder 1993).

Few evaluations of methods for preserving insects so that their DNA is suitable for analysis have been conducted. Storage of insects in liquid nitrogen is probably the best method for preserving large amounts of undegraded DNA, and should be the standard against which other methods are compared. However, it is often useful to analyze DNA from insects preserved in museums that have been stored dried, in alcohol, or in Carnoy's solution for cytogenetic analysis. Post et al. (1993) compared the yield and degree of degradation of DNA extracted from *Simulium damnosum* after the flies were pinned and dried or stored in saline, Carnoy's solution, methanol, or propanol. The highest yields of DNA and the least degradation were obtained with flies preserved in liquid nitrogen, stored in 100% ethanol at 4°C, or dried over silica gel. Pinned specimens and those in saline gave undetectable yields of DNA, and the DNA from flies in Carnoy's solution, methanol, and propanol was degraded to small fragments. Post et al. (1993) concluded that storage in 100% ethanol at 4°C and desiccation over silica gel offer suitable alternatives to storage in liquid nitrogen.

## Precipitating Nucleic Acids

During cloning, or many other projects, it is often necessary to concentrate DNA samples or change the solvent in which a nucleic acid is dissolved. The DNA isolated by phenol extraction contains trace amounts of phenol, which could disrupt the activity of enzymes in subsequent manipulations if it is not purified further. Purification can be achieved by ethanol precipitation, isopropanol precipitation, or several other methods. The most versatile is probably ethanol precipitation, because it can concentrate both DNA and RNA and purify DNA after phenol extractions.

Basically, DNA is precipitated by combining the DNA sample, a salt, and ethanol at −20°C or lower (Table 6.2). The precipitated salt of the nucleic acid is then sedimented by centrifugation, the ethanol supernatant is removed, and the nucleic acid pellet is resuspended in a buffer. The choice of salt or buffer is determined by the nature of the sample and by the planned use for the nucleic acid. Once the DNA is purified, it can be stored at 4°C in TE buffer (pH 8). The EDTA in the TE buffer chelates heavy metal ions which are commonly required for DNase activity. For very long-term storage (5 years or more), the DNA can be frozen at −80°C but should not be subjected to freeze–thaw cycles.

The now-purified DNA can be cut, either by shearing or with a restriction nuclease. Shearing produces random fragments while restriction endonucleases can generate fragments of a desired size and with termini, or ends, appropriate to the annealing and ligation steps of the experiment outlined in Figure 6.1.

## Shearing DNA

A variety of protocols are available to mechanically produce fragmented DNA. DNA in cells can be broken by shear forces in solution. Sonication with ultrasound can produce DNA fragments about 300 nucleotides long. High-speed stirring of cells in a blender at 1500 revolutions/minute for 30 minutes will produce DNA molecules with a mean size of about 8 kb. Breakage occurs

**Table 6.2**

A Sample Protocol for Precipitating DNA with Ethanol[a]

For recovery of DNA from a typical reaction (for 1 μg DNA in 20 μl):

1. To 20 μl aqueous DNA sample in a microcentrifuge tube, add 2 μl 3 M sodium acetate, pH 5.5, and 40 μl ethanol.

2. Mix well by vortexing and immerse the tube in a −70°C bath composed of methanol plus dry ice for 15 minutes. The mixture should freeze or form a slurry.

3. Centrifuge the DNA precipitate in a benchtop microcentrifuge at maximum speed for 10 minutes in a cold room. A whitish pellet of DNA should appear at the bottom of the tube. In general, pellets of 10 μg are visible while pellets of 2 μg will be invisible.

4. Remove the ethanol supernatant using a micropipette taking care not to disturb the pellet or the area of the tube where the pellet should be located.

5. Add 100 μl 70% ethanol (chilled to −20°) to the sample and vortex. This step removes any solute trapped in the precipitate.

6. Reprecipitate the DNA by centrifugation for 2 minutes and remove the supernatant as before.

7. Dry the pelleted DNA for 1–2 minutes in a vacuum desiccator, taking care to release the vacuum gently so as not to dislodge the dried sample.

8. Resuspend the DNA in TE (pH 8) buffer. (TE buffer contains 10 mM Tris–HCl at pH 8; 1 mM $Na_2$ EDTA)

[a]Modified from Berger and Kimmel 1987.

essentially at random with respect to DNA sequence and the broken ends consist of short, single-stranded (ss) regions. These single-stranded termini must be modified before the DNA can be joined to a vector.

## Cutting DNA with Restriction Endonucleases

Most cloning projects use restriction endonucleases to cut DNA. Restriction enzymes were discovered as an outcome of basic research aimed at understanding the control by bacteria of infection by **bacteriophages** (viruses that invade bacteria). Most bacteria contain specific endonucleases that guard against invasion of foreign DNA. These endonucleases make cuts in double-stranded (ds) DNA invading the cell unless the DNA has been modified, usually by methylation. Foreign DNA has an inappropriate methylation pattern but the bacterium's own DNA is protected because its methylation pattern is recognized as appropriate.

More than 1400 restriction enzymes (also called restriction endonucleases) have been found in more than 1117 different microorganisms. Endonucleases with 120 different sequence specificities are commercially available (Kessler and Manta 1990). Restriction enzymes are identified by three-letter abbreviations for the parent organism using a nomenclature proposed by Smith and Nathans (1973). For example, Hin denotes Haemophilus influenzae and Bam stands for Bacillus amyloliquefaciens and are italicized. An additional letter is added if it is needed to identify a specific strain or serotype (Hind or Bam H), but these are not italicized ordinarily. A Roman numeral is added to reflect the order of identification or characterization of the specific endonuclease (Hind III or Bam HI, Table 6.3).

Restriction endonucleases recognize specific sequences in DNA. They cleave the DNA specifically, producing either blunt or staggered cuts (Table 6.4). Most endonucleases recognize sequences 4–6 nucleotides long, but some have 7- and 8-base recognition sites, and a very few recognize 12-base recognition sites. Restriction enzymes cleave DNA to form 5'-phosphate and 3'-hydroxyl termini on each strand (Table 6.4). Some endonucleases produce breaks that are staggered, generating either 5'-phosphate extensions on each strand or 3'-hydroxyl extensions on each strand. Other endonucleases produce breaks that can be opposed, with "blunt" ends (Table 6.4).

It is often desirable to generate DNA fragments of a specific length, with a specific sequence, and with a particular type of end. This precision is possible with well-characterized DNA that has been sequenced. However, with un-

**Table 6.3**

Target Sites for Selected Restriction Endonucleases[a]

| Organism from which endonuclease was isolated | Abbreviation | Recognition sequences[b] \| indicates cleavage site |
|---|---|---|
| Bacillus amyloliquefaciens H | Bam HI | G\|GATCC |
| Bacillus globigii | Bgl II | A\|GATCT |
| Escherichia coli RY13 | Eco RI | G\|AATTC |
| Haemophilus aegyptius | Hae III | GG\|CC |
| Klebsiella pneumonia | Kpn II | GGTAC\|C |
| Nocaardia otitidis-caviarum | Not I | GC\|GGCCGC |
| Providencia stuartii | Pst I | CTGCA\|G |
| Serratia marcescens | Sma I | CCC\|GGG |

[a]Modified from Kessler and Manta 1990.
[b]Recognition sequences are written from 5' to 3' with only one strand given. The vertical bar indicates a cleavage site.

**Table 6.4**

Three Types of Termini, or Ends, Are Created by Restriction Enzyme
Cleavage of Double-Stranded DNA[a]

| 5′ Overhang produced by Eco RI | Termini/enzyme blunt end produced by Pvu II | 3′ Overhang produced by Kpn I |
|---|---|---|
| G\|AATTC | CAG\|CTG | GGTAC\|C |
| CTTAA\|G | GTC\|GAC | C\|CATGG |
| G$^{3'}$ $^{5'}$AATTC | CAG$^{3'}$ $^{5'}$CTG | GGTAC$^{3'}$ $^{5'}$C |
| CTTAA$^{5'}$ $^{3'}$G | GTC$^{5'}$ $^{3'}$GAC | C$^{5'}$ $^{3'}$CATGG |

[a]The vertical bar indicates where cleavage occurs when DNA is restricted, or cut, by three representative enzymes.

characterized DNA, while it is possible to predict whether the ends will be blunt, or with 5′ or 3′ overhangs, it is difficult to plan precisely the length of the DNA fragments that will be generated after DNA is digested with a restriction endonuclease. This is because the frequency of any particular restriction endonuclease cleavage site within an unknown DNA sequence is unpredictable. If a DNA sequence contained 50% guanine and cytosine (G+C) base pairs, and if the bases were distributed randomly, a 4-base sequence recognized by the restriction endonuclease would occur on an average of every 256 bases ($4^4$); a 6-base sequence would occur approximately every 4 kb ($4^6$ or 4096 bases); and an 8-base sequence would occur approximately every 65 kb. However, many segments of DNA are not random in their G+C content. For example, highly repetitive DNA may have several nucleotides repeated millions of times, which obviously would bias restriction site frequencies significantly. The percentage of G+C in DNA from different sources may vary from 22 to 73%.

Many restriction endonucleases are commercially available either in native or cloned form. Most manufacturers provide standardized buffers (high-, medium-, or low-salt) for optimizing the reaction conditions and protocols for the digestions. Among those available, some recognize identical sequences, although they may vary with respect to the site of cleavage, sensitivity to methylation, and cleavage rates. The choice of endonuclease is determined by the goals of the project. Enzymes that produce small segments of a few hundred bases are useful for restriction mapping or for sequencing. Enzymes that produce fragments of 1 to 10 kb are useful for mapping large DNA regions and for cloning whole genes with their introns and control sequences. Generating even larger fragments (5–50 kb) is necessary for cloning into cosmid vectors or for genome walking, as discussed in Chapters 7 and 8.

Restriction endonucleases can be degraded if not properly stored at $-20°C$.

Digestions with restriction endonucleases contain the DNA substrate, the restriction endonuclease(s), Tris buffer, $Mg^{++}$, NaCl, 2-mercaptoethanol, and bovine serum albumin. All endonucleases require $Mg^{++}$ as a cofactor and most are active at pH values ranging from 7.2 to 7.6. The major difference among the endonucleases is their dependence on ionic strength and their temperature optima. Most digestions are done at 37°C, but a few restriction endonucleases perform better at lower temperatures. A few restriction endonucleases, such as *Taq* I, which was isolated from a bacterium growing in hot springs, function well at temperatures as high as 75°C. Endonuclease activity is usually measured with bacteriophage λ DNA as a substrate, but activity of the endonuclease varies greatly with different DNA substrates and can also be modified by the neighboring sequences. Activity rates can vary by a factor of 10- to 50-fold.

The number and variety of endonucleases available for genetic manipulation continue to increase (Pingoud et al. 1993). Endonucleases that recognize longer recognition sequences are particularly useful because large DNA fragments can now be separated by pulse-field electrophoresis. New microbial sources of enzymes are being sought, especially those that tolerate high temperatures. Catalogs obtained from many suppliers contain much useful information on restriction endonuclease activity and their appropriate reaction conditions.

## Joining DNA Molecules

Different DNA fragments cleaved by restriction endonucleases can be joined together, which makes it possible to insert exogenous DNA into plasmid vectors. Two **DNA ligases** are commonly used. One is derived from *E. coli* and the other from the bacteriophage T4. Their requirements for cofactors differ. T4 DNA ligase requires ATP while *E. coli* ligase requires oxidized nicotinamide adenine dinucleotide ($NAD^+$). Both catalyze the joining of a $5'$-phosphate and a $3'$-OH group to form a phosphodiester bond. T4 DNA ligase will catalyze the joining of blunt-ended DNA molecules and cohesive-ended molecules, although more enzyme is required for blunt-ended ligation of DNA molecules.

If the restriction endonuclease employed generates DNA fragment ends with uneven ends or overhangs (Table 6.4), then the sequences of the DNA within the single-stranded regions of the two molecules have to be complementary for ligation to occur. Ligation of 4-base extensions is easier than ligation of 2-base extensions. Extensions that consist of G+C bases ligate more readily than those with A+T base pairs. Blunt ends are more difficult to ligate, requiring 20 to 100 times more T4 DNA ligase and higher DNA concentrations. The surrounding DNA sequences do not affect ligation efficiency, but ligation is negatively influenced by the presence of contaminating endonucleases or by phosphatase.

The optimum temperature for ligating DNA is 37°C but the hydrogen-

**Table 6.5**

Blunt End Ligation when the Vector to Insert Molar Ratio = $3^a$

| | |
|---|---|
| In a siliconized Eppendorf tube in ice mix: | |
| Dephosphorylated vector DNA (~4 kb) | 160 ng |
| DNA fragment (~1 kb) | 13 ng |
| 10× ligase buffer I | 4 μl |
| (250 mM Tris–HCl at pH 7.5, 50 mM $MgCl_2$, 25% w/v poly-ethylene glycol 8000, 5 mM dithiothreitol (DTT), 4 mM ATP) | |
| T4 DNA ligase | 1 Weiss unit |
| Water to a final volume of | 20 μl |

Incubate at 23° for 4 hours and stop the reaction by adding 1 μl of 0.5 M EDTA. Dilute 5-fold before adding the mixture to competent *E. coli* cells for transformation.

$^a$Modified from Cobianchi and Wilson (1987).

bonded joint between sticky ends is unstable at this temperature. The ligation reaction thus is carried out at a temperature which is a compromise between the optimum for the rate of the enzyme action and the association of the termini, and usually is done between 4 and 15°C. Ligation reactions often are allowed to take place overnight at these low temperatures.

The ligation reaction also is promoted by adjusting the ratio of insert DNA and vector DNA. When a linear DNA fragment is produced by a restriction endonuclease from a circular vector, the linear fragment will often recircularize and hydrogen bond to itself, or to other linear vector sequences. To prevent this, the linearized plasmid vector DNA can be treated with alkaline phosphatase to remove 5'-terminal phosphate groups. Alkaline phosphatase prevents recircularization of the plasmid or formation of plasmid dimers. However, the phosphatase must be eliminated if the vector and linear DNA are to be ligated. Circularization of the vector will then occur if the foreign DNA (untreated with phosphatase) joins the ends of the vector.

Only T4 DNA ligase is able to join blunt-ended DNA molecules. A typical blunt end ligation reaction is described in Table 6.5.

## Growth, Maintenance, and Storage of *Escherichia coli*

DNA manipulations require manipulating bacteria, primarily derivatives of *E. coli* K12 strains. An extensive amount of genetic information has been obtained on *E. coli* and over 1000 genes have been mapped. Standard microbial techniques are employed: pure cultures of *E. coli* are obtained by propagating cultures from single, isolated colonies on agar plates. Dilution streaking with an inoculating loop will readily produce isolated colonies and an isolated colony can be restreaked to obtain a pure master plate which can be stored at 4°C for a month. It is important that the cultures be kept pure, that the phenotypes be verified prior to use, and that the cultures be stored properly. For long-term storage, cultures can be stored in stab vials, as frozen glycerol cultures, or as lyophilized cultures.

Overnight cultures of most strains of *E. coli* produce ~4 X $10^9$ bacteria per milliliter (ml) depending upon the medium, degree of aeration, the strain, and the temperature. To determine the cell concentration, dilutions of the culture should be plated. Detailed methods for manipulating *E. coli* are readily available in a number of laboratory manuals, some of which are listed at the end of this chapter.

## Plasmids for Cloning in *Escherichia coli*

Plasmids are widely used in cloning. Many plasmids have undergone extensive genetic engineering to enhance their value as vectors. Wild-type plasmids are small nucleic acid molecules that are stably inherited as extrachromosomal units in many kinds of bacteria. Plasmids are widely found, but usually are not essential to their host. Many plasmids carry genes for antibiotic resistance, antibiotic production, heavy metal resistance, an ability to degrade aromatic compounds, sugar fermentation, enterotoxin production, or hydrogen sulfide production. Most are double-stranded DNA molecules that are covalently closed circles, but some are linear molecules.

Plasmids can be classified into two types, depending upon whether they carry a set of genes that promote **bacterial conjugation**. Plasmids also can be categorized as to whether they are maintained in multiple copies in hosts or in limited numbers per cell. Generally, plasmids that promote bacterial conjugation are relatively large and present in one to three copies per bacterial cell. Plasmids that do not promote bacterial conjugation are smaller and present in multiple copies per cell. Plasmids are called 'promiscuous' if they are capable of promoting their own transfer to a wide range of bacteria and of being stably maintained in their new hosts. Promiscuous plasmids can transfer cloned DNA molecules into a range of different bacteria.

Wild-type plasmids could be used for cloning in *E. coli* but they suffer from a number of disadvantages, and genetically engineered plasmids have been developed that have a number of desirable attributes. The first improvement involved removing excess DNA so that the plasmid is easier to manipulate *in vitro*, resistant to damage by shearing, and readily isolated from bacterial cells. The smaller size is also an advantage because bacterial cells usually can sustain several smaller plasmids, which will increase the yield of the recombinant DNA molecules.

A second improvement is the addition of a **selectable marker** gene to the plasmid, which allow identification of specific cells with a desirable new genotype. The selectable marker allows the experimenter to identify those bacterial cells that have taken up the plasmid during the transformation process. Many selectable markers are antibiotic resistance genes that allow the transformed bacteria to be grown on selective media. A third improvement involves adding unique DNA segments in a **multiple cloning site** that can be cut by restriction endonucleases. The presence of unique restriction or cloning sites is helpful because cloning requires that both the vector and the exogenous DNA be cut with the same endonuclease (or ones that produce the same kinds of ends) so that the ends can be ligated together. If the plasmid had more than one site that was cut by a specific endonuclease, the plasmid vector would be cut into several fragments, resulting in defective vectors.

Vectors have been engineered by sophisticated techniques to perform a variety of defined tasks. Some vectors facilitate expression of proteins; for example, **baculovirus** vectors are used to produce large amounts of foreign proteins in insect cells (Chapter 7). Some vectors help identify regulatory signals that turn genes on or off, some are used for direct selection of recombinants, some have increased stability so that they are not eliminated from their host cells, and others are genetically altered so that high copy numbers per host cell can be maintained. Some vectors are particularly useful for DNA sequencing because they produce a single-stranded copy of DNA rather than a double-stranded molecule (Chapter 8). Some vectors are modified so that proteins are secreted through the host cell wall to facilitate purification of the proteins, and others are modified to produce fusion proteins to facilitate protein purification. Many versatile vectors are produced and are available from commercial sources.

A key to effective genetic engineering is the ability to identify those cells that have been genetically transformed. To aid in this, most plasmid vectors contain at least two selectable markers. If two markers are present, then one is often the site into which the exogenous DNA is cloned. Insertion of exogenous DNA should inactivate the resistance gene, so that the *E. coli* cells with this plasmid can be identified as susceptible, and thus as containing plasmids with exogenous DNA. Another selectable marker is the *lacZ* construct, which allows selection of blue/white bacterial colonies. This construct is described in more detail in Chapter 7.

Plasmid vectors, such as pBR322 and its derivatives, are widely used because they can be produced in multiple copies within a cell, they are easily purified, and can produce large amounts of the cloned gene (Figure 6.2). This plasmid carries both ampicillin and tetracycline resistance genes and an **origin of replication**, which is a sequence at which replication of the DNA molecule is initiated. This plasmid has been completely sequenced and its restriction sites are totally characterized. This means that the exact length of each fragment from a restriction digest can be predicted and these fragments can serve as DNA markers for

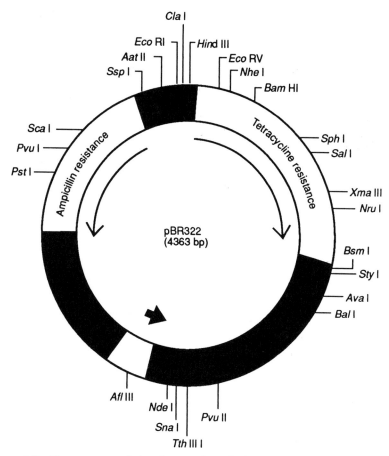

**Figure 6.2** The structure of the cloning plasmid pBR322 showing the unique sites where restriction endonucleases can cleave the DNA. The thin arrows inside the circle indicate the direction of transcription of the ampicillin and tetracycline resistance genes, which serve as selectable markers. The thick arrow shows the direction of DNA replication. (Adapted from Old and Primrose 1980.)

sizing other DNA fragments. pBR322 fragments produced after digestion with restriction enzymes range in size from several bp to the entire 4.3-kb-long plasmid.

Some unique restriction sites occur within both the ampicillin and tetracycline resistance genes of pBR322 (Figure 6.2) and these are very useful in cloning. If exogenous DNA is inserted into a site in the ampicillin resistance gene where a restriction enzyme cuts uniquely, the ampicillin gene will be

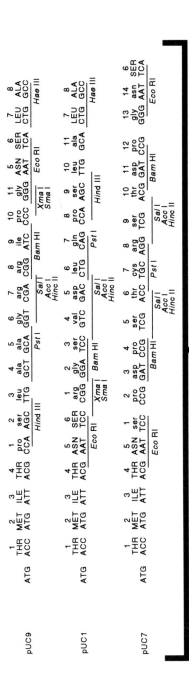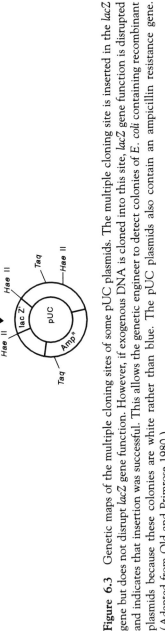

**Figure 6.3** Genetic maps of the multiple cloning sites of some pUC plasmids. The multiple cloning site is inserted in the *lacZ* gene but does not disrupt *lacZ* gene function. However, if exogenous DNA is cloned into this site, *lacZ* gene function is disrupted and indicates that insertion was successful. This allows the genetic engineer to detect colonies of *E. coli* containing recombinant plasmids because these colonies are white rather than blue. The pUC plasmids also contain an ampicillin resistance gene. (Adapted from Old and Primrose 1980.)

inactivated. The recombinant plasmids contained within their *E. coli* host can then be identified by spreading the transformed *E. coli* first onto culture plates with media containing tetracycline and then **replica plating** them onto plates with ampicillin. Replica plating is a procedure in which a particular pattern of colonies on an agar surface is reproduced with a template that contains an image of the pattern. The template is often produced by pressing a piece of sterile velvet upon the original agar surface, which transfers cells from each colony to the cloth and then to another agar surface. Those recombinant *E. coli* colonies that are unable to survive on the ampicillin can then be recovered from the colony growing on the original tetracycline plate. Many derivatives of pBR322 have been constructed to fulfill particular cloning goals (Balbas et al. 1986).

Most vectors, including the pUC series and others developed by commercial producers, have a variety of **polylinkers** or multiple cloning sites in them (Figure 6.3). A polylinker is a short segment of DNA with sites where several different restriction endonucleases can cut. This gives the genetic engineer options as to which restriction enzyme to use. The polylinker usually is placed within a selectable marker gene (*lacZ*); the gene function is disrupted when exogenous DNA is cloned into it at any of the restriction sites. The colonies containing recombinant plasmids then can be identified by their color.

## Transforming *Escherichia coli* with Plasmids

A plasmid carrying exogenous DNA must be inserted into *E. coli* in order to amplify the DNA in most experiments. The process of inserting a plasmid into *E. coli* is called bacterial **transformation**. For many years, efforts to transform *E. coli* were unsuccessful and it was only in 1970 that methods were developed. Transformation required understanding the genetics of *E. coli* and having the ability to manipulate the physiological status of the *E. coli* cells to optimize the transformation reaction.

A simple transformation procedure involves suspending *E. coli* cells that are in the log phase of their growth cycle in an ice-cold solution containing membrane-disrupting agents such as polyethylene glycol (PEG), dimethyl sulfoxide (DMSO), or divalent cations such as calcium chloride (Table 6.6). Plasmid DNA is then added to a small aliquot of these **competent cells** (competent for transformation) and the incubation on ice is continued for another 30 minutes. A heat shock is then administered by putting the cells into 42°C for two minutes. The cells are then transferred to nutrient broth and incubated for 30 to 60 minutes to allow the plasmid to express its phenotypic properties (plasmids often carry antibiotic resistance genes as selectable markers). The cells are then plated out onto agar plates with a selective medium (perhaps containing the antibiotic to which the plasmid gene confers resistance). Only those bacteria that have taken up the plasmid with the selectable marker should survive and reproduce on the selective medium.

**Table 6.6**

Producing Competent *E. coli* and Transforming Them with Plasmid DNA Using $CaCl_2$ [a]

---

Producing competent cells

1. Grow a fresh overnight culture of *E. coli* in LB broth at 37°C.

2. Dilute the cells 40-fold into 1 liter of fresh medium. Incubate at 37° with good aeration until their density produces an absorbency rating at 550 nm of 0.4–0.5.

3. Immediately chill the culture by swirling in an ice-water bath.

4. When the cells are chilled, centrifuge the culture at 4°C at 5000 rpm for 10 minutes.

5. Decant the supernatant and place the pellet in ice.

6. Resuspend the pellet in 500 ml ice-cold 100 mM $CaCl_2$. It is easier to resuspend the pellets if they are vortexed before the $CaCl_2$ is added. The cells can be suspended by sucking them up and down in a 25-ml pipet.

7. Once the cells are resuspended, incubate in ice for 30 minutes with occasional swirling.

8. Pellet the cells once again at 5000 rpm for 10 minutes at 4°C.

9. Resuspend in 40 ml of ice-cold 100 mM $CaCl_2$ and 15% glycerol.

10 Distribute aliquots of 0.2 ml of cells into sterile Eppendorf tubes in ice.

11. Keep in ice at 0–4°C for 12–24 hours. This is essential for maximal competency, although the cells are competent at this stage.

12. Freeze the tubes in ethanol–dry ice or liquid nitrogen and place immediately at −70°. The cells remain competent for months if stored at −70°.

Transformation of competent cells

1. Thaw a tube of frozen competent cells at 4°C.

2. Add DNA in buffer.

3. Incubate in ice for 30 minutes.

4. Heat shock for 2–5 minutes in a 42° water bath.

5. Add 0.4 ml of LB broth at room temperature to each tube and incubate for 1 hour at 37°.

6. Spread on agar plate with appropriate antibiotics. Incubate plates overnight at 37°.

---

[a]Modified from Berger and Kimmel 1987.

The process of transformation is not entirely understood. The various agents affect the bacterial cell wall and, in the case of $CaCl_2$, also may be responsible for binding DNA (the plasmid) to the cell wall. The actual uptake of DNA is stimulated by the brief heat shock. It is known that large DNA molecules are taken up less efficiently than smaller ones. The efficiency of transformation varies with the strain of *E. coli* used, and is typically expressed as the number of transformant cells per microgram ($\mu$g) of plasmid DNA. Various protocols produce efficiencies of $10^7$ or $10^8$ transformants per $\mu$g of plasmid DNA.

**Electroporation** also can be used to insert DNA into bacterial cells. Electroporation involves disrupting the cell membrane briefly by an electrical current so that DNA can be incorporated.

## Purifying Plasmid DNA from *Escherichia coli*

Removing plasmids from *E. coli* is necessary if experiments are to be conducted on the now-cloned DNA. The trick is to lyse the *E. coli* cells just sufficiently so the plasmids can escape without too much contamination by the bacterial chromosomal DNA. If the lysis of the bacterial cell is done gently, most of the bacterial chromosomal material that is released will be of higher molecular weight than the plasmids and can be removed, along with the cell debris, by complexing with detergents and high-speed centrifugation. The plasmid DNA, which is left in the clear liquid remaining, is then extracted by one of two methods.

1. Cesium chloride centrifugation with **ethidium bromide** (EtBr) yields bands in the centrifuge tube that contain chromosomal and plasmid DNA at different levels due to the different densities of linear and supercoiled DNA in the presence of ethidium bromide. Ethidium bromide stains DNA by intercalating between the double-stranded DNA base pairs and in so doing causes the DNA to unwind (Figure 6.4). A plasmid DNA molecule that has not been nicked is a circular double-stranded supercoil which has no free ends and can only unwind to a limited extent, thus limiting the amount of EtBr that it binds. A linear DNA molecule, such as fragmented bacterial chromosomal DNA, can bind more EtBr and become stiffer, extending the molecule and reducing its buoyant density.

2. A second method for extracting and purifying plasmid DNA exploits the observation that within a pH range of 12.0–12.5, linear DNA will completely denature, but closed circular (plasmid) DNA will not. Plasmid-containing bacteria are treated with lysozyme to weaken the cell wall and then lysed with sodium hydroxide and sodium dodecyl sulfate (SDS). The chromosomal DNA is denatured, but upon neutralization with acidic potassium acetate, it renatures and aggregates to form an insoluble network. The high concentration of potassium acetate also causes precipitation of protein–SDS complexes and high-molecular-weight RNA. If the pH of the alkaline denaturation step has been controlled carefully, the plasmid DNA molecules will remain circularized and in solution while the contaminating molecules precipitate. The precipitate can be removed by centrifugation and the plasmid purified and concentrated by ethanol precipitation (Table 6.7).

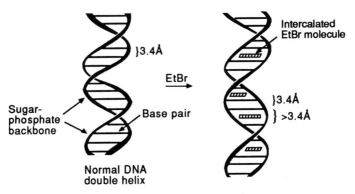

A) Ethidium bromide

B) Intercalation of EtBr into the double helix

**Figure 6.4**   Ethidium bromide (EtBr) can intercalate into DNA and cause the DNA to unwind. DNA with EtBr will fluoresce if exposed to UV radiation. EtBr is used to visualize DNA fragments on gels after electrophoresis. It can also serve as a mutagen.

## Electrophoresis in Agarose and Acrylamide Gels

DNA and RNA molecules are separated by size and visualized by agarose or acrylamide **gel electrophoresis**. Gel electrophoresis provides a powerful method for resolving mixtures of single- or double-stranded nucleic acid molecules. The nucleic acids can be directly visualized *in situ* in the gel and also can be recovered from the gel by a variety of methods determined by the subsequent steps in a particular protocol.

At a pH near neutrality, linear DNA is negatively charged and migrates from cathode to anode in a gel with a mobility dependent on fragment size, voltage applied, composition of the electrophoresis buffer, base composition, gel

**Table 6.7**

Small-Scale Plasmid Preparations (Minipreps)

1. Prepare 5-ml transformed *E. coli* cultures in LB broth containing the appropriate antibiotic. The cultures can be grown in disposable 14-ml plastic centrifuge tubes by picking colonies with a sterile toothpick and dropping the toothpick into the tube. Cap the tube and incubate at 37°C with shaking (250 rpm) for 16 hours.

2. Pellet the cells by centrifugation at 5000 rpm for 5 minutes. Discard supernatant and toothpick.

3. Add 100 μl of 50 mM glucose, 25 mM Tris–HCl (pH 8), 10 mM EDTA, 2 mg/ml lysozyme (freshly prepared). Resuspend pellet and incubate for 10 minutes.

4. Add 200 μl 0.2 N NaOH, 1% SDS. Mix gently. Incubate on ice for 10 minutes. The SDS–NaOH solution must be made just prior to use and kept at room temperature. Mix 3.5 ml water, 1 ml 1 N NaOH and 0.5 ml 10% SDS.

5. Add 150 μl 3 M potassium acetate (pH 4.8). Mix gently. Incubate for 10 minutes in a freezer. A white precipitate will form.

6. Centrifuge for 15 minutes at 15,000 rpm at 4°C.

7. Pour supernatant into Eppendorf tubes and fill with cold ethanol. Incubate in ice for 10 minutes.

8. Pellet the DNA for 1 minute in a microcentrifuge and aspirate off supernatant. Add 0.5 ml cold 70% ethanol and aspirate off.

9. Dry under vacuum. Resuspend in 50 μl distilled water containing 10 μg/ml pancreatic ribonuclease (RNase) to remove RNA.

10. The DNA is suitable for restriction analysis or fragment preparation. Use 5–10 μl per reaction. If the DNA does not cut well, it can be extracted with phenol and precipitated with ethanol.

concentration, and temperature. **Agarose** is used for longer DNA molecules and **polyacrylamide** for shorter. Nondenaturing polyacrylamide gels can be used to separate dsDNA fragments between 6 bp (20% acrylamide) and 1000 bp (3% acrylamide) in length. Nondenaturing agarose gels can separate DNA fragments between 70 bp (3% agarose) and 10,000 bp (0.1% agarose). Single-stranded DNA can be separated by agarose or polyacrylamide gel electrophoresis by including a denaturing reagent in the gel.

DNA molecules up to 2000 kb in length can be resolved in conventional agarose gel and buffers by providing electrical fields in alternately pulsed and perpendicular orientation. By this technique, often called **pulsed field gel electrophoresis**, it is possible to separate intact yeast chromosomes. Unfortunately, intact insect chromosomes appear to be too large to be separated by this technique.

### Agarose Gels

The resolving power of agarose is a function of the concentration of the dissolved agarose. The migration rate of linear nucleic acids through agarose depends upon the molecular size, conformation, and voltage gradient. The ability to detect the conformation of the DNA (circular, nicked circle, or linear plasmids) is often useful. DNA from 60 to 0.1 kb can be detected with gels containing different percentages of agarose. Thus, it is possible to separate DNA that is 0.1–3 kb long in 2% gels, DNA that is 0.8–10 kb long in 0.7% gels, and DNA that is 5–60 kb long in 0.3% gels. Agarose gels are usually electrophoresed at room temperature, except for low percentage agarose gels (<0.5%), which are easier to handle at cooler temperatures, and low melting temperature agarose gels, which may melt if run too fast at room temperature.

Agarose powder for agarose gel electrophoresis comes in grades which vary in purity and melting temperature. Type II agarose is generally used, although it contains contaminants which coelute with DNA and inhibit most commonly used enzymes. This means that DNA must be purified following elution from this gel if it is to be ligated or cut with restriction enzymes. An alternative involves using high-quality, low-melting-temperature agarose, which melts at 65°C and sets at 30°C. Low-melt agarose allows DNA to remain double stranded and also allows many enzymes to be used in the liquid agar.

### Polyacrylamide Gels

Polyacrylamide gels result from polymerization of acrylamide monomers into linear chains and the linking of these chains with N, N'-methylene bis-acrylamide (often called bis). The concentration of acrylamide and the ratio of acrylamide to bis determine the pore size of the resultant three-dimensional network and thus its sieving effect on nucleic acids of different sizes. Polyacrylamide gels can be used to purify synthetic oligonucleotides, isolate or analyze DNA less than 1 kb in size, and resolve small RNA molecules by two-dimensional or pulsed field gel electrophoresis. Polyacrylamide gels are also used for sequencing DNA. Polyacrylamide gel ingredients are highly toxic and should not be inhaled or touched without gloves.

## Detecting, Viewing, and Photographing Nucleic Acids in Gels

Ethidium bromide is the dye of choice for both single- and double-stranded nucleic acids in agarose and polyacrylamide gels (Figure 6.4). Agarose gels are somewhat less sensitive in detecting small amounts of DNA than are polyacrylamide gels. The sensitivity for single-stranded DNA is 5- to 10-fold less. Ethidium bromide is often incorporated into the gel and running buffer during

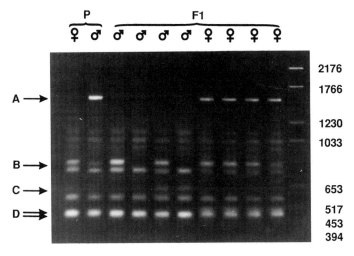

**Figure 6.5**   Photograph of DNA stained with ethidium bromide and illuminated with UV light.

electrophoresis. Alternatively, the gels can be stained after electrophoresis by placing them in buffer containing ethidium bromide for 30 minutes.

As little as 0.05 μg of DNA can be visualized in one band when the gel is exposed to ultraviolet (UV) light (Figure 6.5). The EtBr–nucleic acid complex absorbs UV irradiation at about 260 nanometers (nm) or 300 nm. The fluorescence of ethidium stacking in duplex DNA is 10 times greater than that of free EtBr and the emission is at 590 nm (red-orange). Commercially available UV sources (transilluminators) are available for viewing DNA stained with EtBr in 254- or 366-nm models. While the short-wave model can detect smaller amounts of sample DNA, it damages DNA by nicking or dimerization, making it unsuitable for most cloning work. Safety glasses and a face mask must be worn by the experimenter around UV light sources.

Agarose and polyacrylamide gels can be dried by heating under vacuum in commercially available dryers and stored for reference. Drying must be performed prior to autoradiography, and agarose gels should not be heated above their melting temperature!

## Identifying DNA by Southern Blot Analysis

It is often necessary to identify specific DNA sequences. Southern (1975) developed a method for detecting specific DNA fragments that are complementary to a specific RNA or DNA sequence in an agarose gel. This method provided molecular biologists with an essential tool. Variations on 'Southern blotting' have been developed to identify RNA and proteins in gels. These modifications have been called **Northern** (RNA) and **Western** (protein) **blotting**, respectively.

In Southern blotting, DNA fragments in an agarose gel are denatured into a single-stranded form by alkali treatment (Figure 6.6). The gel is then laid on top of buffer-saturated filter paper. The top of the gel is covered with a nitrocellulose filter membrane. This membrane is then overlaid with dry filter paper. Additional layers of dry filter paper or absorbent papers are stacked on top. Buffer passes out of the gel, drawn by capillary action into the dry filter papers. As the buffer moves from the gel, it elutes some of the denatured DNA from the gel. When the single-stranded DNA comes in contact with the nitrocellulose lying on top of the gel, it binds. The blotting process is carried out over several hours and results in the transfer of part of the DNA from the gel to the nitrocellulose membrane. It produces a pattern of bands on the membrane surface that resembles the original bands in the gel, with a minimal loss of resolution. The stack of filter papers is then removed and the nitrocellulose membrane is baked at 80°C in a vacuum to bind the DNA permanently on the nitrocellulose. An alternative method for transferring the DNA to a membrane involves electroblotting, which requires specialized equipment.

Determining whether the DNA of interest is present on the blot requires probing the DNA on the nitrocellulose membrane with radiolabeled **probes** (Figure 6.6). The probes can be radiolabeled RNA, single-stranded DNA, or a synthetic oligonucleotide which is complementary in sequence to the DNA of interest. The probe must bind specifically to the DNA of interest but not to the nontarget DNA or the nitrocellulose in a nonspecific manner. To prevent nonspecific binding, especially by single-stranded DNA probes which have a high affinity for the nitrocellulose, the nitrocellulose with the bound DNA of interest is pretreated by placing it in a solution containing 0.2% each of Ficoll (an artificial polymer of sucrose), polyvinylpyrrolidone, and bovine serum albumin (also known as Denhardt's solution). The mixture often includes an irrelevant nucleic acid such as salmon sperm DNA, which may act by occupying all the available nonspecific binding sites on the membrane.

The temperature at which Southern blotting is conducted is adjusted to maximize the rate of hybridization of the probe with the immobilized DNA on the nitrocellulose and to minimize the amount of nonspecific binding. After the hybridization step, in which the radiolabeled probe binds to the immobilized DNA on the membrane, the membrane is washed to remove any unbound probe. The temperature at which the washing takes place also determines the stringency of the Southern blot. The regions on the membrane where hybridization took place are detected by placing the membrane in contact with X-ray film. The length of time the X-ray film is exposed to the radioactivity can vary, and is determined by the amount of DNA in the blot and the degree of homo-

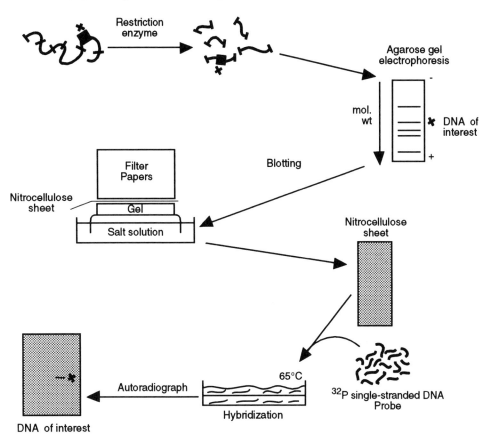

**Figure 6.6**   Outline of a Southern blot procedure. DNA is cut with restriction enzymes, electrophoresed, and blotted onto nitrocellulose by capillary action. The nitrocellulose sheet containing the DNA is baked to bind the DNA to the nitrocellulose. Specific DNA is identified by the binding of a radiolabeled probe (here $^{32}$P-labeled single-stranded DNA) in a hybridization procedure. Excess probe is washed off and the nitrocellulose sheet is then exposed to X-ray film to visualize the location of the DNA that contains sequences homologous to the radiolabeled probe. The conditions under which hybridization occurs can be varied to increase or decrease the specificity of the reaction between DNA and probe. (Adapted from Williams and Patient 1988, by permission of Oxford University Press.)

logy between the DNA and the probe. If there is only a small amount of DNA present, as would be expected for a single-copy gene in a blot of genomic DNA, the film may be exposed to the blot for up to three weeks.

Southern blots can detect single-copy gene sequences in genomic DNA digests. Recent modifications of the Southern blot method have employed nylon membranes as substrates because they are more robust and can be reused. Thus, one probe can be removed by high-temperature washing, and the DNA can be reprobed with a different radiolabeled probe. Another advantage to nylon membranes is that the DNA can be permanently fixed to the membrane by a brief exposure to UV light, which cross-links the DNA and fixes it to the membrane. In contrast, fixing to nitrocellulose requires baking for an hour or two. Another advance has been the use of electrophoretic transfer of DNA from the gel (electroblotting) to the membrane rather than relying on capillary blotting.

## Labeling DNA or RNA Probes

In molecular biology, many DNA manipulation techniques depend on hybridizing a nucleic acid probe to a target DNA or RNA sequence. Probes are required in Southern and Northern blots, dot blots, colony or plaque blots, and *in situ* hybridization. Dot blots can be used to identify unfractionated DNA or RNA molecules that have been immobilized on a nitrocellulose membrane. Plaque or colony blots detect DNA released from lysed bacteria or phage after immobilizing the DNA on a nitrocellulose membrane. **In situ hybridization** is employed to detect DNA or RNA molecules in cytological preparations.

Nucleic acid probes can be radiolabeled by several methods, as described in Chapter 7. Only one method to uniformly label dsDNA probes is described here. **Nick translation** consists of the translation or movement of a nick (or break in one strand) of a dsDNA molecule (Figure 6.7). Nicks are introduced at widely separated, random sites along the DNA molecule by treating the DNA with small amounts of DNase. A nick exposes a free 3'-OH group, and DNA polymerase I of E. *coli* will then remove nucleotides from the 5' side of the nick. The simultaneous removal of nucleotides from the 5' side and the addition of nucleotides to the 3' side by DNA polymerase I results in movement of the nick along the DNA. If dNTPs are radiolabeled with $^{32}P$ and the nicks are random, the duplex DNA molecule will become labeled uniformly along its length as it incorporates radiolabeled dNTPs. The reaction may be carried out with all four dNTPs labeled or only one radiolabeled. Nick translation is particularly useful for producing large amounts of probe for multiple hybridization reactions and where a high probe concentration is required.

Nick translation kits are available from a number of commercial sources and contain instructions, a stock mixture of DNA polymerase I and DNase I, and a series of buffers lacking one or more unlabeled dNTPs. The radioactive dNTPs must be obtained fresh within a few days before the labeling reaction is set up because of the rapid decay of the $^{32}P$-labeled dNTPs.

## Removing DNA from Agarose Gels after Electrophoresis

Several methods have been developed to recover DNA from agarose gels, including (1) electrophoresis onto a DEAE-cellulose membrane, (2) electroelu-

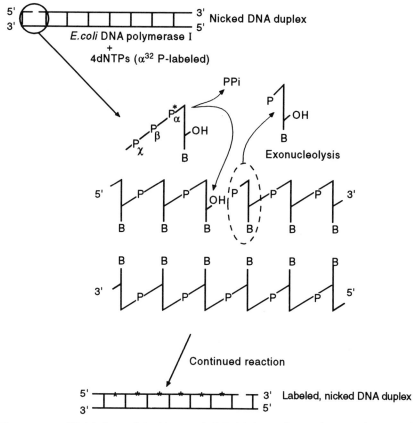

**Figure 6.7** $^{32}$P-labeling of double-stranded DNA by nick translation. The asterisks indicate the location of radiolabeled phosphate groups. (Adapted from Old and Primrose 1980.)

tion into dialysis bags, and (3) use of low-melting-temperature agarose gels. With method 1, fragments of DNA are separated by electrophoresis, a slit is cut in the gel immediately ahead of the DNA fragment of interest, and a sliver of DEAE-cellulose membrane is inserted. Electrophoresis is continued until the DNA in the band has been transferred to the membrane. The membrane is removed from the slit and washed, and the DNA is eluted from the membrane.

Method 2, electroelution, allows recovery of DNA of a wide size range, but is inconvenient. DNA is separated by electrophoresis in agarose gel containing ethidium bromide. The band of interest is located with an ultraviolet lamp, and is cut from the gel with a razor blade. The gel fragment containing the DNA of interest is then placed in a piece of dialysis tubing, sealed, and placed into an electrophoresis tank. Electric current is passed through the bag to elute the

DNA out of the gel and onto the inner wall of the bag. The polarity of the current is reversed to release the DNA from the wall of the bag, the bag is opened, and all the buffer containing the DNA is transferred to a clean tube.

Method 3 uses low-melting-temperature agarose gels. DNA of interest is electrophoresed, the band of interest is detected by staining with ethidium bromide, cut out, and placed into a clean tube. A buffer is added, the mixture is heated to 65°C to melt the agarose, and centrifuged, which will leave the DNA in the aqueous phase and the agarose in the interface. The DNA in the aqueous phase can be purified with phenol. The DNA in the aqueous phase is then precipitated by ammonium acetate and cold ethanol. At this point the DNA is sufficiently pure to be digested by restriction enzymes or modified by ligases.

None of the three methods are fully satisfactory in producing large amounts of pure DNA. Problems include the presence of enzymatic inhibitors in the agarose, which can affect subsequent DNA manipulations. Large fragments ($>5$ kb) of DNA are often inefficiently recovered from agarose gels because these longer fragments bind very tenaciously to purification matrices such as the diethylaminoethyl (DEAE)-cellulose membrane. Small ($<500$ nanograms or ng) amounts of DNA also are recovered inefficiently from gels. The methods are labor intensive so that recovery of DNA from gels is not readily performed on large numbers of samples. Various kits also are available from commercial sources that can purify DNA from gels.

## Restriction Site Mapping

So far in this experiment, DNA has been cloned into a plasmid, amplified in *E. coli*, and a specific sequence(s) have been identified by Southern blot analysis using a radiolabeled probe. Specific bands of DNA have been isolated and the DNA purified from the agarose gel by one of three methods.

Information about the cloned DNA now can be obtained by restriction site mapping, DNA sequencing, and translation of the DNA into proteins using an expression vector. DNA sequencing is described in Chapter 8, and translation and expression of DNA are discussed in Chapter 7. **Restriction site mapping** is a relatively simple technique that provides a physical map of the sites in the DNA at which different restriction enzymes cut.

One method for constructing restriction maps involves digesting the DNA with a series of single restriction endonucleases in separate reactions. The products of each digestion are electrophoresed on agarose or polyacrylamide gels. DNA marker fragments of known size are electrophoresed in lanes adjacent to the DNA being examined to provide estimates of the lengths of the DNA fragments generated. DNA molecular markers of known size are available from commercial sources. The DNA is stained with ethidium bromide and the bands that were produced are examined under ultraviolet light. The banding patterns can be photographed (Figure 6.5).

After the single digestions are done, the DNA can be digested simultaneously with two restriction enzymes. Again, the size of the digestion products is analyzed by gel electrophoresis, using not only size markers but the samples of the first digest for comparison. If the products of the first and second digests are electrophoresed in adjacent lanes on the gel, it is possible to detect small differences in migration rate. Maps are built up from these data by a process of trial and error and basic logic. Based on the sizes of the DNA fragments generated, it is possible to define the order of the restriction sites and the distances between them. The resolution of map distances depends on the accuracy with which the sizes of the DNA fragments can be estimated relative to those of the size markers. However, restriction maps are rarely accurate to less than 100–200 bp. Restriction maps of DNA provide the experimenter with useful information for additional experiments. Furthermore, such restriction site maps can be used as in systematics or population genetics studies (Chapters 13, 14).

Many of the techniques described in this chapter can be used for other purposes, including preparing a genomic library, as described in Chapter 7.

## General References

Ausubel, F. M., R. Brent, R. E. Kingston, D. M. Moore, J. G. Seldman, J. A. Smith and K. Struhl, Eds. *Current Protocols in Molecular Biology*. Greene Publ. Assoc., New York. These two large volumes offer research protocols. The publisher has a subscription update service so that new pages can be inserted into the binder to replace outdated material or expand on existing information.

Berger, S. L. and A. R. Kimmel, Eds. 1987. *Guide to Molecular Cloning Techniques* (Methods Enzymol., Vol. 152). Academic Press, San Diego.

Brown, T. A., Ed. 1991. *Essential Molecular Biology: A Practical Approach*. Vols. I and II. IRL Press, Oxford.

Brown, T. A., Ed. 1991. *Molecular Biology LabFax*. Bios Sci. Publ., Oxford.

Glover, D. M. 1984. *Gene Cloning: The Mechanisms of DNA Manipulation*. Chapman and Hall, London.

Hames, B. D. H. and S. J. Higgins, Eds. 1985. *Nucleic Acid Hybridisation—A Practical Approach*. IRL Press, Oxford.

Old, R. W. and S. B. Primrose. 1989. *Principles of Gene Manipulation: An Introduction to Genetic Engineering*. 4th Ed. Blackwell Scientific, Oxford.

Sambrook, J., E. F. Fritsch and T. Maniatis. 1989. *Molecular Cloning: A Laboratory Manual*. 2nd Ed. Cold Spring Harbor Press, Cold Spring Harbor, NY. Three volumes provide detailed protocols of methods.

Sealey, P. G. and E. M. Southern, Eds. 1982. Gel electrophoresis in DNA Pp. 39–116. In: *Gel Electrophoresis of Nucleic Acids—A Practical Approach*, D. Rickwood and B. D. Hames, Eds. IRL Press, Oxford.

Williams, J. G. and R. K. Patient. 1988. *Genetic Engineering*. IRL Press, Oxford.

## References Cited

Balbas, P., X. Soberon, E. Merino, M. Zurita, H. Lomeli, F. Valle, N. Flores and F. Bolivar. 1986. Plasmid vector pBR322 and its special-purpose derivatives—A review. Gene 50: 3–40.

Beye, M. and U. Raeder. 1993. Rapid DNA preparation from bees and %GC fractionation. Bio-
Techniques 14: 372–373.

Henry, J. M., A. K. Raina and R. L. Ridgway. 1990. Isolation of high-molecular-weight DNA from
insects. Anal. Biochem. 185: 147–150.

Jowett, T. 1986. Preparation of nucleic acids. Pp. 275–286. In: Drosophila: A Practical Approach. D.
B. Roberts, Ed. IRL Press, Oxford.

Kessler, C. and V. Manta. 1990. Specificity of restriction endonucleases and DNA modification
methyltransferases—A review (Edition 3). Gene 92: 1–248.

Pingoud, A., J. Alves and R. Geiger. 1993. Restriction enzymes. Pp. 107–200. In: Methods in
Molecular Biology. Vol. 16. Enzymes of Molecular Biology. M. M. Burrell, Ed. Humana Press,
Totowa, NJ.

Pinol, J., O. Fancino, A. Fontdevila and O. Cabre. 1988. Rapid isolation of Drosophila DNA. Nucleic
Acids Res. 16: 1726.

Post, R. J., P. K. Flook and A. L. Millest. 1993. Methods for the preservation of insects for DNA
studies. Biochem. System. Ecol. 21: 85–92.

Smith, H. O. and D. Nathans. 1973. A suggested nomenclature for bacterial host and modification
and restriction systems and their enzymes. J. Mol. Biol. 81: 419–423.

Southern, E. M. 1975. Detection of specific sequences among DNA fragments separated by gel
electrophoresis. J. Mol. Biol. 98: 503–517.

# Chapter 7

# Cloning and Expression Vectors, Libraries, and Their Screening

## Overview

DNA cloning has five essential components: (1) a method for generating DNA fragments, (2) reactions to join exogenous DNA fragments to a vector, (3) a method to introduce the vector into a host cell where the vector ensures the DNA is replicated, (4) methods for selecting or identifying the recombinant molecules, and (5) methods for analyzing the cloned DNA. This chapter describes how to develop a genomic library in λ phage vectors, or produce a complementary DNA (cDNA) library, and clone into cosmids or phagemids or single-stranded M13 phage. Screening libraries for genes of interest can be accomplished by nucleic acid or antibody hybridizations and chromosome walking. Cloning is feasible because a diverse array of enzymes are available to synthesize, ligate, and modify the ends of DNA molecules. The cloning projects described in this chapter illustrate techniques and approaches used and provide the basis for identifying specific genes, producing DNA copies of mRNA, and, in some cases, producing gene products (proteins) in E. coli by incorporating the DNA into expression vectors. Two insect baculoviruses have been genetically engineered as expression vectors and are used to produce proteins in insect cells in tissue culture or even in intact lepidopteran larvae.

## Introduction

The goal in cloning is to multiply cells that contain a single vector molecule that contains exogenous DNA. A **vector** is the agent used to replicate, or multiply, the exogenous DNA. Vectors are segments of DNA with an **origin of replication** so that the vector is replicated after it is introduced into the host cell. Vectors can be either plasmids, bacteriophages, or hybrid structures called cosmids or phagemids.

Chapter 6 introduced the use of plasmid vectors. This chapter introduces four additional vectors derived from the *E. coli* **bacteriophage** λ; the single-stranded DNA bacteriophage **M13**; engineered, hybrid vectors combining components from bacterial plasmids and λ that are called **cosmids**; and **phagemids**, engineered, hybrid molecules that combine elements of plasmids and M13 vectors. The most commonly used host cell is *E. coli* but other hosts are used, including bacteria such as *Bacillus subtilis* or the yeast *Saccharomyces cerevisiae*. Insect tissue culture cells also can be used as hosts for **baculovirus** expression vectors.

A multitude of sophisticated vectors have been developed for cloning and many are commercially available and designed for special purposes. The choice of an appropriate vector depends upon the goal of the experiment. Developing a new vector requires an extensive knowledge of the biology and genetics of *E. coli*, the plasmid or bacteriophage, and enzymology. It is impossible to describe all existing vectors and their uses in this book and, furthermore, such a description would become obsolete rather rapidly. There are at least 100 commercially available vectors, with new vectors engineered and advertised regularly.

Basically, a vector is a segment of DNA with an origin of replication which allows it to be maintained stably after it is introduced into its host cell. Most vectors contain unique restriction sites in a region that contains nonessential genetic information. The unique restriction sites are where foreign DNA fragments can be inserted into the vector. Often several cloning or restriction sites are combined in a single region, which is called a **polylinker** or multiple cloning site.

Various plasmid vectors are available, including derivatives of pBR322, one of the most widely used vectors. It was described and its structure illustrated in Chapter 6. Another plasmid vector series is the pUC group. pUC plasmids contain a functional segment of the *E. coli lacZ* gene, which produces colonies with a blue color if provided with the appropriate substrate in the agar medium. If exogenous DNA is cloned into the cloning site located within the *lacZ* gene, the *lacZ* gene product is no longer produced, and the *E. coli* colonies are colorless rather than blue. DNA cloned into the pUC plasmids can be subcloned easily because these plasmids have two antibiotic resistance genes that can serve as selectable markers. The pUC plasmids also produce an increased copy number in *E. coli* cells, which results in an increased yield in recombinant DNA molecules compared with the pBR322 series plasmids.

Cloning can be used to produce gene libraries; to develop mutated genes for experiments; to provide single-stranded DNA for sequencing; and to permit eukaryotic genes to be translated in *E. coli*, insect tissue culture cells, or lepidopteran larvae. Figure 7.1 identifies many of the steps and procedures involved in cloning, but a full description of all the techniques employed is beyond the scope

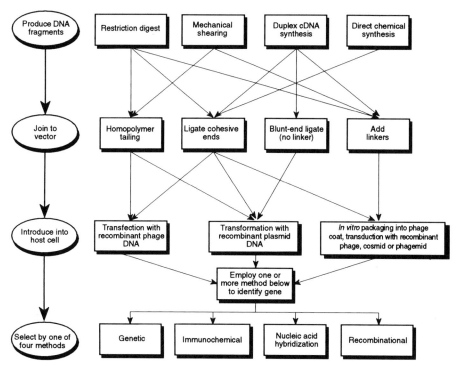

**Figure 7.1**   Generalized scheme outlining the steps employed in cloning DNA in *E. coli*. There are four major components: obtaining DNA fragments, joining them to the vector, introducing the recombinant molecule into an appropriate host cell, and identifying or selecting the recombinant DNA of interest. All of these steps can be achieved in several ways. The choice of an appropriate cloning scheme depends upon the goals of the experiment. (Revised from Old and Primrose 1989.)

of this book. References at the end of the chapter provide additional information and protocols.

## Producing a Genomic Library

The perfect **genomic library** would contain all the DNA sequences representative of the entire genome, in a stable form, and as a manageable number of overlapping cloned segments. The clones would contain sufficiently large DNA segments that they could contain whole genes and their flanking sequences. A library should be easy to construct and easy to screen for sequences of interest. Ideally, the library could be amplified without loss or misrepresentation of sequences, and stored for years without significant loss of information. Unfortunately, no single vector provides all of these desirable attributes.

Genomic DNA can be cloned into bacteriophage λ or into cosmids to make a genomic library. λ is more commonly used because cloning is relatively easy. The relative disadvantage is that the average λ insect library would contain >100,000 clones, each with an insert that averages ~15–20 kb in length. Cosmids have an advantage in that the size of the inserted DNA can be 2- to 3-fold larger and therefore fewer cosmids need to be evaluated to find the gene(s) of interest. Thus, cosmids can provide a significant advantage when it is important to work with an entire gene and its flanking sequences. Unfortunately, cosmids are more difficult to work with.

Genomic libraries constructed from random fragments of DNA are called 'shotgun' libraries. Generating a genomic library for an insect may require $10^5$ to $10^6$ clones to ensure that all sequences in the genome are represented. The formula for estimating the required size of the library is

$$N = \frac{\ln (1 - P)}{\ln (1 - a/b)},$$

in which $N$ is the number of clones required, $P$ is the probability that a given sequence will be present, $a$ is the average size of the DNA fragments inserted in the vector, and $b$ is total size of the genome.

Alternatively, before the DNA is cloned into a library, it may be enriched in some manner to increase the frequency of the desired DNA sequences in the starting material. In any case, the library can be read only if there is a key to open it. The key to libraries is some sort of probe, as described later.

### λ Phage as a Vector

**Lambda** (λ) is a genetically complex, but well-studied, double-stranded DNA bacteriophage of *E. coli*. The entire sequence of the DNA molecule of λ has been determined. Of the 48.5 kb, nearly 40% of the DNA is not essential for propagating the phage in its host. If this nonessential DNA is removed, exogenous DNA can be inserted. At each end of the linear DNA molecule there are short, single-stranded 5′-projections of 12 nucleotides, called **cos sites**, that are complementary to each other in sequence.

The *cos* sites enable the λ chromosome to circularize after the linear phage is injected into its *E. coli* host (Figure 7.2). After replication within the host cell, the λ DNA is in a linear form when it is packaged into a protein coat. The protein coat consists of an icosahedral head and a tail that ends in a tail fiber (Figure 7.2). The infective phage thus consists of the DNA molecule plus a protein head and tail. The protein coat is important because phage particles adsorb to their host cells by the tip of the tail fiber to receptor sites in the outer

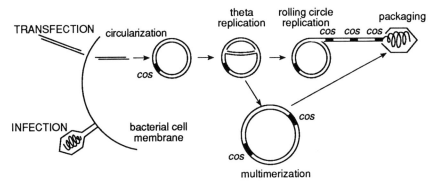

**Figure 7.2** The typical lytic cycle of bacteriophage λ begins when the phage adsorbs to an *E. coli* cell by the tail fiber. The linear DNA molecule is injected into the host cell where it circularizes by annealing the *cos* sites. The λ DNA replicates, producing long, concatenated molecules with the individual chromosomes annealed at the *cos* sites. Later in the infection cycle, phage genes for the protein head and tail are turned on. When these components are produced, the DNA is packaged into the icosohedral head after the DNA is cut at the *cos* sites. If the DNA is too long or too short, it will not be packaged in the head, but will be lost. Lysis of the host cell then occurs, typically releasing approximately 100 progeny phages.

membrane of the cell. Adsorption is temperature independent and also dependent on the presence of magnesium ions. A wild type λ bacteriophage has two different phases to its life cycle. While one stage (the temperate or **lysogenic**) is of little interest to the genetic engineer, λ has been genetically modified into a family of vectors and retains many characteristics of the second, or **lytic**, phase.

Wild-type strains of λ may either be lytic or lysogenic. In the lytic phase, early DNA transcription establishes the lytic process, middle genes replicate and recombine the DNA, and late genes produce protein for packaging this DNA into mature phage particles. Phage DNA is replicated in a "rolling circle" mode, as illustrated in Figure 7.2. Multiple copies of the replicated DNA molecules are assembled in a linear tandem array, with the termini of each molecule joined at the *cos* sites. The *cos* sites form the recognition site of a specific endonuclease which subsequently cuts these DNA molecules during the packaging process so that a single DNA molecule is inserted into the head of the protein coat. In summary, in a lytic infection, the phage takes over the host cell: phage DNA is replicated, head and tail proteins are made, the replicated DNA is packaged, and the host cell is lysed to release about 100 infective particles.

In the temperate or lysogenic phase, most phage functions become repressed and lysis is avoided. In lysogeny, the λ DNA is inserted into the host chromosome by site-specific recombination and the phage genome (prophage) is repli-

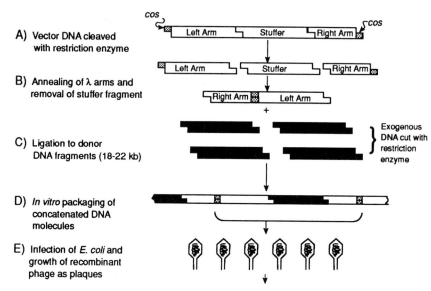

A) Vector DNA cleaved
   with restriction enzyme

B) Annealing of λ arms and
   removal of stuffer fragment

C) Ligation to donor
   DNA fragments (18-22 kb)

D) *In vitro* packaging of
   concatenated DNA
   molecules

E) Infection of *E. coli* and
   growth of recombinant
   phage as plaques

**Figure 7.3** (A) Schematic outline of a bacteriophage λ replacement vector. A linear molecule contains the *cos* sites, a left and right arm, and a 'stuffer' region with nonessential DNA. (B) The vector is digested with an appropriate restriction enzyme, the stuffer fragment is removed, and the two arms anneal. (C) Exogenous DNA that has been cleaved with an appropriate restriction enzyme is added, and the fragment is ligated in. (D) Exogenous DNA fragments 18–22 kb long can be incorporated because these molecules can be successfully packaged into the λ coat by *in vitro* packaging. (E) *E. coli* is infected with the λ and thousands of individual plaques are produced. Each plaque contains many thousands of replicas of a single phage containing exogenous DNA.

cated as part of the *E. coli* chromosome. However, nearly all λ vectors used in genetic engineering lack the ability to enter the lysogenic phase.

λ has been genetically improved as a vector in two ways: (1) Genes in the central region of the λ chromosome that code for events associated with recombination and lysogeny can be deleted or replaced, allowing exogenous DNA to be carried (Figure 7.3). (2) Vectors are engineered to contain DNA sequences that can be cut in specific sites by unique restriction enzymes to facilitate the insertion of exogenous DNA. Engineered versions of λ are of two major types. **Insertion vectors** have a single target site at which foreign DNA can be inserted, while **replacement vectors** have a pair of sites defining a fragment that can be removed and replaced by foreign DNA (Figure 7.3).

Once exogenous DNA has been inserted into the vector λ, this molecule can be multiplied by inserting it into host *E. coli* cells in one of two ways. Naked λ DNA molecules lacking a protein coat can be introduced into *E. coli* cells in a process called **transfection**. Transfection is the infection of bacteria by viral nucleic acid alone. The efficiency of transfection is $>10^4$ recombinant clones per µg of donor DNA. This efficiency would suffice for the construction of genomic libraries from species with small genomes. However, larger genomes, such as those of insects, require a more efficient method of inserting the vector DNA into *E. coli* or the amount of DNA isolated for cloning could become a limiting factor.

The way to increase efficiency in introducing recombinant λ DNA molecules into *E. coli* is called *in vitro* **packaging**. By incorporating the recombinant DNA molecules into phage protein coats, *E. coli* can be infected much more readily, and this increases the likelihood of producing complete genomic libraries. *In vitro* packaging provides the genetic engineer with the efficiency of the intact phage to penetrate its bacterial host. Efficiency of infection of *E. coli* with packaged DNA can be $10^6$ recombinants per µg of vector DNA, an increase in efficiency over transfection by nearly two orders of magnitude.

*In vitro* packaging involves a number of steps and specific conditions (Figure 7.4). One critical condition is the size of the inserted DNA. If λ DNA is to be packaged efficiently into a protein coat, it is important that the amount of exogenous DNA inserted into the vector be regulated carefully. The *cos* sites must be separated by DNA that is 78 to 105% of the length of the wild-type DNA, or approximately 50 kb. There is an upper and a lower limit to how much exogenous DNA can be inserted into λ vectors. In an insertion vector, only 14 kb of DNA or less can be cloned. In a replacement vector, up to 22 kb of DNA can be inserted. In replacement vectors, a pair of restriction sites flank the nonessential central region of the phage DNA called the 'stuffer region'. When the stuffer region is excised and the insert DNA is ligated into this region, a DNA molecule is produced that can be packaged efficiently.

*In vitro* packaging requires the following components: (1) the DNA molecules to be packaged, (2) high concentrations of phage head precursor, (3) proteins that participate in the packaging process, (4) and phage tails. These packaging ingredients are obtained by combining a very concentrated mixture of the lysate from two different λ strains that are lysogenic. One mutant λ strain can progress no further in the packaging process than the prehead stage because it carries a mutation in a gene (gene *D*) and therefore accumulates this precursor. The other mutant λ strain is prevented from forming any head structure by a mutation in a different gene (gene *E*), but can produce the tail component. In the mixed lysate, both head and tail components become available so that a complete phage can be assembled that contains recombinant DNA.

Transfer of DNA into the *E. coli* host involves adsorption of phage to specific receptor sites in the outer membrane of the *E. coli* cell. Because phage will adsorb to dead cells and debris, healthy bacterial cultures should be used for

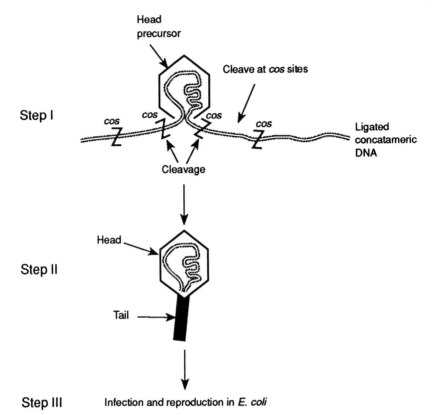

**Figure 7.4**  *In vitro* packaging of cloned DNA in λ involves providing a protein head and tail precursor. The DNA is cut at the *cos* sites, and if the DNA is about 50 kb in length, it will be packaged in the head. The completed phage is then used to infect *E. coli* and amplify the recombinant molecule.

transformation or efficiency will be reduced. Once the *E. coli* have taken up the phage, they are plated out on nutrient agar and allowed to grow at least overnight at 37°C. Bacterial colonies begin to grow, but if they are infected with λ, clear areas (**plaques**) of lysed cells will be seen, surrounded by an opaque background of unlysed bacteria (Figure 7.5). Each plaque represents a bacterial cell that has been infected, ideally only by a single λ. Thus, each plaque should contain multiple copies of a *single* kind of recombinant DNA molecule. Even the smallest plaque is likely to contain sufficient phage DNA to be detectable by **plaque hybridization**, as described later.

A visual method for identifying plaques containing λ with recombinant DNA is available. Many λ vectors carry the *lacZ* gene. This gene codes for part of

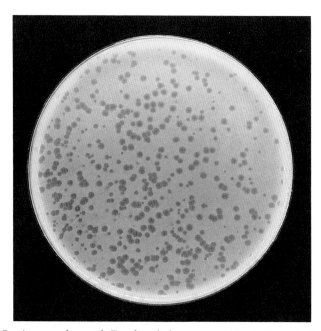

**Figure 7.5** An agar plate with E. *coli* and plaques caused by the bacteriophage λ. Each clear area indicates where a single bacterium initially was infected with λ. After replication, the emerging λ attacked adjacent E. *coli* and lysed them, resulting in a clear "plaque" of killed E. *coli* and λ on the surface of the agar.

the β-galactosidase enzyme, which cleaves lactose to produce glucose and galactose. Inserting exogenous DNA into this gene inactivates synthesis of β-galactosidase, making it easy to identify recombinants in transformed E. *coli* cells growing on agar. To identify E. *coli* colonies containing recombinant phage or plasmids, the agar is made up with a lactose analog called **X-gal** (5-bromo-4-chloro-3-indolyl-β-D-galactopyranoside). The X-gal is cleaved by β-galactosidase into a product that is bright blue. If exogenous DNA has inserted into and disrupted the b-galactosidase gene, plaques will appear white or colorless. Plaques without recombinant vectors will be blue.

Figure 7.6 outlines the steps involved in one strategy for producing a representative genomic library in a λ replacement vector. The genomic DNA and the vector DNA can be prepared simultaneously. In this example, the genomic DNA is cut with a mixture of *Hae* III and *Alu* I in a reaction so that the DNA is only partially digested. The genomic DNA is then sized so that fragments about 20 kb long are available. Meanwhile, the λ vector DNA is digested with the restriction enzyme *Eco* RI. It is also purified so that the stuffer fragments are

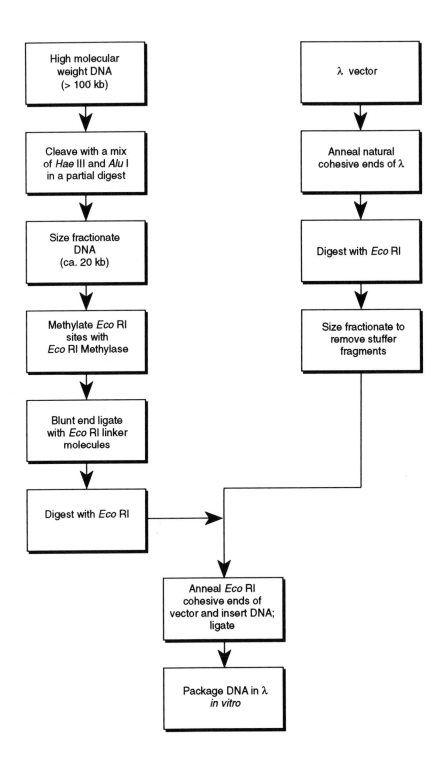

removed. The genomic DNA has linker molecules added to it prior to the annealing reaction. When the genomic DNA and the vector DNA are combined, they anneal at their complementary cohesive ends and are ligated together. The last step involves providing a protein coat for the DNA by *in vitro* packaging.

Commercial cloning kits are also available, which simplify the procedures considerably because such kits provide vectors, enzymes, *in vitro* packaging materials, and detailed protocols. It is even simpler to supply genomic DNA to a company who will provide a complete genomic library for a fee. However, a few points should be made about constructing a genomic library.

The genomic DNA to be cloned must be of high molecular weight and not excessively sheared during its isolation from the insect. High-molecular-weight DNA is needed because the DNA will be partially digested with a restriction enzyme to generate a random collection of DNA fragments and these need to be about 20 kb long. The DNA to be digested must be longer than 20 kb so that after digestion both ends of the fragment will have cohesive ends. DNA fragments with only one cohesive end (and one broken end) cannot be inserted into vectors. DNA shorter than about 20 kb won't be packaged into the phage and will be lost. Thus, the extraction should be carefully executed to avoid damaging or shearing the DNA.

The genomic DNA fragments ideally will be representative of the entire insect genome. If the restriction enzymes that are used cut relatively frequently compared with the desired fragment size, a *partial* digestion will produce a set of overlapping fragments. Ideally these fragments will be a nearly random array of the entire genome. However, it is possible that not all regions of the genome will be represented in the desired size range. Some regions of the genome may lack the appropriate cleavage sites for the enzyme being used, or the DNA may not be cleaved with equal efficiency everywhere, particularly in heterochromatic regions.

After the genomic DNA has been partially digested with an appropriate restriction endonuclease, the DNA fragments are size-fractionated by centrifugation or gel electrophoresis. This separates out the DNA fragments greater than or smaller than approximately 18–22 kb. Larger molecules could give rise to oversized recombinant molecules that cannot be packaged within the phage head, and smaller molecules could give rise to multiple inserts or fail to be packaged at all.

Preparing a representative library requires high-quality vector DNA. Large-

---

**Figure 7.6**   One method for producing a representative genomic library in a λ replacement vector. Two parallel processes are carried out: preparation of the exogenous DNA and preparation of the vector DNA. These are then ligated together and packaged *in vitro*. The specific restriction enzymes and ligation method can be varied.

scale preparation of λ DNA should yield pure preparations that lack the central 'stuffer' region or it can reinsert back into the vector later on. Removal of the central stuffer region is carried out by centrifugation, elution, or electrophoretic separation. However, it is difficult to remove all of the stuffer fragment, so it is important to determine, by appropriate controls, how often the 'stuffer' is reinserted back into the vector.

The ligation reaction must be carefully regulated by optimizing concentrations of vector and exogenous DNA. Because a portion of the DNA molecules to be ligated will have damaged ends, the ratio of vector DNA to insert DNA will probably have to be determined empirically in small reactions. It is desirable to produce long concatenated molecules that can be cut at the *cos* sites and packaged. Likewise, the appropriate ratios of ingredients used for *in vitro* packaging will have to be determined empirically. Once the DNA has been packaged, the phage can be stored at 4°C for years. Alternatively, the phage can be amplified by multiplication in *E. coli*.

Commonly used vectors derived from λ include the gt and EMBL series. λ gt10 was designed for cloning short DNA fragments, especially cDNA. λ gt11 is used to construct cDNA libraries, as described later. DNA properly aligned with the *lacZ* gene in λ gt11 will be expressed in *E. coli* as a **fusion protein**. EMBL vectors are a family of replacement vectors that provide a high level of reproduction in *E. coli*, polylinker cloning sites, and the ability to select for recombinant phage. EMBL3 and EMBL4 vectors, or their derivatives, are particularly useful for constructing genomic libraries.

### Cloning with Cosmids

**Cosmids** are engineered vectors that combine characteristics of both plasmids and phage. They have been constructed to include a fragment of λ DNA that includes the *cos* site, which is the recognition site for packaging λ DNA into the protein coat (Figure 7.7). Cloning into cosmids is similar to cloning in λ. It involves digesting exogenous DNA with a restriction enzyme, cutting the cosmid vector with a compatible restriction enzyme, combining the two, and ligating them.

Once the exogenous DNA is inserted into the cosmids, cosmids are packaged in a manner similar to that employed with λ, employing the *cos* sites. Packaging the cosmid recombinants into phage coats provides a useful method for selecting the size of the inserted DNA. What is significant about cloning with cosmids is the fact that 32–47 kb of foreign DNA can be inserted into the vector and still be packaged. Because cosmids can carry such large DNA fragments, they are particularly good vectors for constructing libraries for organisms, such as insects, with large genomes. After *in vitro* packaging, cosmids are used to infect a suitable *E. coli* strain. Infection of *E. coli* involves injection and circularization of the cosmid DNA but no phage protein is produced. Transformed *E. coli* cells are identified on the basis of their resistance to a specific antibiotic.

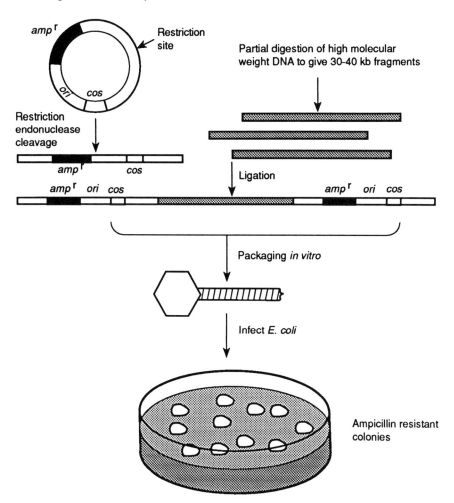

**Figure 7.7**  Outline of procedures used in cloning with a cosmid vector. This vector contains a *cos* site, a restriction site for inserting exogenous DNA, and a gene for ampicillin resistance. Exogenous DNA is cut with an appropriate restriction enzyme, as is the vector. The vector and exogenous DNA are ligated together, producing a recombinant molecule 37–52 kb long, which can be packaged in λ by *in vitro* packaging. The packaged vector infects *E. coli*, injecting its DNA into the host, where it circularizes and multiplies. *E. coli* cells that receive the cosmid are distinguished from cells that are not infected by their ability to survive on media containing ampicillin.

While having a large capacity for DNA fragments is a benefit in cloning with cosmids, it can be a detriment. If, during a partial digestion with restriction enzymes, two or more genomic DNA fragments join together in the ligation reaction, a clone could be created with fragments that were not initially adjacent to each other. This could be a problem if the researcher is interested in the relationship between a gene of interest and the DNA surrounding it on the chromosome. The problem can be overcome by size fractionating the partial digest. However, even then, cosmid clones could be produced that contain noncontiguous DNA fragments ligated to form a single insert. This problem can be solved by dephosphorylating the foreign DNA fragments to prevent them from ligating together, but this makes cosmid cloning very sensitive to the exact ratio of insert and vector DNAs. If the ratio is unbalanced, vector DNAs could ligate together without containing any exogenous DNA insert. This is resolved by treating the vector to two separate digestions, which generate vector ends that are incapable of ligating to each other after phosphatasing.

Commonly used cosmid vectors include the pJB8 and the pcosEMBL family. pJB8 is probably most useful for constructing genomic libraries. Because this vector is small, fragments up to 45 kb can be inserted into it and packaged into phage heads. The pcosEMBL family was designed to simplify isolation of specific recombinants from cosmid libraries and speed up isolation of large regions of complex genomes in an ordered array of overlapping clones (i.e., chromosome walking, which is described later). The vectors in this family differ by having different cloning sites and different numbers of *cos* sites. In our theoretical experiment, recombinant cosmids can be identified by rearing *E. coli* in the presence of the antibiotic ampicillin (Figure 7.7).

### Cloning in the Filamentous Phage M13

**M13** is a filamentous phage of *E. coli* that contains a circular single-stranded DNA molecule that is 6407 nucleotides long. M13 only infects strains of *E. coli* that have **F pili** because the site where these phage adsorb appears to be at the end of the F pilus (Figure 7.8). It is not known how the phage DNA gets from the end of the F pilus to the inside of the *E. coli* cell. Replication of M13 DNA does not result in host cell lysis. However, the infected cells grow and divide more slowly and extrude virus particles. Up to 1000 M13 phage particles may be released into the medium by each infected cell.

The life cycle of the M13 phage is different from that of λ. First, the M13 phage contains single-stranded DNA. During infection, single-stranded M13 phage DNA enters the *E. coli* cell. The DNA then is converted from a single-stranded form to a double-stranded or replicative form (RF). The double-stranded RF multiplies until about 100 RF molecules are produced within the cell (Figure 7.8). Then, the replication of the RF becomes asymmetric due to the accumulation of a viral-encoded binding protein that is specific to single-

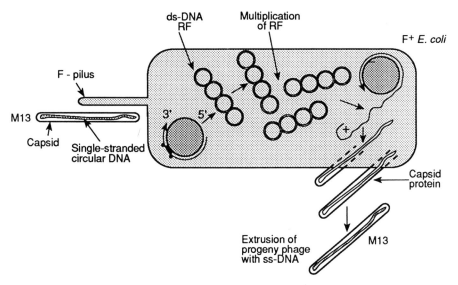

**Figure 7.8** An outline of the infection cycle of the bacteriophage M13. The single-stranded phage attaches to the F-pilus of E. *coli*, injects its DNA into the host, and begins to produce approximately 100 copies of double-stranded (RF) molecules. DNA replication then shifts to producing ssDNA molecules, which are extruded through the host cell wall, during which time they are coated with a capsid protein coat. M13 has been engineered as a vector and is used to produce ssDNA molecules, which is particularly useful in DNA sequencing reactions.

stranded DNA. The binding protein binds to the M13 DNA and prevents synthesis of a complementary strand. Subsequently, only single-stranded viral DNAs are synthesized and these are extruded from the host cell. As the single strands of M13 DNA move through the E. *coli* cell membrane, the DNA binding protein is removed and the M13 DNA is coated with capsid protein.

M13 has been engineered into a family of cloning vectors because it has a number of advantages. First, single-stranded DNA is required in a number of applications, including the dideoxy DNA sequencing method (described in Chapter 8), and site-directed mutagenesis. Second, the use of single-stranded vectors allows the genetic engineer to combine cloning, amplification, and strand separation of a double-stranded DNA fragment into one operation. Third, because the phage DNA is replicated in a double-stranded circular (RF) intermediate stage, the RF DNA can be purified and manipulated just like a plasmid. Fourth, both RF and single-stranded DNA will transfect competent E. *coli* cells and yield either plaques or infected colonies. Fifth, it is possible to package DNA up to six times the length of the wild-type M13 DNA. Sixth, the orientation of an insert can be determined, either by restriction analysis or hybridization of two single-stranded clones to form a double-stranded DNA fragment, which can then be detected by gel electrophoresis.

The M13 phage does not contain excess DNA that can be removed. However, there is a 507-bp region which contains the origins of replication for the viral DNA and its complementary strand. In most vectors derived from M13, foreign DNA has been inserted at this site. M13 vectors also have been modified by inserting polylinkers and the *lacZ* gene so that white plaques are formed instead of blue if exogenous DNA has been inserted into the *lacZ* cloning site.

## Phagemids

A **phagemid** is an engineered vector that contains plasmid and M13 components. Phagemids provide another method for obtaining ssDNA. Phagemids carry two replication origins, one a standard plasmid origin and the other derived from M13. The M13 origin is crucial in the synthesis of ssDNA. Production of ssDNA also requires enzymes and coat proteins coded by phage genes, which are lacking in phagemids. As a result, cells containing a phagemid vector must be coinfected with a helper phage if ssDNA is desired. The helper phage converts the phagemids into ssDNA molecules, which are then assembled into phage particles and secreted from the cell.

Many of the new commercially available vectors are phagemids and these may eventually supersede M13 cloning systems. Foreign DNA is cloned into pUC118 and propagated as dsDNA. There is a 476-bp fragment of M13 in these vectors, as well as a gene for ampicillin resistance. If cells carrying the phagemid are infected with the helper virus M13K07, then phage particles are produced containing ssDNA.

## cDNA Cloning

**Complementary DNA (cDNA)** is DNA that is complementary to the mRNA that is transcribed from the gene. Because the mRNA is processed, it will lack the introns usually found in insect genes. Thus, a cDNA clone will contain the DNA sequence of the protein of interest, but will lack introns and probably not contain the control sequences that regulate gene expression. cDNA cloning is used to produce a **cDNA library** or to produce probes for screening genomic libraries. A cDNA library allows the genetic engineer to clone *only those genes that are active at a specific time or in specific tissues*. Genes that are not actively being transcribed into mRNA will not be represented in the cDNA library. Thus, a cDNA library usually contains fewer clones than a genomic library. The gene of interest may occur in a frequency of one in $10^3$ or one in $10^4$. By contrast, a single-copy gene may be present in a genomic library in a frequency of only one in $10^5$ to one in $10^6$. Another benefit of a cDNA library is that it is possible, if an appropriate **expression vector** is used for cloning, to express a gene in a host such as E. *coli* or yeast. This enables the genetic engineer to produce large amounts of a specific gene product.

The quality of a cDNA library depends on the quality of the mRNA used as the template and the fidelity with which it can be **reverse transcribed** into cDNA. Messenger RNA, together with a suitable primer, and a supply of deoxyribonucleoside triphosphates can be converted into a double-stranded DNA molecule with the enzyme reverse transcriptase (Figure 7.9).

The cloning process involves two steps: (1) the first strand of cDNA is produced, and (2) a strand that is complementary to the first strand is synthesized, so that a double-stranded DNA molecule is produced. The primer is usually an oligonucleotide consisting of deoxythimidine residues (dT) because it can hybridize to the 3' poly(A) tails of template mRNA and thus give rise to full-length copies of double-stranded DNA. Once the double-stranded cDNA molecule has been synthesized, it can then be inserted into a plasmid or phage vector that is capable of replicating in E. *coli* and, in some cases, of being translated into a protein.

A key to producing cDNA is the enzyme **reverse transcriptase**. Reverse transcriptase is capable of two functions *in vitro*: a polymerase activity and a ribonuclease H activity. The polymerase activity requires (1) a template RNA molecule hybridized to a DNA primer with a 3'-OH group, and (2) all four dNTPs to synthesize a DNA molecule which is a faithful complement of the RNA.

Cloning a cDNA library is much more complex than cloning a genomic library into λ or cosmids. Before beginning the process, the goals of the project must be carefully considered and the basic approach chosen after deciding how the cDNA library will be screened to identify the gene(s) of interest (Kimmel and Berger 1987). For example, if antibodies will be used to identify clones capable of synthesizing specific peptides, the cDNA should be cloned into expression vectors such as the pUC family of plasmids or λ gt11 phage.

The cloning techniques vary in the type of primer used, the method for second-strand synthesis, and methods for coupling the cDNA to the vector, which can be either a plasmid or λ. Reverse transcriptases are commercially available that can synthesize copies of mRNA sequences that are more than 3 kb long. However, the transcripts often terminate prematurely and the clones containing the 5'-end of the mRNA will be rare.

Figure 7.9 outlines the synthesis of double-stranded cDNA from mRNA. Messenger RNA is often prepared for cDNA cloning by affinity chromatography on oligo(dT) cellulose. The reaction is preceded by a brief heat denaturation of the mRNA to eliminate its secondary structure, because reverse transcriptase is inhibited if the mRNA exhibits a secondary structure. The polyadenylated mRNA, the primer, and the reverse transcriptase are combined. The primer in

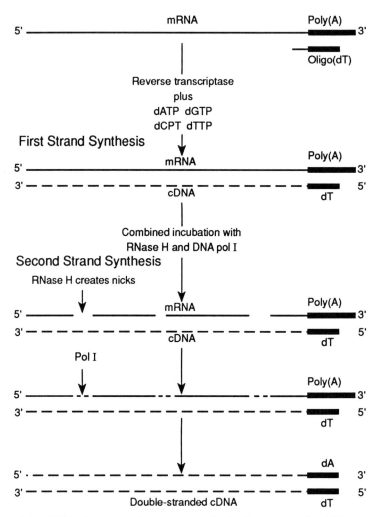

**Figure 7.9**   cDNA cloning involves two steps. In the first step an oligo(dT) anneals to the poly(A) region of mRNA. The enzyme reverse transcriptase and dATPs, dCTPs, dGTPs, and dTTPs are provided to produce the first cDNA strand. Synthesis of the second strand of the cDNA involves additional incubation with RNAse H and DNA polymerase I. The double-stranded cDNA produced is a complement to the mRNA, and thus lacks introns or regulatory sequences.

this case is a short sequence of (dT) residues. The product of the first strand synthesis is a hybrid of mRNA and the synthesized cDNA. The first strand is used as a template-primer complex to make the second strand of DNA. The enzyme RNase H is used to introduce gaps in the mRNA strand. At the same time, DNA polymerase I enzyme uses the primer–template complexes formed by RNase H to synthesize a double-stranded DNA.

Once the double-stranded DNA is synthesized, it is inserted into a vector. To do this, the synthesized molecule needs to have ends that can be ligated into the vector. One option is to make the cDNA blunt ended by end filling with the Klenow fragment of DNA polymerase I and then ligating it into a vector that has been cut with a restriction enzyme that produces a blunt end. Another option involves the addition of cohesive ends to the cDNA so that it will ligate into the vector more easily. There are three methods to add cohesive ends to the double-stranded DNA: (1) tailing with terminal transferase, (2) adding linkers, and (3) adding adaptors. The details of carrying out these procedures can be found in many cloning protocols.

There are about 10,000 *different* mRNA molecules in an average insect cell. At least 200,000 cDNA clones should be generated to be sure that a representative cDNA library is constructed. If the desired clone is a single-copy gene, then it will be very rare, so that powerful screening methods will be required to isolate the clone of interest.

Isolating RNA is more difficult than isolating DNA. Preparation of mRNA requires the absolute elimination of ribonucleases from glassware, pipettes, tips, and solutions. Anything that might contaminate the reactions with RNase must be eliminated, including hair, dust, and sneezes. Even fingerprints contain enough RNase to degrade the RNA. Furthermore, RNase is a very hardy enzyme and difficult to eliminate, even by autoclaving. Phenol extraction followed by ethanol precipitation is the most common technique for isolating RNA. Once RNA has been isolated, it must be evaluated for quality, often by agarose gel electrophoresis.

## Enzymes Used in Cloning

A number of enzymes used in genetic engineering have been mentioned in this and previous chapters. Table 7.1 summarizes their name, principal activity(ies), source, and functions in genetic manipulations. Enzymes used to synthesize DNA include terminal transferase, DNA polymerase I, and reverse transcriptases. Enzymes that modify DNA include S1 nuclease, exonuclease III, Bal31 nuclease, and DNase I. There are more than 1400 restriction endonucleases that can cleave DNA in a predictable manner. T4 and *E. coli* DNA ligases join DNA sequences. Calf intestinal phosphatase (CIP) and T4 polynucleotide kinase are used to modify the 5′ ends of DNA molecules.

**Table 7.1**

Enzymes Useful for DNA Manipulation

| Enzyme and source | Functions in genetic engineering |
|---|---|
| Enzymes that synthesize DNA | |
| DNA polymerase I (*E. coli*) | 5′ to 3′ DNA synthesis of template DNA with a primer; exonuclease functions (5′ to 3′ and 3′ to 5′) used to correct errors in DNA synthesis *in vivo*; generate radiolabeled DNA probes by nick translation; synthesize second strand of cDNA. |
| Klenow fragment of DNA polymerase I (*E. coli*) | DNA synthesis without 5′ to 3′ exonuclease ability; makes dsDNA from ssDNA; used in dideoxy sequencing; DNA labeling by random priming or end filling; converts 5′-overhangs of DNA cut with restriction enzymes to blunt ends. |
| T4 DNA polymerase (phage T4) | Exonuclease in 3′ to 5′ direction; fills in overhanging ends of DNA cut with restriction enzymes. |
| T7 DNA polymerase (phage T7) | 3′ to 5′ exonuclease activity used in DNA end labeling; converts 3′ overhangs to blunt ends. |
| *Taq* DNA polymerase (*Thermus aquaticus*) | DNA synthesis at 60–70°C in polymerase chain reaction. |
| Reverse transcriptases (from several RNA tumor viruses) | Synthesize copies of DNA from ss mRNA or DNA with template; cDNA synthesis most important. |
| Terminal transferase (mammalian thymus) | Adds residues to any free 3′-terminus; used to add poly (dG) and (dC) to 2 DNA molecules to be joined. |
| Enzymes that degrade DNA | |
| S1 Nuclease (*Aspergillus*) | Degrades ssDNA endonucleolytically; removes projecting 3′ regions of ssDNA in cloning and S1 mapping. |
| Exonuclease III (*E. coli*) | Degrades 1 of 2 strands of dsDNA from 3′ end of a blunt-ended double helix or from a projecting 5′ end. |
| Bal 31 nuclease (*Alteromonas espejiana*) | Degrades both strands of dsDNA with blunt ends. |

(*continues*)

**Table 7.1**  (*Continued*)

| Enzyme and source | Functions in genetic engineering |
| --- | --- |
| DNase I (pancreas) | Introduces random nicks in dsDNA prior to labeling by nick translation; produces random fragments for shotgun cloning and sequencing in M13; study of chromatin structure; study of DNA-protein complexes. |
| **Enzymes that join DNA** | |
| T4 DNA ligase | Seals ss nicks between adjacent nucleotides in dsDNA molecule, requires ATP; used to ligate 2 restriction fragments of DNA together in cloning. |
| E. coli DNA ligase | Seals ss nicks between adjacent nucleotides in dsDNA molecule, requires ATP; used to ligate 2 restriction fragments of DNA together in cloning but requires $NAD^+$. |
| **Enzymes that modify the 5′ ends of DNA** | |
| Calf intestinal phosphatase (CIP) | Removes 5′-phosphate groups to generate an OH-terminus; prevents unwanted ligation of DNA fragments during cloning; used for end labeling DNA probes. |
| T4 polynucleotide kinase | Adds phosphates to 5′-OH ends; used in chemical cleavage method of DNA sequencing; used to add linkers or adapters in cloning. |
| **Enzymes that cut DNA** | |
| Restriction endonucleases (many bacteria) | Types I, II, and III, >1400 types known. (see Table 6.3 for partial list); cleaves DNA; produces predictable termini, either blunt, 5′-overhang, or 3′-overhang. |

## Isolating a Specific Gene from a Library

The production of a genomic or a cDNA library is only a first step. The information in a library can be obtained only if the library can be screened. Screening identifies specific genes and provides information about genome organization or gene regulation. The ability to screen a library is dependent upon the availability of a probe. As pointed out in Chapter 6, a probe often is a molecule labeled with radioactive isotopes.

To identify a specific gene from among the thousands of clones in a library requires a way to identify a specific DNA sequence in the gene of interest. There are four ways to obtain a suitable probe. (1) The amino acid sequence of the protein is known for the species being studied, or a related species, and can be used to predict and synthesize the sequence of an oligonucleotide hybridization probe. Because the genetic code is degenerate and the third base can vary without altering the amino acid sequence, the probe used may actually incorporate a mixture of oligonucleotides with optional bases, especially in the third site of the codon. (2) The gene of interest has already been cloned from a related organism, so that it can be used as a heterologous hybridization probe. For some genes, particularly the housekeeping genes, conservation of functional domains in proteins has been extensive, so that probes from diverse species can be used effectively. (3) The protein is abundant in a particular tissue so the relevant clone can be identified by its relative abundance in a tissue-specific cDNA library. (4) If the protein has been purified, it can be used to generate an antibody against it. The antibody can be used to identify recombinant cells that have the specific enzymes.

Once a probe is obtained and labeled, it is used in DNA hybridization experiments to identify those clones that contain the DNA of interest. Hybridization involves immobilizing DNA samples from different clones on a solid support (either a nitrocellulose or nylon membrane) and then probing the unknown DNA with a DNA or RNA probe to identify a number of clones that contain the DNA of interest. Identification is possible because the labeled probe is able to base-pair with the desired DNA so that it can be detected by autoradiography or other methods. There are a number of different DNA hybridization techniques. Southern blot analysis, described in Chapter 6, is one such technique. Another is **plaque screening**.

Figure 7.10 illustrates plaque screening of *E. coli*. First, *E. coli* cells are infected with λ phages that contain exogenous DNA and grown on an agar substrate. A nitrocellulose filter is laid onto the *E. coli* lawn and plaques. The precise orientation of the filter is marked in relation to the plaque pattern on the plate. Some of the phages in the plaques are adsorbed onto the filter, where they release their DNA. The DNA is denatured by an alkali treatment and then brought to a neutral pH. After denaturation, the single-stranded DNA on the filter is incubated with a radiolabeled probe. The probe base-pairs with a specific nucleotide sequence from the gene of interest, but not with DNA from plaques containing other genes. The point at which the probe hybridized to the immobilized DNA on the filter is located by autoradiography. The filter and the original agar substrate are then compared and the corresponding plaque is located on the original agar substrate. A few phages are picked from each plaque that yielded a spot on the X-ray film. These phages are used to infect individual new *E. coli* cultures to produce multiple copies of each phage.

Plaque hybridization allows several hundred thousand plaques to be screened at once and so a single copy gene can be isolated from among thousands

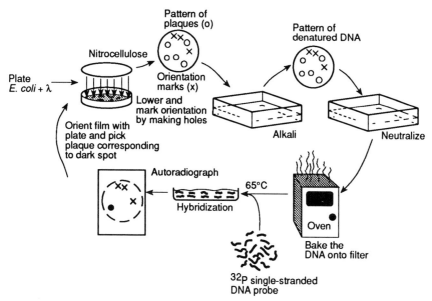

**Figure 7.10**   Plaque screening can locate specific genes. It involves *in situ* hybridization of *E. coli* that have been transformed with a λ vector. *E. coli* infected with recombinant phage are plated out. A nitrocellulose filter is laid on top of the bacterial lawn and plaques. Some of the phage in a plaque adsorb to the filter. The filter is treated with an alkali to denature the phage DNA, neutralized, baked in an oven to immobilize the DNA, and placed in a solution with a radiolabeled DNA probe. The probe base-pairs with sequences in the DNA that are complementary and identifies plaques that may contain the gene of interest. It is then possible to go back to the plate, pick a few phages from those plaques, and multiply them in *E. coli*.

of clones. Because the DNA that was inserted into the λ vector was cut at random, it is likely that more than one clone (plaque) will be identified that contains DNA of interest. Ideally, at least one clone isolated by the probe will contain the complete gene, but this can only be determined after the DNA has been sequenced, a technique that is described in Chapter 8.

Another, more advanced, technique employed to identify specific DNA sequences is called **chromosome walking** (Figure 7.11). It is particularly useful with *Drosophila* but less useful with other insects for which less genetic information is available. Chromosome walking is used to isolate a gene of interest for which no probe is available. The gene of interest *must* be linked to a marker gene that has been identified and cloned. This marker gene is used as a probe to screen a genomic library. All fragments containing the marker gene are selected

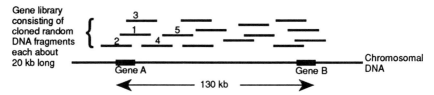

**Figure 7.11**   Chromosome walking is used to identify a gene of interest when a probe is not available. It can only be carried out when it is known that the target gene is linked to another gene which has been cloned and sequenced. First a clone containing gene A is isolated in fragment 1. This fragment is sequenced and new probes are synthesized that contain sequences from each end of the fragment. The new probes are used to identify overlapping DNA clones in the library on each side of fragment 1, that is, clones 2, 3, and 4. Clone 4 can, in turn, be sequenced, and used as a probe to identify clone 5, and so on until gene B is reached.

and sequenced. The fragments are then aligned and those cloned segments farthest from the marker gene in both directions are subcloned for the next step. The subclones are used as probes to screen the genomic library again to identify new clones containing DNA with overlapping sequences. As the process is repeated, the nucleotide sequences of areas farther and farther away from the marker gene are identified and eventually the gene of interest will be found. As shown in Figure 7.11, the goal is to identify gene B, for which no probe is available. However, sequences of a nearby gene (A) are available in cloned fragment 1. In a large and random genomic DNA library, many overlapping cloned fragments will be present. Thus, clone 1 sequences can be used as a probe to identify overlapping clones 2, 3, and 4. Clone 4 subsequently can be used as a probe to identify clone 5, and so on, until gene B is reached.

Once a gene has been identified in a genomic library, its DNA sequence can be determined, as described in Chapter 8. However, a DNA sequence by itself is of limited value. If nothing is known about the gene product, it may be difficult to determine unambiguously which sequences are the coding regions and which are introns. Intron boundaries may be established by their similarities to sequences of known introns (consensus sequences). If the gene product is unknown, it may be possible to identify the sequenced gene's function by comparing the DNA sequence with other sequences in DNA databases, although a fully convincing match is not always found. Thus, going from clone to DNA sequence to gene product is one of the most difficult problems in molecular genetic research, particularly if genes are being studied for which there are no clearly associated gene products.

One solution is to attempt to express the gene in order to obtain a gene product. To be expressed, genes require a promoter and, often, upstream control

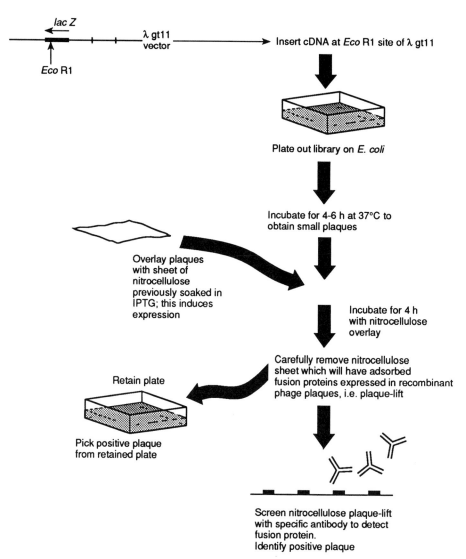

**Figure 7.12** Specific genes also can be identified by immunochemical screening. A λ gt11 library in *E. coli* is screened in a manner similar to plaque hybridization except that the genes of interest are identified by binding of a specific antibody to a fusion protein.

sequences. A variety of expression vectors have been developed which allow the expression of cloned genes in *E. coli*. Such vectors require the presence of *E. coli* promoters if the eukaryote sequence is to be expressed. A detailed description of *E. coli* expression vectors is provided by Pouwels (1991), and will not be given here. Of particular interest to entomologists, however, is the use of baculovirus expression vectors to express many insect genes, as well as other eukaryotic genes, as described later.

cDNA libraries can be screened by hybridization screening or by expression screening. If a protein of interest has been purified and part or all of the protein sequence has been obtained, then it is possible to predict the sequence of synthetic oligonucleotides that can be used as a hybridization probe to detect the appropriate cDNA clone(s). Alternatively, if an antibody to the protein is available, it can be used to identify the clone(s) of interest if the cDNA library is cloned into an expression vector such as λ gtll (Figure 7.12). This vector contains the *lacZ* gene of *E. coli*, and has a unique *Eco* RI restriction site near the end of the gene. cDNA cloned into this site in the correct orientation and reading frame will produce a fusion protein. Upon lysis of the *E. coli* cells, the protein is released and picked up on nitrocellulose in just the same way as in plaque screening. The plaque containing the interesting cDNA clone can be detected by incubating the filter with a specific antibody.

It is also possible to determine the difference in abundance between two different mRNA populations. Thus, mRNA produced from different tissues from the same organism or mRNA produced from a tissue before and after a specific induction signal can be compared by differential screening. A cDNA library is prepared from one of the two mRNA populations and the two copies are immobilized on filters. The filters are then screened twice, once with highly labeled cDNA prepared from one of the two mRNA populations and once with a probe from the other mRNA population. By comparing the signals produced on the two filters probed with the different probes, it is possible to determine whether mRNA sequences are present in one population, but are absent or rare in the alternative mRNA population.

## Labeling Probes by a Variety of Methods

**Nucleic acid hybridization** is used for many different purposes in molecular genetics. Nearly all phases of cloning and characterization or analysis of DNA involve hybridizing one strand of nucleic acid to another. Nucleic acid hybridization relies on the fact that two single-stranded nucleic molecules with complementary bases (DNA with DNA and DNA with RNA) are able to pair up via hydrogen bonds. The strength of the hybridization is determined by the length of the homologous sequences, various experimental conditions, and the degree of sequence homology. Chapter 6 described one application of nucleic acid hybridization, the Southern blot, and one method for labeling probes, nick translation. Other nucleic acid hybridization techniques include colony or plaque hybridization, and Northern blot analysis, in which the immobilized nucleic acid is RNA.

A number of labeling methods other than nick translation are available, as outlined in the following paragraphs. The success of nucleic acid hybridization relies on methods to introduce radioactive label into cloned segments of DNA or RNA. Each labeling technique has optimal sizes, efficiency, and different amounts of required nucleic acid template required. One measure of the efficiency of labeling is the **specific activity** of the label. Specific activity refers to the amount of radioactivity per μg of probe DNA. The specific goals of the project will determine which labeling technique is employed. Detailed protocols are available in a variety of laboratory manuals and kits. Nonradioactive probes using biotin and chemiluminescent labels also are available in kits.

### Synthesis of Uniformly Labeled DNA Probes by Random Primers

Short oligonucleotides can serve as primers for DNA synthesis by DNA polymerases on single-stranded templates. If the primers used are random in sequence, they will form hybrids at many different locations along the template strand so that the strand being synthesized will incorporate a radiolabeled dNTP randomly along its length. Such a single-stranded DNA probe will have a very high specific activity. If reverse transcriptase is used for synthesis, the template can be RNA. If DNA is the template, then the Klenow fragment of DNA polymerase I is used.

### Synthesis of Probes by Primer Extension

Primer extension is used to synthesize probes from denatured double-stranded DNA. It can be used to produce probes from denatured, closed-circular DNA or from denatured, linear, double-stranded DNA. The purified DNA is mixed with random primers and denatured by boiling. Radiolabeled dNTPs and the Klenow fragment of DNA polymerase I then are added to carry out synthesis of the probe. Random primers anneal to the denatured DNA and the radioactive product is synthesized by primer extension. Probes prepared by random priming are usually 400–600 nucleotides long.

### End-Labeled Probes

A variety of methods are available to introduce labels at either the 3' or 5' ends of linear DNA. Usually only a single label is introduced at one end of the molecule, so the specific activities produced by such techniques are lower than those obtained by the uniform labeling methods described above. Both DNA and RNA can be end labeled. The advantages to end labeling are that the location of the labeled group is known and very small fragments of DNA can be labeled, including restriction digest fragments.

### Single-Stranded Probes

Single-stranded DNA, cDNA, or RNA probes have an advantage over double-stranded probes because more probe is available to hybridize with the nucleic acid of interest. Single-stranded probes do not anneal to themselves to become unavailable as probes. RNA probes have the advantage that they do not need to be denatured before being used because they are already single stranded. As a result, RNA probes have higher specific activity (the ratio of radioactive to nonradioactive molecules of the same kind) than DNA probes. Double-stranded DNA probes must be denatured before being used, which produces two strands. If only one of the DNA strands has been labeled, the unlabeled strand can dilute the reaction mixture.

Single-stranded probes are prepared from DNA templates by synthesizing radiolabeled DNA that is complementary to sequences cloned in a bacteriophage vector such as M13 or a phagemid. RNA probes can be produced by transcription of double-stranded DNA in a vector with a powerful promoter derived from E. coli bacteriophages T7 and T3 using a bacteriophage DNA-dependent RNA polymerase. The labeled transcript produced is complementary to one of the two template strands. The probe can therefore be used as a strand-specific probe in hybridization reactions. cDNA probes are used to isolate cDNA clones of genes that are expressed in specific cells or tissues.

### Synthetic Probes

Knowledge of the sequence of a few amino acids in a protein will allow a specific gene to be isolated with a synthetic probe. Automated machines are available to synthesize short segments of single-stranded DNA in which the sequences in these synthetic probes are defined precisely. Alternatively, probes can be produced that consist of alternative sequences, as determined by the degeneracy of the genetic code. Because there are 64 possible codons and only 20 amino acids, most amino acids are coded for by more than one codon. Thus, a probe that consists of a mixture of degenerate sequences can be used to screen libraries (degenerate probe).

## Baculovirus Vectors for Expressing Foreign Polypeptides in Insect Cells

E. coli cells infected with plasmid or phage expression vectors have been used to express foreign genes. However, it is sometimes difficult to obtain complex polypeptides derived from eukaryotes in a biologically active form from the prokaryote *E. coli*. This is because post-translational changes are often made to proteins in eukaryotes that cannot be performed in *E. coli*. These changes include **glycosylation** and **phosphorylation**. As a result, eukaryotic expression vectors have been developed for use in a variety of eukaryotic cells, including yeast and insect cells.

The most effective expression vectors for use in insect cells have been engineered from baculoviruses (O'Reilly et al. 1992). Baculoviruses are viruses with double-stranded, circular DNA genomes contained within a rod-shaped protein coat. The Baculoviridae are divided into three subgroups—nuclear poly-hedrosis viruses (NPV), granulosis viruses, and nonoccluded viruses. Most NPVs primarily infect lepidopterans, where they produce nuclear inclusion bodies in which progeny virus particles are embedded. Polyhedrin is the protein compo-nent of the crystalline matrix that protects the viral particles when they are outside their insect host. Several NPVs have been used as biological pesticides in pest management programs. Perhaps the most extensively studied baculovirus is *Autographa californica* NPV. It has a relatively broad host range, infecting over 28 different lepidopteran species. The life cycle of wild-type baculovirus begins when caterpillars eat the protein matrix (polyhedrin) and release the virus particles. Virus replication occurs within host insect cells, but the protein matrix is not produced early in the infection. However, when the caterpillar is near death, the virus resumes polyhedrin production until approximately 20% of the insect host cell proteins consist of polyhedrin.

Two baculoviruses, the *Autographa californica* nuclear polyhedrosis virus (AcNPV) and the silk moth *Bombyx mori* nuclear polyhedrosis virus (BmNPV), have been exploited as vectors to carry exogenous DNA into insect cells in cell culture or into living silk moth larvae in order to produce foreign proteins (Maeda 1989). The productivity of the baculovirus as a vector is based on the extremely high efficiency of its promoter for the polyhedrin gene. The poly-hedrin gene can be deleted from the vector so that exogenous DNA coding for a protein can be inserted into this site. The polyhedrin promoter enables the production of very large amounts of the desired polypeptides. These engineered vectors are used to express foreign genes either *in vitro* in insect cell cultures or *in vivo* in the silk moth larva. The level of expression of foreign gene products varies in efficiency, depending on the specific foreign protein being produced.

The commercial production of proteins using baculovirus expression vectors is one of the most exciting developments of the biotechnology industry. Proteins that elicit an important biological response are often produced in tiny quantities in an organism and their isolation is thus not economical. Furthermore, some proteins, particularly those from higher eukaryotes, cannot be produced in *E. coli* in a form in which they will be biologically active. A number of mammalian proteins have been produced using *A. californica* NPV in cultured *Spodoptera frugiperda* cells. The first such protein, human interferon, was produced by *A.*

*californica* NPV in insect cells in 1983. Since then, many additional genes have been expressed, including growth factors, tumor antagonists, blood clotting or anticlotting factors, protein hormones, antibodies, and vaccines against hepatitis B, acquired immune deficiency syndrome (AIDS), and malaria (Miller 1988, O'Reilly et al. 1992).

Biotechnology companies are developing proteins expressed in insect cells as commercial products. Companies are conducting research to learn how to scale up the culture of insect cells in commercial-sized fermentor systems (Maiorella et al. 1988). However, some sceptics believe the baculovirus expression system is less useful than that of yeast because insect cells are more difficult and more expensive to grow. Other scientists think that intact insects can be used as bioreactors. They argue that a single *S. frugiperda* caterpillar theoretically can produce up to 3 mg of a pharmaceutically important protein at a maintenance cost of less than a penny. Many problems remain before these products can be widely used. There are concerns about the fact that little is known about the insect glycosylation system, which could affect biological activity of these proteins in mammals. Furthermore, getting these products through the regulatory process remains to be achieved.

The baculovirus expression system offers outstanding research opportunities. It can be used to study the physiology, biochemistry, and endocrinology of insects. If the genes coding for insect proteins have been cloned, baculovirus expression vectors can be used to produce both small and large proteins such as neuropeptides and eclosion or prothoracicotropic hormones (Maeda 1989). Genes cloned from *D. melanogaster* or other insects have been expressed in *S. frugiperda* cells using *A. californica* NPV vectors. For example, the *Kruppel* gene product of *D. melanogaster*, and cecropin, a bactericidal peptide secreted from the fat body of the moth *Hyalophora cecropia* in response to bacterial infection, have been produced.

The ability to produce abundant amounts of specific proteins by baculovirus expression vectors will promote dramatic advances in basic and applied knowledge of insect physiology and development. Improvements will continue to be made in baculovirus vectors that will further enhance their ability to express proteins (Luckow and Summers 1988). For example, research is under way to develop vectors that function early in the infection cycle of the insect cell so that stable transformation of insect cells can be achieved and continuous production of proteins can be obtained (Jarvis et al. 1990). Stable transformation is desirable because the older baculovirus vectors produce high levels of foreign gene products in insect cells only transiently before the virus kills the host cells.

In the next chapter, we discuss how to sequence DNA.

## General References

Ausubel, F. M., R. Brent, R. E. Kingston, D. M. Moore, J. G. Seldman, J. A. Smith and K. Struhl, Eds. 1987. *Current Protocols in Molecular Biology*. Greene Publ. Assoc., New York. These two ring-bound volumes offer detailed research protocols. The publisher has a subscription update service so that new pages can be inserted into the binder to replace outdated material or to expand existing information.

Berger, S. L. and A. R. Kimmel, Eds. 1987. *Guide to Molecular Cloning Techniques* (Methods Enzymol., Vol. 152). Academic Press, San Diego.

Brown, T. A., Ed. 1991. *Essential Molecular Biology: A Practical Approach.* Vols. I and II, IRL Press, Oxford.

Glover, D. M., Ed. 1985. *DNA Cloning.* Vol. I. *A Practical Approach.* IRL Press, Oxford.

Old, R. W. and S. B. Primrose. 1989. *Principles of Gene Manipulation: An Introduction to Genetic Engineering.* 4th Ed. Blackwell Scientific, Oxford.

O'Reilly, D. R., L. K. Miller and V. A. Luckow. 1992. *Baculovirus Expression Vectors: A Laboratory Manual.* Freeman, New York.

Ptashne, M. 1986. *A Genetic Switch: Gene Control and Phage Lambda.* Blackwell Scientific, Oxford.

Sambrook, J., E. F. Fritsch and T. Maniatis. 1989. Molecular *Cloning: A Laboratory Manual.* 2nd Ed. Cold Spring Harbor Press, Cold Spring Harbor, NY. Volumes 1, 2, and 3 provide detailed protocols for a large number of techniques.

Williams, J. G. and R. K. Patient. 1988. *Genetic Engineering.* IRL Press, Oxford.

## References Cited

Jarvis, D. L., J. G. W. Fleming, G. R. Kovacs, M. D. Summers and L. A. Guarino. 1990. Use of early baculovirus promoters for continuous expression and efficient processing of foreign gene products in stably transformed lepidopteran cells. Bio/Technology 8: 950–955.

Kimmel, A. R. and S. L. Berger. 1987. Preparation of cDNA and the generation of cDNA libraries: Overview. Methods Enzymol. 152: 307–316.

Luckow, V. A. and M. D. Summers. 1988. Trends in the development of baculovirus expression vectors. Bio/Technology 6: 47–55.

Maeda, S. 1989. Expression of foreign genes in insects using baculovirus vectors. Annu. Rev. Entomol. 34: 351–372.

Maiorella, B., D. Inlow, A. Shauger, D. Harano. 1988. Large-scale insect cell-culture for recombinant protein production. Bio/Technology 6: 1406–1410.

Miller, R. E. 1988. Baculoviruses as gene expression vectors. Annu. Rev. Microbiol. 42: 177–199.

Old, R. W. and S. B. Primrose. 1989. *Principles of Gene Manipulation: An Introduction to Genetic Engineering.* 4th Ed. Blackwell Scientific, Oxford.

O'Reilly, D. R., L. K. Miller and V. A. Luckow. 1992. *Baculovirus Expression Vectors: A Laboratory Manual.* Freeman, New York.

Pouwels, P. H. 1991. Survey of cloning vectors for *Escherichia coli.* Pp. 179–239. In: *Essential Molecular Biology: A Practical Approach.* Vol. I. T. A. Brown, Ed. IRL Press, Oxford.

# Chapter 8

# DNA Sequencing and Genome Analysis

## Overview

There are two basic methods for sequencing DNA, plus several automated methods. For both basic methods, sequencing involves four procedures: (1) cloning and preparing template DNA, (2) performing the sequencing reactions, (3) gel electrophoresis of the samples, and (4) compilation and interpretation of the data. Sequencing is carried out on genomic DNA, cDNA, or mitochondrial DNA to evaluate regulatory sequences as well as coding and noncoding regions. DNA sequences are used to reconstruct phylogenies, identify taxonomic groups, and study population ecology and genetics. DNA to be sequenced can be cloned by traditional or polymerase chain reaction (PCR) methods. Automated sequencing methods are being developed to carry out the very large-scale sequencing necessary for projects in which the goal is to sequence an entire genome. Information obtained from a *Drosophila* genome project now under way will revolutionize both fundamental and applied aspects of insect genetics.

The most commonly used manual sequencing method is the Sanger or dideoxy chain-terminating method, developed in 1977. DNA is synthesized *in vitro* on a single-stranded template using a primer, a mixture of radioactively labeled deoxynucleotide triphosphates (dNTPs) and dideoxynucleotide triphosphates (ddNTPs). The reaction is terminated at the position at which the ddNTP is incorporated in the growing DNA chain instead of the dNTP. Four different reactions are carried out, one for each base. The DNA fragments from the four reactions are then separated by acrylamide gel electrophoresis and the base sequence is identified by autoradiography of the banding patterns.

In the 'chemical' sequencing method developed by Maxam and Gilbert in 1977, single-stranded DNA, derived from double-stranded DNA and labeled at the 5' end with $^{32}$P or $^{35}$S, is subjected to chemical cleavage protocols that selectively make breaks on one side of a specific base. Fragments from the

184

reactions are then separated according to size by acrylamide gel electrophoresis and the sequences are identified by autoradiography.

## Introduction

DNA sequencing is an important component of many insect molecular genetics projects. Sequencing often is a necessary component of a cloning project, while in other cases it is the desired end point and the sequences are used in taxonomic, ecological, or evolutionary studies. Advances in technology are providing an opportunity to sequence entire genomes, which will revolutionize both basic and applied knowledge of gene structure and function. In identifying the sequences of promoters, protein coding sequences, and noncoding regions of DNA in genomic or mitochondrial DNA, it is possible to deduce relationships between organisms and reconstruct their evolutionary history. The development of extensive computerized databases of DNA and protein sequences allows hypotheses to be constructed regarding the structure and function of the proteins and their secondary structures. All these options became possible only after DNA sequencing methods were developed in the late 1970s.

Two basic DNA sequencing methods were developed at about the same time—the chemical or Maxam-Gilbert method (1977, 1980) and the chain-terminating method of Sanger et al. (1977). Both use the same basic procedures: (1) cloning and preparing the DNA templates, (2) performing the sequencing reactions on the DNA templates, (3) gel electrophoresis of the samples, and (4) compiling and interpreting the data. The Maxam and Gilbert sequencing method is described briefly, but more details are given for the chain-terminating method because it is more readily employed by novices.

DNA to be sequenced can be either genomic DNA, mitochondrial DNA, cDNA, or DNA amplified by the polymerase chain reaction. Because cDNA typically ranges in size from 500 to 8000 bp, it can be cloned and sequenced with fewer steps than genomic DNA. However, because cDNA lacks introns and regulatory structures, it provides less information than genomic DNA. A project to sequence only cDNA would probably miss some genes that are expressed at very low levels or in a tissue-specific or time-dependent manner. Sequencing genomic DNA has the disadvantage that effective computer tools are needed to discover the sequences that actually code for a gene, because up to 90% of these sequences are noncoding DNA. Some DNA sequences are associated with centromeres or telomeres and others have no known function. Computer programs have been developed to search DNA sequence data and identify possible regulatory sequences, potential start or stop codons, open reading frames, and sequences that may indicate the location of intervening sequences or introns.

Before DNA can be sequenced, it is cloned, and a series of clones must be identified that contain the DNA of interest, because it is time consuming and

**G A T C G T A C**

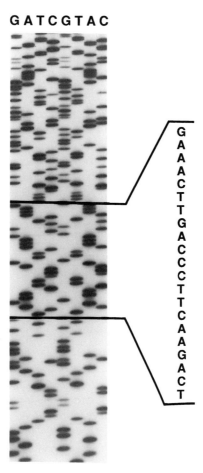

**Figure 8.1**   Autoradiograph of a sequencing gel obtained by the dideoxy chain-terminating method. Four different reactions are carried out, in which DNA synthesis of a complementary DNA chain is terminated by incorporating a radiolabeled dideoxy base (ddATP, ddCTP, ddGTP, or ddTTP) rather than a deoxy base (dATP, dCTP, dGTP, dTTP). The resulting fragments of synthesized DNA are visualized by acrylamide gel electrophoresis and autoradiography. The sequence is read by determining which lane contains each succeeding DNA segment, as determined by a band. Thus, reading from the bottom, the shortest fragment has a band in the T lane (just above the line). The next band is a C, then the next is an A, and so on. Bands at the very top and bottom of the gel are not read. This sequence is a portion of the *mariner* transposable element cloned from the predatory mite *Metaseiulus occidentalis* (A. Jeyaprakash and M. Hoy, unpublished observations).

expensive to sequence the entire genome of eukaryotic species. Sometimes the goal is to sequence an entire gene, but in other cases it is to sequence a specific region of a gene.

The length of DNA that can be sequenced by a single reaction is limited to 200 to a maximum of 1000 bases (Figure 8.1). Thus, the cloned DNA must be converted into smaller segments or **subclones** and cloned into vectors that are specialized for sequencing such as M13 or plasmid sequencing vectors. Cloning vectors can contain DNA inserts ranging in size from 100 to 1,000,000 bp. For example, yeast artificial chromosomes can contain inserts up to 1 million bp long, cosmids can contain inserts 30,000 to 45,000 bp long, and λ can contain inserts 10,000 to 20,000 bp long. These inserts are too long to be sequenced in an intact form with current methods, however.

Sequencing is performed on single-stranded DNA, so having the DNA fragment inserted into a vector, such as M13, that yields ssDNA is a great advantage. The sequencing vectors derived from M13 can contain inserts ranging in size from 500 to 2000 bp. Double-stranded plasmid sequencing vectors can contain inserts 1000 to 5000 bp long. Thus, if the goal is to sequence long genomic DNA regions, such as entire genes, a series of sequencing reactions is carried out to produce overlapping DNA sequences of smaller regions, which are then ordered.

## The Dideoxy or Chain-Terminating Method

Briefly, the **dideoxy** or **chain termination sequencing** method developed by Sanger involves *de novo* synthesis of a series of labeled DNA fragments from a single-stranded DNA template. Two methods are employed to produce a single-stranded DNA template: (1) denaturing double-stranded DNA by heating it, and (2) cloning it into a vector that produces single-stranded DNA. The single-stranded DNA segment to be sequenced serves as the template for the synthesis, by complementary base-pairing, of a new radiolabeled strand of DNA.

Because DNA synthesis is employed in the chain-terminating method, the sequencing reaction requires a DNA polymerase, radiolabeled deoxyribonucleotides (dNTPs), a primer, and dideoxyribonucleotides (ddNTPs) (Table 8.1). Because several different DNA polymerases could be used (Klenow fragment, Sequenase 2, or thermophilic DNA polymerases such as *Taq*), different protocols are available. DNA sequencing kits that contain the necessary enzymes and components can be purchased, with specific protocols provided. The dNTPs can be labeled either with $^{32}$P or $^{35}$S, but $^{35}$S labeling produces sharper bands and improves the resolution of the autoradiogram. The structure of a dNTP that has been labeled with $^{35}$S is shown in Figure 8.2.

The dideoxy or chain-terminating reaction is begun by adding a short oligonucleotide primer that is complementary to a region of DNA adjacent to the DNA segment to be sequenced (Table 8.1; Figure 8.3). The primer is nor-

**Table 8.1**

An Example of a Dideoxy (Sanger) Sequencing Protocol Using Modified Bacteriophage
T7 DNA Polymerase (Sequenase)[a]

---

Sequencing reagents
    Annealing buffer (5× concentrate)
        200 mM Tris–HCl, pH 7.5
        100 mM $MgCl_2$
        250 mM NaCl
        Dithiothreitol (DDT) 0.1 M
    Labeling nucleotide mixture (for use with radiolabeled dATP)
        1.5 mM dGTP, 1.5 mM dCTP, and 1.5 mM dTTP
    Termination nucleotide mixtures (one for each dideoxynucleotide)
        Each mixture contains 80 mM dGTP, 80 mM dATP, 80 mM dTTP, 80 mM
          dCTP, and 50 mM NaCl
        In addition, the 'G' mixture contains 8 mM dideoxy-dGTP; the 'A' mix, 8 mM
          ddATP; the 'T', 8 mM ddTTP; and the 'C', 8 mM ddCTP
    Stop solution
        95% formamide, 20 mM EDTA, 0.05% bromphenol blue, and 0.05% xylene
          cyanol FF

Procedures
  1. Annealing template and primer
    For each template, a single annealing (and subsequent labeling) reaction is used.
      Combine the following:
      Primer                                0.5 pmol
      DNA                                   0.5–1.0 pmol (1 mg of M13)
      Annealing buffer                      2 ml
      Water to bring total volume to 10 ml
    Warm the capped tube to 65°C for 2 minutes and allow the mixture to cool slowly to
    room temperature over a period of about 30 minutes.

  2. Labeling reaction
    To the annealed template-primer add the following:
      DTT (0.1 M)                           1 ml
      Labeling nucleotide mix               2 ml
      [α-$^{35}$S] or [α-$^{32}$P]dATP      5 mCl (typically 0.5 ml)
      Sequenase                             3 units
    Total volume should be approximately 15 ml; mix thoroughly and incubate for 2–5
    minutes at room temperature.

  3. Termination reactions
    Label 4 tubes 'G', 'A', 'T', and 'C'. Fill each with 2.5 ml of the appropriate dideoxy
    termination mixture.
    When the labeling reaction is complete, transfer 3.5 ml of it to the tube prewarmed
    to 37°C, labeled 'G'. Do the same for each of the other three tubes (A, T, and C).

---

(continued)

**Table 8.1**    (*Continued*)

> After 2–5 minutes at 37°C, add 4 ml of Stop Solution to each termination reaction, mix, and store on ice. To load the gel, heat the samples to 75–80°C for 2 minutes or more and load 2–3 ml in each lane.
>
> 4. Polyacrylamide gel electrophoresis
> 5. Analysis of sequences

*a*Modified from Tabor and Richardson (1987).

mally 15–30 nucleotides long and is annealed to the template in a preincubation step. Four separate reactions are set up to determine the position, respectively, of the A, T, G, and C bases in the template DNA. Each reaction requires a mixture of DNA polymerase, primers, dNTPs, ddNTPs, and the template DNA. The ddNTPs are derivatives of dNTPs that do not contain a hydroxyl group at the 3' position of the deoxyribose ring (Figure 8.4).

When the ddNTPs are incorporated into the growing DNA chains instead of dNTPs, that particular DNA molecule is terminated at that point. All four dNTPs are present in each tube with each ddNTP but the ratio is adjusted so that ddNTPs are less frequent than dNTPs. This makes the incorporation of a ddNTP a random event. The newly synthesized DNA molecules in a specific reaction tube therefore are a *mixture* of DNA fragments of different lengths, each with a fixed *starting point* (determined by the primer). Thus, for example, in the reaction in which the chain is terminated when thymines are incorporated, a ddNTP does not always get incorporated into the first site where a thymine occurs. Nor does a ddNTP necessarily get incorporated into the chain where the second thymine occurs. However, over the 200 to 400 nucleotides being se-

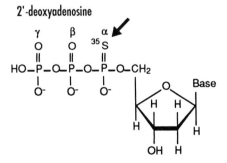

**Figure 8.2**    Structure of [α-$^{35}$S]deoxynucleoside triphosphate. Labeling with $^{35}$S results in sharper bands, which increases the resolution of the sequencing gels.

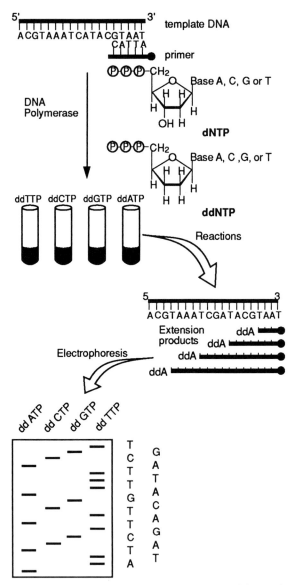

**Figure 8.3** A diagram of the steps involved in the Sanger dideoxy chain-terminating method of DNA sequencing. (Modified from Hunkapiller et al., 1991, *Science*, 254, p. 60; copyright 1991 by the AAAS.)

Normal deoxynucleoside triphosphate ( i.e. 2'-deoxynucleotide )

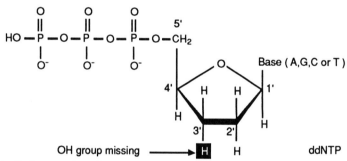

**Figure 8.4**  Dideoxynucleoside triphosphates (ddNTPs) act as chain terminators because they lack a 3'-OH group found on normal deoxynucleotides (dNTPs).

quenced, each site where a thymine is incorporated will have ddNTPs incorporated into the molecules, so that a series of molecules of different length is produced (Figure 8.1). Thus, *populations* of synthesized molecules are produced in which the chain is terminated at the sites where thymines occur. These DNA molecule populations are visualized by gel electrophoresis.

The base sequence is visualized by running the radiolabeled DNA fragments from the four reactions on an acrylamide gel in four adjacent lanes. Each reaction tube will produce a series of bands, with each band representing a population of molecules at which the DNA molecule was terminated by incorporating a ddNTP. The banding pattern in the four lanes is visualized on an X-ray film (Figure 8.1). Ideally, the two strands of the DNA should be sequenced independently to reduce errors generated by artifacts in the sequence reactions or inadequate resolution of regions of the sequence on the gel.

## DNA Sequences Are Analyzed on Polyacrylamide Gels

Both DNA sequencing methods generate sets of DNA molecules that share a common end but vary in their length at the other end, depending on the specific nucleotide sequence at that end. Both methods also employ radioactive labeling to visualize the results. Once the DNA is electrophoresed, the gel is dried down onto paper and exposed to X-ray film. The autoradiogram produced displays a ladder of bands representing the migration position of the different radiolabeled DNA segments. For example, the sequence of 20 nucleotides can be read to the right of the four lanes in Figure 8.1. The first base at the top is a T, the second a C, the third a G, and so on.

Both sequencing methods use very thin polyacrylamide gels to discriminate between nucleic acid molecules that may differ in length by only one nucleotide. Polyacrylamide gels are cast between two sealed glass plates held approximately 0.3 mm apart and run in a vertical electrophoresis tank. It is important to have a uniform electrical field across the width of the gel or the resulting banding pattern will be distorted (the 'smile' effect) or fuzzy. Gels are usually 20 or 40 cm wide and range in length from 40 cm to up to 2 meters long. A gel 40 cm long will allow resolution of approximately 200 nucleotides and take about two hours to run.

A sequencing gel is a high-resolution gel containing 6–20% polyacrylamide and 7 M urea. The DNA to be analyzed is denatured before electrophoresis by heating it to 80°C in a buffer containing formamide, which lowers the melting temperature of double-stranded DNA molecules. The urea minimizes DNA secondary structure, which could affect mobility of the DNA through the gel. The gel is run using sufficient power so that it is heated to about 50°C, which also minimizes DNA secondary structure. The rate of migration of an individual DNA molecule is determined by its length. The shorter fragments migrate more quickly than the longer, thus ending up nearer the bottom of the gel. Protocols for gel electrophoresis of DNA are widely available (Davies 1982, Sealey and Southern 1982, Howe and Ward 1989).

Gels are sometimes difficult to interpret because of a phenomenon called compression. Compressions often occur at GC-rich regions in the sequence and result in anomalous bands. Compressions usually are due to hairpin loop structures (secondary structures) at the ends of the DNA chains.

## Variations on Dideoxy Sequencing Methods

There are a number of different protocols for sequencing DNA by the dideoxy chain-termination method (Ambrose and Pless 1987, Barnes 1987, Mierendorf and Pfeffer 1987, Howe and Ward 1989, Sambrook et al. 1989, Tabor and Richardson 1987). Each protocol has its virtues and benefits. Variables include such factors as whether the DNA to be sequenced is double- or single-stranded and whether the DNA is sequenced directly from double-stranded plasmid DNA, after subcloning into a single-stranded bacteriophage such as M13, or from polymerase chain reaction products.

Double-stranded DNA can be sequenced by denaturing the DNA into single strands using alkali. However, during the sequencing reaction, the two strands can renature and this may generate some artifacts in the sequencing and reduce the number of bases that can be read in a single gel. As a result, ssDNA templates are usually better for sequencing because more bases may be read and fewer artifacts are generated. Large amounts of ssDNA templates can be produced if the DNA is cloned into vectors containing the origin of replication of a single-stranded bacteriophage such as M13 (Chapter 7).

Sequencing reactions require a primer. If the template DNA is a subclone that was inserted into the multiple cloning site of a vector, then the primer almost always is an oligonucleotide complementary to sequences flanking the multiple cloning site. This allows any fragment cloned into the multiple cloning sites to be sequenced using the same primer. Most vectors have the *lacZ* sequences flanking the multiple cloning sites. Thus, an oligonucleotide primer directed to this sequence, which is 16–17 nucleotides long, is commonly used and is called the **M13 universal primer**. This primer is designed to be complementary to the strand of DNA packaged into M13 or into any plasmid vector containing an M13 origin of replication. Methods also have been developed for direct sequencing from denatured plasmid DNA (Mierendorf and Pfeffer 1987), which eliminates the need to isolate or subclone DNA fragments. Direct sequencing from plasmid DNA is particularly convenient when analyzing large numbers of constructions.

## Sequencing Strategies for Longer DNA Segments

Because sequence data can only be obtained for several hundred to 1000 nucleotides from a single reaction set, one of two sequencing strategies must be adopted when the goal is to sequence long regions of DNA: directed or random (Figure 8.5).

Directed strategies provide direct and sequential analysis of a large DNA segment from one end to the other by walking, deletion, or transposon insertion (Sambrook et al. 1989). Random methods are often called 'shotgun' sequencing strategies (Figure 8.5B). In the 'shotgun approach', the DNA can be digested

## A)  Directed strategy

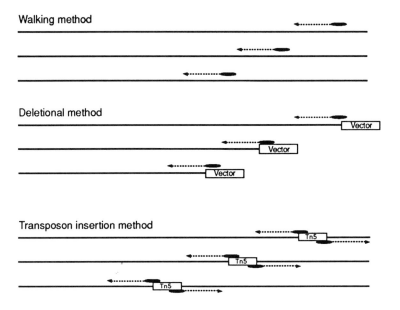

**Walking method**

**Deletional method**

**Transposon insertion method**

## B)  Random strategy

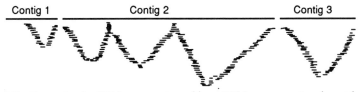

Contig 1                Contig 2                Contig 3

**Figure 8.5**   Strategies for DNA sequencing of long DNA segments involve either directed (A) or random (B) methods. Directed methods include walking, deletional sequencing, or transposon insertion. (Adapted from Hunkapiller et al., 1991, *Science*, 254, p. 61; copyright 1991 by the AAAS.)

with a restriction endonuclease and subfragments cloned and sequenced. The nucleotide sequence of the various inserts (200–500 nucleotides) is established and a computer is used to determine how the various fragments (contigs) overlap and to establish the entire sequence of the original DNA fragment used to generate the subclones. Disadvantages to the 'shotgun' method are that it may underrepresent some fragments; sequencing is redundant to ensure that the majority of the larger sequences have been included in one or more subclones; and there is no way to identify specific fragments.

## Sequencing DNA by the Polymerase Chain Reaction

DNA fragments can be amplified by the **polymerase chain reaction** (PCR) directly from genomic or cloned DNA. In this procedure, a specific segment of DNA can be amplified by one million-fold or more using *Taq* DNA polymerase on other thermophilic DNA polymerases. The PCR eliminates the need to prepare large amounts of DNA for sequencing and subcloning steps. The use of ***Taq* DNA polymerase** in DNA sequencing is described in Chapter 9. Because *Taq* polymerase tolerates high temperatures, DNA can be denatured at higher temperatures to reduce any secondary structures that would inhibit the sequencing reaction.

Conventional PCR requires primers of known sequence that flank the region to be amplified. Several techniques have been developed for sequencing double-stranded DNA produced by the PCR (Chapter 9). Asymmetric PCR allows the preferential amplification of only one of the complementary DNA strands so that single-stranded DNA template is produced for dideoxy sequencing.

## The Maxam and Gilbert Sequencing Method

The **Maxam and Gilbert DNA sequencing** method is also called the chemical-cleavage method (Maxam and Gilbert 1977, 1980). It uses chemical reagents to generate base-specific cleavages of the DNA to be sequenced (Figure 8.6, Table 8.2). It is less used today than the Sanger method, in part because the chemicals used are toxic and the methods are more labor intensive. The primary advantage of the Maxam–Gilbert method over the Sanger chain-terminating method is that DNA sequences are obtained from the original DNA molecule and not from a synthesized copy. Thus, one can analyze DNA modifications such as methylation and study DNA secondary structure and the interaction of proteins with DNA by the Maxam–Gilbert method.

To start, one needs pure DNA that has been cut by restriction endonucleases to generate DNA fragments of specific length and with known sequences at one end. Each DNA fragment then must be radioactively labeled at one end with a $^{32}$P-phosphate group in sufficient quantity that at least four different chemical reactions can be carried out. Next, specific bases in the DNA fragment are altered in at least four separate chemical reactions (Table 8.2). For example, guanine is methylated by dimethyl sulfate. Each reaction is carried out in a manner that limits the reaction so that, for example, only one guanine base is modified per several hundred guanine nucleotides. The altered base is then removed, in a subsequent step, by cleavage at the modification points with piperdine. The result is a set of end-labeled fragments of different lengths that will show up as a ladder of DNA bands on the gel because the reaction was

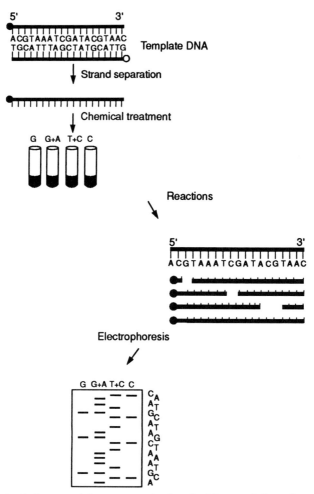

**Figure 8.6** A diagram of the steps involved in the Maxam–Gilbert chemical cleavage method of DNA sequencing. (Modified from Hunkapiller et al., 1991, *Science*, 254, p. 60; copyright by the AAAS.)

limited and not all the guanines were altered in the reaction. Four different reaction samples (one for A>C, G, T+C, and C) are then run side by side on a sequencing gel and the results are visualized by autoradiography (Figure 8.6).

## Analyzing DNA Sequence Data

DNA sequencing projects that generate substantial amounts of data will require computer assistance to analyze them (Howe and Ward 1989, Doolittle

**Table 8.2**

Reagents for Maxam–Gilbert DNA Sequencing

| Base specificity | Base-modifying reagent | Base-modifying reaction |
| --- | --- | --- |
| G | Dimethyl sulfate | Methylation of N7 makes C8-C9 bond susceptible to cleavage by base |
| G+A | Formic acid | Weakens bond of A+G |
| T+C | Hydrazine | Opens pyrimidine rings, which makes them susceptible to removal |
| A | Hydrazine + NaCl | Opens cytosine rings |

1990, Weir 1990, Gribskov and Devereux 1991). Software packages are available for all common laboratory computer systems and, depending on the size of the computer, can analyze the sequences in greater or lesser detail. Sequence data can be put into the computer in several ways. Gels can be read visually by the scientist, digitized and entered, or scanned by automated laser scanners which can enter the data directly. Software has been developed that can interpret ambiguities in the autoradiograph.

The automated methods provide speed and help to minimize clerical errors that can occur. The computer can compare readings from several sequencing runs, search for and identify overlaps and contiguities between runs, compare results from sequencing the complementary strands of the DNA, and identify possible errors. Once the sequence data have been entered into the computer and possible clerical or experimental errors resolved, the next step is to analyze the data.

In a random or shotgun sequencing project, the DNA is broken into fragments of about 400 nucleotides, which are cloned and sequenced. The relationships between the cloned fragments are determined by comparing their sequences. DNA segments that are related to one another by a partial overlap are called **contigs**. If a sequence overlaps with another, then the two contigs can be joined. The process of comparing sequences and aligning them is continued until it is possible to produce a continuous sequence for the DNA of interest.

A variety of questions can be asked. The sequence can be searched for all known restriction endonuclease target sites and the computer can generate a comprehensive and precise restriction map for the DNA. The sequences can be searched for interesting structures such as **tandem repeats** and **inverted repeats**. The sequences can be used to predict which proteins are coded for based on the sequences in the possible open reading frames on each of the two strands. (Both strands must be interpreted because it is not known in advance which is the coding strand and which is the noncoding strand.) An **open reading frame** is a segment of DNA that does not include a termination codon, and may indicate that a polypeptide coding region exists in the sequence. Finally, the DNA sequence itself, or the deduced polypeptide sequence, may be compared with a data bank of other sequences. Often, because of the degeneracy of the DNA code, similarities are found between two polypeptide sequences which would not have been apparent if the comparison had been carried out at the DNA sequence level alone.

DNA sequence similarities may be present because of **convergent evolution** or through **homology**. Convergent evolution implies that the two sequences did not have a common ancestral sequence, but that selection for a particular function in two different lineages has converged on a particular structure or related structures. The term homology has been used differently by different people. Traditionally, similar structures in different organisms have been called homologous if the organisms have descended from a common ancestor. However, some molecular biologists have used the term 'percent homology' to imply a degree of *similarity* in DNA sequence, which may or may not be due to descent from a common ancestor. That use is misleading and should be discontinued.

A search of DNA or protein sequence banks for similarities with any newly discovered sequence may turn up interesting information. Amazing degrees of sequence similarities have been discovered. For example, in D. *melanogaster*, **homeotic genes**, genes that direct cells in different segments to develop in particular patterns, have been cloned. The **homeobox**, a segment of approximately 180 bp, was found to be characteristic of homeotic genes, and probes using the homeobox have been used to isolate previously unknown homeotic genes in other insects. Even more interesting, sequences homologous to the homeobox have been isolated from various vertebrates, including mice and humans, indicating that similar genetic mechanisms may control some aspects of development in higher organisms. The high degree of conservation between the homeobox sequences of *Drosophila*, frogs, mice, and humans indicates that these sequences have been conserved over more than 500 million years, which is when invertebrates and vertebrates are thought to have diverged. More than 400 genes have already been identified that are similar in the human and *Drosophila* (Koshland 1991). Thus, information obtained from the Human Genome Project and from the *Drosophila* Genome Project is expected to provide significant advances in our understanding of evolution and developmental biology.

The DNA sequence data banks are expanding rapidly and are important resources for the research community. There are three major DNA sequence databases—the DNA Data Bank of Japan (DDBJ), the European Molecular Biology Laboratory Nucleotide Sequence Data Library (EMBL), and the GenBank Genetic Sequence Data Bank (GenBank). There were over 50 million nucleotides of DNA sequences in Genbank by 1991 and the entries are growing exponentially (Cinkosky et al. 1991). Efforts are being made to ensure that the data entered are accurate, but errors apparently are common in published accounts in both the data banks and scientific journals. Errors, particularly in noncoding regions, may arise from outright sequencing errors or clerical errors. Submission of a sequence in a machine-readable form is becoming a prerequisite for publishing in many journals and can be accomplished by using electronic computer mail systems.

## Future Developments in DNA Sequencing Methods

The demand for improved DNA sequencing methods for the human genome project has spurred the development of innovative new approaches for high-speed DNA sequencing methods (Hunkapiller et al. 1991). The new technologies strive to improve efficiency, reduce costs, and reduce errors. Despite improved methods, current automated techniques are still considered inadequate for truly large-scale sequencing projects. They can turn out only 2000 sequenced bases per day, a rate that would require 1.5 million worker-days to sequence the entire human genome. As in small-scale projects, large-scale sequencing projects involve DNA cloning, physical mapping, subcloning, sequencing, and information analysis, but because of the complexity and volumes of data generated, the different processes must be coordinated and automated. Improvements in methods and organization must be achieved over the next 10 years by 100- to 1000-fold if the Human Genome Project is to be completed rapidly and efficiently (Hunkapiller et al. 1991). DNA sequencing currently costs from $2 to $10 per base, but improved methods may result in a 100-fold decrease in cost.

Some new sequencing techniques use different fluorescent dyes to mark specific nucleotides which can then be scanned by multiple-wavelength fluorescence. Others use methods that allow analysis of sequences from very small amounts of DNA, which can eliminate a time-consuming step in sequencing— the amplification of DNA to be sequenced. Improvements also are being attempted in electrophoresis. For example, capillary gel electrophoresis is being explored as a method to speed up the separation and improve the resolution of DNA molecules (Smith 1991). Sequence analysis by computerized scanning methods is already available and is being improved. Software for analysis and statistical comparisons with DNA sequences in data banks will continue to improve.

A variety of new sequencing technologies are being investigated, including mass spectrometry, sequencing by hybridization, sequencing of single molecules, and atomic probe microscopy using scanning tunneling microscopes and atomic force microscopes. Research on sequencing methods is extensive and intensive, and Hunkapiller et al. (1991) suggested that within the next ten years there is a 50% chance that large-scale DNA sequencing methods will use entirely new techniques.

## The *Drosophila* Genome Project

One of the largest initiatives in biology today involves plans to sequence entire genomes. These mammoth undertakings have generated controversy and excitement. Concerns have been raised about the costs, time, and goals. The Human Genome Project is admitted to be difficult and expensive. The difficulty is demonstrated by a discussion of the numbers of bases to be sequenced. There are 23 human chromosomes, containing 30,000 to 90,000 genes. These genes are found in a total of 3000 million nucleotides. A reasonable Ph.D. project, until recently, would have involved the isolation, cloning, and sequencing of a single gene, taking approximately three years. This implies that at least 100,000 work-years of effort would be required to sequence the functional genes. The amount of nonfunctional DNA, which is 30 to 100 times as great, would require even more time to sequence. Thus, the Human Genome Project is looking for faster and cheaper methods for sequencing. It also is looking at sequencing the genome of organisms with smaller amounts of DNA.

The U.S. Human Genome Program has identified five organisms as models for the study of the human genome, including *E. coli*, the yeast *Saccharomyces cerevisiae*, the nematode *Caenorhabditis elegans*, *D. melanogaster*, and the laboratory mouse (NIH 1991). Analysis of the $1.65 \times 10^8$ bp of DNA in the *D. melanogaster* genome is considered particularly valuable because it contains large amounts of nonprotein coding DNA, introns, transposable elements, heterochromatin, centromeres, and complex gene structures, all of which probably have interesting roles in genome function and evolution. More important to entomologists, however, is the revolutionary effect this project will have on our knowledge of insect genetics.

Knowledge of the structure and function of the *Drosophila melanogaster* genome already is far greater than that for any other multicellular organism (Kafatos et al. 1991). Since 1970, more than 12,000 papers have been published on *D. melanogaster*. To date, approximately 3800 different genes, perhaps a fourth of the total, have been mapped by recombination studies. Many of these 3800 genes have been associated cytogenetically with one of the 5000 bands of the polytene salivary gland chromosomes. Approximately 3000 transcription units have been placed on the cytogenetic map by using *in situ* hybridization to localize the DNA on specific polytene chromosomes. Approximately 2% of the unique sequence DNA of *Drosophila* has been sequenced and is in data banks. It is now possible, although tedious, to clone and sequence any gene for which a mutant is available (Kafatos et al. 1991).

The *Drosophila* Genome Project is expected to support research in the following areas: (1) Develop a high-resolution **physical map**, which will serve as a basis for DNA sequencing and detailed functional studies. A physical map is a series of overlapping clones for which information is available on the sequences at their ends and knowledge of their physical location. The physical map should be placed within a database that has cross-references to the genetic information (banding patterns, mutations, etc.) already available for *D. melanogaster*. With such an integrated database of the *Drosophila* genome, it would be easy to recover any gene of interest by accessing the pertinent clones from a central distribution center. Once the physical map is available, it should be possible to develop a more thorough understanding of the structure, function, organization, and evolution of genes and chromosomes in this model insect. (2) Conduct feasibility studies for large-scale DNA sequencing projects, especially for regions containing DNA of great biological interest. Large-scale studies are defined as those that attempt to determine three megabase pairs of contiguous DNA sequence within three years. (3) Develop new techniques to identify coding sequences in genomic DNA and to obtain high-quality cDNA libraries that are representative of the complete coding information for the genomic DNA.

The *Drosophila* Genome Project has begun (Merriam et al. 1991). The availability of a physical map is an important starting point for sequencing the *Drosophila* genome and several laboratories are working on different aspects of this. For example, Siden-Kiamos et al. (1990) are producing overlapping cosmid clones that are assigned to specific chromosome sites, and a master library of 20,000 cosmids already has been used to map a portion of the X chromosome. Likewise, Hoheisel et al. (1991) report efforts to develop reference libraries to physically map the *D. melanogaster* genome.

## References Cited

Ambrose, B. J. B. and R. C. Pless. 1987. DNA sequencing: Chemical methods. Methods Enzymol. 152: 522–538.

Barnes, W. M. 1987. Sequencing DNA with dideoxyribonucleotides as chain terminators: Hints and strategies for big projects. Methods Enzymol. 152: 538–556.

Davies, R. W. 1982. DNA sequencing. Pp. 117–172. In: *Gel Electrophoresis of Nucleic Acids.* D. Rickwood and B. D. Hames, Eds. IRL Press, Oxford.

Cinkosky, M. J., J. W. Fickett, P. Gilna and C. Burks. 1991. Electronic data publishing and GenBank. Science 252: 1273–1277.

Doolittle, R. F., Ed. 1990. *Molecular Evolution: Computer Analysis of Protein and Nucleic Acid Sequences* (Methods Enzymol., Vol. 183), Academic, New York.

Gribskov, M. and J. Devereux, Eds. 1991. *Sequence Analysis Primer.* Stockton Press, New York.

Hoheisel, J. D., G. G. Lennon, G. Zehetner and H. Lehrach. 1991. Use of high coverage reference libraries of *Drosophila melanogaster* for relational data analysis: A step towards mapping and sequencing of the genome. J. Mol. Biol. 220: 903–914.

Howe, C. J. and E. S. Ward, Eds. 1989. *Nucleic Acids Sequencing: A Practical Approach.* IRL Press, New York.

Hunkapiller, T., R. J. Kaiser, B. F. Koop and L. Hood. 1991. Large-scale and automated DNA sequence determination. Science 254: 59–67.

Kafatos, F. C., C. Louis, C. Savakis, D. M. Glover, M. Ashburner, A. J. Link, I. Siden-Kiamos and R. D. C. Saunders. 1991. Integrated maps of the *Drosophila* genome: Progress and prospects. Trends Genet. 7: 155–161.

Koshland, D. E. 1991. Flying into the future. Science 254: 173.

Maxam, A. M. and W. Gilbert. 1977. A new method for sequencing DNA. Proc. Natl. Acad. Sci. USA 74: 560–564.

Maxam, A. M. and W. Gilbert. 1980. Sequencing end-labeled DNA with base-specific chemical cleavages. Methods Enzymol. 65: 499–560.

Merriam, J. M. Ashburner, D. L. Hartl and F. C. Kafatos. 1991. Toward cloning and mapping the genome of *Drosophila*. Science 254: 221–225.

Mierendorf, R. C. and D. Pfeffer. 1987. Direct sequencing of denatured plasmid DNA. Methods Enzymol. 152: 556–562.

National Institutes of Health. 1991. Request for applications: Mapping the *Drosophila* Genome RFA HG-91-05. National Center for Human Genome Res., Bethesda, MD.

Sambrook, J., E. F. Fritsch and T. Maniatis. 1989. DNA sequencing. Chapter 13. In: *Molecular Cloning: A Laboratory Manual*. 2nd Ed. Cold Spring Harbor Laboratory Press, Cold Spring Harbor, New York.

Sanger, F., S. Nicklen and A. R. Coulson. 1977. DNA sequencing with chain-terminating inhibitors. Proc. Natl. Acad. Sci. USA 74: 5463–5467.

Sealey, P. G. and E. M. Southern. 1982. Electrophoresis of DNA. Pp. 39–76. In: *Gel Electrophoresis of Nucleic Acids*. D. Rickwood and B. D. Hames, Eds. IRL Press, Oxford.

Siden-Kiamos, I., R. D. C. Saunders, L. Spanos, T. Majerus, J. Treanear, C. Savakis, C. Louis, D. M. Glover, M. Ashburner and F. C. Kafatos. 1990. Towards a physical map of the *Drosophila melanogaster* genome: Mapping of cosmid clones within defined genomic divisions. Nucleic Acids Res. 18: 6261–6270.

Smith, L. M. 1991. High-speed DNA sequencing by capillary gel electrophoresis. Nature 349: 812–813.

Tabor, S. and C. C. Richardson. 1987. DNA sequence analyses with a modified bacteriophage T7 DNA polymerase. Proc. Natl. Acad. Sci. USA 84: 4767–4771.

Weir, B. S. 1990. *Genetic Data Analysis*. Sinauer Assoc., Sunderland, MA.

# Chapter 9

# DNA Amplification by the Polymerase Chain Reaction: Molecular Biology Made Accessible

## Overview

Occasionally a technique is developed that offers the possibility of changing the kinds of questions that can be answered in biology. Within a few short years, the polymerase chain reaction (PCR) has become just such a powerful method for solving myriad basic and applied problems. The PCR is a method for amplifying small amounts of DNA or RNA. It can be used to isolate DNA fragments, end-label DNA, clone cDNA and genomic DNA, sequence DNA, mutate specific DNA sequences, alter promoters, and quantitate the amount of RNA and DNA. Modifications of the PCR continue to be developed and these lead to additional new applications.

The PCR requires DNA polymerase, dNTPs, a template DNA molecule, and primers. The sequences at each end of the DNA to be amplified, or sequences of flanking DNA at each end, are needed in order to synthesize appropriate primers. With PCR applications in which two specific primers are used, amplification of DNA is geometric, producing large quantities of specific sequences of DNA suitable for sequencing, cloning, or probing. PCR techniques that use single primers, such as RAPD-PCR, also can result in DNA amplification. RAPD-PCR is a technique that uses short, randomly chosen primers to amplify DNA segments for which no genetic information is available. The resulting banding patterns provide information about genetic variation within the entire genome of insects.

The power of the PCR to amplify DNA is dramatic; theoretically even a single molecule can be amplified to produce any amount of copies. However, this power creates formidable problems with contamination and requires careful organization of PCR experiments and the use of adequate controls. The relative ease with which the PCR can be used by novices in molecular biology has made it possible for a diverse group of biologists to apply a new technology to study molecular systematics, evolution, ecology, behavior, and development.

Recently, other methods, including the ligase chain reaction (LCR) and several other amplification techniques, have become available to amplify and/or detect specific nucleic acids.

## Introduction

The **polymerase chain reaction (PCR)** is an *in vitro* or cell-free method for synthesizing specific DNA or RNA sequences in nearly any amount required. It is one of the most accessible and versatile techniques available to entomologists interested in both basic and applied problems. The PCR is recognized to be one of the most powerful new techniques in biology. It is used to isolate specific DNA fragments, end-label DNA, mutagenize specific DNA fragments, clone cDNA and genomic DNA, sequence DNA, quantitate RNA and DNA, and alter a variety of sequences to study gene expression. DNA polymerase and the PCR technique were designated the Molecule of the Year by *Science* magazine in 1989 (Guyer and Koshland 1989). The PCR has become a standard laboratory method in an extraordinarily short time since it was invented in 1985 (K. Mullis, U.S. patent 4,683,195, July 1987; U.S. patent 4,683,202, July 1987). Between 1985 and 1989, the PCR was cited in more than 1000 publications and in 1993 Kary Mullis received the Nobel Prize in chemistry for his work on the PCR.

Improvements in and optimization of the PCR have led to its use by numerous scientists. The PCR has rapidly become a common procedure in forensics and diagnostics, and is revolutionizing studies of basic biology, ecology, and evolution. This relatively simple technique has given scientists who have limited experience in molecular biology the opportunity to apply molecular genetic techniques to diverse problems (Arnheim et al. 1990).

While the PCR technique is relatively simple, the process is, in fact, not completely understood. A number of parameters can be modified to optimize the PCR (Carbonari et al. 1993). The PCR involves complex kinetic interactions between the template or target DNA, product DNA, oligonucleotide primers (polymers made up of 10–30 nucleotides), deoxynucleotide triphosphates (dNTPs), buffer, and enzyme (DNA polymerase). These relationships change during the course of the reaction.

The PCR works well with most DNA targets, but adjustments are usually needed in reaction parameters in order to improve specificity and yield. These include altering the reaction buffer (particularly the $MgCl_2$ concentration); relative concentrations of template DNA, primer, dNTPs, and enzyme; annealing time and temperature; and extension time and temperature. No single protocol is appropriate for all situations and each new application will require optimization. For example, amplifying a 100-bp DNA fragment is not equivalent to amplifying a 10-kb fragment.

This chapter introduces the basic principles of the PCR, discusses some of the modifications of the basic method, identifies some of the current methods and applications, and provides references for additional information. PCR technology is changing rapidly and new applications and methods of significance to entomologists will no doubt become available.

## The Basic Polymerase Chain Reaction

The PCR involves combining a DNA sample (the template) with oligonucleotide primers, deoxynucleotide triphosphates (dNTPs), and a **DNA polymerase** in a buffer. The specificity of the basic PCR depends on base-pairing by the two oligonucleotide primers to the target DNA that serves as the template. The primers, which are sequences that flank the DNA template to be amplified, anneal to the strands of the denatured template DNA. Repeated PCR cycles involve heat denaturation to separate the template DNA strands, annealing of primers to the complementary DNA sequences, and extension of the new DNA strands by DNA polymerase, with the base sequence of the new strands determined by the template DNA. The primers are constructed so that DNA synthesis proceeds across the region *between* the annealed primers (Figure 9.1). This mixture is repetitively heated and cooled until the desired amount of amplification of the template DNA is achieved (Figure 9.1).

The first few cycles in the PCR are particularly critical for accurate and efficient amplification of DNA sequences (Ruano et al. 1991). All cycles of the PCR begin with denaturing of the template DNA and any previously synthesized product so that the template DNA is single stranded. As the temperature is lowered, the primers anneal to the complementary sequences of the template DNA. The annealing step of the early cycles requires the primers to 'scan' the genomic DNA template for the correct target sequences to which they can anneal. Because much of the genomic DNA will not have the correct complementary sequence, annealing during early cycles may not be as efficient as it is during the middle cycles. In the first few cycles, interactions of primers with genomic DNA template can lead to products that are not specific. The PCR will be specific only if the two primers bind to sites on the complementary strand of the DNA and these sites are less than 10 kb apart.

During the middle cycles, the previously synthesized product is the preferred template for the primers, so the target template is perfectly demarcated (Figure 9.1). Finally, in the late cycles, amplified products in high concentration will hybridize to themselves, thus blocking the primers from their complementary sites. DNA sequences up to about 10 kb in length can be synthesized, although sequences of 2 kb or less are more readily obtained.

The power of the PCR is based on the fact that the products of one replication cycle can serve as a template for the next. Each successive cycle essentially doubles the amount of DNA synthesized in the previous cycle, which

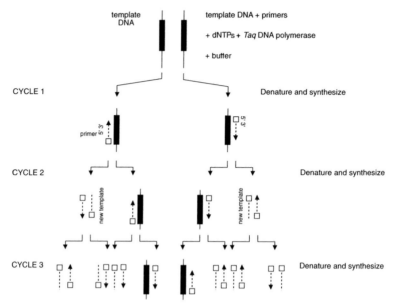

**Figure 9.1**   The standard PCR method. Template DNA is isolated and mixed with primers, dNTPs, and *Taq* DNA polymerase in a buffer with $Mg^{++}$. The double-stranded template DNA is heated to denature it so that the primer can anneal to single-stranded target DNA. *Taq* synthesizes a new, single strand of complementary DNA using the primer to initiate synthesis. The dNTPs are added in a sequence determined by the template DNA strand. The process of denaturation, annealing of the primer to the template, and DNA synthesis (or extension) is called a cycle. The DNA is once again heated to denature it, and all strands can serve as templates for DNA synthesis. As a result, the synthesized DNA increases in a geometric manner as it doubles each cycle. The DNA is synthesized from the 5' to the 3' direction. The primer is indicated by the small rectangle, while the new DNA strand is a dashed line. The solid lines represent the original DNA strands. After three cycles, the two template strands (one double-stranded molecule) have become 16 (8 double-stranded molecules).

results in the exponential accumulation of the specific target DNA at approximately $2^n$, where n is the number of cycles. The ends of the discrete DNA product are determined by the 5' ends of the primers (Figure 9.1).

Table 9.1 illustrates a procedure suitable for amplifying genomic DNA from *Drosophila* and demonstrates the relative simplicity of the technique, once specific conditions have been optimized for a particular problem. Table 9.2 lists some of issues that must be considered in setting up new PCR reactions.

**Table 9.1**

Example of Standard PCR Reaction Protocol for Amplifying *Drosophila* DNA

---

1. Set up a 100-$\mu$l reaction in a 0.5-ml microfuge tube, mix, and overlay with 75 $\mu$l of mineral oil:
   Template DNA ($10^5$ to $10^6$ target molecules)
   20 pmol each primer (each primer 18–30 nucleotides long)
   100 mM Tris–HCl (pH 8.3 at 20°C)
   10 mM $MgCl_2$
   0.05% Tween-20
   50 $\mu$M each dNTP
   2 units of *Taq* DNA polymerase

2. Perform 25 to 35 cycles of PCR using the following temperature profile:
   Denaturation at 96°C, 15 seconds (a longer initial time is usually desirable)
   Primer annealing at 55°C, 30 seconds
   Primer extension at 72°C, 1.5 minutes

3. Cycling should conclude with a final extension at 72°C for 5 minutes. Stop reactions by chilling to 4°C and/or adding EDTA (ethylene dinitrilotetraacetic acid, a chelating agent) to 10 mM.

---

The PCR is performed in commercially available temperature cyclers that allow the programming of the three fundamental reaction conditions: denaturing, annealing, and extending (Figure 9.2). A typical amplification cycle would involve denaturing the template DNA at 94°C for 20 seconds, annealing the primers to the template DNA at 55°C for 20 seconds, and extension (or synthesis) of the DNA at 72°C for 30 seconds. Because the instruments require time to heat and cool to a specific temperature, the actual cycle time may require 10 minutes or more, depending upon the machine used. If 25 cycles are performed, then the total time required will be approximately four hours and the target DNA will have been amplified approximately one millionfold.

### DNA Polymerases

The PCR, as first described in 1985, used the **Klenow fragment** of *Escherichia coli* DNA polymerase I at 37°C to produce copies of target DNA (Mullis and Faloona 1987, Saiki et al. 1985). Because the Klenow fragment is heat sensitive, fresh enzyme had to be added to each cycle, making the PCR an exceedingly tedious procedure.

The efficiency and fidelity of the PCR was dramatically improved by employing a heat-resistant DNA polymerase (**Taq DNA polymerase**) so that the procedure could be carried out at high temperatures (Saiki et al. 1988, Eckert and Kunkel 1991, Taylor 1991). *Taq* DNA polymerase was isolated from the bacterium *Thermus aquaticus*, which was collected from a hot spring in Yellowstone National Park. Because *Taq* DNA polymerase can withstand repeated exposures to temperatures up to 94–95°C, its use greatly increased the ease with which multiple amplifications of template DNA could be achieved. *Taq* DNA polymerase needs to be added to the reaction only once.

*Taq* DNA polymerase is commercially available in both native and cloned form. A more thermostable recombinant DNA polymerase called the Stoffel fragment also is available. The Stoffel fragment persists longer at 97.5°C and exhibits optimal activity over a broader range of magnesium ion concentrations, as well as lacking intrinsic 5′ to 3′ exonuclease activity. A recombinant version of another enzyme originally isolated from *Thermus thermophilus* can be used to reverse transcribe RNA and to synthesize cDNA in a procedure called RNA-PCR. A method of purifying *Thermus aquaticus* DNA polymerase also has been published (Engelke et al. 1990).

*Taq* DNA polymerase is a 94-kilodalton (kDa) protein with a temperature optimum of approximately 75–80°C. It can extend more than 60 nucleotides per second at 70°C with a GC-rich 30-oligomer primer. In a PCR mixture, *Taq* DNA polymerase retains 50% of its activity after about 40 minutes at 95°C. Use of *Taq* has increased the specificity and yield of PCR over that possible with the Klenow fragment because the primers can be annealed and extended at higher temperatures, which eliminates much of the nonspecific amplification. Also, longer PCR products are produced, because the secondary structure of the template DNA is eliminated at higher temperatures. Fragments about 500 bp long can be synthesized with the Klenow fragment, but fragments up to 10 kb in length can be produced with *Taq* DNA polymerase.

The accuracy of the PCR has been estimated (Mullis et al. 1986, Keohavong and Thilly 1989). *Taq* DNA polymerase has no 3′ to 5′ exonuclease (or proofreading) activity, but has a 5′ to 3′ exonuclease activity during polymerization (Erlich et al. 1991). This means that initial estimates of a misincorporation rate of $10^{-4}$ nucleotides per cycle by *Taq* during the PCR were too high. Current estimates of misincorporation rates are $10^{-5}$ nucleotides per cycle in a PCR reaction with optimized conditions.

### Primers

While all the components of a PCR are important, primers are crucial to achieve a successful PCR (Table 9.2). Many of the modifications of the PCR have involved modifying the number, size, and specificity of the primers used, as described later.

What is a primer and where do you get them? A primer is a short (10–30 nucleotides) polymer of oligonucleotides. The basic PCR method requires either that the sequence, or the flanking sequences, of the DNA targeted for amplification by the PCR be known in order to synthesize a primer. Primers anneal to the target DNA by complementary base pairing, with A annealing to T, and C to G. The primers determine the length, specificity, and nature of the amplified DNA produced by PCR. Standard PCR requires a pair of primers that flank the target DNA to be amplified. DNA extension occurs from each 3' OH end of the primer, so that the ends of the amplified DNA are defined by the 5' ends of the primers. The length of the DNA generated during the PCR is equal to the sum of the lengths of the two primers plus the distance in the template DNA located between the primers. For some PCR applications, single primers can be used, as explained later.

Standard primers can, in some cases, be purchased from commercial enterprises in purified form ready for use. Most primers must be synthesized to order based on the specific goal of the experiment. Primers are usually synthesized by phosphoramidate chemistry on a DNA synthesizer. A number of commercial suppliers will provide this service, with the price based on the number of bases in the oligonucleotides. Primers are sometimes called 10-mers or 20-mers, based on their length.

Selecting efficient and specific primers for the standard PCR remains somewhat empirical, although computer programs have been developed to aid in designing appropriate primers (Table 9.2). It is desirable, where possible, to select primer pairs with an average G+C content of around 50% and a random base distribution. It is important to avoid complementary 3' ends of the primer pairs to avoid primer-dimer artifacts that will reduce the yield of the desired DNA. Runs of three or more Cs or Gs at the 3' ends of primers may promote mispriming at G+C-rich sequences. Palindromic sequences (that is, sequences that read the same from either end) within primers also should be avoided. Sequences that will yield a significant secondary structure should be avoided. Computer programs can assist in determining whether a particular primer sequence will have any of these problems.

Most standard primers are 20 to 30 nucleotides long. The specific sequence of bases in the traditional primer pair is based on the known DNA sequence being amplified. This usually means that the DNA being amplified is from a gene that is characterized and has been sequenced, or is from a related species. Primers can be constructed that contain extensions so that restriction enzyme sites, regulatory codons, or labels can be added to the amplified DNA. These sequences will be incorporated into the 5' end of the target sequence.

### Preparing DNA Samples

The PCR is so sensitive that relatively crude DNA preparations may be used as templates. Ideally, $10^5$ to $10^6$ target molecules of DNA will be available as template DNA, although successful PCRs have been achieved with one or a few DNA molecules from a single cell. There must be at least one intact DNA strand that includes the region to be amplified, and impurities must be sufficiently

**Table 9.2**

Optimizing Standard PCR Involves Optimizing Reaction Components

| PCR component | Issues to consider |
|---|---|
| Primer | 1. Select primers with a random base distribution and GC content similar to those of template DNA being amplified.<br>2. Avoid primers with stretches of polypurines and poly-pyrimidines or other unusual sequences.<br>3. Avoid sequences with a secondary structure, especially at the 3' end.<br>4. Check primers for complementarity and avoid primers with 3' overlaps to reduce primer-dimer artifacts.<br>5. Construct primers 20–30 nucleotides long.<br>6. Optimize amount of primers used for each reaction. |
| Template DNA | 1. Template DNA should be free of proteases that could degrade the *Taq* DNA polymerase.<br>2. Template DNA with high levels of proteins or salts should be diluted or cleaned up to avoid inhibition of *Taq* activity.<br>3. Highly concentrated template DNA may yield non-specific product or inhibit the reaction. |
| PCR buffer | 1. $MgCl_2$ concentration is very important.<br>2. Excess $Mg^{++}$ promotes production of nonspecific product and primer-dimer artifacts.<br>3. Insufficient $Mg^{++}$ reduces yields.<br>4. Presence of EDTA or other chelators can affect the availability of $Mg^{++}$. |
| *Taq* polymerase | 1. Excessive *Taq* concentrations can yield nonspecific products and reduce yield. Recommended concentrations are between 0.5 and 2.5 units per 100 µl reaction. Add the *Taq* at 94°C and mix thoroughly.<br>2. Stringency can be increased by increasing the annealing temperature, adjusting dNTP concentrations, and minimizing incubation time. |
| dNTPs | 1. dNTP concentrations should be equivalent to minimze misincorporation errors.<br>2. Low dNTP concentrations minimize mispriming, but if too low, can reduce the amount of product produced. |
| Cycle parameters<br>  Incubation | 1. Time varies with length of target being amplified; 1 minute per kb is average. |

*(continues)*

**Table 9.2**  (*Continued*)

| PCR component | Issues to consider |
|---|---|
| | 2. Ramp time (time to change from one temperature to another) should be minimized to improve specificity. Check your machine to be sure. |
| | 3. Insufficient heating for denaturing step is a common problem; 94°C results in complete separation, but excess time can cause denaturation of *Taq* polymerase. |
| Annealing | 1. Annealing temperature depends on length and GC content of primers; 55°C good for primers 20 nucleotides long (50% GC). |
| | 2. Higher annealing temperatures may be needed to increase primer specificity. |
| Cycle number | 1. Optimum number varies with starting concentration of template DNA, and all of the above parameters. |
| | 2. Too many cycles increases amount of nonspecific product, while too few results in a low yield. |

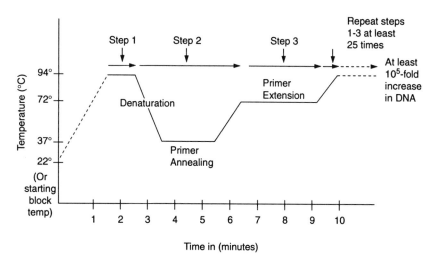

**Figure 9.2**  Example of a typical PCR protocol. Step 1 involves denaturating the double-stranded DNA template at 94°C. Step 2 involves annealing the primer to the single-stranded target DNA by base-pairing. Step 3 involves synthesis of new DNA from the 3′ end of the primer by *Taq* DNA polymerase using dNTPs in a sequence determined by the template DNA. Steps 1–3 are a cycle, and approximately 25 cycles will yield an increase in DNA content by a factor of approximately one million ($2^{25}$). (Figure used with permission of Perkin-Elmer Corp. 1988.)

dilute so that polymerization is not inhibited (Table 9.2). Concentrations of 0.05 up to 1.0 μg of genomic DNA typically are used for amplification of single-copy genes. Less DNA (0.5 to 2 ng) can be used to amplify multiple-copy genes such as nuclear ribosomal RNA genes, because these genes are repeated about 200 to 500 times in the genome of eukaryotes.

It is possible to amplify DNA in crude lysates, particularly if the amount of material is small. Potential inhibitors of the PCR can be eliminated by boiling or dilution. Lysing cells in boiling water is a quick and effective method of preparing DNA for PCR, although only a small volume can be used or the accumulation of cellular debris will inhibit the reaction. Cells in complex biological fluids or cells resistant to lysis require additional processing. Template DNA should be free of exogenous and endogenous proteases which could degrade the *Taq* DNA polymerase, nucleases, DNA binding proteins, and high levels of heat-precipitable proteins.

Table 9.3 describes a protocol for preparing genomic DNA suitable for the PCR from a single *Drosophila melanogaster* (Jowett 1986). Other techniques or variations are possible, and simpler (Table 9.4). For example, Singer-Sam et al. (1989) reported that Chelex, a polyvalent chelating agent in resin form, prevents degradation of DNA heated in low-ionic-strength buffers, probably by chelating heavy metal ions that may serve as catalysts in the breakdown of DNA. Adding Chelex during boiling appears to increase the amount of DNA produced from samples containing low levels of template DNA. This preparation method has been used to isolate DNA from hundreds of individual insects or mites either for the standard PCR or RAPD-PCR, which is discussed later (Edwards and Hoy 1993).

### PCR Automation

The polymerase chain reactions consist of repeated cycles involving at least two temperatures and generally three (Figure 9.2). The high temperature is needed to denature the DNA template and subsequent product. The low temperature should allow annealing of the primer to the DNA template and extension of the DNA by the DNA polymerase. A third, intermediate temperature near the optima for DNA polymerase extension is used for the extension phase. These temperature cycles can be achieved by moving samples between two or more water baths. However, this is extremely tedious and it is difficult to maintain precise control over timing. The annealing temperatures should be neither too low nor too high for efficient amplification. An algorithm has been developed that is useful for determining the optimal annealing temperature for a given pair of oligonucleotides and a template DNA sequence (Rychlik et al. 1990). This algorithm can be computed easily with a computer program (Osborne 1992).

A variety of commercial thermal cyclers controlled by microprocessors auto-

**Table 9.3**

Extracting Genomic DNA from a Single *Drosophila melanogaster*[a]

---

1. In a 1.5-ml microfuge tube freeze a fly in liquid nitrogen. (Store at −70°C until needed.)

2. Thaw and add 100 μl of 10 mM Tris–HCl (pH 7.5) 60 mM NaCl, 50 mM EDTA, 0.15 mM spermine, 0.15 mM spermidine.

3. Grind fly with a yellow pipette tip.

4. Add 100 μl of 1.25% SDS, 0.3 M Tris–HCl, 0.1 M EDTA, 5% sucrose, 0.75% freshly added diethylpyrocarbonate (DEP).

5. Mix and incubate 30–40 minutes at 60°C to ensure lysis of the nuclei and release of DNA.

6. Cool and add 30 μl of 8 M potassium acetate.

7. Cool for 45 minutes on ice.

8. Spin for 1 minute in a microfuge.

9. Remove supernatant, avoiding the lipid on the surface, and add 2 volumes of ethanol.

10. Leave at room temperature.

11. Spin for 5 minutes and pour off the supernatant.

12. Wash the pellet with 70% ethanol.

13. Dry under vacuum.

14. Take up pellet in 25 μl of TE (10 mM Tris, 1 mM EDTA).

The method lyses the nuclei once the tissue is broken up. The SDS and protein form complexes. DEP is a protein denaturant and nuclease inhibitor. The protein–SDS complexes are precipitated by adding potassium, leaving the DNA in solution. The final DNA is contaminated with RNA, which can be removed by adding RNase to a concentration of 100 μg/ml.

---

[a] Adapted from Jowett 1986.

mate the rapid and precise heating and cooling required for maximum efficiency of the PCR. There are three basic categories of commercial temperature cyclers in which the samples are (1) heated and cooled by fluids, (2) heated by electric resistances and cooled by fluids, and (3) heated by electric resistances and cooled by semiconductors. Accuracy and reproducibility in temperature control should be a concern when choosing a temperature cycler. Most temperature cyclers are designed for use with 0.5-ml centrifuge tubes which are placed in holes in the heating and cooling blocks. A close fit between the walls of the block and the tube wall is critical in maintaining accurate temperatures, and 50 μl of glycerol or mineral oil in the well may assist in this. Temperature cycles also can be achieved in an oven, which will allow nonstandard containers to be used for the

**Table 9.4**

A Rapid Method for Extracting DNA from a Single Insect or Mite Using Chelex 100 Chelating Resin[a]

1. Add a single insect or mite to a microcentrifuge tube. Insects can be alive or frozen at −80°C.

2. Rap tube sharply to move the insect to the bottom of the tube. If the insect or mite is difficult to detect visually, add a small amount of buffer and spin in a microfuge tube to ensure the specimen is at the bottom.

3. Immerse the bottom of each microfuge in liquid nitrogen. Freeze a plastic pestle in liquid nitrogen. (Pestles are prepared in advance by melting the ends of 200-μl pipette tips and molding them to the bottoms of microcentrifuge tubes.

4. Macerate the frozen specimen five to ten times within the tube with the frozen pestle.

5. Add 200 μl of a 5% (w/v) Chelex solution (Bio-Rad Laboratories).

6. Vortex the solution vigorously to thaw it.

7. Remove the pestle and place tube into a temperature cycler and heat to 56°C for 15 minutes.

8. Centrifuge the sample (>100 g) for 15 seconds in a nanofuge to allow removal of the DNA solution from the top of the tube. Avoid removing any Chelex resin from the bottom of the tube. The DNA can be used for both traditional and RAPD-PCR. The DNA is not suitable for cutting with restriction enzymes, ligation reactions, or DNA sequencing.

[a]Adapted from Edwards and Hoy 1993.

PCR. Thermocyclers should be checked periodically to determine how accurate they are and calibrated if needed; differences of even 1–2°C can be significant in the PCR.

Commercially available temperature cyclers cost between U.S. $3000 and $8000. The instruments differ in the design of the cooling system, control of ramping time between temperature steps, memory capacity for program storage, sequential linking of programs, and capacity of the heating block to hold different numbers of microcentrifuge tubes. Thermocyclers that do not provide a uniform temperature across a heating block can lead to variation in outcomes from the reactions taking place in different microfuge tubes. Different models or brands of temperature cyclers, while ostensibly programmed to produce the same temperature profiles, may not be not equivalent and also can alter the outcomes of the PCR.

### Specificity of the PCR

The specificity (or fidelity) of the PCR based on DNA synthesis by the Klenow enzyme is not high. The use of the *Taq* polymerase not only simplifies the PCR procedure but increases the specificity and overall yield of the reaction. The higher temperature optimum for the *Taq* polymerase (ca. 75°C) allows the use of higher temperatures for primer annealing and extension, which increases the stringency of the reaction and minimizes the extension of primers that are mismatched with the template.

The increased specificity of the PCR with the use of *Taq* DNA polymerase also results in an increased yield of the target fragment because competition by nontarget products for DNA polymerase and primers is reduced. In the later cycles, the amount of DNA polymerase is no longer sufficient to extend all the annealed primer–template complexes in a single cycle period, which results in reduced efficiency and a 'plateau' in the amplification reaction. This plateau is reached after approximately 30 cycles when *Taq* is used rather than after 20 cycles when the Klenow fragment is used. *Taq* also allows amplification of longer fragments than the Klenow fragment.

The standard PCR with *Taq* DNA polymerase can be enhanced by a variety of methods. Stringency of the annealing step can be controlled by adjusting the annealing temperature; high-temperature annealing and extension (greater than 55°C) and a balanced ratio of $Mg^{++}$ and dNTP concentrations give the greatest fidelity in the final PCR product (Table 9.2).

Minimizing the incubation time during the annealing and extension step limits the amount of mispriming and extension that occurs. Reducing primer and *Taq* polymerase concentration also reduces mispriming. Changing the $MgCl_2$ concentration can increase specificity by allowing a higher annealing temperature, which increases the stringency of the reaction, and by direct effects on the *Taq* polymerase itself. Limiting the amount of one of the four dNTPs may promote misincorporation.

Estimates of the specificity or fidelity of *Taq* polymerase vary (Kwiatowski et al. 1991, Landgraf and Wolfes 1993). For example, the products of six PCR amplifications of a 2.5-kb fragment containing the *Sod* gene from *Drosophila melanogaster* were cloned and sequenced (Kwiatowski et al. 1991). Four substitutions and two single-nucleotide frameshifts were found, or an error rate of $3.0 \times 10^{-4}$ for substitutions and $1.5 \times 10^{-4}$ for frameshifts. The error rates per nucleotide polymerized are $1.5 \times 10^{-5}$ and $0.7 \times 10^{-5}$ for substitutions and frameshifts, respectively. *Taq* polymerase appears to extend a mismatched primer–template significantly less efficiently than a correct primer–template. Misincorporated bases cannot be removed and this can promote termination of the extending DNA chain, which will prevent propagation of the errors in the subsequent PCR cycles. Because the accumulation of mutations is proportional to the number of replications of the DNA, the fewer PCR cycles that are required to provide an adequate yield of DNA, the better. Starting with adequate amounts of template reduces the number of PCR cycles needed to produce a specific amount of product, and hence the amount of misincorporations.

## Detecting Primer Artifacts

Low-molecular-weight DNA products may be produced during the PCR and are most obvious if the PCR is carried out with high primer concentrations, too much enzyme in early cycles, with small amounts of template DNA, and many cycles. These unwanted low-molecular-weight DNA products may be **primer-dimers** or other artifacts derived from the primers. Primer-dimers occur when the enzyme makes a product reading from the 3' end of one primer across to the 5' end of the other primer. Since each primer serves as both primer and template, a sequence complementary to each primer is produced, which upon denaturation is a perfect template for further primer binding and extension. As the number of cycles is increased over thirty, the probability of mispriming events increases, as does the amount of artifact product formed. The accumulation of a large amount of such PCR product depletes primers and dNTPs from the reaction mixture and competes for enzyme with the desired target DNA. If a second amplification is necessary to obtain adequate product, it is best done with 1 μl of the first reaction using a fresh reaction mixture.

## Cycle Number

The optimum number of cycles will vary with the starting concentration of the DNA, as well as all the other parameters (Table 9.2). Too many cycles can increase the amount of nonspecific background products. Too few cycles will give a low product yield. The number of DNA target molecules and cycles needed to give a good yield have been estimated:

| Number of target DNA molecules | Number of cycles |
|---|---|
| $3 \times 10^5$ | 25 to 30 |
| $1.5 \times 10^4$ | 30 to 35 |
| $1 \times 10^3$ | 35 to 40 |
| 50 | 40 to 45 |

The efficiency of DNA amplification declines in late cycles. This is called the amplification plateau because the product stops accumulating exponentially and enters a linear or stationary phase. The plateau appears to be unavoidable. The amplification plateau is possibly due to insufficient *Taq* DNA enzyme concentration or polymerization time in the late cycles, depletion of the primer, depletion of dNTPs, and the presence of product in such high concentration that reassociation with product occurs in preference to association with primers. Other possibilities include competition for reaction components by nonspecific products and product reassociation so that the product can't serve as a template for additional amplification. In general, it is better to set up multiple reactions to obtain more DNA than to try to avoid the plateau.

### Reducing the Evils of Contamination

Because the PCR is a very sensitive technique and can generate copies of a DNA sequence from very small amounts of DNA (theoretically from a single molecule), it is crucial that laboratory techniques be meticulous to prevent contamination of the laboratory, supplies, and equipment with target DNA. Contamination can be an enormous problem. Because amplified DNA sequences are produced in large quantities, carryover of tiny quantities can lead to false positives in subsequent reactions.

It is very crucial to separate physically the amplification reactions from the site where the evaluation of the amplification products is conducted (Lindahl 1993). Ideally, two separate sites should be available, one for amplification and one for analysis. Each separate room or containment unit should have a separate set of supplies and pipetters (Table 9.5). Amplified DNA should not be brought into an area where DNA is being prepared for amplification. Reagents and supplies should never be taken from an area where PCR analyses are being performed. Pipetters should not be taken into the preparation area after use with amplified material.

Reagents should be aliquotted to minimize the possibility of contamination. All reagents used in the PCR should be prepared, aliquotted, and stored in an area that is free of PCR-amplified DNA products. Similarly, oligonucleotides used for amplification should be synthesized and purified in an environment free of the PCR products.

Barrels of pipetters often become contaminated by DNA aerosols. To reduce contamination of samples by pipetters, positive displacement pipettes with disposable tips and plungers are recommended. These plugged tips and plungers are completely self-contained and will reduce contamination.

**Table 9.5**

Laboratory Activities for the PCR Ideally Should be Separated from Each Other in at Least Two Locations, With One for Pre-PCR Work and the Other for Post-PCR Work

| Pre-PCR work | Post-PCR work |
|---|---|
| Weigh chemicals | Electrophorese PCR products |
| Prepare buffer | Process PCR products |
| DNA extraction | Sequence PCR products |
| Prepare stock solutions | Analyze restriction digests |
| Prepare PCR reaction | Load products for second-stage amplification |

Contamination also can be reduced by changing gloves frequently, wearing laboratory coats, uncapping tubes carefully to reduce aerosol formation, and minimizing handling of DNA samples (Kitchin et al. 1990). Components of the PCR (mineral oil, dNTPs, primers, buffer, and enzyme) can be added to the tubes before the target DNA. Contamination of the tube will be reduced if each tube is capped before adding DNA to the next tube. Contamination can come from electrophoresis equipment, dot-blot apparatus, razor blades, micro-centrifuges, water baths, and other equipment, as well as from airborne bacteria, or sputum and skin cells from the person performing the PCR.

Ultraviolet irradiation at 254 nm has been thought to be effective in inactivating contaminating DNA and has been widely used. However, UV irradiation may be only partially effective (Dwyer and Saksena 1992). PCR mixtures with dNTPs are highly absorbent at that wavelength, and have been shown to reduce the efficiency of UV decontamination (Frothingham et al. 1992). Nor is autoclaving useful in eliminating previously amplified PCR products (Dwyer and Saksena 1992). Meticulous attention to the physical separation of pre- and post-PCR manipulation is more critical in minimizing contamination or carryover problems.

Multiple controls are required for the PCR. The choice of controls will allow detection of contamination and false positives. For a positive control, DNA should be used that amplifies weakly but consistently. Using DNA that produces strong positive responses will generate large amounts of amplified DNA, which could cause contamination. Negative controls should be well characterized. In addition, multiple reagent controls should be included with each amplification. A small number of molecules of PCR product in the reagents would lead to sporadic positive results. Thus, it is important to carry out reagent controls *repeatedly* so that rare contaminants will be detected. Reagent controls should contain all the necessary components for the PCR except for the template DNA, which will provide an extremely sensitive method for detecting contamination. In situations in which primers are being used for the first time, it is desirable to test the primers to determine if the sequences chosen are appropriate. As one test, if the PCR using only one primer or the other produces a product, the results probably are an artifact.

## Some Modifications of the PCR

What can you do if you are interested in amplifying DNA from an arthropod for which little genetic information is available? What if you want single-stranded DNA rather than double-stranded DNA? What if you want to find the sequence of DNA upstream or downstream from a specific gene? Some solutions to these problems have been achieved by modifying the types and numbers of primers used in the PCR (Table 9.6), as discussed later. Other new applications of the PCR are being developed rapidly.

**Table 9.6**

Modifications of the PCR Use Different Types of Primers

| PCR type, primer number, and size | Potential uses |
|---|---|
| **Standard** Paired, 15–30 nucleotides each | Amplify DNA for which sequence information is available. |
| **Anchored** One known primer, second is made | Amplify DNA when only one sequence is known for a primer. Synthesis of cDNA with the known primer is carried out using mRNA; a poly(G) tail is added to the cDNA. The second primer is made by synthesizing a primer with a poly(C) sequence, which allows amplification of a second DNA strand that is complementary to the cDNA. |
| **Arbitrary** Single, 18–30 nucleotides, arbitrary sequence | Amplify regions of DNA internal to regions to which arbitrary primers (M13 sequencing primer, M13 reverse sequencing primer, or T3 sequencing primer) anneal on opposite strands. One or more DNA fragments will be produced and these can be used to generate genome maps or discriminate among individuals, populations, or species. |
| **Asymmetric** Paired, 10–30 nucleotides, primers in a 1 : 50 to 1 : 100 ratio | Amplify single-stranded DNA for sequencing. |
| **DAF-PCR** Single, 5 nucleotides, arbitrary sequence | DNA Amplification Fingerprinting (DAF-PCR) by amplification of genomic DNA using a single short primer of arbitrary sequence. |
| **Degenerate** Multiple types, 15–30 nucleotides | Amplify DNA from genes related to those for which the sequence is known, or for which only part of the sequence is known, or for members of a gene family. The degeneracy of the DNA code for amino acids determines how many primer types are included in the reaction. |

*(continues)*

**Table 9.6**  (*Continued*)

| PCR type, primer number, and size | Potential uses |
|---|---|
| **Inverse** | |
| Paired, 15–30 nucleotides, inverse orientation | Amplify regions of DNA of unknown sequence flanking known sequences; used for identifying upstream and downstream sequences, chromosome walking, library screening, determining insertion sites of transposable elements, etc. Primers are oriented so DNA synthesis occurs away from the core DNA with a known sequence. |
| **PCR-RFLP** | |
| Paired, 18–30 nucleotides | Nuclear or mitochondrial DNA is amplified by the standard PCR, then cut with restriction enzymes. Banding patterns are visualized on a gel after staining with ethidium bromide. Standard primers are used; if none are available, sequences are obtained from DNA cloned in a genomic library. |
| Quantitation of mRNA | Several methods are known: (1) Two different cDNAs are amplified and the absolute level of one is calculated if the other is known. (2) The sample is spiked with a known amount of control DNA, and target and control DNA are amplified and compared to estimate amount of target DNA. |
| **RAPD-PCR** | |
| Single, 10 nucleotides, random sequence | Random Amplified Polymorphic DNA-PCR. Amplify regions of DNA that are flanked by the random primer sequences. Multiple DNA fragments may be produced and used as markers for genome mapping or identifying individuals, populations, or species. |
| RNA amplification | mRNA is reverse transcribed and the cDNA is amplified by PCR using primers 18–22 nucleotides long. |
| SSP-PCR | Single-Specific-Primer PCR amplifies double-stranded DNA when sequence information is available for one end only. A second generic vector primer is used at the unknown end. The primers are used in a PCR after ligating the unknown end to a generic vector. This method also can be used to obtain sequence data and restriction site data in unknown regions outside the known sequence. |

## Anchored PCR

In situations in which only one sequence is known that is suitable for a primer (rather than two), **anchored PCR** can be used to amplify DNA. The procedure involves synthesis of cDNA with the known primer from mRNA (Collasius et al. 1991, Dorit et al. 1993). A poly(G) tail is added to the cDNA. The second primer is developed by synthesizing a primer with a poly(C) sequence, which allows amplification of a second DNA strand that is complementary to the cDNA. Subsequent cycles yield amplified DNA from both strands.

## Arbitrary Primers

Ecologists, evolutionary biologists, and geneticists often wish to develop genetic markers for insects for which little genetic information is available. **Arbitrarily primed PCR (AP-PCR)** can produce a characteristic pattern for any genome, which could be useful in developing markers for breeding programs, genetic mapping, population genetics, or epidemiology (Welsh and McClelland 1990, Welsh et al. 1992).

AP-PCR involves two cycles of low-stringency amplification, followed by amplification at higher stringency using a single primer designed for other purposes. The term **stringency** refers to the PCR conditions such as the annealing temperature. If a high annealing temperature is employed, then the primers will only anneal to the template DNA if a high proportion of the sequences match. Lower annealing temperatures allow the annealing of primers to template DNA sequences with some mismatches. The sequence of the primer is chosen arbitrarily. Full-length primers (20 to 34 nucleotides long) that have been used include the universal M13 sequencing primer, the M13 reverse sequencing primer, and the T3 sequencing primer.

At low temperatures, an arbitrary primer can anneal to many sequences with some mismatches. By chance, some primers will be able to anneal to the target DNA within a few hundred bases of each other and on opposite strands. Sequences between these positions will then be amplified by the PCR. The extent to which sequences amplify will depend on the efficiency of priming at each pair of annealing sites and the efficiency of extension. During early cycles, those that prime most efficiently will predominate. During later cycles, those that amplify most efficiently will predominate.

AP-PCR has been used to generate reproducible and discrete DNA patterns with a diverse array of species. Between three and twenty DNA products were produced from the genomes sampled, which allowed differentiation between closely related strains of some species (Welsh et al. 1990). However, the polymorphisms revealed segregate as dominant markers, and extensive standardization of methods are required to obtain reproducible results (Black 1993).

## Asymmetric PCR

Single-stranded DNA can be produced by **asymmetric PCR**. By providing an excess of primer for one of the two strands, typically in ratios of 50:1 to 100:1, amplification results in a product that is primarily single stranded. Early in the reaction, both strands are produced, but as the low-concentration primer is depleted, the strand primed by the abundant primer accumulates arithmetically. Such ssDNA is particularly useful for sequencing by the dideoxy chain-terminating method (Chapter 8).

## DAF-PCR

Amplification of genomic DNA can be directed by only one primer of an arbitrary sequence to produce a characteristic spectrum of short DNA products of varying complexity. This approach, **DNA amplification fingerprinting** (**DAF**), does not rely on having DNA sequence information. It can generate fingerprints from a variety of species (Caetano-Anolles et al. 1991). Primers as short as five nucleotides can produce complex banding patterns, allowing differentiation among individuals as well as among different strains.

### Degenerate Primers and Standard PCR

**Degenerate primers** may be necessary when only a limited portion of a protein sequence is known for a target gene. Degenerate primers also may allow detection of new or uncharacterized sequences in a related family of genes, or may be useful when the goal is to amplify known members of a gene family. A mixture of oligonucleotides varying in base sequence, but with the same number of bases, can be substituted for primers with a defined sequence.

Designing degenerate primers for the PCR requires several considerations. Recall that the genetic code is degenerate, with more than one codon coding for most amino acids. Methionine and tryptophan are encoded by a single codon, but the other amino acids may be encoded by two to six different codons. When designing degenerate primers, it is useful to chose a segment of the protein in which there are amino acids with minimal degeneracy. The lower the degeneracy in the primers, the higher the specificity. The degeneracy of the primer may be restricted by considering which codons are most often used in a particular species (codon bias), if it is known for a particular organism. Furthermore, degeneracy may be reduced if primers containing fewer (15 to 20) nucleotides are used. Since a single mismatch may prevent *Taq* from extending the primer, degeneracy at the 3′ end should be avoided. Empirical testing of primers may be necessary and modifications made to ensure that DNA products are synthesized.

### Inverse PCR

**Inverse PCR** allows amplification of an unknown DNA sequence that flanks a 'core' region with a known sequence (Ochman et al. 1990). Inverse PCR can be used to produce probes to identify adjacent and overlapping clones from a DNA library. Inverse PCR also can generate end-specific DNA fragments for chromosome walking.

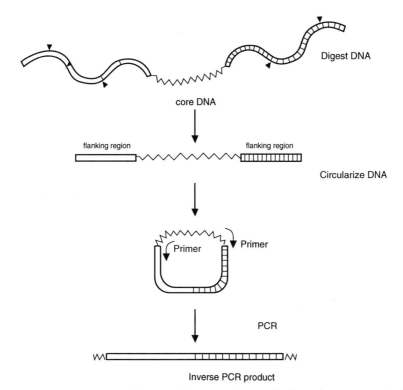

**Figure 9.3** Inverse PCR allows amplification of DNA flanking the 'core' DNA, for which sequence information is available. The template DNA is digested with an appropriate restriction enzyme to produce fragments approximately 2–4 kb long, with the core DNA in the middle. The DNA is then circularized by ligation. Primers, dTNPs, and *Taq* DNA polymerase are added and the PCR is carried out. Primers are oriented so that synthesis of DNA occurs away from the core DNA into the flanking regions.

The basic method for inverse PCR involves digesting template DNA, circularizing the digested DNA, and amplifying the flanking DNA outside the 'core' region with the primers oriented in the opposite direction of the usual orientation (Silver 1991). Primers for inverse PCR are synthesized in the opposite orientation and are homologous to the ends of the core region so that DNA synthesis proceeds across the *uncharacterized* region of the circle rather than across the characterized core region (Figure 9.3).

DNA containing the core sequence is digested with an appropriate restriction enzyme to produce a fragment about 3 to 4 kb long. Preliminary Southern blot analyses may be necessary to identify restriction enzymes that produce end fragments of suitable size for circularization and amplification. If an enzyme is used that cleaves within the core region, either the upstream or downstream segment of DNA will be amplified. Enzymes that do not cleave within the core region allow amplification of both flanking sequences, with their junction determined by the restriction enzyme and type of circularization method used. Once the DNA has been digested, the ends of the fragment are ligated to form a circular molecule. Ligation is performed with T4 DNA ligase in a dilute DNA concentration to favor formation of monomeric circles.

Inverse PCR has been modified to enable the isolation of cDNA clones. For example, using this procedure, a scorpion (*Leiurus quinquestriatus hebraeus*) cDNA family encoding α-neurotoxins was isolated (Zilberberg and Gurevitz 1993). The procedure involves ligating molecular adapters to the cDNA followed by circularization by self-ligation. Two back-to-back oligonucleotide primers designed for a known region of the corresponding polypeptide are used to amplify the circularized cDNA by inverse PCR. The PCR product thus contains a divided coding sequence separated by the molecular adapter that is flanked by 5′ and 3′ noncoding cDNA sequences. The linear PCR product is treated with T4 DNA polymerase, phosphorylated, and circularized again. The circular PCR product is digested at a unique restriction site within the adapter, forming a linear clonable molecule. Unfortunately, inverse PCR can be difficult to carry out, primarily because self ligation of the ends of linear DNA fragments in complex genomes is inefficient (Arnheim and Erlich 1992).

## PCR-RFLP

Many of the disadvantages of traditional restriction fragment length polymorphism (RFLP) analysis for analyzing population variation using individual insects are eliminated by using **PCR-RFLP** (Karl and Avise 1993). If no primers are available from the literature, a genomic DNA library is constructed and clones are isolated. Clones with inserts of 500 to 2000 bp are chosen and sequences of the first 100–150 nucleotides from both ends are obtained so that PCR primers can be derived. Nuclear DNA is amplified by the PCR using these primers and digested with appropriate restriction enzymes. The cut DNA is visualized after electrophoresis by staining with ethidium bromide. The advantage to PCR-RFLP is that DNA extracted from a single individual is sufficient after PCR amplification to provide DNA bands that can be visualized without having to be hybridized with radiolabeled probes.

### Quantitation of mRNA

Several methods have been developed to use PCR to quantitate the amount of specific mRNAs (Arnheim and Erlich 1992, Siebert and Larrick 1992). The amount of product depends on the initial amount of mRNA, the efficiency of reverse transcription, and the efficiency of the PCR. Two approaches have been developed. In one, two different cDNAs are amplified and, based on the ratio of the two, the absolute levels of one can be calculated if the other is known. This assumes that the two cDNA targets are amplified at similar efficiencies. Coupled **reverse transcription and PCR (RT-PCR)** is remarkably sensitive and can, in theory, detect single copies of a particular RNA in a highly complex sample, although the relative inefficiency of reverse transcriptase means that detection of ten or more copies is more realistic (Foley et al. 1993).

The second approach involves adding to the sample a known amount of a control DNA sequence. The control sequence should be as similar as possible to the cDNA target sequence so they will have similar amplification efficiencies using the same pair of primers. The target and control DNA can differ by as little as a single nucleotide substitution that alters a restriction enzyme site. The absolute amount of the target sequence can be determined by comparing the amounts of the control and target DNA produced in the same PCR. To quantitate the efficiency of reverse transcription, the sample can be spiked with an RNA transcript from a plasmid containing a modified target DNA. A known amount of the RNA transcribed from the plasmid is added to the sample as a control (Arnheim and Erlich 1992, Raeymaekers 1993).

### Random Primers and RAPD-PCR

A method similar to AP-PCR was developed in 1990, when Williams et al. demonstrated that genomic DNA from a diverse group of organisms could be amplified using a single short (9 or 10 nucleotides long) primer composed of an arbitrary sequence of oligonucleotides. The primers can be designed without having any nucleotide sequence information for the organism being tested. The only constraints are that the primers should have 50 to 80% G+C content and no palindromic sequences. Different random primers used with the same genomic DNA produce different numbers and sizes of DNA fragments, but these can be reproduced if appropriate care is employed (Ellsworth et al. 1993, Kernodle et al. 1993, Meunier and Grimont 1993, MacPherson et al. 1993, Williams et al. 1993). DNA segments can be amplified from a wide variety of species. The amplified DNA can be detected as bands in ethidium bromide-stained agarose gels (Figure 9.4). The authors called this modified PCR method RAPD because it produced Random Amplified Polymorphic DNA.

**RAPD-PCR** has been used to develop genetic maps and to identify molecular markers in populations (Hadrys et al. 1992, Tingey and del Tufo 1993), as well as to determine paternity in dragonflies (Hadrys et al. 1993), identify polymorphisms in parasitic wasps (Edwards and Hoy 1993, Roehrdanz et al. 1993), and analyze parentage in communally breeding burying beetles (Scott et al. 1992, Scott and Williams 1993). RAPD-PCR makes it possible to identify hundreds of new genetic markers in a short time, which allows genetic maps to

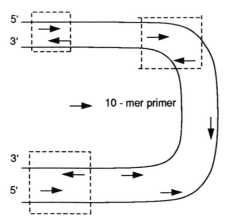

**Figure 9.4**  RAPD-PCR provides a 'random' sample of the genome. A single 10-primer that is chosen arbitrarily is added to the genomic DNA. It anneals to complementary sequences throughout the genome. If two primers anneal in the opposite direction from each other within a distance of less than approximately 2 kb, as in the three areas shown in the dashed boxes, DNA fragments of different size are amplified and can be visualized on an agarose gel stained with ethidium bromide. DNA regions that are not in the boxes will not amplify, even though one primer has annealed with the DNA in that region.

be developed rapidly (Ragot and Hoisington 1993). RAPD-PCR is particularly valuable for genome mapping in those species for which other genetic markers are lacking or rare.

Genomic DNA sequences differing by only a single base may not be amplified in the RAPD protocol, or may result in a complete change in the number and size of the amplified DNA segments. Thus RAPD may be able to detect small differences in the genomes of individual insects or mites, different insect populations, or species. It is possible that RAPD markers will be useful for mapping genomes in a wide variety of insect species, or can be used to identify individuals, populations, and species. RAPD-PCR is a particularly useful technique for species for which little genetic information is available. Two or more primers have been employed simultaneously in RAPD-PCR to generate reproducible DNA fragments that are different from those obtained with each single primer (Micheli et al. 1993).

Unfortunately, RAPD-PCR is particularly sensitive to initial DNA concentration during amplification so bands may vary in their intensity even when present in different individuals. Another difficulty with RAPD-PCR is that all bands are inherited as dominant alleles, which means that heterozygotes cannot

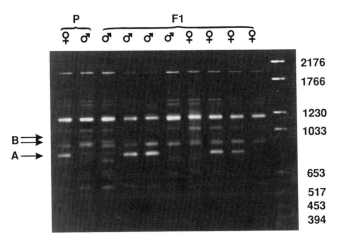

**Figure 9.5** Photograph of a gel showing mode of inheritance of RAPD-PCR DNA fragments from the parasitic wasp *Trioxys pallidus*. A single 10-mer primer anneals to different regions of the genome and if two primers anneal in the opposite orientation, amplification of several DNA sequences occurs. The size of the bands (bp) is indicated by the size marker lane on the right. The arrow marked A indicates that the band in the mother is inherited in about half of her haploid male and diploid female progeny, which is consistent with the hypothesis that the mother was heterozygous for the band. The arrows marked B identify faint bands that are not inherited in a Mendelian fashion and are not used in analyses. (From Edwards and Hoy 1993.)

be identified unless progeny testing is conducted (Black 1993). This is not an issue, however, with haplodiploid species where haploid males can be evaluated. Another problem is that comigration of similar sized bands with different sequences can occur but may not be detected. The role of RAPD in population ecology and genetics studies is discussed further in Chapter 14.

### RNA Amplification

Messenger RNA can be reverse transcribed and the resultant cDNA can then be amplified by the PCR using *Taq* DNA polymerase. This procedure will allow detection of gene expression in small numbers of specific cells or tissues. The reactions have been carried out with RNA isolated from as few as 10 to as many as 1000 cells.

The process involves (1) isolation of mRNA, (2) reverse transcription of mRNA into cDNA, and (3) amplification of cDNA by PCR. Primers for the amplification should be about 18–22 nucleotides long, and should occur in separate exons to inhibit amplification of any contaminating genomic DNA in the RNA preparation.

### Simultaneous Amplification and Detection

Specific PCR products can be detected by adding ethidium bromide to the PCR tube. The fluorescence of the ethidium bromide increases in the presence of double-stranded DNA, so that an increase in fluorescence in such a PCR indicates a positive amplification, which can be monitored by UV photography of PCR tubes, or by use of a fluorometer (Higuchi et al. 1992). The PCR amplification can even be monitored continuously, which makes possible the development of instruments for automated DNA amplification in situations in which a large number of samples need to be processed.

### Single Specific Primer PCR

The **Single Specific Primer-PCR** (SSP-PCR) method allows the amplification of double-stranded DNA even though only the sequence for one end of a specific region is known, and thus only one specific primer can be obtained. SSP-PCR also allows genome walking from known into unknown regions of the chromosome (Shyamala and Ames 1989, 1993).

SSP-PCR involves (1) digesting chromosomal DNA with one or two restriction enzymes; (2) ligating the unknown end of the restricted chromosomal DNA to an oligomer of known sequence that is sufficiently long to serve as a PCR primer (generic primer), or ligating the unknown end to a vector, in which case the vector sequence can be used to design a PCR primer; (3) amplifying the ligation mixture with the generic primer annealing to the generic oligomer or to the vector and a specific primer annealing to the known sequence. The exponential accumulation of the product resulting from the combination of the generic primer and the specific primer will dominate over the linear accumulation of nonspecific products from the generic primer alone.

## Other Thermostable DNA Polymerases

*Taq* DNA polymerase is not the only DNA polymerase that can be used in the PCR. Genetically engineered variants of *Taq* DNA polymerase are being developed and evaluated (Erlich et al. 1991). Other thermostable DNA polymerases isolated from other bacterial species collected from hot springs or from deep ocean vents also are being evaluated for PCR. VENT DNA polymerase, isolated from a deep ocean vent bacterium, *Thermococcus litoralis*, is commercially available. A thermostable enzyme has been isolated from *Thermus thermophilus*, which can also reverse-transcribe RNA efficiently at high temperatures. The thermostability of this enzyme may minimize the importance of secondary structure in the RNA template and allow efficient cDNA synthesis at high temperatures. DNA polymerase has been isolated from the archebacterium *Sulfolobus acidocaldarius*, which carries out polymerization at 100°C; this polymerase could facilitate the amplification of DNA regions with a high level of secondary structure and enhance specificity (Arnheim and Erlich 1992). Polymerases from

*Thermoplasma acidophilum, Thermococcus litoralis,* and *Methanobacterium thermoautotrophicum* have 3′ to 5′ exonuclease activities, which means that they have the ability to proofread, which may reduce the rate of misincorporation over that of *Taq* DNA polymerase.

## Some Research Applications

The PCR can be applied to a diverse array of both basic and applied problems (Table 9.7). Protocols for the different methods are available in several books (Innis et al. 1990; Erlich 1989, Erlich et al. 1989, Ausubel et al. 1991, McPherson et al. 1991) or individual journal papers (Carbonari et al. 1993, Erlich and Arnheim 1992, White et al. 1989). In addition, many different PCR applications are described in the journals *BioTechniques* and *PCR Methods and Applications.* The following examples provide evidence of the versatility of the PCR, but are only an abbreviated introduction to the diversity of PCR applications. New modifications of the PCR continue to be made in order to solve a diversity of diagnostic, ecological, evolutionary, genetic, and developmental biology questions.

### Amplifying Ancient or Aged DNA

Amplification of ancient DNA is possible (Paabo 1990, 1991, Jackson et al. 1991, Cano et al. 1993a, 1993b). Museum specimens and archeological specimens thus can be evaluated. Old DNA is usually heavily modified or degraded, but PCR allows the extraction of the few copies of intact DNA present. DNA from pathogens contained within museum specimens of arthropods can be used for the PCR. For example, Lyme disease spirochete (*Borrelia burgdorferi*) DNA extracted from the midgut of ticks (*Ixodes dammini*) stored for 50 years in 70% alcohol was successfully amplified by the PCR (Persing et al. 1990). Individual tick specimens were processed as follows: specimens were removed from 70% ethanol with flame-sterilized forceps and air-dried on filter paper discs for 5 minutes. Two hundred microliters of 0.5-mm glass beads were incubated in 1 ml of 1% bovine serum albumin in distilled water at 37°C for 30 minutes and then washed twice in 1-ml of distilled water. Ticks were placed whole into 0.6-ml microcentrifuge tubes containing a slurry (20 μl) of the treated glass beads. Specimens were crushed into the beads with a disposable plastic dowel for 30 to 45 seconds to liberate the midgut contents, and 25 μl of PCR buffer was added. Samples were boiled for 5 minutes, then cooled on ice; 5-μl portions of the supernatant fluid were used for the PCR.

An even simpler protocol was used by Azad et al. (1990) to determine whether individual ticks or fleas were infected with rickettsia. Individual ticks or fleas were placed in 100 μl of BHI (brain heart infusion broth) and boiled for 10 minutes. The PCR was carried out with 10 μl of the suspension.

Because the PCR can be applied to frozen or formalin-fixed tissues, dried

**Table 9.7**

Some Entomological Problems and Potential PCR Applications(s)

| Problem | PCR technique(s) |
|---|---|
| Amplify ancient DNA | Standard PCR |
| Amplify mRNA | RNA PCR |
| Chromosome walking | Inverse PCR; SSP-PCR |
| Cloning a gene | Blunt end cloning; sticky-ended cloning; anchored PCR; PCR with degenerate primers; SSP-PCR |
| Constructing a genetic map | AP-PCR; RAPD; inverse PCR |
| Constructing a phylogeny | Standard PCR with primers having polylinkers for cloning and sequencing; asymmetric PCR and sequencing; PCR-RFLP |
| Detecting gene expression | RNA PCR |
| Detecting mutations | Standard PCR; RAPD; AP-PCR; PCR-RFLP |
| Detecting pathogens in arthropod vectors | Standard PCR |
| Detecting integrated DNA in transgenic arthropods | Standard PCR; Ligation-mediated PCR; Inverse PCR; anchored PCR |
| Engineering DNA | |
|    Introduce restriction sites into DNA fragments | Attach sequences to 5′ end of primers and conduct standard PCR |
|    Label DNA with $^{32}$P or biotin for sequencing, probes, or isolating DNA strands on a column | |
|    Assemble overlapping DNA segments to make synthetic DNA | |
|    Introduce substitutions, deletions, or insertions in product DNA | Alter primer sequence when synthesizing, then standard PCR |
| Evolutionary analyses | Standard PCR; RAPD; AP-PCR; DNA sequencing; PCR-RFLP |
| Identify species | Standard PCR; RAPD; AP-PCR; PCR-RFLP |
| Identify strains, races, or biotypes | Standard PCR; RAPD; AP-PCR; PCR-RFLP? |
| Identifying upstream and downstream sequences | Inverse PCR; SSP-PCR |
| Monitoring dispersal of individuals | Standard PCR; RAPD; AP-PCR |
| Sequencing a gene | Asymmetric PCR to produce ssDNA; dideoxynucleotide chain-termination sequencing method with *Taq* polymerase |

museum specimens and alcohol-preserved specimens, the PCR will be very useful in field studies by reducing potential dangers involved in maintaining and transporting infected disease vectors. In addition, detection of pathogens by the PCR appeared to be significantly more sensitive than enzyme-linked immunosorbent assays (ELISA). A DNA hybridization technique called the dot-blot also can be used to assay individual insects. Dot-blot assays have been found to be 100-fold more sensitive than agarose gel visualization of DNA with ethidium bromide, making it possible to detect both live and dead rickettsia and densities of as few as a single rickettsia. It seems likely that, because the PCR is exquisitely sensitive, its application could allow studies of transovarial transmission or quantitation of the persistence of microorganisms in vectors that cannot be accomplished in any other way.

The PCR has been used to amplify DNA from tissues preserved in formalin followed by paraffin embedding. Specimens up to 40 years old have yielded DNA up to 800 bp in length (Wright and Manos 1990). The integrity of the DNA and the duration of fixation do affect the length of the product that can be amplified.

The PCR has been used to amplify DNA fragments from insects preserved in ancient amber. DNA was extracted from a fossil termite, *Mastotermes electrodominicus*, preserved in amber estimated to be 25 to 30 million years old (DeSalle et al. 1992). Fragments of 16S ribosomal DNA from mitochondria and 18S rDNA from the nuclear genome were amplified by the PCR and used to construct a molecular phylogenetic analysis in which living termites, cockroaches, and mantids were compared and found to be monophyletic. Both 18S rRNA and 16S rRNA were extracted and amplified by the PCR from an extinct 120–130-million year old conifer-feeding weevil (Coleoptera: Nemonychidae) isolated from Lebanese amber (Cano et al. 1993a) or from a 25- to 40-million-year-old bee from Dominican amber (Cano et al. 1993b).

### Amplifying RNA

The PCR can be used to amplify messenger RNA sequences from complementary DNA (Kawasaki 1990). This allows analysis of gene expression during development, quantitation of mRNA from specific tissues, rearrangements of DNA during cell differentiation, and RNA processing.

### Chromosome Walking

Inverse PCR provides an effective method for detecting overlapping and adjacent clones in a library. Inverse PCR or SSP-PCR also can be used to provide information on promoters and other regulatory structures.

### Cloning a Gene

The PCR can generate microgram quantities of a specific DNA fragment, which simplifies the procedures involved in cloning single-copy genes from genomic DNA. The cloning of PCR products is often necessary, especially when characterizing a heterogeneous array of PCR products. Cloning a PCR product can be achieved in several ways (Scharf 1990, Dietmaier et al. 1993). The DNA can be cloned directly into a plasmid or M13 vector. Restriction sites frequently are inserted at the 5′ end of each primer so that the amplified DNA can be cloned directly into a vector after digestion of the amplified DNA (Kaufman and Evans 1990). It also is possible to clone PCR products by blunt-end and sticky-end cloning methods, although blunt-ended cloning is notoriously difficult (Scharf 1990). Various commercial kits are available that allow direct cloning of PCR products by taking advantage of the fact that the terminal transferase activity of *Taq* polymerase results in the addition of a single A at the 3′ ends of the amplified fragment, thus generating a one-nucleotide sticky end. This results in an overhang at each end of the double-stranded DNA, and cloning vectors with nucleotides complementary to these can be used to directly clone the PCR product. Modifications of this tactic also are available (Dietmaier et al. 1993).

The isolation of a gene requires some prior knowledge of the gene (Clackson et al. 1991). The nature and extent of that knowledge determine the cloning strategy to be used. If a probe (primer) is available from another species, genomic DNA can be screened by the PCR using standard or degenerate primers (McPherson et al. 1991, Clackson et al. 1991). Because many *Drosophila* genes have been cloned and extensive sequence information is available for this insect, entomologists may be able to apply this information to develop degenerate primers to obtain genes of interest from other arthropods.

This success of this approach is illustrated by the cloning of a sodium channel gene from *Drosophila* and the house fly, *Musca domestica*, using degenerate primers (Knipple et al. 1991). Several vertebrate sodium channel genes were cloned and comparisons of the inferred amino acid sequences of the α-subunits of sodium channels from rat brain and rat skeletal muscle with those of the electric eel revealed a 70% similarity when conservative substitutions were taken into account. Two sodium channel genes (*para* and *DSC1*) cloned from *Drosophila* were found to be homologous to the vertebrate sodium channel genes. Using this information, it was possible to generate DNA primers designed to isolate a segment of the gene homologous to *para* from the house fly.

The PCR was performed on genomic house fly DNA using degenerate primers. The 5′-end primer consisted of a 256-fold degenerate sequence 20 nucleotides long. The 3′-end primer consisted of a 64-fold degenerate sequence 21 nucleotides long. Both had additional sequences appended to their 1′ ends to provide a *Hind* III and *Xba* I restriction enzyme recognition sequence, respectively, to facilitate the cloning of the amplification products. The PCR product was 104 bp long, consisting of 87 bp of coding sequence plus the flanking sequences attached to the 5′ ends of the primers. To confirm that the PCR-generated DNA was derived from the house fly, amplified DNA was labeled with $^{32}$P and used as a hybridization probe of genomic Southern blots containing digests of house fly, *Drosophila*, and mouse DNA. Because the only specific hybridization signal after high stringency washing was to the house fly DNA, it was concluded that the amplified DNA was from the house fly and not an artifact or contaminant. The PCR products were also cloned and sequenced. The *para*-homologous sequence isolated from the house fly differed from that of *Drosophila* at only 16 nucleotides (81.6% similarity). The substitutions, primarily in the third base position of codons, had no effect on the amino acid sequence of this region.

Doyle and Knipple (1991) subsequently used the same degenerate (mixed sequence) primers to amplify DNA from seven insect species and an arachnid, including the tobacco budworm *Heliothis virescens*, the mosquito *Aedes aegypti*, the diamondback moth *Plutella xylostella*, the gypsy moth *Lymantria dispar*, the cabbage looper *Trichoplusia ni*, the Colorado potato beetle *Leptinotarsa decemlineata*, the American cockroach *Periplaneta americana*, and the two-spotted spider mite *Tetranychus urticae*. Following amplification, the PCR products were sequenced and only 5 of 60 clones sequenced were not derived from *para* homologs. This study, and others, suggests that degenerate primers derived from conserved segments of characterized *D. melanogaster* genes can be used to clone genes from a diverse array of other arthropods. Interestingly, intraspecific polymorphisms were found in nucleotide sequences from three moth species, which could reflect the presence of duplicated genes or allelic variants in the populations from which the genomic DNA was isolated.

Doyle and Knipple (1991) suggested that, because the PCR can be used to analyze specific segments of DNA from individuals, the *para* gene could be used to analyze insect populations for pesticide resistance. If the sodium channel is a target site of DDT and pyrethroid pesticides, genetic variants of the sodium channel may be important in resistance to these pesticides. The probes could potentially lead to the development of highly specific methods for monitoring at least one type of pesticide-resistant genotype in pest populations.

### Detecting Exogenous DNA in Transgenic Arthropods

Determining whether exogenous DNA has integrated into eukaryotic chromosomes is difficult because the copy number of the specific sequence of interest is limited and the amount of DNA for analysis by Southern blot often is too small. The presence of chromosomal junction fragments formed by integrated exogenous DNA and the genomic flanking sequences is indicative that exogenous DNA has integrated into the chromosomes. These junctions can be detected by inverse PCR, panhandle PCR, and ligation-mediated PCR (Izsvak and Ivics 1993).

### Detecting Pathogens in Vector Arthropods

Arthropod vectors such as ticks, fleas, and mosquitoes are involved in maintaining and transmitting pathogens to humans and other vertebrates. The detection of pathogenic microorganisms within vector arthropods is important in conducting epidemiological studies and developing control strategies. A number of antigen-detection techniques have been developed, including direct or indirect immunofluorescence tests and ELISA using polyclonal or monoclonal antibodies. Other techniques involve recovery of the microorganisms from vectors by culture in embryonated eggs or tissue culture cells or by experimental infections in laboratory animals. The recovery of pathogenic microorganisms by these methods requires either live or properly frozen specimens. These techniques are expensive, time-consuming, and may involve dissection and preparation of specimens from live arthropods, which can be hazardous.

The PCR offers a new approach to detecting and identifying pathogenic microorganisms if sequence information is available to design appropriate primers (Wise and Weaver 1991). The PCR can be carried out with material from dead specimens, is more sensitive than most immunological techniques, and is more rapid. For example, rickettsia have been identified in fleas and ticks using primers that amplify a 434-bp fragment of the 17-kDa protein antigen from the *Rickettsia rickettsii* genome (Azad et al. 1990, Gage et al. 1992). Malarial DNA has been detected in both infected blood and individual mosquitoes by the PCR (Schriefer et al. 1991), and the Lyme disease agent *Borrelia burgdorferi* has been identified in ticks by species-specific amplification of the flagellin gene (Johnson et al. 1992).

### Developmental Biology

It is possible to detect the presence of specific mRNAs in tissues or cells by reverse transcription and DNA amplification by the PCR. The increased sensitivity of the PCR will allow advances in understanding when specific genes are expressed during development.

### Engineering DNA

It is possible to genetically engineer DNA fragments by the PCR in several ways. One method involves adding sequences to the 5' end of primers. Such sequence changes are readily accepted, even though these add-ons do not base-pair with the template DNA. The DNA being synthesized contains the add-on because the primers are incorporated in the synthesized DNA fragment. For example, it is possible to add a restriction site sequence to DNA being amplified by the PCR by attaching the restriction site sequence to the primers (Figure 9.6). Such restriction sites facilitate subsequent manipulations of the final PCR product.

Another DNA sequence that can be added on is the T7 promoter, which is located at the 1' end of one primer. This promoter allows RNA copies to be generated from the DNA synthesized in the PCR. Although the add-on sequences are mismatched and don't base pair to the template DNA, in most cases

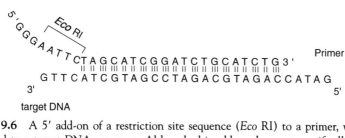

**Figure 9.6** A 5' add-on of a restriction site sequence (*Eco* RI) to a primer, which is annealed to a target DNA sequence. Although this add-on does not specifically match the template DNA, this does not significantly affect the PCR. The extra bases that are added 5' to the *Eco* RI site ensure that the efficiency of the restriction enzyme cleavage is maintained.

they have little effect on the specificity or efficiency of the amplification. Specificity is apparently imparted most significantly by the 3' end of the primer.

This technique also can be used to specifically label one PCR product strand or the other, or both, with a radioactive, biotin, or fluorescent label (Chehab and Kan 1989). DNA sequences also can be altered at any position by modifying primers so that substitutions, additions, or deletions are made in the amplified DNA.

### Evaluating Efficacy of Disease Control

In regions of the world where malaria is endemic, the use of bed nets impregnated with a synthetic pyrethroid insecticide has been proposed as a method to control the mosquito vector of malaria (Gokool et al. 1992). The PCR has been used to determine whether the use of bed nets prevents mosquito-based transmission of malaria. PCR primers were used to amplify and fingerprint the human DNA contained within a mosquito blood meal. DNA fingerprints also were obtained from the blood of individuals sleeping under the bed nets and compared with DNA fingerprints obtained from the mosquitoes (Gokool et al. 1992). The many hypervariable regions of the human genome can produce individual-specific patterns of DNA fragment lengths. The DNA banding patterns indicated that few mosquitoes had fed on individuals protected by treated bed nets.

### Evolutionary Analyses

Analysis of evolution involves reconstruction of phylogenies and analysis of populations. A variety of methods have been used in the past, including biochemical methods to compare protein or gene sequences, electrophoresis of proteins, protein sequencing, immunological evaluations, and DNA–DNA hybridization. Restriction mapping of nuclear and organelle DNAs also has been useful, although restriction maps have limited resolving power. However, it has been impractical to obtain DNA sequence data from large numbers of individuals because of the great effort and expense required to screen large numbers of clones and map and sequence them.

The relative ease and simplicity of the PCR makes molecular studies more accessible to evolutionary biologists, and examples will be provided in Chapters 13 and 14. Analyses of molecular sequences and functions by the PCR should enhance our understanding of evolution. The PCR makes it possible to directly sequence amplified fragments of genes from individuals and populations, thus increasing the resolving power and phylogenetic range of comparative studies. For most applications of the PCR, it is necessary to know the sequence to synthesize primers. However, by choosing sequences that are highly conserved among widely divergent species, it is possible to design 'universal primers' to amplify a particular nuclear or organelle gene fragment from many members of a major taxonomic group. This allows comparisons of sequences from classes or phyla for taxonomic work as well as enhancing population studies that involve identifying individuals and biotypes (Chapters 13, 14).

Universal primers have been developed for both nuclear and mitochondrial genes. For example, primers are available that amplify a region of approximately 515 bp from the 18S rDNA from many fungi, protozoa, algae, plants, and animals. The primers are based on conserved sequences among the 18S rDNA, but they do not amplify bacterial or mitochondrial rRNA genes. Mitochondrial gene sequences are useful for many problems in evolutionary and population biology. Because mitochondria are inherited maternally, sequence analyses of mitochondrial DNA allow construction of maternal phylogenies. Mitochondrial DNA evolves at a high rate, which makes it useful for population studies within species and for species identifications.

### Genomic Footprinting

The PCR can increase the concentration of rare DNA fragments in a complex mixture and has been adapted to analyze DNA–protein interactions (genomic footprinting), which is distinct from DNA footprinting. The goal of genomic footprinting is to determine the contacts between DNA bases and specific proteins in a living cell. DNA footprinting determines these interactions in vitro. For a review, see Arnheim and Erlich (1992).

### Mutation Detection and Analysis

The PCR can be used to identify new mutations because it is easier to analyze PCR-amplified DNA from organisms with a specific mutation than to clone each gene. Mutations can be detected either by gel electrophoresis or by chemical cleavage (Arnheim and Erlich 1992). Two double-stranded DNA fragments that differ by a single nucleotide substitution may be detected by their different thermal denaturation profiles, as shown by denaturing-gradient gel electrophoresis. The two DNA samples are run in adjacent lanes of a polyacrylamide gel with a gradient of DNA denaturant along the axis of DNA migration. Their mobility should be retarded when the DNA reaches the position in the gel where the concentration of denaturant is sufficient to cause melting of the DNA. A modification of this method involves attaching a GC-rich segment to one of the PCR primers, which enhances the ability to detect all possible single-base substitutions.

## Sequencing DNA

Techniques have been developed to sequence DNA by both the Maxam–Gilbert and the Sanger method using the PCR (Ausubel et al. 1991, Ellingboe and Gyllensten 1992, Kocher 1992, Olsen et al. 1993). Both methods permit the rapid determination of sequences without the need to construct a library or screen the library for the gene(s) of interest. In addition, sequencing of the PCR products can be either direct or after amplification and cloning (Olsen et al. 1993). Direct sequencing of PCR products means that only a single sequence needs to be determined, which will represent an average of the sequences of the target DNAs in solution. When PCR products are cloned and sequenced, a minimum of four or five different sequences should be determined to distinguish the positions of point mutations as well as to determine if errors occurred during the PCR. In theory, it is possible to obtain sequence information from as little as one copy of target DNA, which was impossible before the development of the PCR.

A common protocol for sequencing DNA involves carrying out asymmetric PCR to produce single-stranded DNA from M13 phage plaques, λ plaques, plasmid vectors in bacterial colonies, or genomic DNA. A total of 30–40 cycles of PCR may be required for asymmetric PCR because it is less efficient than the standard PCR. Once a single-stranded DNA molecule is produced, *Taq* DNA polymerase is highly effective in the Sanger chain-terminating method of DNA sequencing. The basic sequencing protocol involves annealing an oligonucleotide primer to a single-stranded template, labeling the primer in a short, low-temperature PCR in the presence of one labeled dNTP and three unlabeled dNTPs, all at low concentration, and extending the labeled primer in four separate base-specific, high-temperature reactions, each in the presence of higher concentrations of all dNTPs and one chain-terminating ddNTP. As with any other sequencing project, the products of the reactions are separated by high-resolution polyacrylamide gel electrophoresis and visualized by autoradiography.

If all the PCR products are analyzed on a gel or used as a hybridization probe, or sequenced, the problems introduced by errors in the PCR are limited.

However, sequences derived from a *single clone* may be incorrect, because the PCR produces a 0.3 to 0.8% error rate after 20 to 30 cycles. Ideally, several independent clones will be sequenced, and these will be from separate reactions.

## Ligase Chain Reaction and Other Amplification Methods

While the PCR is by far the most important technique currently available to amplify DNA or RNA *in vitro*, it has several limitations. The PCR requires the use of thermal cycling equipment and heat-stable DNA polymerases. It only works with DNA, although if RNA can be converted to cDNA in a separate step, it can then be amplified by PCR. Several other techniques have been developed that either involve amplifying a nucleic acid sequence in a sample or use nucleic acids to amplify the signal from a detection reaction (Landegren 1993).

The **ligase chain reaction (LCR)** or ligation amplification reaction (**LAR**) is an *in vitro* method for amplifying nucleic acids that involves a cyclical accumulation of the ligation products of two pairs of complementary oligonucleotides (Figure 9.7). The LCR uses oligos that completely cover the target sequence (Barany 1991). One set of oligonucleotides is designed to be completely complementary to the left half of a sequence being sought, and a second set matches the right half. Both sets are added to the test sample, along with ligase cloned from thermophilic bacteria. If the target sequence is present, the oligos blanket their respective halves of that stretch, with their ends barely abutting at the center. The ligase interprets the break between the ends as a nick in need of repair and produces a covalent bond. This creates a full-length segment of DNA complementary to the target sequence. The solution is then heated to separate the new full-length strand from the original strand, and both then serve as targets for additional oligos.

After sufficient cycles, the full-length ligated oligos are identified in one of several ways, using radioisotopes, fluorescence, or immunological methods. The LCR is particularly useful because the ligase won't join the adjoining oligos if the oligos and the target are not perfectly matched; thus this procedure is highly specific. However, the PCR is faster and requires fewer copies of target DNA than the LCR. Combining both the PCR and LCR takes advantage of the best attributes of both reactions (Landegren 1993a,b).

**Strand displacement and amplification (SDA)** involves digesting a DNA sample with restriction enzymes and then subjecting the sample to an initial denaturation. A specific primer is then hybridized to the end of one strand of the fragment of interest. DNA polymerase extends the primer and also extends the template molecule by copying the 5' end of the primer. During the reaction, modified dATP nucleotides are incorporated which contain sulfur rather than phosphorus in that position. Because the incorporated A residues have a sulfur

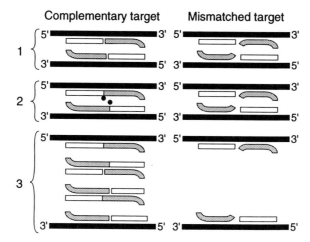

**Figure 9.7**  The ligase chain reaction (LCR) uses primers that anneal to a specific, complementary sequence on the target DNA to be amplified. LCR uses two primers that completely cover the target sequence; one primer is complementary to the left half of a sequence and the second primer matches the right half. Both primers are added to the target DNA along with ligase. If the target sequence is present, the primers blanket the DNA fragment with the ends of the primers abutting in the center. The ligase interprets the tiny break between these ends as a nick in need of repair and ligates them together with a covalent bond, which creates a full-length stretch of DNA complementary to the target. Amplification subsequently is achieved in a manner similar to the PCR. The solution is heated to separate the new, full-length DNA strand from the original and both then serve as targets for additional primers.

in the α position, the primer, but not its DNA copy, is nicked by a restriction enzyme that recognizes a site at the 5′ end of the primer. DNA synthesis is initiated at this nick, but the DNA polymerase used lacks the 5′ exonucleolytic activity required for nick translation. The downstream fragment is released by strand displacement during DNA synthesis. Successive cycles of hemirestriction and strand displacement follow. The result is an exponential amplification of the target DNA because displaced strands are recruited as templates for an analogous reaction on the other strand of the target DNA.

## Concluding Remarks

Several other DNA and RNA amplification techniques have been developed for specific objectives, and each technique has advantages and disadvantages (see Landegren, 1993b, for a review). However, the speed, specificity, versatility, and sensitivity of the PCR already has had a significant impact on genetics, immunology, forensic science, evolutionary biology, ecology, and population biology (Arnheim and Erlich 1992). The PCR provides a way to obtain information on RNA and DNA sequences present in individual cells, and to carry out retrospective studies on DNA from insects in amber. The PCR and other nucleic acid amplification methods provide molecular genetic tools that promise to revolutionize many of these fields in as-yet-unforeseen ways. In only a few years, the PCR has changed the way in which research is conducted. End users and commercial developers are developing sophisticated techniques for integrating the PCR into many applications and techniques, ranging from diagnostics to *in situ* PCR, in which the PCR process is applied to fixed cells (Haff 1993). The PCR is becoming an important tool in systematic, evolutionary, and ecological studies of insects, as will be discussed in Chapters 13 and 14.

## General References

### Books

Ellingboe, J. and U. B. Gyllensten, Eds. 1992. *The PCR Technique: DNA Sequencing*. Eaton Publ., Natick, MA.

Erlich, H. A., Ed. 1989. *PCR Technology, Principles and Applications of DNA Amplification*. Stockton Press, New York.

Erlich, H. A, R. Gibbs, and H. H. Kazazian, Jr., Eds. 1989. *Polymerase Chain Reaction* (Current Communications in Molecular Biology). Cold Spring Harbor Laboratory Press, Cold Spring Harbor, NY.

Innis, M. A., D. H. Gelfand, J. J. Ninsky, and T. J. White, Eds. 1990. *PCR Protocols: A Guide to Methods and Applications*. Academic Press, San Diego.

McPherson, M. J., P. Quirke and G. R. Taylor, Eds. 1991. *PCR: A Practical Approach*. IRL Press, Oxford.

### Journals

*PCR Methods and Applications* is published by Cold Spring Harbor Laboratory Press, Cold Spring Harbor, NY.

*BioTechniques* is published by Eaton Publishing, Natick, MA, and contains many articles about the PCR and related nucleic acid amplification techniques.

## References Cited

Arnheim, N. and H. Erlich. 1992. Polymerase chain reaction strategy. Annu. Rev. Biochem. 61: 131–156.

Arnheim, N., T. White and W. E. Rainey. 1990. Application of PCR: Organismal and population biology. BioScience 40: 174–182.

Ausubel, F. M., R. Brent, R. E. Kingston, D. M. Moore, J. G. Seldman, J. A. Smith and K. Struhl, Eds. 1991. The polymerase chain reaction. *Current Protocols in Molecular Biology*. Greene Publ. Assoc., New York.

Azad, A. F., L. Webb, M. Carl and G. A. Dasch. 1990. Detection of Rickettsiae in arthropod vectors by DNA amplification using the polymerase chain reaction. Ann. N.Y. Acad. Sci. 590: 557–563.

Barany, F. 1991. The ligase chain reaction in a PCR world. PCR Methods Appl. 1: 5–16.

Black, W. C., IV. 1993. PCR with arbitrary primers: Approach with care. Insect Mol. Biol. 2: 1–6.

Caetano-Anolles, G., B. J. Bassam and P. M. Gresshoff. 1991. DNA amplification fingerprinting using very short arbitrary oligonucleotide primers. Bio/Technology 9: 553–557.

Cano, R. J., H. N. Poinar, N. J. Pieniazek, A. Acra and G. O. Poinar, Jr. 1993a. Amplification and sequencing of DNA from a 120–135-million-year-old weevil. Nature 363: 536–538.

Cano, R. J., H. N. Poinar, D. W. Roubik and G. O. Poinar, Jr. 1993b. Enzymatic amplification and nucleotide sequencing of portions of the 18S rRNA gene of the bee *Proplebeia dominicana* (Apidae: Hymenoptera) isolated from 25–40 million year old Dominican amber. Med. Sci. Res. 20: 619–622.

Carbonari, M., D. Sbarigia, M. Cibati and M. Fiorilli. 1993. Optimization of PCR performance. Trends Genet. 9:42–43.

Chehab, F. F. and Y. W. Kan. 1989. Detection of specific DNA sequences by fluorescence amplification: A color complementation assay. Proc. Natl. Acad. Sci. USA 86: 9178–9182.

Clackson, T., D. Gussow and P. T. Jones. 1991. General application of PCR to gene cloning and manipulation. Pp. 187–214. In: *PCR: A Practical Approach*. M. J. McPherson, P. Quirke and G. R. Taylor, Eds. IRL Press, Oxford.

Collasius, M., H. Puchta, S. Schlenker and G. Valet. 1991. Analysis of unknown DNA sequences by polymerase chain reaction (PCR) using a single specific primer and a standardized adaptor. J. Virol. Methods 32: 115–119.

Dallman, M. J. and A. C. G. Porter. 1991. Semi-quantitative PCR for the analysis of gene expression. Pp. 215–224. In: *PCR: A Practical Approach*. M. J. McPherson, P. Quirke and G. R. Taylor, Eds. IRL Press, Oxford.

DeSalle, R., J. Gatesy, W. Wheeler and D. Grimaldi. 1992. DNA sequences from a fossil termite in Oligo-Miocene amber and their phylogenetic implications. Science 257: 1933–1936.

Dietmaier, W., S. Fabry and R. Schmitt. 1993. DISEC-TRISEC: di- and tri-nucleotide-sticky-end cloning of PCR-amplified DNA. Nucleic Acids Res. 21: 3603–3604.

Dorit, R. L., O. Ohara and W. Gilbert. 1993. One-sided anchored polymerase chain reaction for amplification and sequencing of complementary DNA. Methods Enzymol. 218: 36–47.

Doyle, K. E. and D.C. Knipple. 1991. PCR-based phylogenetic walking: Isolation of *para*-homologous sodium channel gene sequences from seven insect species and an arachnid. Insect Biochem. 21: 689–696.

Dwyer, D. E. and N. Saksena. 1992. Failure of ultra-violet irradiation and autoclaving to eliminate PCR contamination. Mol. Cell. Probes 6: 87–88.

Eckert, K. A. and T. A. Kunkel. 1991. The fidelity of DNA polymerases used in the polymerase chain reactions. Pp. 225–245. In: *PCR: A Practical Approach*. M. J. McPherson, P. Quirke and G. R. Taylor, Eds. IRL Press, Oxford.

Edwards, O. R. and M. A. Hoy. 1993. Polymorphisms in two parasitoids detected using random amplified polymorphic DNA (RAPD) PCR. Biol. Control. 3: 243–257.

Ellingboe, J. and U. B. Gyllensten, Eds. 1992. *The PCR Technique: DNA Sequencing*. Eaton Publ., Natick, MA.

Ellsworth, D. L., K. D. Rittenhouse and R. L. Honeycutt. 1993. Artifactual variation in randomly amplified polymorphic DNA banding patterns. BioTechniques 14: 214–217.

Engelke, D. R., A. Krikos, M. E. Bruck and D. Ginsburg. 1990. Purification of *Thermus aquaticus* DNA polymerase expressed in *Escherichia coli*. Anal. Biochem. 191: 396–500.

Erlich, H. A. and N. Arnheim. 1992. Genetic analysis using the polymerase chain reaction. Annu. Rev. Genet. 26: 479–506.

Erlich, H. A., D. Gelfand and J. J. Sninsky. 1991. Recent advances in the polymerase chain reaction. Science 252: 1643–1651.

Foley, K. P., M. W. Leonard and J. D. Engel. 1993. Quantitation of RNA using the polymerase chain reaction. Trends Genet. 9: 380–385.

Frothingham, R., R. B. Blitchington, D. H. Lee, R. C. Greene and K. H. Wilson. 1992. UV absorption complicates PCR decontamination. BioTechniques 13:208–210.

Gage, K. L., R. D. Gilmore, R. H. Karstens and T. G. Schwan. 1992. Detection of *Rickettsia rickettsii* in saliva, hemolymph and triturated tissues of infected *Dermacentor andersoni* ticks by polymerase chain reaction. Mol. Cell. Probes 6: 333–341.

Gokool, S. D., F. Smith and C.F. Curtis. 1992. The use of PCR to help quantify the protection provided by impregnated bednets. Parasitol. Today 8: 347–350.

Guyer, R. L. and D. E. Koshland, Jr. 1989. The molecule of the year. Science 246:1543–1546.

Hadrys, H., M. Balick and B. Schierwater. 1992. Applications of random amplified polymorphic DNA (RAPD) in molecular ecology. Mol. Ecol. 1: 55–63.

Hadrys, H., B. Schierwater, S. L. Dellaporta, R. DeSalle and L. W. Buss. 1993. Determination of paternity in dragonflies by random amplified polymorphic DNA fingerprinting. Mol. Ecol. 2: 79–87.

Haff, L. 1993. PCR: Applications and alternative technologies. Bio/Technology 11: 938–940.

Higuchi, R., G. Dollinger, P. S. Walsh and R. Griffith. 1992. Simultaneous amplification and detection of specific DNA sequences. Bio/Technology 10: 413–417.

Innis, M. A., D. H. Gelfand, J. J. Ninsky, and T. J. White, Eds. 1990. *PCR Protocols: A Guide to Methods and Applications.* Academic Press, San Diego.

Izsvak, Z. and Z. Ivics. 1993. Two-stage ligation-mediated PCR enhances the detection of integrated transgenic DNA. BioTechniques 15: 814–816.

Jackson, D. P., J. D. Hayden and P. Quirke. 1991. Extraction of nucleic acid from fresh and archival material. Pp. 29–50. In: *PCR. A Practical Approach.* M. J. McPherson, P. Quirke and G. R. Taylor, Eds. IRL Press, Oxford.

Johnson, B. J. B., C. M. Happ, L. W. Mayer and J. Piesman. 1992. Detection of *Borrelia burgdorferi* in ticks by species-specific amplification of the flagellin gene. Am. J. Trop. Med. Hyg. 47: 730–741.

Jowett, T. 1986. Preparation of nucleic acids. Pp. 275–286. In: Drosophila: A *Practical Approach.* D. B. Roberts, Ed. IRL Press, Oxford.

Karl, S. A. and J. C. Avise. 1993. PCR-based assays of Mendelian polymorphisms from anonymous single-copy nuclear DNA: Techniques and applications for population genetics. Mol. Biol. Evol. 10: 342–361.

Kaufman, D. L. and G. A. Evans. 1990. Restriction cleavage at the termini of PCR products. BioTechniques 8: 304–306.

Kawasaki, E. S. 1990. Amplification of RNA. Pp. 21–28. In: *PCR Protocols: A Guide to Methods and Applications.* Academic Press, San Diego.

Keohavong, P. and W. G. Thilly. 1989. Fidelity of DNA polymerases in DNA amplification. Proc. Natl. Acad. Sci. USA. 86: 9253–9257.

Kernodle, S. P., R. E. Cannon and J. G. Scandalios. 1993. Concentration of primer and template qualitatively affects products in random-amplified polymorphic DNA PCR. BioTechniques 14: 362–364.

Kitchin, P. A., Z. Szotyori, C. Fromholc and N. Almond. 1990. Avoidance of false positives. Nature 344: 201.

Knipple, D. C., L. L. Payne and D. M. Soderlund. 1991. PCR-generated conspecific sodium channel gene probe for the house fly. Arch. Insect Biochem. Physiol. 16: 45–53.

Kocher, T. D. 1992. PCR, direct sequencing, and the comparative approach. PCR Methods Appl. 1: 217–221.

Kwiatowski, J. D. Skarecky, S. Hernandez, D. Pham, F. Quijas and F. J. Ayala. 1991. High fidelity of the polymerase chain reaction. Mol. Biol. Evol. 8: 884–887.

Landegren, U. 1993a. Ligation-based DNA diagnostics. BioEssays 15: 761–765.

Landegren, U. 1993b. Molecular mechanics of nucleic acid sequence amplification. Trends Genet. 9: 199–204.

Landgraf, A. and H. Wolfes. 1993. *Taq* polymerase (EC 2.7.7). Pp. 31–58. In: *Methods in Molecular Biology*. Vol. 16. *Enzymes of Molecular Biology*. M. M. Burrell, Ed. Humana Press, Totowa, NJ.

Lindahl, T. 1993. Recovery of antediluvian DNA. Nature 365: 700.

MacPherson, J. M., P. E. Eckstein, G. J. Scoles and A. A. Gajadhar. 1993. Variability of the random amplified polymorphic DNA assay among thermal cyclers, and effects of primer and DNA concentration. Mol. Cell. Probes 7: 293–299.

McPherson, M. J., K. M. Jones and S. J. Gurr. 1991. PCR with highly degenerate primers. Pp. 171–186. In: *PCR: A Practical Approach*. M. J. McPherson, P. Quirke and G. R. Taylor, Eds. IRL Press, Oxford.

Meunier, J. R. and P. A. D. Grimont. 1993. Factors affecting reproducibility of random amplified polymorphic DNA fingerprinting. Res. Microbiol. 144: 373–379.

Micheli, M., R. R. Bova, P. Calissano and E. D'Ambrosio. 1993. Randomly amplified polymorphic DNA fingerprinting using combinations of oligonucleotide primers. BioTechniques 15: 388–390.

Mullis, K. B. 1987. Process for amplifying nucleic acid sequences. US Patent 4,683,202. Filed 10–25–85; issued 7–28–87.

Mullis, K. B. and F. Faloona. 1987. Specific synthesis of DNA *in vitro* via a polymerase-catalyzed chain reaction. Methods Enzymol. 155: 335.

Mullis, K., F. Faloona, S. Scharf, R. Saiki, G. Horn and H. Erlich. 1986. Specific enzymatic amplification of DNA in vitro: The polymerase chain reaction. Cold Spring Harbor Symp. Quant. Biol. 51: 263.

Ochman, H., M. M. Medhora, D. Garza and D. L. Hartl. 1990. Amplification of flanking sequences by inverse PCR. Pp. 219–227. In: *PCR Protocols: A Guide to Methods and Applications*. Academic Press, San Diego.

Olsen, D. B., G. Wunderlich, A. Uy and F. Eckstein. 1993. Direct sequencing of polymerase chain reaction products. Methods Enzymol. 218: 79–92.

Osborne, B. I. 1992. HyperPCR: A Macintosh Hypercard program for the determination of optimal PCR annealing temperature. CABIOS 8: 83.

Paabo, S. 1990. Amplifying ancient DNA. Pp. 159–166. In: *PCR Protocols: A Guide to Methods and Applications*. Academic Press, San Diego.

Paabo, S. 1991. Amplifying DNA from archeological remains: A meeting report. PCR Methods Appl. 1: 107–110.

Persing, D. H., S. R. Telford III, P. N. Rys, D. E. Dodge, T. J. White, S. E. Malawista and A. Spielman. 1990. Detection of *Borrelia burgdorferi* DNA in museum specimens of *Ixodes dammini* ticks. Science 249: 1420–1423.

Raeymaekers, L. 1993. Quantitative PCR: Theoretical considerations with practical implications. Anal. Biochem. 214: 582–585.

Ragot, M. and D. A. Hoisington. 1993. Molecular markers for plant breeding: Comparisons of RFLP and RAPD genotyping costs. Theor. Appl. Genet. 86: 975–984.

Roehrdanz, R. L., D. K. Reed and R. L. Burton. 1993. Use of polymerase chain reaction and arbitrary primers to distinguish laboratory-raised colonies of parasitic Hymenoptera. Biol. Control 3: 199–206.

Ruano, G., D. E. Brash and K. K. Kidd. 1991. PCR: The first few cycles and amplifications. A Forum for PCR Users. Perkin Elmer Cetus, Norwalk, CT.

Rychlik, W. W. J. Spencer and R. E. Rhoads. 1990. Optimization of the annealing temperature for DNA amplification *in vitro*. Nucleic Acids Res. 18: 6409–6412.

Saiki, R. K., S. J. Scharf, F. Faloona, K. B. Mullis, G. T. Horn, H. A. Erlich and N. Arnheim. 1985. Enzymatic amplification of β-globin genomic sequences and restriction site analysis for diagnosis of sickle cell anemia. Science 230: 1350–1354.

Saiki, R. K., D. H. Gelfand, S. Stoffel, S. J. Scharf, R. Higuchi, G. T. Horn, K. B. Mullis and H. A. Erlich. 1988. Primer-directed enzymatic amplification of DNA with a thermostable DNA polymerase. Science 239: 487–491.

Scharf, S. J. 1990. Cloning with PCR. Pp. 84–91. In: *PCR Protocols: A Guide to Methods and Applications*. Academic Press, San Diego.

Schriefer, M. E., J. B. Sacci, Jr., R. A. Wirtz and A. F. Azad. 1991. Detection of polymerase chain reaction-amplified malarial DNA in infected blood and individual mosquitoes. Exp. Parasitol. 73: 311–316.

Scott, M. P. and S. M. Williams. 1993. Comparative reproductive success of communally breeding burying beetles as assessed by PCR with randomly amplified polymorphic DNA. Proc. Natl. Acad. Sci. 90: 2242–2245.

Scott, M. P., K. M. Haymes and S. C. Williams. 1992. Parentage analysis usind RAPD PCR. Nucleic Acids Res. 20: 5493.

Shyamala, V. and G. Ferro-Luzzi Ames. 1989. Genome walking by single-specific-primer polymerase chain reaction: SSP-PCR. Gene 84: 1–8.

Shyamala, V. and G. Ferro-Luzzi Ames. 1993. Single specific primer-polymerase chain reaction (SSP-PCR) and genome walking. Pp. 339–348. In: *Methods in Molecular Biology*. Vol. 15. *PCR Protocols: Current Methods and Applications*. B. A. White, Ed. Humana Press, Totowa, NJ.

Siebert, P. D. and J. W. Larrick. 1992. Competitive PCR. Nature 359: 557–558.

Silver, J. 1991. Inverse polymerase chain reaction. Pp. 137–146. In: *PCR: A Practical Approach*. M. J. McPherson, P. Quirke and G. R. Taylor, Eds. IRL Press, Oxford.

Singer-Sam, J., R. L. Tanguay and A. D. Riggs. 1989. Use of Chelex to improve the PCR signal from a small number of cells. Amplifications 1989, No. 3, p. 11.

Taylor, G. R. 1991. Polymerase chain reaction: Basic principles and automation. Pp. 1–14 . In: *PCR: A Practical Approach*. M. J. McPherson, P. Quirke and G. R. Taylor, Eds. IRL Press, Oxford.

Tingey, S. V. and J. P. del Tufo. 1993. Genetic analysis with random amplified polymorphic DNA markers. Plant Physiol. 101: 349–352.

Welsh, J. and M. McClelland. 1990. Fingerprinting genomes using PCR with arbitrary primers. Nucleic Acids Res. 18: 7213–7218.

Welsh, J., C. Petersen and M. McClelland. 1990. Polymorphisms generated by arbitrarily primed PCR in the mouse: Application to strain identification and genetic mapping. Nucleic Acids Res. 19: 303–306.

Welsh, J., K. Chada, S. S. Dalal, R. Cheng, D. Ralph and M. McClelland. 1992. Artibrarily primed PCR fingerprinting of RNA. Nucleic Acids Res. 20: 4965–4970.

White, T. J., N. Arnheim and H. A. Erlich. 1989. The polymerase chain reaction. Trends Genet. 5: 185–189.

Williams, J. G. K., A. R. Kubelik, K. J. Livak, J.A. Rafalski and S. V. Tingey. 1990. DNA polymorphisms amplified by arbitrary primers are useful as genetic markers. Nucleic Acids Res. 18:6531–6535.

Williams, J. G. K., M. K. Hanafey, J. A. Rafalski and S. V. Tingey. 1993. Genetic analysis using random amplified polymorphic DNA markers. Methods Enzymol 218: 704–740.

Wise, D. L. and T. L. Weaver. 1991. Detection of the Lyme disease bacterium, *Borrelia burgdorferi*, by using the polymerase chain reaction and a nonradioisotopic gene probe. J. Clin. Microbiol. 29: 1523–1526.

Wright, D. K. and M. M. Manos. 1990. Sample preparation from paraffin-embedded tissues. Pp. 153–158. In: *PCR Protocols: A Guide to Methods and Applications*. Academic Press, San Diego.

Zilberberg, N. and M. Gurevitz. 1993. Rapid isolation of full length cDNA clones by "inverse PCR": Purification of a scorpion cDNA family encoding α-neurotoxins. Anal. Biochem. 209: 203–205.

# Chapter 10

# P Elements and P-Element Vectors for Transforming *Drosophila*

## Overview

The engineering of P-transposable elements to serve as vectors of exogenous DNA has revolutionized *Drosophila* genetics. P-element vectors serve as effective tools for research on insect development and analyses of gene structure, function, regulation, and position effects. The P-transposable element is found in certain strains (called P) of *Drosophila melanogaster*, but is lacking in others (M strains). When P-strain males and M-strain females are crossed, the $F_1$ progenies exhibit a condition called hybrid dysgenesis because the P-transposable elements present in the chromosomes of the $F_1$ progenies are no longer prevented from transposing. The resultant insertions of P elements into new chromosomal sites lead to germ line mutations and sterility. P elements may have invaded *D. melanogaster* during the past 40 years from another *Drosophila* species. One possible method by which the P element could have been transferred to *D. melanogaster* is by a mite. It has been hypothesized that P elements were obtained from eggs of one *Drosophila* species by a parasitic mite feeding on their eggs. Subsequent feeding by this 'infected' mite on *D. melanogaster* eggs might have resulted in the mechanical transfer of P elements to *D. melanogaster* and their subsequent spread in field populations around the world.

When P-element vectors containing cloned genes are microinjected into early stage *Drosophila* embryos, some of the P-element vectors integrate into the chromosomes in the germ line tissues. If the newly inserted DNA is transmitted to the progeny of the injected embryos, a stable transformation has occurred and the new strain can be used for a variety of genetic analyses. A technique called transposon tagging, using P-element vectors, facilitates the identification and cloning of genes from *Drosophila*. When a single P-element vector inserts into a gene and causes a visible mutation, that gene is 'tagged' and can be isolated and cloned. P-element vectors also can be used to repair gaps left in chromatids when P elements excise, and this offers the possibility of inserting exogenous DNA into targeted, rather than random, sites in the *Drosophila* chromosome.

## Introduction

The **P element** was first genetically modified to serve as a vector of exogenous DNA in 1982 (Spradling and Rubin 1982, Rubin and Spradling 1982). A variety of different P-element vectors now are used routinely to introduce exogenous DNA into *Drosophila melanogaster*. P-element-mediated transformation of *D. melanogaster* has revolutionized how geneticists study gene structure, function, regulation, position effects, dosage compensation, and development in this eukaryotic organism. P-element-mediated transformation also has allowed geneticists to begin to unravel the genetic basis of behavior, development, and sex determination in *Drosophila*, as described in Chapters 11 and 12. This chapter describes P elements and hybrid dysgenesis, and the methods employed in introducing P-element vectors into the germ line of *D. melanogaster*. This approach to germ line transformation has inspired entomologists to attempt to engineer insects other than *Drosophila*, as described in Chapter 15.

## P Elements and Hybrid Dysgenesis

Intact P elements are transposable elements that are 2907 bp long and code for a single polypeptide that has transposase activity (Figure 10.1). There are four exons (numbered from 0 to 3) and these are flanked by inverted repeats 31 bp long. In addition, there are internal inverted repeats that are 11 bp long, at sequence 126–136 and 2763–2773. The presence of intact inverted repeats is required if the P element is to transpose or move. Multiple copies of P elements (30 to 50) normally are dispersed throughout the genome of P strains of *D. melanogaster*, but do not cause hybrid dysgenesis because transposition is suppressed by unknown factors in the P cytotype.

The movement of P elements causes mutations by inactivating genes, altering rates of transcription, and altering developmental or tissue-specific gene expression. P-element movements also cause chromosome breakage and nondisjunction that can lead to chromosome rearrangements and germ cell death. Transposition of P elements in somatic cells can reduce the life span of *D. melanogaster* males, which is probably due to genetic damage from P element insertion and excision (Woodruff 1992).

P elements initiate a syndrome called **hybrid dysgenesis** in *Drosophila melanogaster* (Kidwell et al. 1977, Bingham et al. 1982). Hybrid dysgenesis occurs when male flies from a strain with chromosomes that contain P elements (P males) are mated with females from a strain lacking P elements (M females). Their $F_1$ progenies exhibit a high rate of mutation, chromosomal aberrations

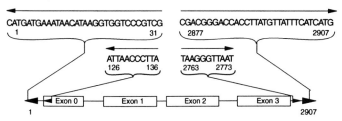

**Figure 10.1** Structure of an intact P element. There are four exons (0–3), separated by short introns (thin line). The 31-bp inverted terminal repeats (sequences 1–31 and 2877–2907) are indicated by the filled arrows. There are also II bp of inverted repeats at sequences 126–136 and 2763–2773. (Modified from Engels 1989.)

and, sometimes, complete sterility, caused by movements (transposition) of P elements in their germ line chromosomes. The reciprocal cross does not generate hybrid dysgenesis because the female's cytotype suppresses transposition of the P elements.

Transposition of P elements occurs in germ line tissues in *D. melanogaster* but does *not* occur in somatic tissues, because of tissue-specific splicing of the mRNA (Laski et al. 1986), although Kitamura et al. (1993) found that this splicing occurred in somatic tissues during embryogenesis. Three short introns have to be cut out of the original RNA transcript to produce the mRNA that codes for active transposase. Both somatic and germ line cells transcribe the transposase gene, and the somatic cells can remove the first two introns. However, only the portion of the pole cells that reach the gonads and differentiate into primordial germ cells can remove all three introns to produce a functional 87-kilodalton (kDa) transposase (Rio et al. 1986, Kobayashi et al. 1993). In somatic cells, a 66-kDa protein is produced that functions as a repressor of P-element activity. Several models for the molecular mechanisms of P cytotype repression have been proposed by (Gloor et al. 1993, Lemaitre et al. 1993, Misra et al. 1993, Ronsseray et al. 1993).

## P Element Structure Varies

Many P elements in the *Drosophila* genome are defective. Some have internal deletions and are unable to produce their own transposase but, if they retain their 31 bp terminal repeats, they can move if supplied with transposase by intact elements. However, P elements are unable to mobilize other transposable element families in *D. melanogaster* (Eggleston et al. 1988). P elements with defective 31 bp repeats are unable to transpose, and the 31 bp repeats are thought to be the site of action of the transposase. The frequency of transposition depends on the size of the P element; smaller elements are able to move more readily than larger ones. The location of the P element in the chromosome also is important in determining the frequency of transposition. While transposition is more or less 'random', P elements tend to insert near the region of a gene that codes for the start of transcription.

## Transposition Method

P elements appear to move from site to site in the genome (jump) by a 'cut-and-paste' method (Engels et al. 1990, Gloor et al. 1991, Sentry and Kaiser 1992). When a P element jumps, it leaves behind a double-stranded gap in the DNA that must be filled in. The gap is repaired by using a matching sequence as a template. This matching sequence can occur on the sister chromatid or elsewhere in the genome. If the transposition occurs in an individual that is *heterozygous* for the P insertion, and the matching site on the homologous chromosome is used as the template for DNA replication and repair, there can be a precise *loss* of the P-element sequence in the original site, although there is no *net loss* in the genome because the P element has simply changed locations.

However, if a P element jumps after the chromosomes have duplicated, but before the cell divides, one of the sister chromatids will still have the P element in its original position. In this situation, this homologous P element may serve as the template for filling in the hole left when the P element moved to a new position elsewhere in the genome. Under these circumstances, the number of P elements in the genome is *increased* by one. The P element is *replaced* in its original site by gap repair *and* it is also now present in a new site in the genome.

The cut-and-paste mechanism of transposition implies that P elements do not have to confer an advantage on the organism to invade and persist in the genome. In fact, a mathematical simulation model indicates that P elements can become fixed in populations even when fitness is reduced by 50% (Hickey 1982) and many laboratory studies have shown that colonies can change from M to P strains relatively rapidly. The cut-and-paste model also has been the conceptual basis of **targeted gene replacement** in *Drosophila* (Engels et al. 1990, Gloor et al. 1991, Sentry and Kaiser 1992), which is described in the following paragraphs.

## Origin of P Elements

P elements are relatively new to *D. melanogaster*. Surveys indicate that laboratory strains of *D. melanogaster* collected before 1950 lack P elements, but most colonies collected from the wild within the past 40 years have P elements (Anxolabehere et al. 1988, Engels 1989, 1992). Surveys also have indicated that very closely related transposable elements are found in populations of many *Drosophila* species, including *D. paulistorum* (Daniels et al. 1984), *D. guanche* (Miller et al. 1992), and *D. bifasciata* (Hagemann et al. 1992). Anxolabehere and Periquet (1987) found P elements in two other dipteran families (Opomyzidae and Trixoscelididae). Simonelig and Anxolabehere (1991) found that a P element isolated from *Scaptomyza pallida*, a drosophilid distantly related to *D. melanogaster*, is able to transpose in *D. melanogaster* and to mobilize a *D. melanogaster* P element that is defective. The genetic similarity of P elements from distantly related species suggests that P elements have invaded *D. melanogaster* recently.

The donor species that provided a P element to *D. melanogaster* is thought to have come from species of the *willistoni* group, which is not closely related to *D. melanogaster* (Daniels et al. 1984, Lansman et al. 1985, Daniels and Strausbaugh 1986). DNA sequences of P elements from *melanogaster* and *willistoni* are nearly identical although these species diverged from each other about 50 million years ago. Thus, there should have been sufficient time for considerable sequence divergence in the P elements if they had been present in both genomes at that time. Both *D. melanogaster* and *D. willistoni* overlap in their geographical ranges in Florida and in Central and South America, but they apparently are unable to interbreed. Thus, it has been hypothesized that P elements invaded *D. melanogaster* populations recently, probably by **horizontal transfer** from *D. willistoni* or a species closely related to *D. willistoni* (Kidwell 1992).

Two mechanisms have been proposed to explain how the P element could have infected *D. melanogaster*. As indicated, one involves horizontal transfer and the other involves interspecific crosses. Horizontal transfer could have been effected by a viral, bacterial, fungal, protozoan, spiroplasmal, mycoplasmal, or small arthropod vector [Hymenoptera or Acari (mites)]. One candidate for vector may be a semiparasitic mite, *Proctolaelaps regalis* (Houck et al. 1991, Kidwell 1992). *P. regalis* is found associated with both *Drosophila* species; it has been found in laboratory colonies and in the field associated with fallen or rotting fruit, which is the natural habitat for *Drosophila*. Laboratory observations indicate that *P. regalis* feeds on fly eggs, larvae, and pupae, and can make rapid thrusts of its mouth parts (chelicerae) into a series of adjacent hosts. This brief feeding on multiple hosts might allow it to pick up DNA from one egg and inject it into another.

Houck et al. (1991) removed mites from colonies of *Drosophila* with P elements in their genome and analyzed them by PCR and Southern blot analysis. The analyses indicated the mites carried both P element and *Drosophila* ribosomal DNA sequences. Mites isolated from M-strain colonies (which lack P elements) lacked P element DNA sequences.

For *P. regalis* to have transferred P elements to *D. melanogaster* from *D. willistoni*, a number of conditions had to occur in the proper sequence (Houck 1993). Females of *D. melanogaster* and *D. willistoni* had to deposit their eggs in close proximity and mites had to feed sequentially on one and then the other, in the correct order. The recipient fly egg had to be less than three hours old (prior to blastoderm stage); the germ line of the recipient embryo had to incorporate a complete copy of the exogenous P element; the transformed individual had to survive to adulthood; and the adult had to reproduce. Full confirmation of horizontal transfer of P elements between *Drosophila* species by this method has not yet been obtained. However, if horizontal transfer occurs in wild populations by mites or other vectors, our thoughts about genome evolution will be revolutionized.

The second mechanism for horizontal transfer of P elements involves interspecific crosses. Crosses between the sibling species *Drosophila simulans* and *D. mauritiana* produce sterile males, but fertile females. When $F_1$ females are back-crossed to males of either species, a few fertile males are produced. To determine whether interspecific transmission of P elements might occur, the two species were crossed (Montchamp-Moreau et al. 1991). The hybrid progeny were evaluated by *in situ* hybridization of larval salivary glands and by Southern blot analysis. The results indicated that the P element is able to pass from one species to another through interspecific crosses when the postmating sterility barrier is incomplete. Hybridization, although rare, occurs between some *Drosophila* species. Although *D. melanogaster* and *D. willistoni* apparently are unable to cross, interspecific crosses may have allowed the transfer of other types of transposable elements between *Drosophila* species.

## P-Element Vectors and Germ Line Transformation

### Protocols

After P elements were cloned (Rubin et al. 1982), they were genetically engineered to serve as vectors for inserting exogenous DNA into the germ line of *D. melanogaster* (Rubin and Spradling 1982, Spradling and Rubin 1982). A number of different vectors with different characteristics have been produced for specific purposes. The following example provides a model for the steps and procedures involved in P-element-mediated transformation of *Drosophila* (Table 10.1, Figure 10.2):

1. Construct or choose an appropriate P-element vector containing the DNA and marker gene(s) of interest. In this example, the marker gene is a wild type version (*rosy*+) of the mutant gene for *rosy* eyes. The wild-type allele is dominant over *rosy*, so if a single copy of *rosy*+ is present, the fly will have normal eyes. This vector with the *rosy*+ gene is unable to insert into the chromosome because it cannot produce transposase. Select a **helper plasmid**, such as pπ25.7wc, that contains a complete DNA sequence coding for transposase. This vector is unable to insert into *Drosophila* chromosomes because it lacks 23 bp of one terminal repeat, hence the designation wc for 'wings clipped'.

2. Microinject both the vector and helper plasmid DNAs into embryos ($G_0$)

**Table 10.1**

Procedures Involved in Transformation of D. *melanogaster* with P-Element Vectors[a]

| Generation | Step |
|---|---|
| $G_0$ | 1. Host strain embryos are injected with a mixture of helper P elements and a P transposon with a selectable marker. |
| | 2. Flies surviving to adults are mated individually to noninjected flies to obtain a maximum number of progeny. |
| $G_1$ | 3. Offspring of each $G_0$ fly are examined for the transformed phenotype. Transformed progeny from individual $G_0$ flies are mated together to establish transformed parent lines. |
| $G_2$ and beyond | 4. Parent lines are examined by DNA blot analysis to determine the number of P transposons in each line. |
| | 5. Representatives of each parent line are mated to various balancer stocks to determine which chromosome contains the P-element inserts. |
| | 6. Lines are evaluated for stability and expression level, which may vary, depending upon insertion site of the P element. |

[a]Modified from Karess 1985, by permission of Oxford University Press.

from an appropriate host strain. Embryos should be in the preblastoderm stage, when the embryo is a syncytium.

3. Mate the injected $G_0$ individuals that survive to adulthood with males or females that are homozygous for rosy. If the wild-type gene was inserted into the chromosome, then the progeny will have wild-type, rather than rosy, eyes. If no transformation occurred, then their progeny will have rosy eyes.

4. Mate $G_1$ progeny with wild type (rosy$^+$) flies to produce the next generation ($G_2$) with wild-type eyes.

5. Select individual $G_2$ lines with wild-type eyes.

6. Identify possible transformants containing single transposon insertions at unique sites (single-insert lines) and verify insert structure.

7. Analyze the properties of the transformed lines, including level of expression of the inserted DNA and stability of the transformed line.

8. Cross the most useful lines to balancer stocks to enable the lines to be maintained in a stable condition.

Insertion of exogenous DNA into *Drosophila* chromosomes is enhanced if preblastoderm embryos are microinjected. At that stage, the cleavage nuclei are in a **syncytium** (lacking nuclear membranes) and exogenous DNA can more easily be inserted into the chromosomes. The preblastoderm embryos are forming the **pole cells** that will give rise to the germ line tissues (ovaries and testes).

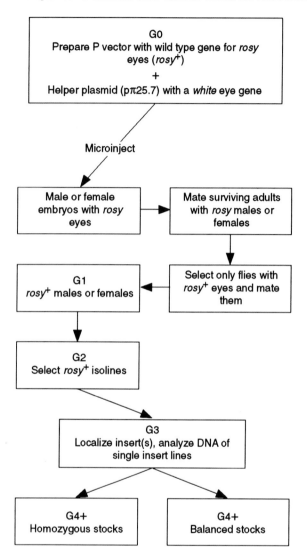

**Figure 10.2**   Steps in transforming *D. melanogaster* with P-element vectors. See text for details. (Adapted from Spradling 1986, by permission of Oxford University Press.)

Insertion of exogenous DNA into the chromosomes of the germ line results in stable transformation of that strain of flies. If only somatic cells contain the exogenous DNA, the flies cannot transmit the desired trait to their progeny. Such flies may exhibit the trait, but are only transiently transformed and do not transfer it to their progeny.

Insertion of P elements into chromosomes apparently occurs more or less at random, although heterochromatic segments of the chromosomes are less amenable to incorporation than euchromatic regions. The entire injected P-element vector, which usually consists of bacterial plasmid and P-element sequences into which exogenous DNA has been cloned, does not insert into the chromosome. Rather, the DNA inserted consists of the P sequences contained within the inverted terminal repeats. The plasmid DNA outside the inverted repeats does not insert and is lost during subsequent development.

Once transformed fly lines are obtained, the lines typically are stable unless transposase is provided in some manner. Sometimes an experimenter wants to induce movements of inserted DNA in stable colonies, and such **secondary transposition** can be induced if transposase is introduced by injecting helper elements into the preblastoderm embryo.

Isolating pure *Drosophila* stocks containing a single P-element insertion requires a sequence of steps over several generations (Figure 10.2). In this example, a gene for wild-type eye color is introduced into a mutant strain of flies with *rosy* eyes. Initially, a P-element vector containing a wild-type version of the *rosy* gene (*rosy*+) is mixed with a helper P element that can provide the necessary transposase (pπ25.7wc) (Figure 10.3). The mixture is microinjected into *rosy* embryos prior to the formation of pole cells. Within the embryo, a small amount of the microinjected DNA is included in the cytoplasm that is budded off by the forming pole cells. During the development of the germ line from the pole cells, these exogenous DNAs are transcribed. Because the helper element can transcribe mRNA to produce transposase, transposase enters the nuclei of germ line cells, where it promotes transposition of the defective P element carrying the wild-type gene.

During embryogenesis and development of the injected larvae (which are the $G_0$ generation), transcription and translation of the wild-type gene can produce sufficient xanthine dehydrogenase to influence the eye color of the adults. ($G_0$ indicates the generation that is injected, $G_1$ indicates their progeny, and so on.) $G_0$ flies with a normal eye color do not necessarily have the *rosy*+ gene inserted into the germ line chromosomes because the injected DNA may be transcribed and translated while in the cytoplasm, or the DNA may only be inserted into the chromosomes of somatic cells (**transient transformation**).

The next generation of flies ($G_1$) is the crucial generation to be screened for wild-type eyes because these flies should have wild-type eyes *only* if the *rosy*+ gene inserted into the chromosomes of the germ line. $G_1$ flies have never been

A) Structure of P-element derivative: Carnegie 20

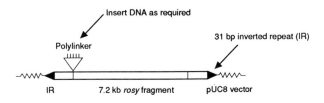

Lacks own transposase gene

B) Structure of helper P-element derivative: pπ 25.7 wings clipped

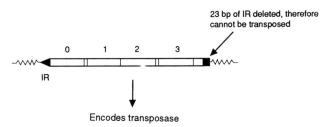

Encodes transposase

**Figure 10.3** Examples of modified P-element vectors. (A) The Carnegie 20 vector contains a 7.2-kb segment of DNA coding for the *rosy* gene. It contains a polylinker for inserting exogenous DNA and retains the 31-bp inverted repeats (IR, dark arrows). This vector cannot transpose without a helper element because it cannot make transposase. (B) The helper element, pπ25.7, wings clipped, produces transposase, but 23 bp of inverted repeat have been deleted at one end so this vector cannot insert into the chromosome.

shown to have unintegrated DNA carried over from their parents, so the presence of one or more progeny with normal eye color in this generation indicates that **stable transformation** occurred.

Individual $G_1$ flies may contain *multiple insertions* of the P element. Also, the P element may have inserted into *different sites* in different $G_1$ flies. As a result, colonies derived from single flies must be screened to identify those with a single insert before the transformed lines can be studied in detail (Figure 10.2).

To determine how many P elements inserted into the chromosomes of each colony and their location, DNA from $G_2$ adult flies is prepared from each isoline and evaluated by Southern blot analysis. (See Chapter 6 for a description of

Southern blot analysis.) DNA is cut with restriction endonucleases and probed with radiolabeled P sequences to determine the number of insertions. Lines containing multiple insertions should be discarded because these lines will be difficult to analyze. $G_3$ lines with single inserts are then crossed to *Drosophila* stocks containing appropriate **balancer chromosomes**. Balancer chromosomes function to prevent crossing-over between homologous chromosomes and thus help to maintain stable stocks. The location of the transposon in each single-insert line can be determined by *in situ* hybridization to salivary gland chromosomes.

Transformation success rates vary from experiment to experiment and experimenter to experimenter. Usually, it is important to obtain about ten single insert lines containing a transposon of interest. This may require microinjection of 600 or more embryos because survival of embryos after microinjection averages 30–70%, and of these, only 50–60% survive to adulthood ($G_0$). Even after $G_0$ adults are obtained, damage caused by microinjection may result in early death or sterility in 30 to 50%.

Transformation does not take place in all the germ line cells in an injected embryo. Usually only a small fraction of the germ line cells of a $G_0$ individual contains integrated P elements and produces transformed $G_1$ progeny, perhaps because integration of the P-element vector occurs late during germ line development. Thus, it is important to maximize the recovery of $G_1$ progeny from each $G_0$ individual to increase the probability of detecting progeny in which integration of P elements into a chromosome occurred. The size of the introduced P element is another factor that may influence transformation success; the larger the construct, the less frequent the insertion.

Detailed information on the life history and culture of *Drosophila* is available in a variety of references (Roberts 1986, Ashburner 1989), as are detailed protocols for transforming *Drosophila* with P-element vectors (Karess 1987, Spradling 1986). These protocols provide complete information on the appropriate equipment for microinjection, how to stage and dechorionate embryos, align them on slides, desiccate them, and inject them in the region that contains the pole cells. Directions are available for preparing the DNA for injection and for pulling the needles.

### P-Element Vectors

P elements have been engineered to provide an array of vectors with different characteristics and functions (for examples, see Rubin and Spradling 1983, Karess and Rubin 1984, Cooley et al. 1988, Sass 1990, Handler et al. 1993b). Generally, the vectors contain restriction sites for cloning, and usually contain a selectable marker gene. For example, the P element has been cloned into a bacterial plasmid vector such as pUC8 (Figure 10.3). Most of the P element has been replaced by a wild type *rosy*[+] gene, but the terminal repeats essential for transposition are kept. In addition, a polylinker site for cloning exogenous DNA is present, which allows up to 40 kb of DNA to be inserted. This vector is injected into *Drosophila* embryos along with a helper element that provides transposase, but which cannot transpose itself because it has a deletion in one of its terminal inverted repeats (Figure 10.3). This specific helper element, pπ25.7wc, has been called 'wings clipped' because it cannot insert into the *Drosophila* chromosome.

### Characterizing Transformants

Identification of transformed flies is achieved in several ways. If a visible marker, such as an eye color, is included in the vector, then putatively transformed flies can be determined visually. Alternatively, selectable markers can be used, such as resistance to neomycin. Fly larvae reared on a diet containing lethal concentrations of neomycin will survive only if they have an active neomycin gene inserted into their chromosomes. These markers are insufficient to prove stable transformation, however. Ideally, DNA from putatively transformed lines will be extracted and analyzed by Southern blot analysis. If large numbers of fly lines need to be characterized, dot blot analysis can be done. *In situ* hybridization of larval salivary gland chromosomes will allow a determination to be made of the number of insertions and their location(s). It is desirable to identify lines that carry only a single insertion if the timing and level of expression is to be determined. Different lines are likely to have different levels of expression because of position effects (Spradling and Rubin 1983, Levis et al. 1985).

# Uses for P-Element Vectors

### Transposon Tagging

The insertion of transposons into genes allows the isolation and cloning of specific genes *if* the altered gene exhibits an altered phenotype. However, because many P strains contain 30 to 50 copies of P elements, transposon tagging has been of limited value in *D. melanogaster*. This is because the new strains are unstable, and it is very tedious to identify the locus (or loci) that has been interrupted by the transposition of a specific element among the many present in the genome.

An efficient method for **transposon tagging** was developed by Cooley et al. (1988) that makes it possible to identify and clone any gene that alters a phenotype in *D. melanogaster* after a P element has inserted into it. The mutated gene can then be cloned because it has been 'tagged' with the P-element sequences. This technique provides a method for identifying the location and function of a significant fraction of the genes that alter the phenotype of *D. melanogaster*.

**A)**

**B)**

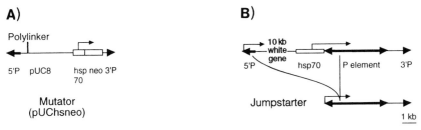

**Figure 10.4**   Two P-element vectors, mutator (A) and jumpstarter (B), were developed to facilitate insertion of a single P element to identify and clone genes in D. melanogaster. Jumpstarter encodes transposase and can therefore mobilize mutator. Mutator is able to transpose and carries ampicillin and neomycin resistance genes to facilitate identification and subsequent cloning of the Drosophila gene into which it has inserted. (Modified from Cooley et al. 1988.)

The technique relies on the development of two specially designed P-element vectors. The goal is to introduce a *single* P element into the germ line of flies *lacking* P elements. One vector, called jumpstarter, encodes transposase and mobilizes a second vector, called mutator, to transpose (Figure 10.4). The structure of the mutator element facilitates identifying and cloning genes because it carries ampicillin and neomycin resistance genes, which provide selectable markers.

Flies containing the mutator element are crossed to flies with a single jumpstarter element ($F_0$ generation). $F_1$ male flies that have both transposons are individually mated to normal females ($F_1$ generation) in the presence of G418 (an analog to neomycin). Transposase produced by the jumpstarter element will allow transposition of the mutator element to a new site on one of the autosomes in a portion of the germ line cells of the $F_1$ males. All $F_2$ female progeny should inherit the X-linked neomycin resistance from their fathers and survive. The $F_2$ males will lack the neomycin resistance gene and die *unless* they develop from one of the rare germ line cells containing a mutator element that has transposed to the autosomes. Half the males surviving will not have inherited the chromosome containing jumpstarter, so the new insertion will be completely stable.

Saving one such $F_2$ son per cross guarantees that each will contain an independent insertion event. These fly lines then can be analyzed for abnormal phenotypes; the DNA from the fly lines can be cloned, and the DNA flanking the P insertion can be identified by selecting clones that carry ampicillin resistance. The unique genomic DNA isolated from the flanking regions can be used as a probe to identify cosmids from a genomic library containing the gene of interest.

## Expressing Exogenous Genes

Genetic engineering techniques permit the expression of exogenous genes in a variety of organisms, and the availability of a transformation method for *Drosophila* makes it possible to express interesting genes in this insect. For example, Rancourt et al. (1990) obtained expression in *D. melanogaster* of two antifreeze protein genes isolated from the Atlantic wolffish, *Anarhichas lupus*. The two genes were cloned into a P-element vector with *Drosophila* yolk protein gene promoters. These highly active promoters were expressed in *Drosophila* females shortly after eclosion and remained active for several weeks. Transformed adult *Drosophila* females produced 1.5–5 mg/ml of antifreeze protein in their hemolymph. The antifreeze activity of the purified protein was determined by measuring freezing point depression and shown to have full biological activity.

## Evaluating Mammalian Drug Metabolism and Mutagenicity

Many human cancers are caused by changes in the genetic material as a result of mutation and chromosomal rearrangements, and *Drosophila* has been used to evaluate the mutagenicity of some drugs. However, *Drosophila* is unable to metabolize some classes of drug, such as polycyclic aromatic hydrocarbons, in the same manner as mammals. As a result, mice or rats are more often used for testing the toxicity and mutagenicity of drugs.

Transgenic strains of *Drosophila* have been developed which express mammalian genes (such as cytochrome P-450) for different drug metabolizing enzymes (Jowett 1991). These *Drosophila* have been used to assay the effect of chemicals on mutation rate, somatic recombination, nondisjunction, and aneuploidy. Whether transgenic *Drosophila* could provide an alternative to using mammalian systems for *in vivo* genotoxicity tests remains to be determined, but this approach has the advantage of exploiting actual human genes rather than rodent equivalents.

## Evaluating Position Effects

**Transposon jumping** can be employed to move stably inserted P elements to other sites within the genome. This allows researchers to explore the effects of position on gene expression. To induce transposon jumping, embryos from a transformed strain are injected with 'wings clipped' helper plasmids that transcribe transposase. The transposase interacts with the terminal repeats of the inserted P element, causing it to transpose to a new site. The helper element does not integrate, so the new strain will be stable until transposase is again supplied.

## Targeted Gene Transfer

The ability to replace or modify genes in their normal chromosomal locations, **targeted gene transfer**, has not been possible with *Drosophila* until recently

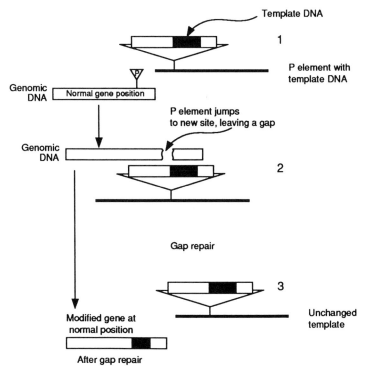

**Figure 10.5**  Targeted gene mutation in *D. melanogaster* is based on the gap repair hypothesis. If a P element jumps out of a normal gene, it will leave a gap that must be repaired. Repair is thought to involve using DNA with homologous ends from within the genome as a template for DNA repair. If a new P element with a modified gene structure is present, the sequence in the gap can be filled in using the modified gene as the template, leading to a targeted gene alteration. (Modified from Gloor et al. 1991.)

(Ballinger and Benzer 1989, Kaiser and Goodwin 1990, Gloor et al. 1991, Sentry and Kaiser 1992). The cut-and-paste model of P element transposition provided a scheme for inserting a gene into the gap left behind by a P element (Figure 10.5).

As noted above, P transposition is thought to leave a double-stranded gap in the original insertion site, and this gap may be extended by exonucleases. The gap can be repaired, using a template provided by a sister chromatid, a homologous chromosome containing a homologous DNA sequence, or an extrachromosomal element. If the sister chromatid or homologous chromosome has a second copy of the P element, the P-element sequences will be restored in the gap, giving the impression that transposition has been replicative.

Engels et al. (1990) proposed a method for **site-directed mutagenesis** (**targeted mutagenesis** or targeted gene replacement). The first step is to obtain a

P-element insertion in the gene of interest, preferably close to the site to be modified. This is feasible because many different colonies of *Drosophila* have been developed that contain P elements, and the precise location of these elements is known. The next step is to transfer the desired *replacement* gene into a second colony with a P-element vector. Then, individuals from the first colony are crossed with the second. A source of transposase is added to promote transposition and targeted gene replacement. The replacement gene serves as the template to fill in the gap left when the P element transposes. The result is that the repaired DNA is converted to the sequence that has been introduced.

Targeted gene replacement of the X-linked *white* locus was achieved by Gloor et al. (1991). They worked with a colony that carries a P element in one of the *white* exons. This P element cannot transpose because it lacks transposase. Transposase was added by crossing flies from this colony with a fly carrying an immobile P element on chromosome 3 that can produce transposase. A vector, called P[walter], was injected that carried an altered *white* gene with 12 bp substitutions. These base pair substitutions add or remove a restriction enzyme site, which provides an efficient way to determine whether the replacement *white* allele was used to repair the gap induced by P element transposition ( = gene conversion). About 1% of fly lines were identified with different amounts of gene conversions, ranging from a few base pairs to alterations of at least 2790 bp.

A 1% gene conversion rate is sufficiently frequent to make targeted gene transfer a practical method for systematically altering genes in their normal locations to see how their function is modified. Another advantage to targeted gene transfer is that it is possible to insert genes longer than 40 kb by this method. P-element-mediated transposition is limited to inserting DNA segments less than 40 kb in length.

Nassif and Engels (1993) investigated the length and stringency of homology required for repair of double-strand DNA breaks in *Drosophila* germ cells using the targeted gene transfer system. They found that a relatively short match (of a few hundred base pairs) of homologous sequence on either side of the target is sufficient to promote gap repair. However, the gap repair was sensitive to single base mismatches within the homologous regions. Interestingly, the data also suggest that the ends of a broken chromosome can locate a single homologous template anywhere in the genome by using a short stretch of closely matching sequence. How this occurs remains mysterious, but the search is sufficiently efficient that up to several percent of the progeny exhibited gene conversion events (targeted gene replacement) at the *white* locus. This high rate of gene conversion is considered to be unlikely if the process were dependent upon *random* collisions between homologous DNA sequences.

Transposable elements have been identified in most organisms investigated (Chapters 4, 5). If other elements transpose by a mechanisms similar to that of P elements, then targeted gene conversion could be feasible in insects and mites (Sentry and Kaiser 1992). Targeted gene insertion would make it possible to introduce new genetic information into specific chromosomal sites, or to modify existing genes in directed ways rather than to rely on random insertions of the exogenous DNA. Such random insertions can cause lethality if they occur into essential genes, or can result in poor levels of expression if insertion occurs into heterochromatic or other unsuitable regions. Targeted genetic manipulation of organisms other than *Drosophila* has not yet been achieved (Chapter 15).

## Transformation of Other Species by P Elements

DNA from *D. melanogaster* has been introduced into *D. simulans* with P-element vectors. Such transposition events in *D. simulans* can produce a syndrome of hybrid dysgenesis similar to that found in *melanogaster* (Daniels et al. 1989). P-element vectors also have been used to transform the more distantly related *D. hawaiiensis* (Brennan et al. 1984). Thus, P-element vectors can integrate and transpose in several *Drosophila* species other than *melanogaster*. Rio et al. (1988) suggested that *Drosophila* P-element transposase was active even in mammalian cells and yeast, and this elicited optimism about the possibility of using P elements for genetic engineering of other organisms. Unfortunately, efforts to use P-element vectors to transform arthropod species outside the genus *Drosophila* have failed to date (O'Brochta and Handler 1988, Handler and O'Brochta 1991, Handler et al. 1993a) (Chapter 15). As a result, the potential for using other transposable elements from *Drosophila*, such as *hobo* and *mariner*, as vectors for transformation of other insects is being investigated. Preliminary results are promising. Atkinson et al. (1993) found that the *hobo* element of *Drosophila* can excise in the house fly *Musca domestica*, and Garza et al. (1991) have introduced the *mariner* element from *D. mauritiana* into the germ line of *D. melanogaster*.

## References Cited

Anxolabehere, D. and G. Periquet. 1987. P-homologous sequences in Diptera are not restricted to the Drosophilidae family. Genet. Iber. 39: 211–222.

Anxolabehere, D., M. G. Kidwell and G. Periquet. 1988. Molecular characteristics of diverse populations are consistent with the hypothesis of a recent invasion of *Drosophila melanogaster* by mobile P elements. Mol. Biol. Evol. 5: 252–269.

Ashburner, M. 1989. Drosophila: *A Laboratory Handbook*. Cold Spring Harbor Laboratory Press, Cold Spring Harbor, NY.

Atkinson, P. W., W. D. Warren and D. A. O'Brochta. 1993. The *hobo* transposable element of *Drosophila* can be cross-mobilized in houseflies and excises like the Ac element of maize. Proc. Natl. Acad. Sci. USA 90: 9693–9697.

Ballinger, D. G. and S. Benzer. 1989. Targeted gene mutations in *Drosophila*. Proc. Natl. Acad. Sci. USA 86: 9402–9406.

Bingham, P. M., M. G. Kidwell and G. M. Rubin. 1982. The molecular basis of P-M hybrid dysgenesis: The role of the P element, a P-strain-specific transposon family. Cell 29: 995–1004.

Brennan, M. D., R. G. Rowan and W. J. Dickinson. 1984. Introduction of a functional P element into the germ-line of *Drosophila hawaiiensis*. Cell 38: 147–151.

Cooley, L., R. Kelley and A. Spradling. 1988. Insertional mutagenesis of the *Drosophila* genome with single P elements. Science 239: 1121–1128.

Daniels, S. B. and L. D. Strausbaugh. 1986. The distribution of P element sequences in *Drosophila*: The *willistoni* and *saltans* species groups. J. Mol. Evol. 23: 138–148.

Daniels, S. B., L. D. Strausbaugh, L. Ehrman and R. Armstrong. 1984. Sequences homologous to P elements occur in *Drosophila paulistorum*. Proc. Natl. Acad. Sci. USA 81: 6794–6797.

Daniels, S. B., A. Chovnick and M. G. Kidwell. 1989. Hybrid dysgenesis in *Drosophila simulans* lines transformed with autonomous P elements. Genetics 121: 281–291.

Eggleston, W. B., D. M. Johnson-Schlitz and W. R. Engels. 1988. P-M hybrid dysgenesis does not mobilize other transposable element families in *D. melanogaster*. Nature 331: 368–370.

Engels, W. R. 1989. P-elements in *D. melanogaster*. Pp. 437–484. In: *Mobile DNA*. D. E. Berg and M. M. Howe, Eds. Am. Soc. Microbiol., Washington, DC.

Engels, W. R. 1992. The origin of P elements in *Drosophila melanogaster*. BioEssays 14: 681–686.

Engels, W. R., D. M. Johnson-Schlitz, W. B. Eggleston and J. Sved. 1990. High-frequency P element loss in *Drosophila* is homolog dependent. Cell 62: 515–525.

Garza, D., M. Medhora, A. Koga and D. L. Hartl. 1991. Introduction of the transposable element *mariner* into the germline of *Drosophila melanogaster*. Genetics 128: 303–310.

Gloor, G. B., N. A. Nassif, D. M. Johnson-Schlitz, C. R. Preston and W. R. Engels. 1991. Targeted gene replacement in *Drosophila* via P element-induced gap repair. Science 253: 1110–1117.

Gloor, G. B., C. R. Preston, D. M. Johnson-Schlitz, N. A. Nassif, R. W. Phillis, W. K. Benz, H. M. Robertson and W. R. Engels. 1993. Type I repressors of P element mobility. Genetics 135: 81–95.

Hagemann, S. W., J. Miller and W. Pinsker. 1992. Identification of a complete P-element in the genome of *Drosophila bifasciata*. Nucleic Acids Res. 20: 409–413.

Handler, A. M. and D. A. O'Brochta. 1991. Prospects for gene transformation in insects. Annu. Rev. Entomol. 36: 158–183.

Handler, A. M., S. P. Gomez and D. A. O'Brochta. 1993a. A functional analysis of the P-element gene-transfer vector in insects. Arch. Insect Biochem. Physiol. 22: 373–384.

Handler, A. M., S. P. Gomez and D. A. O'Brochta. 1993b. Negative regulation of P element excision by the somatic product and terminal sequences of P in *Drosophila melanogaster*. Mol. Gen. Genet. 237: 145–151.

Hickey, D. A. 1982. Selfish DNA: A sexually transmitted nuclear parasite. Genetics 101: 519–535.

Houck, M. A. 1993. Mites as potential horizontal transfer vectors of eukaryotic mobile genes: Proctolaelaps regalis as a model. Pp 45–69. In: *Mites: Ecological and Evolutionary Analyses of Life-History Patterns*. M. A. Houck, Ed. Chapman and Hall, New York.

Houck, M. A., J. B. Clark, K. R. Peterson and M. G. Kidwell. 1991. Possible horizontal transfer of *Drosophila* genes by the mite *Proctolaelaps regalis*. Science 253: 1125–1129.

Jowett, T. 1991. Transgenic *Drosophila* as an *in vivo* model for studying mammalian drug metabolism. BioEssays 13: 683–689.

Kaiser, K. and S. F. Goodwin. 1990. "Site-selected" transposon mutagenesis of *Drosophila*. Proc. Natl. Acad. Sci. USA 87: 1686–1690.

Karess, R. E. 1987. P element mediated germline transformation of *Drosophila*. Pp. 121–141. In: *DNA Cloning. Vol. II. A Practical Approach*. D. M. Glover, Ed. IRL Press, Oxford.

Karess, R. E. and G. M. Rubin. 1984. Analysis of P transposable element functions in Drosophila. Cell 38: 135–146.

Kidwell, M. G. 1992. Horizontal transfer of P elements and other short inverted repeat transposons. Genetica 86: 275–286.

Kidwell, M. G., J. F. Kidwell and J. A. Sved. 1977. Hybrid dysgenesis in *Drosophila melanogaster*: A syndrome of aberrant traits including mutation, sterility, and male recombination. Genetics 86: 813–833.

Kitamura, T., S. Kobayashi and M. Okada. 1993. Developmentally regulated splicing of the third intron of P element in somatic tissues in *Drosophila* embryos. Dev. Growth Differ. 35: 67–73.

Kobayashi, S., T. Kitamura, H. Sasaki and M. Okada. 1993. Two types of pole cells are present in the *Drosophila* embryo, one with and one without splicing activity for the third P-element intron. Development 117: 885–893.

Lansman, R. A., S. M. Stacey, T. A. Grigliatti and H. W. Brock. 1985. Sequences homologous to the P mobile element of *D. melanogaster* are widely distributed in the subgenus *Sophophora*. Nature 318: 561–63.

Laski, F. A., D. C. Rio and G. M. Rubin. 1986. Tissue specificity of *Drosophila* P element transposition is regulated at the level of mRNA splicing. Cell 44: 7–17.

Lemaitre, B., S. Ronsseray and D. Coen. 1993. Maternal repression of the P element promoter in the germline of *Drosophila melanogaster*: A model for the P cytotype. Genetics 135: 149–160.

Levis, R., T. Hazelrigg and G. M. Rubin. 1985. Effects of genomic position on the expression of transduced copies of the *white* gene of *Drosophila*. Science 229: 558–561.

Miller, W. J., S. Hagemann, E. Reiter and W. Pinsker. 1992. P-element homologous sequences are tandemly repeated in the genome of *Drosophila guanche*. Proc. Natl. Acad. Sci. USA 89: 4018–4022.

Misra, S., R. M. Buratowski, T. Ohkawa and D. C. Rio. 1993. Cytotype control of *Drosophila melanogaster* P element transposition: Genomic position determines maternal repression. Genetics 135: 785–800.

Montchamp-Moreau, C., G. Periquet and D. Anxolabehere. 1991. Interspecific transfer of P elements by crosses between *Drosophila simulans* and *Drosophila mauritiana*. J. Evol. Biol. 4: 131–140.

Nassif, N. and W. Engels. 1993. DNA homology requirements for mitotic gap repair in *Drosophila*. Proc. Natl. Acad. Sci. USA 90: 1262–1266.

O'Brochta, D. A. and A. M. Handler. 1988. Mobility of P elements in drosophilids and non-drosophilids. Proc. Natl. Acad. Sci. USA 85: 6052–6056.

Rancourt, D. E., I. D. Peters, V. K. Walker and P. L. Davies. 1990. Wolffish antifreeze protein from transgenic *Drosophila*. Bio/Technology 8: 453–457.

Rio, D. C., F. A. Laski and G. M. Rubin. 1986. Identification and immunochemical analysis of biologically active *Drosophila* P element transposase. Cell 44: 21–32.

Rio, D. C., G. Barnes, F. A. Laski, J. Rine and G. M. Rubin. 1988. Evidence for P element transposase activity in mammalian cells and yeast. J. Mol. Biol. 200: 411–415.

Roberts, D. B., Ed. 1986. Drosophila: A *Practical Approach*. IRL Press, Oxford.

Ronsseray, S., B. Lemaitre and D. Coen. 1993. Maternal inheritance of P cytotype in *Drosophila melanogaster*: A 'pre-P cytotype' is strictly extra-chromosomally transmitted. Mol. Gen. Genet. 241: 115–123.

Rubin, G. M. and A. C. Spradling. 1982. Genetic transformation of *Drosophila* with transposable element vectors. Science 218: 348–353.

Rubin, G. M. and A. C. Spradling. 1983. Vectors for P element-mediated gene transfer in *Drosophila*. Nucleic Acids Res. 11: 6341–6351.

Rubin, G. M., M. G. Kidwell, and P. M. Bingham. 1982. The molecular basis of P-M hybrid dysgenesis: The nature of induced mutations. Cell 29: 987–994.

Sass, H. 1990. P-transposable vectors expressing a constitutive and thermoinducible *hsp82–neo* fusion gene for *Drosophila* germline transformation and tissue-culture transfection. Gene 89: 179–186.

Sentry, J. W. and K. Kaiser. 1992. P element transposition and targeted manipulation of the *Drosophila* genome. Trends Genet. 8: 329–311.

Simonelig, M. and D. Anxolabehere. 1991. A P element of *Scaptomyza pallida* is active in *Drosophila melanogaster*. Proc. Natl. Acad. Sci. USA 88: 6102–6106.

Spradling, A. C. 1986. P element-mediated transformation. Pp. 175–97. In: Drosophila: A *Practical Approach*. D. B. Roberts, Ed. IRL Press, Oxford.

Spradling, A. C. and G. M. Rubin. 1982. Transposition of cloned P elements into *Drosophila* germline chromosomes. Science 218: 341–347.

Spradling, A. C. and G. M. Rubin. 1983. The effect of chromosomal position on the expression of the *Drosophila* xanthine dehydrogenase gene. Cell 34: 47–57.

Woodruff, R. C. 1992. Transposable DNA elements and life history traits. I. Transposition of P DNA elements in somatic cells reduces the life span of *Drosophila melanogaster*. Genetica 86: 143–154.

# Part III

# Applications to Entomological Problems

# Chapter 11

# Sex Determination in Insects

## Overview

A single model has been proposed suggesting that sex determination in all insects is based on modifications of the *Drosophila* scheme, but the evidence for this remains controversial. A resolution of the molecular genetics of sex determination in arthropods will have both theoretical and applied applications, resulting in the improved genetic control of pests or useful genetic modifications of beneficial biological control agents. The molecular genetics of development in *Drosophila* is controlled by a hierarchy of regulatory genes, and the determination of sex is a special aspect of such hierarchical development. Sex determination in *D. melanogaster* has three components: (1) dosage compensation, (2) somatic sexual development, and (3) germ line sexual development. The primary cue for determining sex in *Drosophila* is the number of X chromosomes relative to autosomes in a cell (X:A ratio). This ratio determines somatic sex, germ line sex, and dosage compensation by regulating functions of sets of regulatory genes. Sex determination is based on a double switch mechanism, in which a key gene ($Sxl^+$) must be ON to determine the female pathway. If $Sxl^+$ is OFF, the male pathway is the 'default' developmental process. Sex is subsequently determined by the differential splicing of messenger RNAs. At the end of the pathway, sex determination is influenced by a DNA binding regulatory protein coded for by the $dsx^+$ gene. While sex determination is clearly determined by chromosomal and genetic processes, environment also plays a role in sex determination in some arthropods. Several different types of infectious agents also are able to modify sex determination or sex ratio in a wide array of insects and mites.

## Introduction

Sexual reproduction results in genetic exchange, variation, and diversity. In a sexually reproducing species, genes from different individuals can be combined in a single descendant. This results in the construction of new genotypes from preexisting variability by the mechanisms of segregation and recombination. In segregation, homologous chromosomes separate during meiosis to yield haploid gametes. Recombination results in a different assortment of alleles from those found in either of the gametes from which it was derived. Recombination occurs during crossing over between homologous chromosomes, as well as from the independent assortment of maternal and paternal chromosome homologs. Theories for the evolution of sex have focused on the advantages of the combined effects of segregation and recombination. In a sexual population, advantageous mutations that arise at two different loci in two parents can be combined in one individual in later generations.

An understanding of sex determination in insects has both fundamental and applied applications. The evolutionary advantages of sexual reproduction and the reasons why organisms vary the sex ratio of their progeny are among the most discussed topics in evolutionary biology (Bull 1983, Charnov 1982, Colwell 1981, Hamilton 1967, Hartl and Brown 1970, Maynard Smith 1978, Michod and Levin 1988, Thornhill and Alcock 1983, Williams 1975, King 1992, Haig 1993a,b, Howard and Lively 1994, Wrensch and Ebbert 1993). A detailed understanding of the mechanisms of sex determination could provide insights into the regulation of development of a significant character in eukaryotes. Such knowledge also could provide useful tools for the genetic improvement of arthropod natural enemies of pest arthropods and weeds, genetic control of 'Africanized' honey bees, or it could improve the methods by which sterile males are mass reared for genetic control programs (Shirk et al. 1988, Stouthamer et al. 1992, LaChance 1979). See Chapter 15 for a discussion of genetic manipulation of pest and beneficial arthropods.

Sex determination mechanisms in insects appear to be quite diverse (Lauge 1985, Retnakaran and Percy 1985, White 1973, Wrensch and Ebbert 1993). There are insects that have a **genetic sex determination** system by which genetic differences determine maleness or femaleness. Other insects appear to have **environmental sex determination**, in which there are no genetic differences between males and females but environmental conditions determine sex. In a few insects, the hemolymph of the mother determines the sex of the offspring. Ploidy levels sometimes are important in sex determination: both sexes of some species are diploid (diplo-diploidy), while others have haploid males and diploid females (**arrhenotoky**). Other species consist primarily of diploid females (**thelytoky**), and haploid males rarely are produced.

A particularly interesting system has been found in the sciarid fly *Sciara coprophila* (White 1973). Their genome consists of three autosomes, sex chromosomes, and 'L' chromosomes, which are germ line limited and similar to B chromosomes. Both males and females have three pairs of autosomes, two X chromosomes, and generally two L chromosomes. Two kinds of X chromosomes exist (one is termed X and the other is X') and there is no Y chromosome. Spermatogenesis in *S. coprophila* is unusual, with sperm produced that has two sister chromatids of a single X, a haploid set of autosomes, and usually two L chromosomes. As a result, the zygote after fertilization has three X chromosomes, two identical Xs derived from the father and one derived from the mother. Two kinds of females result from these zygotes; one type produces only male progeny and one type produces only female progeny. The female-producing mothers are always XX' and the male producers are always XX. Males are always XX; this is due to the loss of the L chromosomes and two of the three X chromosomes in the male somatic cells so that they are XO. The two X chromosomes are always paternally derived. The L chromosomes also are eliminated in the females, but only one X is eliminated, which results in females with two X chromosomes in the soma. The evolution of such a complex chromosomal system has been attributed to intragenomic conflict associated with different genes having different patterns of inheritance (Haig 1993b).

A basic aspect of sex determination is **dosage compensation** of the X chromosomes, and it, too, varies in its mechanism. Dosage compensation in *Drosophila melanogaster* is achieved by hypertranscription of the single X chromosome in males. Thus, males with one X chromosome produce *equivalent* amounts of gene product as females with two X chromosomes. By contrast, dosage compensation in at least one other insect, the mole cricket *Gryllotalpa fossor*, is analogous to that in mammals; one of the two X chromosomes in females is transcriptionally *inactivated*, with the inactivation occurring randomly within each cell (Rao and Padmaja 1992). *G. fossor* males are XO and females are XX, and one of the two X chromosomes in female cells is late replicating and transcriptionally silent.

Sex-determining systems may vary even within species. For example, *Drosophila* and *Sciara* use the ratio of X chromosomes to autosomes as the key first step to determine sex. In some populations of the house fly *Musca domestica* or the mosquito *Anopheles*, the Y chromosomes determine maleness. In other strains of *M. domestica*, dominant male determiners are found, and in yet other strains dominant female determiners are found.

Sex determination involves both the soma and the germ line tissues. Sexual dimorphism in adult insects is often extreme, with differences in setal patterns, pigmentation, external genitalia, internal reproductive systems, and behavioral patterns.

How do sexually determined differences in soma and germ line arise? The details are becoming clear for *Drosophila*. Despite the diversity in sex-determining mechanisms mentioned above, a model has been proposed in which sex determination in all insects can be based on a single scheme (Nothiger and Steinmann-Zwicky 1985). Whether this single sex-determination model is valid for all insects remains controversial (Bownes 1992).

The most detailed information on sex determination in insects is based on studies of *D. melanogaster*, although some information is available for *Musca domestica*, and a few other economically important insects such as mosquitoes. First, we review the basic sex determination system in *D. melanogaster*. Then, sex determination in other insects is compared against the model proposed by Nothiger and Steinmann-Zwicky (1985). Finally, examples are provided that illustrate the importance of extrachromosomal and microbial genes in modifying sex in arthropods.

## Sex Determination in *Drosophila melanogaster*

An understanding of sexual determination in *D. melanogaster* has relied on identifying a relatively few spontaneous mutants (Table 11.1). This suggests that the number of genes involved in sex determination is not high (Belote et al. 1985, Slee and Bownes 1990). Sex determination in *D. melanogaster* involves three aspects: dosage compensation, somatic cell differentiation, and germ line differentiation. More is known about dosage compensation and somatic cell differentiation than about germ line differentiation.

### Dosage Compensation and Somatic Cell Differentiation

The relative number of X chromosomes and autosomes in *D. melanogaster* is responsible for the primary step in sex determination. Cells with two X chromosomes and two sets of autosomes (2X:2A, or a ratio of 1.0) develop into females, while diploid cells (with 1X:2A, or a ratio of 0.5) develop into males (Figure 11.1). Somehow *Drosophila* is able to 'count' X chromosomes, probably through the influence of two X-linked genes, *sisterless-a*[+] and *sisterless-b*[+] (Table 11.1). It is not the absolute number of X chromosomes that is important, however, but the *ratio* that is critical for determining sex. Thus, flies with equal numbers of X chromosomes and autosomes (XX:AA, XXX:AAA, or X:A, or a ratio of 1.0) develop as females.

Flies with an intermediate X:A ratio (XX:AAA) develop as intersexual flies that appear to be mosaics of discrete patches of male and female tissues. Distinct boundaries between cells exist in *Drosophila* and sex is determined in individual cells (**cell autonomous determination**). No diffusible substances, such as sex hormones, influence sexual development in *Drosophila*.

The Y chromosome does not determine sex in *D. melanogaster*, although it is required for normal spermatogenesis. The Y chromosome is important in sex determination in some insects—*Anopheles*, *Calliphora*, and *Musca* (Table 11.2).

Once the X:A ratio is assessed, the activities of a relatively small number of major regulatory genes are triggered that ultimately lead to male or female differentiation in both the soma and germ line. This is regulated through a cascade of sex-specific events in which RNA transcripts are differentially processed in males and females (Figures 11.1 and 11.2, Table 11.1).

**Table 11.1**

Genes Involved in Somatic Sex Determination in *Drosophila melanogaster*

| Gene | Phenotype | Function |
|---|---|---|
| *daughterless*+ (*da*+) | *da/da* lethal in female embryos | *da*+ activates *Sxl*+ in female embryos. |
| *sisterless a*+ and *b*+ | Lethal in female embryos if mutated | Involved in activating *Sxl*+ in females and 'counts' number of X chromosomes. |
| *Sex-lethal*+ (*Sxl*+) | XX cells | Active in females to determine somatic differentiation by activating *tra*+, necessary for oogenesis but not spermatogenesis. |
| *transformer*+ (*tra*+) | XX flies → pseudomales if mutated | Active with *tra-2*+ in regulating *dsx*+ in females. |
| *transformer 2*+ (*tra-2*+) | XX flies → pseudomales if mutated | Active in females to induce female-specific *dsx*+ expression and repress male-specific *dsx*+ expression. Inactive in males, but needed for spermatogenesis. |
| *doublesex*+ (*dsx*+) | XX, XY flies → intersex if mutated | Active in males to repress female differentiation; in females it represses male differentiation; loss-of-function mutants result in intersexes in both males and females; the pivotal terminal differentiation switch. |
| *intersex*+ (*ix*+) | XX flies → intersex if mutated | Active in females with *dsx*+ product to repress male differentiation; not needed in males. |
| *maleless*+ *msl-1*+ *msl-2*+ *msl-3*+ | Dosage compensation in males | X chromosome transcription rate in males; required in males, but no known function in females. |

<sup>a</sup>Derived from Bownes (1992).

Dosage compensation in *Drosophila* is required to ensure that equal amounts of gene products from the X chromosome are present in both males and females. Males (X:AA) are basically **aneuploid** for an X chromosome, which is a large fraction of the total genome. Aneuploidy (when the chromosomal composition in a cell is not an exact multiple of the haploid set) is normally lethal in organisms, unless equalization of the expression of the X-linked genes is

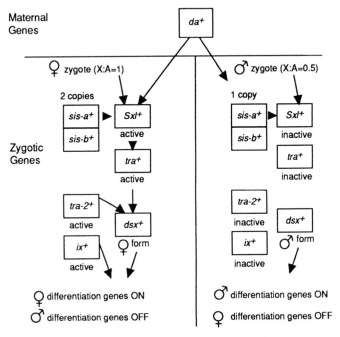

**Figure 11.1** A model for sex determination in *Drosophila melanogaster*. Both maternal and zygotic genes are involved. The product of the *da*⁺ locus is produced by the mother and stored in the egg and zygote. The ratio of X chromosomes to autosomes (A) is 1 in female eggs. This X:A ratio, as well as products of the *da*⁺, *sis-a*⁺ and *sis-b*⁺ genes, is used to activate *Sxl*⁺ in females. The *sis-a*⁺ and *sis-b*⁺ genes are located on the X chromosome, so are present in two copies in females. *Sxl*⁺ is active in females but inactive in males. Because *Sxl*⁺ is active in females, *tra*⁺ and *tra-2*⁺ are active, which leads to the *dsx*⁺ gene being active in females. The *dsx*⁺ locus is a bifunctional gene, producing one product in females and a different one in males. The female *dsx*⁺ product, along with the *ix*⁺ gene product, represses male differentiation genes in females, leading to expression of female differentiation genes. The ratio X:A is 0.5 in males. Thus, *Sxl*⁺ is not activated because insufficient *sis-a*⁺ and *sis-b*⁺ products are made from the single copies present on the single X chromosome in males. Because *Sxl*⁺ is inactive, *tra*⁺ is inactive and *tra-2*⁺ has no function. The combination results in the expression of *dsx*⁺ in the male mode, which represses female differentiation genes and leads to the expression of male differentiation genes. Sex determination in the soma of *D. melanogaster* involves a cascade of genes that are activated and inactivated autonomously in each cell. (Redrawn from Bownes 1992.)

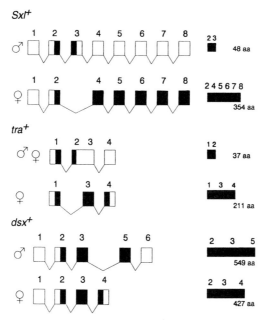

**Figure 11.2** RNA of *Sxl+*, *tra+*, and *dsx+* is processed in a sex-specific manner in *Drosophila melanogaster*. An open square is an exon and a filled square is a coding region. The V indicates introns. RNA splicing is shown at the left, with the splicing patterns indicated by a V. The final protein is shown at the right with the exons and number of amino acids in the protein indicated for each sex. The Sxl and tra proteins are nonfunctional in males. *dsx+* is a double-switch gene with alternative splicing patterns in males and females and two different functional protein products. (Redrawn from Bownes 1992.)

achieved by dosage compensation. Dosage compensation in *D. melanogaster* is achieved by **hypertranscription** of the single X chromosome in males so that an equal amount of gene products is present in both males and females.

Both sexual differentiation and dosage compensation are regulated through a key X-linked gene, *Sex-lethal+* (*Sxl+*), which must be ON in females and OFF in males (Figure 11.1). If, for some reason, *Sxl+* is OFF in females or ON in males, the flies die because of aberrant dosage compensation. Once the X:A ratio is read and the *Sxl+* gene is turned ON or OFF early in embryonic development, the developmental path chosen is stable.

*Sxl+* is transcribed in both sexes, but only female transcripts make the full-length active Sxl protein. *Sxl+* is 23 kb in size and is transcribed in a complex manner. A variety of splicing patterns generate multiple, overlapping transcripts, including male-specific and female-specific size classes (Figures 11.2, 11.3). The earliest *Sxl+* transcripts differ from later transcripts, and may be involved in

Regulation of *tra*⁺ splicing by *Sxl*⁺ products

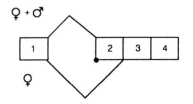

Regulation of *dsx*⁺ splicing by *tra*⁺ and *tra-2*⁺ products

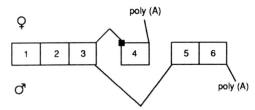

**Figure 11.3**   Splicing patterns of the *tra*⁺ and *dsx*⁺ loci are different in male and female *Drosophila melanogaster*. The solid circle indicates that a repressor (a product of *Sxl*⁺) is important in determining which splicing pattern occurs. When the *tra*⁺ transcript is spliced in the female, the repressor binds at the acceptor site at exon 2 so that binding occurs at the start of exon 3 instead of exon 2. Splicing of the *dsx*⁺ transcript is determined by *tra*⁺ and *tra-2*⁺ products. In females, the products of *tra-2*⁺ and *tra*⁺ (activator, shown as a small square) bind to the acceptor site in exon 4. If no activator is bound to exon 4, the splice acceptor site in exon 5 is used and produces a male-specific product. The default splice pattern results in a male. (Modified from Bownes 1992.)

autoregulation. Male transcripts are inactive because they include an extra exon that stops the translation process. *Sxl*⁺ expression is under **autoregulatory control**, but initiation of *Sxl*⁺ expression requires the action of maternal genes, including *daughterless*⁺ (*da*⁺).

**Maternal genes** influence development of progeny by one of two methods: either the mother produces a gene product that is transferred to and stored in the egg, or the mother's messenger RNA is transferred to and stored in the eggs and subsequently is translated by the embryo. Female progeny of *da* mothers fail to activate *Sxl*⁺ and die as embryos. Male progeny of *da* mothers survive because they do not require *Sxl*⁺. The da protein is probably a transcription factor. In addition to this maternal gene, two X-linked genes, *sisterless-a*⁺ (*sis-a*⁺) and *sisterless-b*⁺ (*sis-b*⁺), appear to be involved in measuring the X:A ratio in the soma (Figure 11.1). Extra doses of *sis*⁺ result in female development.

In *Drosophila* males, the hypertranscription of the single X chromosome requires the functions of at least four autosomal genes, collectively called male-specific lethal genes, which are under the control of the $Sxl^+$ gene (Table 11.1). These genes are *maleless*$^+$(*mle*$^+$), *male-specific lethal 1*$^+$ (*msl 1*$^+$), *male-specific lethal 2*$^+$(*msl 2*$^+$), and *male-specific lethal 3*$^+$(*msl 3*$^+$). Males with mutations at any of these four loci die as late larvae or early pupae, while mutant females are viable and fertile. These four loci are under the control of the $Sxl^+$ gene. In males, the $Sxl^+$ gene is OFF, and these four autosomal genes are ON, a combination which leads to male somatic sexual differentiation. Males with mutations of *mle*$^+$, *msl 1*$^+$, or *msl 2*$^+$ have decreased levels of X-linked gene products, and males with mutations of *mle*$^+$ or *msl 3*$^+$ exhibit decreased transcription of the entire X chromosome. Details of how these four loci function remain obscure (Kuroda et al. 1991; Gorman et al. 1993). The *maleless*$^+$ locus codes for a protein with similarity to RNA and DNA helicases, and the protein is found associated with hundreds of sites along the length of the X chromosome in the male, but not the female. The four genes (*msl 1*$^+$, *msl 2*$^+$, *msl 3*$^+$, and *mle*$^+$) have no known functions in females.

Once $Sxl^+$ is ON in females, a second series of regulatory genes is important in differentiating between the alternative pathways in somatic cell development. These secondary regulatory genes include *transformer*$^+$ (*tra*$^+$, *transformer-2*$^+$(*tra-2*$^+$), *intersex*$^+$ (*ix*$^+$), and *doublesex*$^+$ (*dsx*$^+$) (Figure 11.1). The *tra*, *tra-2*, and *ix* mutations affect somatic sex determination in females, but are not needed for male somatic differentiation (Table 11.1). Although *tra-2*$^+$ is not needed for male differentiation, it is critical for normal spermatogenesis in males.

The *dsx*$^+$ locus is needed for both male and female development. The *dsx*$^+$ gene is a *double switch*, with only one switch functioning in each sex. If *dsx*$^+$ is inactivated, both male and female genes are active within a cell, which results in an intersexual phenotype at the cellular level. When *dsx*$^+$ is active in males, it represses female differentiation. When it is active in females, it represses male development.

A maternally derived gene product, daughterless, is necessary if $Sxl^+$ is to be active in both males and females. The gene (or genes, R-for repression) that specify maleness by repressing $Sxl^+$ has not been identified, but must be located on the autosomes, because a genotype with one X chromosome and one set of autosomes (X:A) is female. When another set of autosomes is added (X:AA), male development occurs. For normal diploid males (X:AA) and females (XX:AA), it is likely that both sexes contain the same limited amount of the product of the R gene or genes, but that two X chromosomes bind all repressor (R) so that $Sxl^+$ can be transcribed. One of the repressors probably is encoded by a gene called *deadpan* (Younger-Shepherd et al. 1992). However, if only one X chromosome is present, some R molecules are able to repress $Sxl^+$. Male development is the 'default pathway' executed in the absence of the $Sxl^+$ gene.

In addition to the sex-determination genes there are genes whose products are directly responsible for the structure and function of sexually dimorphic somatic tissues. A number of structural genes are controlled by the sex-determination regulatory pathway, such as the yolk polypeptide genes, which are expressed in the fat body in a female-specific manner.

## Germ Line Determination

Germ-line-specific genes are required for the development of germ-line tissues in D. melanogaster. Less is known about sex determination in the germ line than in the soma, but it appears that the process of differentiation is different (Pauli and Mahowald 1990, Granadino et al. 1993, Steinmann-Zwicky 1993). Pole cells in the embryo are segregated into the posterior pole of the insect embryo before blastoderm formation, and include the progenitors of the germ cells.

Components of the germ plasm (pole plasm) are synthesized in the mother during oogenesis by a cluster of 15 nurse cells, which are connected to the oocyte at its anterior by cytoplasmic bridges. Pole plasm components are transported into the oocyte and translocated to the posterior pole of the egg. At least eight maternally active genes are important in the production of pole cells, including cappuccino+, spire+, staufen+, oskar+, vasa+, valois+, mago nashi+, and tudor+ (Ephrussi and Lehmann 1992). These genes also are important in the formation of normal abdomens in Drosophila.

During embryogenesis, prospective male and female germ cells are indistinguishable, but differentiation is begun during the larval stage, with male gonads larger than female gonads because they contain more germ cells. The gonads themselves are products of somatic cells of mesodermal origin (Steinmann-Zwicky 1992).

There is controversy as to whether germ line sex determination is cell autonomous, as it is in somatic sex determination. A model (Figure 11.4) has been proposed for germ line sex determination in which determination is incompletely cell autonomous and requires signals from the soma (Pauli and Mahowald 1990, Steinmann-Zwicky 1992).

If sex of the germ cells is determined by both inductive and cell-autonomous components, the following model may apply. Sex determination in the germ line in 1X:2A cells is at first cell autonomous and subsequently is influenced by the soma. In 2X:2A germ cells, sex is determined according to the surrounding soma, but later differentiation depends on intrinsic factors. At least four genes are involved in germ line sex determination—Sxl+, snf+, otu+, and ovo+—but their roles in males and females remain speculative (Pauli and Mahowald, 1990). The ovo+ gene appears to be sensitive to somatic signals and the otu+ gene is necessary at several steps of female germ cell development, but it is not clear how otu+ interacts with the other genes. Ovo+ mutants lacking a gene product do not produce viable female germ cells, but in males the germ cells develop normally.

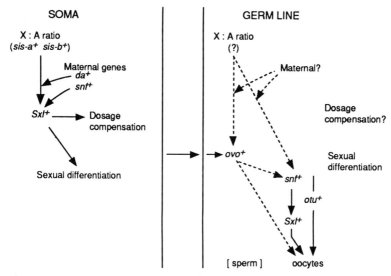

**Figure 11.4** A model suggesting that sex determination in the soma and germ line of *Drosophila melanogaster* are different. In 1X:2A germ line cells, sex determination is initially cell-autonomous, but subsequent differentiation is influenced by the soma. In 2X:2A germ line cells, the surrounding soma influences the sex of the cells, but further differentiation depends on intrinsic factor(s). Solid arrows represent known genetic interactions while interrupted lines represent possible interactions. The (?) indicates that no 'counting elements' equivalent to the somatic *sis-a*+ and *sis-b*+ loci have been found in the germ line. The *ovo*+ gene appears to be sensitive to somatic signals and acts either upstream or independently of the *snf*+ and *Sxl*+ loci. The *otu*+ locus is necessary at several steps in female germ cell development, but the details are unclear. [sperm] indicates it is not known which genes are involved during early sperm development and whether sex determination in the male germ line is the default status when *ovo*+, *Sxl*+, *snf*+, or *otu*+ are not active. (Redrawn from Pauli and Mahowald 1990.)

## Sex Determination and Alternative Splicing

As shown in the preceding discussion, the determination of sex during embryogenesis in *D. melanogaster* is transmitted through a hierarchy of regulatory genes to the terminal differentiation genes, whose products are responsible for the sexually dimorphic traits of the adult fly. The different activities of the regulatory genes in males and females are largely due to sex-specific differences in RNA splicing that lead to the production of functionally different transcripts in the two sexes. The individual genes in this regulatory hierarchy are not only themselves controlled at the level of RNA splicing, but specify in turn the splicing pattern of the transcripts of the genes that function downstream in the hierarchy, producing a cascade of RNA splicing reactions (Baker 1989). Thus, RNA processing has been shown to be an important regulatory mechanism in a significant developmental pathway.

The $Sxl$, $dsx$, $tra$, and $tra-2$ genes have been cloned and the importance of splicing for their function in sex determination has been elaborated (Baker 1989, Bownes 1992). $Sxl^+$ has been found to have two functions in female flies: It positively regulates its own expression and provides each cell with a memory of its sex (Fig. 11.1), and second, it controls other regulatory genes that govern somatic sex determination, germ line sex determination, and dosage compensation (Figure 11.4).

$Sxl^+$ produces two distinct sets of transcripts in males and females during embryogenesis (Figure 11.2). One transcript is present early in embryogenesis and little is known about it. The second set consists of female-specific and male-specific transcripts somewhat later in embryogenesis. These transcripts are continuously present from then on. All the male-specific $Sxl^+$ transcripts contain an exon (3) that is absent in the female-specific transcripts. This male-specific exon introduces a STOP codon, which shortens the open reading frame and results in short, nonfunctional transcripts in males. The transcript in females is larger, but lacks exon 3, and results in a protein containing 354 amino acids.

In females, the $Sxl^+$ gene regulates somatic sexual determination by its control of the $tra^+$ gene (Figure 11.1). Splicing of the $tra^+$ gene is also sex-specific, with a protein produced in females that consists of 211 amino acids (Figure 11.2). The $tra^+$ protein in males consists of 37 amino acids, and is nonfunctional. The first exon of the $tra^+$ primary transcript can be spliced to either of two acceptor sites in the complete primary transcript. In females, the second site is used. The alternative transcript site is used in males and about half the female transcripts (Bownes 1992).

The $dsx^+$ gene is active in both males and females (Figure 11.1). In females, $dsx^+$, along with $ix^+$, acts to repress the genes involved in the terminal differentiation specific to males. The genes involved in female-specific terminal differentiation are not repressed, and so female differentiation occurs. The converse occurs in males. Thus, $dsx^+$ is a double switch and has active, but opposite, regulatory functions in both sexes. The molecular basis for the different functions in the two sexes is again the sex-specific processing of its messenger RNA (Figure 11.2). The primary $dsx^+$ transcript in both sexes has the first three exons in common, but different 3' exons and polyadenylation sites. In females, the three exons that are not sex-specific are spliced to an immediately adjacent, female-specific, fourth exon. In males, this exon is spliced out and the first three exons are joined instead to two male-specific exons located further downstream. The male-specific exons are eliminated from female transcripts by polyadenylation and cleavage at a site at the end of the fourth female-specific exon. The resulting sex-specific $dsx^+$ mRNAs encode proteins with common amino-terminal sequences up to residue 397, but the carboxyl termini consist of 30 female-specific amino acids in females and 152 male-specific amino acids in males.

Male differentiation is a default state of development in *Drosophila* and there is no active transmission of information between the different levels of regulatory genes in the sex-determination hierarchy (Figure 11.1). Thus, male-specific splicing patterns occur by default in the absence of female-specific gene products. Females mutant for $Sxl^+$, $tra^+$, or $tra$-$2^+$ genes process the $dsx^+$ transcript into the male mRNA, also implying that females have the RNA processing machinery needed to carry out the male-specific pattern of processing the $dsx^+$ pre-mRNA and that this corresponds to the default pattern of RNA splicing. Females thus differ from males because the products of the $Sxl^+$, $tra^+$, and $tra$-$2^+$ genes modify the specificity of the RNA splicing machinery so that the female-specific messenger RNAs of $Sxl^+$, $tra^+$, and $dsx^+$ are produced.

# A General Model for Sex Determination in Insects?

The sex-determining systems in insects (and mites) are variable (White 1973, Lauge 1985, Wrensch and Ebbert 1993). For example, the apterygote and many pterygote insects exhibit male heterogamety (males are XO, XY, XXO, XXY, or XYY and females are XX), but in some of the higher insects (Trichoptera, Lepidoptera) females may be heterogametic. Some insects are arrhenotokous, such as many of the Hymenoptera, where males are haploid and females are diploid, although haplo-diploidy *per se* does not necessarily determine sex. In fact, three different models have been proposed to explain sex determination in the haplodiploid Hymenoptera. The sex-determining systems of relatively few Hymenoptera have been analyzed experimentally (Cook 1993a,b).

In the haplo-diploid honey bee *Apis mellifera* and the parasitic wasp *Bracon* (*Habrobracon*) *hebetor*, sex is determined by a series of alleles at a single locus (**single-locus, multiple-allele model**) (Whiting 1943). Haploid males are hemizygous, and diploid females are heterozygous for these factors. If *Bracon* is inbred, diploid individuals are produced that are homozygous for the sex-determining alleles and are therefore males (Figure 11.5). In this model, inbreeding typically will produce diploid males. In fact, the haplo-diploid turnip sawfly *Athalia rosae ruficornis*, when inbred, produced both diploid and triploid males, indicating that sex in this hymenopteran species is also determined by the single-locus, multiple-allele system (Naito and Suzuki 1991).

In other haplodiploid Hymenoptera, sex is determined by a number of alleles at a series of loci. According to this second model, females must be heterozygous at one or more loci, while haploid males are hemizygous (Crozier 1971). After inbreeding, some diploid individuals are produced and these are

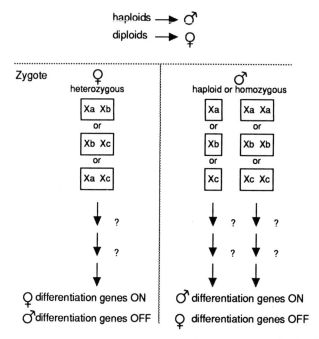

**Figure 11.5**   Multiple alleles at a single locus determine sex in the haplodiploid hymenopteran *Athalia rosae*. Under normal conditions, males are haploid and females are diploid, with females heterozygous for the sex-determining locus (X) while males are hemizygous. If individuals become homozygous for an allele of X, perhaps through inbreeding, they become diploid males. (Redrawn from Bownes, 1992.)

males if they are homozygous for all loci. The first and second models can be combined if the assumption is made that the single-locus, multiple-allele model is a special case of the **multiple-locus, multiple-allele model**. Under this assumption, only one locus has an effect in the first model, and the other loci have little or no effect.

A third model suggests that sex in haplodiploid Hymenoptera is determined by a balance (**genic balance model**) between nonadditive male-determining genes and additive female-determining genes scattered throughout the genome (daCunha and Kerr 1957). In this model, maleness genes ($m$) have noncumulative effects but femaleness genes ($f$) are cumulative. Thus sex is determined by the relationship between $f$ and $m$. In haploid individuals $m>f$, which results in a male. In diploids $ff>mm$, which results in a female.

Within a single species, several different sex-determining mechanisms may occur. The house fly, *Musca domestica*, has five pairs of autosomes and a pair of

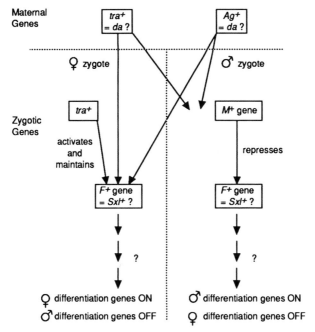

**Figure 11.6** Sex determination in the house fly *Musca domestica*. *tra*+ may be equivalent to *da*+. F+ may be equivalent to *Sxl*+. *tra*+ and *Ag*+ gene products are produced by the mother and stored in the egg. In the female zygote these products activate the F+ gene. The zygote's *tra*+ gene must be active to maintain the function of F+. This leads to expression of female differentiation genes, but the genes lower in the hierarchy are unidentified at present. The M+ gene product is present in males, which represses the F+ gene function, so that female differentiation genes are repressed and unknown male differentiation genes are activated. (Modified after Bownes 1992.)

heterochromatic sex chromosomes; females are XX and males are XY (Figure 11.6). Sex is determined by the presence or absence of the Y chromosome, which carries an epistatic male-determining factor, M, whereas the X chromosome plays no important role in sex determination. In some other strains of M. *domestica*, however, *both* males and females are XX and have a special autosome that may carry a male-determining factor $A^M$. The $A^M$ component is located on at least four different linkage groups in different populations. The presence or absence of the M factor seems to be the primary signal for sex determination in these strains.

Interestingly, in other strains of M. *domestica* both males and females have the M factors in the homozygous state, and the presence or absence of a female-determining dominant factor (F) determines sex. Finally, a dominant maternal-effect mutant, *Arrhenogenous* (*Ag*), has been found in M. *domestica* that causes genotypic female progeny carrying neither M nor F to develop into fertile males. A recessive maternal-effect mutant, *transformer*, causes genotypic female progeny carrying no M factors to follow the male pathway of sexual development to varying degrees. This suggests that the normal *tra*+ gene product is necessary for female determination and/or differentiation and that the gene is expressed during oogenesis and also in zygotes (Inoue and Hiroyoshi 1986). Experiments suggest that M acts early in embryogenesis to suppress a key gene, perhaps F, whose activity is required continuously for development of females, as is $Sxl^+$ in *Drosophila* (Hilfiker-Kleiner et al. 1993).

In the phorid fly *Megaselia scalaris*, the sex-determining linkage group is not fixed; different chromosomes serve as the sex-determining pair in different populations. Traut and Willhoeft (1990) estimate that the male-determining factor moved to a different linkage group with a frequency of at least 0.06%, which is consistent with the hypothesis that the sex-determining factor is moving by transposition. An alternative explanation is that mutations at multiple sex loci in the genome result in males; however, the high rates of change (0.06%) are compatible with known rates of transposition, but are higher than expected if due to mutation.

Environmental conditions can influence sex determination in some arthropods. Many haplodiploid insects adjust the sex ratio of their progeny in response to environmental factors (King 1992). For example, females of many aphelinid wasps in the genus *Encarsia* develop as 'autoparasitoids' of whiteflies (which are considered the primary hosts), but males of the same species develop as parasitoids of female pupae of their own species, which are considered the secondary hosts. Virgin females deposit unfertilized eggs to produce haploid sons on secondary hosts (females of their own species), but typically do not oviposit in primary hosts (whiteflies), even if they are the only hosts available. When a virgin female does deposit haploid male eggs in a primary host, these eggs do not develop, for unknown reasons. However, an unusual population of *Encarsia pergandiella* was found in which males could develop on the primary host. It appears that these haploid males started out as fertilized diploid eggs, but became haploid males after the loss of the paternal set of chromosomes shortly after fertilization. This aberrant chromosomal behavior perhaps was caused by a supernumerary chromosome (Hunter et al. 1993).

In the blowfly *Chrysomya rufifacies* (Diptera, Calliphoridae), females produce *either* female progeny only (**thelygenic** females) *or* male progeny only (**arrhenogenic** females) (Clausen and Ullerich 1990). Thelygenic females are heterozygous for a dominant female-determining maternal effect gene (F') while arrhenogenic females and males are homozygous for the recessive allele (f). This species lacks differentiated sex chromosomes. DNA sequence homology between the D. *melanogaster da*+ gene and a polytene band in the genetic sex chromosome pair of C. *rufifacies* has been observed by *in situ* hybridization, suggesting that there is a functional equivalence between the sequences. The results also suggest that F in C. *rufifacies* and *da+* in D. *melanogaster* are equivalent (Clausen and Ullerich 1990).

Given the brief sample above of some of the different sex-determining systems found in insects, is it really probable that a single model can describe sex determination in all insects as proposed by Nothiger and Steinmann-Zwicky (1985)? They proposed that all the sex determination mechanisms in insects are variations upon a theme (Table 11.2, Figure 11.7). In their model, there is a gene equivalent to *Sxl+*, a repressor (R) which inactivates *Sxl+*, a gene which activates *Sxl+*, and a gene which is equivalent to *dsx+* and which is expressed in two alternative forms to interact with one or the other of the two sets of male and female differentiation genes that are lower in the hierarchy.

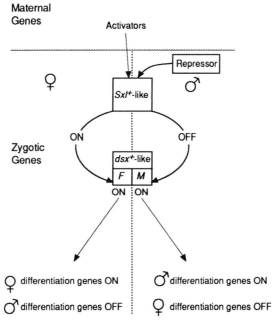

**Figure 11.7**   Is there a general model for sex determination in insects? This model assumes that activators are produced by the mother that activate an *Sxl+*-like gene in the zygote. Other activators may be produced by the zygote, and the combination results in a functional *Sxl+*-like product in females. Male zygotes produce a repressor of the *Sxl+*-like gene and no functional product is produced in males. Next, a *dsx+*-like gene is turned on in both females and males but different products are produced in the two sexes. The different *dsx+* gene products turn on a subsequent series of genes that result in the differentiation of either males or females. Evidence for this general model is fragmentary at this time. (Redrawn from Bownes 1992.)

**Table 11.2**

Modifications of a Basic Model for Sex Determination in Insects[a]

| Genotype and sex | Phenomenon | R | Sxl+ Allelic state | Sxl+ State of activity | Selected examples |
|---|---|---|---|---|---|
| | | | Interpretation[b] | | |
| M/m, male | M, dominant male determiner; locus of M varies from strain to strain | +/−  −/−  M = R  R on a transposable element? | +/+ | Inactive | Aedes, Calliphora, Culex, Chironomus, Megaselia, Musca |
| m/m, female | | | +/+ | Active | |
| Y/X, male | Y, dominant male determiner; Y carries R, heteromorphic | +/−  −/− | +/+ | Inactive | Anopheles, Calliphora, Musca |
| X/X, female | | | +/+ | Active | |
| X:AA, male | X:A ratio; R produced by autosomes | +/+  +/+ | +/+ | Inactive | Drosophila, Sciara |
| XX:AA, female | | | +/+ | Active | |
| Haploid, male | Multiple complementing alleles; mutations at Sxl+, inactive product | −  −/− | a or b | Inactive product | Apis, Habrobracon, Mormoniella |
| Diploid, female | | | a/b | Active product | |

| | | | | | |
|---|---|---|---|---|---|
| MMM $f/f$, male | Several dominant male determiners M; one dominant female determiner F; F epistatic over M; M = R on several chromosomes; F = $Sxlc$, not repressible | +/+ | +/+ | Inactive | Musca |
| MMM $F/f$, female | | +/+ | c/+ | Active | |
| $f/f$, male | F, dominant female determiner, M of other strains epistatic over F; $f = Sxl^-$ $F = Sxl^+$ | −/− | −/− | Inactive | Chironomus |
| $F/f$, female | | −/− | +/− | Active | |
| $f/f$, female arrhenogenic | F', maternal factor, female determiner $f = da^-$ $F' = da^+$ | −/− | +/+ | Inactive | Chrysomya |
| $F'/f$, female thelygenic | | −/− | +/+ | Active | |
| Males and females have same genotype | Environment determines sex—temperature, nutrition; R or $Sxl^+$ conditional | s/s | s/s | depends on conditions | Aedes, Heteropeza |

[a] Modified from Nothiger and Steinmann-Zwicky (1985).

[b] R is a signal gene; $Sxl^+$ is a key switch gene; $da^+$ produces a maternal factor necessary for switching $Sxl^+$ on.

285

Bownes (1992) used their model to compare the sex determination system in the house fly (*Musca domestica*) (Figure 11.5). As noted earlier, different populations of M. *domestica* exhibit apparently different mechanisms of sex determination. M. *domestica* has both X and Y chromosomes, but they are completely heterochromatic, with the Y chromosome of some strains carrying only a male-determining factor M. In other strains, both sexes are XX and have M located on an autosome, but which autosome contains the M varies among populations because M appears to be located on a transposable element. M/m flies are males and m/m are female. Yet, some strains have been found in which both males and females are M/M, and another factor ($F^D$) located on autosome 4 is dominant and epistatic to M and determines femaleness. Maternal genes called *Arrhenogenic*$^+$ (*Ag*$^+$) and *transformer*$^+$ (*tra*$^+$) have been found that encode female-determining factors. *Ag/Ag*$^+$ and *tra/tra* mothers produce fertile pseudomales or intersexual progeny.

According to the Nothiger and Steinmann-Zwicky (1985) model, the male-determining factor (M) in M. *domestica* would correspond to the repressor (R) of *Sxl*$^+$. The genes *tra*$^+$ and *Ag*$^+$ may be equivalent to *da*$^+$ in *Drosophila*. The F gene of M. *domestica* could be equivalent to *Sxl*$^+$. *Sxl*$^+$ is involved in dosage compensation in *Drosophila*, but dosage compensation is not needed in species such as M. *domestica* with heterochromatic sex chromosomes (which usually contain few coding regions) or no sex chromosomes. As a result, insects with heterochromatic sex chromosomes or no sex chromosomes can survive mutations of *tra*$^+$, *Ag*$^+$, and F; such mutations can alter sex determination, but are not lethal to one sex. Genes equivalent to those lower than *Sxl*$^+$ in the *Drosophila* hierarchy have not been identified in M. *domestica* or other insects, although some mutations in mosquitoes may be comparable (Bownes 1992).

Chironomid flies have no differentiated sex chromosomes and sex is determined by a dominant male-determining factor, which is linked in some species to a specific chromosomal inversion. In other species, the location of the male-determining factor is variable; only a few species have morphologically distinguishable sex chromosomes, and the male-determining factor behaves as if it were associated with a transposable element. A female-determining factor has been found in some strains, but the male-determining factor is epistatic to it. Bownes (1992) concluded that sex determination in *Chironomus* is similar to that in M. *domestica*, except that the female-determining factor is epistatic to the male-determining factor in M. *domestica*. Kraemer and Schmidt (1993) found that the sex-determining region of *Chironomus thummi* is associated with highly-repetitive DNA and transposable elements and are attempting to identify the gene or genes that serve as the dominant male sex determiner in *Chironomus*.

Different genera of mosquitoes have several different sex determination systems, but may also conform to the Nothiger and Steinmann-Zwicky model (Bownes 1992). *Anopheles gambiae* and A. *culicifacies* have XY males and XX females. Sex in *Aedes* is determined by a dominant male-determining factor. Intersex flies with phenotypes similar to the *ix*, *dsx*, and *tra* mutants of *Drosophila* have been found in *Aedes aegypti* and *Culex pipiens*. Sex in *Culex* is determined by a single gene on an autosome; *Culex* gynandromorphs have been found, suggesting that sex determination is cell autonomous, as it is in *Drosophila*. The sex of some northern strains of *Aedes* depends upon the temperature at which they are reared, with males transformed into intersexes at higher temperatures. This suggests that an allele equivalent to $ix^+$ is temperature sensitive in these populations. In *Culex pipiens*, a sex-linked gene *cercus* (*c*) changes females into intersexes; these intersexes are sterile and fail to take blood meals. It is possible that *cercus*$^+$ is similar to *tra*$^+$, *ix*$^+$, or *dsx*$^+$ of *Drosophila*, although further study is needed to place these genes in a hierarchical sequence equivalent to that of *D. melanogaster* (Bownes 1992).

Nothiger and Steinmann-Zwicky's (1985) scheme also may account for sex determination in the haplo-diploid Hymenoptera (Table 11.2). According to this model, the multiple alleles which must be heterozygous to confer female development are thought to be mutations of $Sxl^+$. These mutations produce an inactive gene product and lead to male development when they are in the homozygous or hemizygous state. However, the different alleles must complement each other in some way so that when two different mutations are present in a diploid, they are able to make a functional product to produce a female. Bownes (1992) expressed skepticism that this gene is equivalent to $Sxl^+$, and suggested that the gene may be more like *sis-a*$^+$ and *sis-b*$^+$ of *Drosophila*, and is upstream of $Sxl^+$, so that products of this locus would interact with an $Sxl^+$-like gene.

## Meiotic Drive Can Distort Sex Ratios

Mendel's Law of Segregation proposes that equal numbers of paternally- and maternally-derived alternative alleles are distributed to eggs and sperm during meiosis. However, this law is sometimes violated due to a phenomenon called meiotic drive. **Meiotic drive** involves altering the equal assortment of chromosomes during meiosis so that certain chromosomes containing meiotic distortion genes are inherited more frequently than expected (>50%). Meiotic drive most frequently is observed affecting sex chromosomes in the heterogametic sex, so that it alters the sex ratio. Whether meiotic drive mechanisms actually occur more frequently on the sex chromosomes, or whether they are only more frequently discovered there because sex ratio changes are readily observed is unknown (Lyttle 1993). Two examples of meiotic drive will be described, including Segregation Distorter (SD) in *Drosophila* species, and Male Drive (MD) in the mosquitoes *Aedes aegypti* and *Culex quinquefasciatus*.

The SD phenotype is present in low but stable frequencies in most field-collected populations of *D. melanogaster*. *D. melanogaster* males heterozygous for the segregation distorter (SD) chromosome (SD/SD$^+$) may produce virtually only progeny with the SD chromosome instead of half with SD and half without SD$^+$, owing to the failure of sperm with the SD$^+$chromosome to mature (Ashburner 1989). The nuclei of the sperm with the normal SD$^+$ chromosome fail to condense normally at maturation.

The SD phenotype is determined by a complex of three loci—*Sd*, *E(SD)*, and *Rsp*—which are located near the centromeric region of chromosome 2. The product of the *Sd* gene apparently interacts with *sensitive* alleles at the responder locus (*Rsp$^s$*), carried by non-SD chromosomes, causing sperm to develop abnormally. SD chromosomes carry an *insensitive* form of *Rsp* (*Rsp$^i$*), so they are protected from destruction by the Sd product. The three loci are closely linked. A 5-kb tandem duplication was found associated with the *Sd* locus, which may be the cause of that mutation. The *Rsp* locus may be a repeated element, with larger copy numbers present in *Rsp$^s$* (>700 copies) than in *Rsp$^i$* chromosomes (ca. 25 copies). The *Enhancer* locus of SD, *E(SD)*, is required for the full expression of meiotic drive. As noted, the segregation distortion condition affects heterozygous males (SD/SD$^+$), but has no effect on females.

Meiotic drive has been described in *Aedes aegypti* and *Culex quinquefasciatus*. In both species, a Y-linked gene results in the production of excess males (Wood and Newton 1991). Excess males are produced because X chromosomes are broken during meiosis in males and thus fewer X chromosomes than Y chromosomes are transmitted. The production of excess males in *A. aegypti* is thus due to the production of fewer female embryos. The *Distorter* gene (*D*) is linked closely to the sex locus *m*/*M* and causes the chromosome breakage. Additional genes are also involved, and sensitivity to *Distorter* is controlled by *m*, the female-determining locus. In some strains, sensitivity is influenced by a second sex-linked gene *t*. Yet another sex-linked gene, *A*, enhances the effect of *Distorter*. *Distorter* has been found in mosquito populations from Africa, America, Australia, and Sri Lanka, but resistance to it is widespread.

Meiotic drive operates as an evolutionary force which can cause an increase in the population frequency of the allele or chromosome that is favored in transmission, even if it confers a disadvantage on its carriers. It has been proposed that meiotic drive might be used to introduce new genes (such as cold-sensitive lethal genes, insecticide-susceptibility genes, or behavior-altering genes that would reduce the impact of mosquitoes on humans) into natural populations of mosquitoes as a method to achieve control of these important vectors of disease (Wood and Newton 1991).

When different species are crossed, the progeny sometimes show altered sex ratios, with one sex absent, rare, or sterile. The missing or sterile sex is usually the heterogametic sex. This is also known as 'Haldane's Rule'. However, Hurst and Pomiankowski (1991) suggest that Haldane's Rule only occurs in taxa with sex chromosome-based meiotic drive, such as the Lepidoptera and Diptera. Thus, Hurst and Pomiankowski argue that Haldane's Rule may be accounted for in some insects by a loss of suppression of sex ratio distorters when in the novel nuclear cytotype of the hybrid. (An alternative explanation is that Haldane's Rule is based on the faster evolution of genes located on sex chromosomes.) Sex ratio distorters that result in unisexual sterility in crosses between hybrids of different species have been found in many species of *Drosophila*, the dipterans *Musca domestica* and *Glossina morsitans*, the hemipteran *Tetraneura ulmi*, and Lepidoptera (*Acraea encedon*, *Maniola jurtina*, *Danaus chrysippus*, *Philudoria potatoria*, *Mylothris spica*, *Abraxus grossulariata*, *Talaeporia tubulosa*) (Hurst and Pomiankowski 1991). In addition to chromosomally located genes that alter sex ratios, cytoplasmic factors can also alter normal sex ratios or sex determination in insects.

A new class of selfish genes, *Medea*, was found in the flour beetle *Tribolium castaneum* (Beeman et al. 1992). *Medea* is a maternal-effect dominant that causes the death of zygotes that do not carry it. If a mother carries *Medea*, any of her offspring who fail to carry this gene die before they pupate. Females that are heterozygous for *Medea* lose half their progeny if they mate with a wild-type male and one-fourth of their progeny when mated to a heterozygous male. The gene has been mapped to an autosome but details about its genetic structure are not known (Beeman et al. 1992). It is hypothesized that these genes could lead to reproductive isolation and speciation in *Tribolium*. *Medea* is unusual because it segregates as a typical chromosomal gene and it kills diploid progeny after fertilization, whereas other selfish genes are nonchromosomal or have an effect prior to fertilization (as in typical segregation distortion complexes).

## Cytoplasmic Distortion of Normal Sex Ratios

A number of cytoplasmically transmitted organisms (bacteria, viruses, protozoans) have been identified in insects and mites that alter the 'normal' sex-determining mechanism(s) (Table 11.3). Most sex ratio distorters are inherited primarily through the mother and thus are cytoplasmically inherited (Hurst and Majerus 1992, Hurst et al. 1993a). Cytoplasmic agents that can manipulate their host's sex ratio (usually increasing the proportion of females or, occasionally, of males) and promote their own spread are called **cytoplasmic sex ratio distorters**. The spread of a cytoplasmic sex ratio distorter often reduces the fitness of its host and can drive populations to extinction. Sex ratio distorters are usually suspected if crosses produce a heavily female-biased sex ratio, although meiotic drive and hybrid dysgenesis agents are also possible mechanisms (Hurst 1993).

Ebbert (1991, 1993) pointed out that in at least 50 cases, endosymbionts have been shown to alter sex ratios in insects or mites. Examples are found in the Diptera, Heteroptera, Coleoptera, Lepidoptera, and Acari (mites), and may be

**Table 11.3**

Nonnuclear Influences on Sex Determination or Sex Ratio in Arthropods

| Microorganism or factor[a] | Arthropod species | Effect and references |
|---|---|---|
| *Arsenophonus nasoniae* (gram-negative bacterium) | *Nasonia vitripennis* | Kills male eggs (Gherna et al. 1991) |
| Bacterial male killing | *Adalia bipunctata* | Males killed as embryos (Hurst et al. 1992, 1993b) |
| Maternal sex ratio (MSR) A cytoplasmic factor | *Nasonia vitripennis* | Results in nearly all female progeny (Beukeboom and Werren 1992, Werren and Beukeboom 1993) |
| Paternally transmitted sex ratio factor (PSR): A supernumerary B chromosome that is mostly heterochromatic | *Nasonia vitripennis* | Fertilized eggs lose paternal chromosomes; converts females to male progeny; transmitted only by sperm (Werren et al. 1987, Nur et al. 1988, Beukeboom and Werren 1993, Reed 1993) |
|  | *Encarsia pergandiella ?* | Mechanism speculative (Hunter et al. 1993) |
| Sex ratio condition (viral?) | *Drosophila bifasciata* | Death of male embryos (Leventhal 1968) |
| Sex ratio condition (spiroplasmas) | *D. willistoni* and related neotropical species | Males die as embryos (Ebbert 1991) |
| Sex ratio condition | *Oncopeltus fasciatus* | (Leslie 1984) |
| Sterility in male progeny (streptococcal L-form bacteria) | *D. paulistorum* | Induction of semispecies? (Somerson et al. 1984) |
| Thelytoky lost (*Wolbachia* ?) | *Encarsia formosa* | Males produced after antibiotic treatment; microorganisms restore diploidy to unfertilized eggs, resulting in all-female progeny (Zchori-Fein et al. 1992, Kajita 1993) |
| Thelytoky lost; maternally inherited microorganisms (*Wolbachia*) | *Trichogramma* species | Cure with antibiotics results in bisexual populations and male progeny (Stouthamer et al. 1990, Stouthamer and Werren 1993, Stouthamer and Luck, 1993) |

[a]*Wolbachia* also causes cytoplasmic incompatibility.

widespread, but undetected, in many other insect and mite species. Generally, the microorganisms are inherited maternally, with maternal transmission rates typically high, although a few daughters may fail to become infected. The altered sex-ratio conditions generally are found in natural populations at low but stable frequencies of about 10%. The infections appear to reduce fitness of their hosts, with egg hatch or larval survival reduced by half in the progeny of infected females. For example, populations of the two-spot ladybird beetle *Adalia bipunctata* are polymorphic for a cytoplasmically inherited element which produces female-biased sex ratios by causing male offspring to die during embryogenesis. Six of 82 females collected near Cambridge, England, exhibited both a female-biased sex ratio in their progeny and low egg hatch rates. The population sex ratio was 1.15:1, and female-biased (Hurst et al. 1993b). Additional specific examples are described in the following paragraphs.

### Spiroplasmas

The sex ratio condition of *Drosophila willistoni* and related neotropical species is due to a spiroplasma (Ashburner 1989). The spiroplasmas are maternally inherited and transovarially transmitted and cause male embryos to die. Spiroplasmas can be transmitted between species by injecting hemolymph, but spiroplasmas from different species are different. A virus is associated with the spiroplasmas which is different for each spiroplasma. When spiroplasmas from different species are mixed, they clump because the viruses lyse the spiroplasmas of the other species. Study of the spiroplasma from *D. willistoni* is made easier because it is now possible to culture the spiroplasma in a defined medium (Hackett et al. 1986).

### L-Form Bacteria

The *Drosophila paulistorum* complex contains six semispecies (subgroups derived from a single species that are thought to be in the process of speciation) that do not normally interbreed. When these semispecies are crossed in the laboratory, fertile daughters and sterile sons are produced. Streptococcal L-forms (bacteria) have been isolated and cultured in artificial media that are associated with the sterility in these semispecies (Somerson et al. 1984). The L-forms are transferred through the egg cytoplasm and each semispecies is associated with a different microorganism. The L-forms can be microinjected into nonhybrid females and can produce the male sterility expected. This suggests that an L-form has a normally benign relationship with its own host; however, if it is transferred to a closely related host, sterility is induced.

### Thelytoky Microorganisms

Thelytoky (production of females only) in the parasitic wasp *Trichogramma* sp. (Hymenoptera) has been found to be caused by microorganisms in some cases

(Stouthamer et al. 1990, Stouthamer and Werren 1993). When *Trichogramma* with microorganisms are treated with antibiotics, such as rifampin or tetracycline hydrochloride, or are reared at high temperatures (>30°C), strains can be cured, and both male and female progeny are produced. In some *Trichogramma* species, thelytoky is not reversible, and these species lack the microorganisms. Microorganisms are not found in field-collected bisexual (arrhenotokous) populations of *Trichogramma*. The microorganisms can be assayed using a lacmoid stain, and Stouthamer and Werren (1993) found that the number of microorganisms per *Trichogramma* egg ranged from 256 to 668.

Thelytoky in the aphelinid parasitoid *Encarsia formosa* was considered to be obligatory. However, both male and female progeny were produced after females were treated with antibiotics (Zchori-Fein et al. 1992). The males mated and produced sperm, but insemination did not occur.

Other cases of thelytoky in parasitoid wasps also may be induced by microorganisms. Many asexual populations of parasites in both the laboratory and field have produced both male and female progeny after treatment at high temperatures (>30°C), which is consistent with the hypothesis that the microorganisms are inactivated by this treatment. Colonies subsequently reared at lower temperatures may revert to thelytoky, possibly due to regrowth of the microorganisms (Hurst 1993).

### Wolbachia

Rickettsia in the genus *Wolbachia* are one of the most commonly described cytoplasmically inherited microorganisms. *Wolbachia* are small, gram-negative rods that cannot be cultured easily outside their hosts. They are widespread in insects (O'Neill et al. 1992), and have been implicated as both the cause of alterations in sex ratio (resulting in thelytoky) and **cytoplasmic incompatibility** (Rousset and Raymond 1991, Stouthamer et al. 1993). Sequence analysis of DNA encoding a portion of the 16S ribosomal RNA of microorganisms associated with cytoplasmic incompatibility from diverse insect species indicates the microorganisms share >95% sequence similarity, suggesting that these microorganisms are a monophyletic group that specializes in manipulating chromosome behavior and reproduction in insects (O'Neill et al. 1992; Rousset et al. 1992, Stouthamer et al. 1993).

Cytoplasmic incompatibility caused by *Wolbachia* appears to be widespread in insects. Cytoplasmic incompatibility can be partial or complete, and can cause unidirectional reductions in progeny production when males and females from different colonies are crossed. Sometimes incompatibility is found in both reciprocal crosses (bidirectional incompatibility). Cytoplasmic incompatibility can occur when infected males and uninfected females mate. It can also occur when infected individuals mate, suggesting that the incompatibility may be due to the presence of different strains of *Wolbachia* in each arthropod strain. Incompatible crosses usually result in aborted syngamy and haploid embryos die unless they are haplodiploid Hymenoptera.

Sequence data for the 16S ribosomal RNA gene of *Wolbachia* from six different insect genera exhibited 0.2 to 2.6% differences (O'Neill et al. 1992). Whether these differences are due to different strains or species of *Wolbachia* is unresolved, but these differences may not fully explain differences in the cytoplasmic incompatibility found between species and populations. Rather, the variability may be due to specific interactions between the insect host and the *Wolbachia*.

Large numbers of *Wolbachia* have been found in both the ovaries and testes of populations with cytoplasmic incompatibilities. In some cases they can be transmitted by both males and females, while in other cases they are transmitted only by females. Incompatible strains sometimes may be converted to compatible by treating the colonies with heat or antibiotics such as tetracycline. *Wolbachia* was transferred by microinjecting infected egg cytoplasm into uninfected eggs of both *Drosophila simulans* and *D. melanogaster* to produced infected populations. Transinfected strains of *D. simulans* and *D. melanogaster* with high levels of *Wolbachia* exhibited cytoplasmic incompatibilities at high levels, but those with low levels exhibited low levels of incompatibility, suggesting that a threshold level of infection is required for normal expression and that host factors determine the density of the symbiont in the host (Boyle et al. 1993). In the case of *D. simulans*, the *Wolbachia* reduced egg hatch, but did not alter sex ratio. In *D. simulans*, the *Wolbachia* are found in the reproductive organs of both male and female flies but are inherited through the egg cytoplasm. Bressac and Rousset (1993) showed that while *Wolbachia* are present in testes of males, the bacteria are lost by sperm cells during development.

Stouthamer et al. (1993) showed that parthenogenesis-associated bacteria in different parasitic Hymenoptera, including six populations of three species of *Trichogramma* and *Muscidifurax uniraptor*, are up to 99% similar in their DNA sequences. *Wolbachia* are not present in the eggs of bisexual (arrhenotokous) strains, but colonies with microorganisms consisted only of females (thelytoky). The microorganisms causing incompatibility and parthenogenesis alter host chromosome behavior during early mitotic divisions of the host egg. Bacteria causing incompatibility interfere with paternal chromosome incorporation in fertilized eggs, whereas bacteria causing parthenogenesis prevent segregation of chromosomes in unfertilized eggs.

The way in which *Wolbachia* is maintained in populations has considerable theoretical and practical importance (Chapter 15). *Wolbachia pipientis* lowers the fitness of infected males by making them reproductively incompatible with uninfected females (Table 11.4). Crosses between infected individuals have lower fitness than crosses between uninfected individuals. Stevens (1993) evaluated the effect of *Wolbachia pipientis* on reproductive success in populations of the

**Table 11.4**

The Pattern of Partial Reproductive Isolation Caused by *Wolbachia pipientis* Lowers the Fitness of Infected Males by Making Them Reproductively Incompatible with Uninfected Females

|  | Female parent | |
|---|---|---|
| Male parent | Uninfected | Infected |
| Uninfected | Uninfected progeny | Infected progeny |
| Infected | No progeny | Infected progeny |

[a]Modified from Stevens (1993).

flour beetle *Tribolium confusum*. Evolutionary theory suggests that hosts could modify their reproductive behavior in response to selection imposed by the *Wolbachia*, or the *Wolbachia* could modify the host's behavior to their own advantage. Stevens (1993) showed that uninfected *Tribolium* females do not preferentially mate with and/or fertilize eggs with sperm from uninfected males. Rather, reproduction occurred in a way that benefited the *Wolbachia*, which is consistent with the hypothesis that parasites manipulate host behavior to enhance their rate of spread through a population (Table 11.4).

## Paternal Sex Ratio Chromosomes and Cytoplasmic Incompatibility in *Nasonia*

Sex ratios in the parasitic wasp *Nasonia vitripennis* can be altered by at least two different mechanisms. Some natural populations of the haplodiploid parasitoid wasp *Nasonia vitripennis* carry a supernumerary or **B chromosome** that causes a condition called **paternal sex ratio** (**PSR**). B chromosomes are nonvital chromosomes that are found in many plant and animal species. Most B chromosomes are thought to be derived from one of the normal chromosomes and are transmitted at higher rates than expected from Mendelian inheritance, thus exhibiting 'drive' (Jones and Rees 1982).

The PSR condition is only carried by male *N. vitripennis* and is transmitted by sperm to fertilized eggs. After an egg is fertilized by a PSR-bearing sperm, the paternally derived chromosomes condense into a chromatin mass, and subsequently are lost, leaving only the maternal chromosomes. The PSR chromosome itself survives, disrupting normal sex determination by changing fertilized diploid (female) eggs into haploid PSR males. PSR is the only known B chromosome of its kind, and is unusual in its ability to destroy the complete genome of its carrier each generation (Werren et al. 1987, Nur et al. 1988, Beukeboom and Werren 1992, 1993; Beukeboom et al. 1992).

The PSR chromosome constitutes approximately 5.7% of the genome of the carrier males (Reed 1993) and contains three families of related, tandemly repetitive DNAs which are not present in the normal chromosomes (Eickbush et al. 1992). In addition, a tandem repeat family is present on the PSR chromosome, as well as on at least one of the normal chromosomes. Whether these repetitive DNA sequences play a role in PSR function or amplification of the tandem arrays is unknown. The population genetics of PSR is not fully resolved, but has been modeled by Werren and Beukeboom (1993), who found that PSR cannot persist in small subdivided populations.

Cytoplasmic incompatibility caused by microorganisms also is associated with incompatibility between *N. vitripennis* and a closely related sympatric species, *N. giraulti* (Breeuwer and Werren 1990, 1993a,b). Different strains of each species differ in the degree of *interspecific* premating isolation. Matings between the two species occur, but yield only all-male progeny. Cytogenetic analyses of eggs derived from interspecific crosses showed there were differences in mitoses within the diploid eggs. The paternally derived set of chromosomes formed a diffuse tangled mass, which was associated with the presence of microorganisms, and only male progeny were produced in these interspecific crosses. Laboratory crosses were made using males lacking microorganisms and females from the other species; these crosses yielded hybrid female progeny, which were fertile. The bacterial density is correlated with compatibility differences between males and females. Strains with high numbers are incompatible with females from strains with lower numbers. Partial incompatibility can occur in populations with reduced bacterial numbers, perhaps due to incomplete elimination of the paternal chromosome (Breeuwer and Werren 1993a). Thus, the microorganisms may play a part in the process of speciation by preventing gene flow between populations.

## Hybrid Sterility in *Heliothis*

An unusual approach to obtaining sterile males for genetic control programs involves producing sterile backcross males (Makela and Huettel 1979). Viable hybrid progeny are produced when *Heliothis virescens* male and *H. subflexa* female moths mate; all the $F_1$ females are fertile when backcrossed to males from either parental species. However, most of the $F_1$ males are sterile, and if the hybrid females are backcrossed through successive generations to *H. virescens* males, in later generations all male progeny will be sterile. Maternal inheritance is involved in the hybrid male sterility, but the precise mechanism remains controversial.

One potential mechanism is based on incompatibility between sperm mitochondria in backcross males and the cytoplasm, which is derived from the mothers (Miller et al. 1986). A second possible mechanism is that incompatibility is caused by interactions between maternally inherited microorganisms and the paternal genetic material in the nucleus. Evidence for this hypothesis is based on PCR analysis of the 16S rRNA gene sequences from microorganisms associated with *H. virescens* and *H. subflexa* (Krueger et al. 1993). The sequences obtained from the two species were different, and the size of the PCR-amplified 16S rDNA suggested that microorganisms in the sterile backcross males originated from *H. subflexa* while the nuclear genome was essentially derived from *H. virescens*.

## Sex and the Sorted Insects

Genetic control of pest insects represents an attractive alternative to chemical control in terms of safety, specificity, and the limited negative impact it has upon the environment. The screwworm (*Cochliomyia hominivorax*) eradication campaign in the southern U.S.A. demonstrated what can be achieved with mass releases of males sterilized by irradiation (LaChance 1979). The control of screwworm has been an economic success and the principle of sterile insect releases has been applied to other pest insect species, including the Mediterranean fruit fly (*Ceratitis capitata*), tsetse flies (*Glossina palpalis* and *G. morsitans*), mosquitoes (*Anopheles albimanus*), codling moth (*Cydia pomonella*), and ticks (LaChance 1979).

Sterile insect release programs usually require only males, but both sexes must be reared. Not only is it expensive to rear large numbers of 'useless' females, but, in the case of species that are vectors for disease or annoy or bite humans or domestic animals, it is undesirable to release any females, sterile or not! Slight differences in size or color of pupae have been used to sort out undesirable females during mass rearing. Ideally, however, a genetic sexing method should be available that could produce only males of high quality and vigor to compete with wild males for female mates. Since an all-male colony would be difficult to maintain (!), this character ideally would be a conditional trait, perhaps dependent upon temperature or some other environmental cue.

The resolution of the molecular genetics of sex determination in arthropods will have both theoretical (Parkhurst and Ish-Horowicz 1992) and applied applications, and could lead to improved genetic control of pests or useful genetic modifications of beneficial biological control agents.

## References Cited

Ashburner, M. 1989. Drosophila: A *Laboratory Handbook*. Cold Spring Harbor Laboratory Press, Cold Spring Harbor, NY.

Baker, B. S. 1989. Sex in flies: The splice of life. Nature 340: 521–524.

Beeman, R. W., K. S. Friesen and R. E. Denell. 1992. Maternal-effect selfish genes in flour beetles. Science 256: 89–92.

Belote, J. M., M. B. McKeown, D. J. Andrew, T. N. Scott, M. F. Wolfner and B. S. Baker. 1985. Cold Spring Harbor Symp. Quant. Biol. 50: 605–614.

Beukeboom, L. W. and J. H. Werren. 1992. Population genetics of a parasitic chromosome: Experimental analysis of PSR in subdivided populations. Evolution 46: 1257–1268.

Beukeboom, L. W. and J. H. Werren. 1993. Deletion analysis of the selfish B chromosome, paternal sex ratio (PSR), in the parasitic wasp Nasonia vitripennis. Genetics 133: 637–648.

Beukeboom, L. W., K. M. Reed and J. H. Werren. 1992. Effects of deletions on mitotic stability of the paternal-sex-ratio (PSR) chromosome from Nasonia. Chromosoma 102: 20–26.

Bownes, M. 1992. Molecular aspects of sex determination in insects. Pp. 76–100. In: Insect Molecular Science. J. M. Crampton and P. Eggleston, Eds. Academic Press, London.

Boyle, L. S., L. O'Neill, H. M. Robertson and T. L. Karr. 1993. Interspecific and intraspecific horizontal transfer of Wolbachia in Drosophila. Science 260: 1796–1799.

Breeuwer, J. A. J. and J. H. Werren. 1990. Microorganisms associated with chromosome destruction and reproductive isolation between two insect species. Nature 346: 558–511.

Breeuwer, J. A. J. and J. H. Werren. 1993a. Cytoplasmic incompatibility and bacterial density in Nasonia vitripennis. Genetics 135: 565–574.

Breeuwer, J. A. J. and J. H. Werren. 1993b. Effect of genotype on cytoplasmic incompatibility between two species of Nasonia. Heredity 70: 428–436.

Breeuwer, J. A. J., R. Stouthamer, S. M. Barns, D. A. Pelletier, W. G. Weigburg and J. H. Werren. 1992. Phylogeny of cytoplasmic incompatibility microorganisms in the parasitoid wasp genus Nasonia (Hymenoptera: Pteromalidae) based on 16S ribosomal DNA sequences. Insect Mol. Biol. 1: 25–36.

Bressac, C. and F. Rousset. 1993. The reproductive incompatibility system in Drosophila simulans: Dapi-staining analysis of the Wolbachia symbionts in sperm cysts. J. Invertebr. Pathol. 61: 226–230.

Bull, J. J. 1983. Evolution of Sex Determining Mechanisms. Benjamin/Cummings, London.

Charnov, E. 1982. The Theory of Sex Allocation. Princeton University Press, Princeton, NJ.

Clausen, S. and F. H. Ullerich. 1990. Sequence homology between a polytene band in the genetic sex chromosomes of Chrysomya rufifacies and the daughterless gene of Drosophila melanogaster. Naturwissenschaften 77: 137–138.

Colwell, R. K. 1981. Group selection is implicated in the evolution of female-biased sex ratios. Nature 290: 401–404.

Cook, J. M. 1993a. Experimental tests of sex determination in Goniozus nephantidis (Hymenoptera: Bethylidae). Heredity 71: 130–137.

Cook, J. M. 1993b. Sex determination in the Hymenoptera: A review of models and evidence. Heredity 71: 421–435.

Crozier, R. H. 1971. Heterozygosity and sex determination in haplo-diploidy. Amer. Natur. 105: 339–412.

daCunha, A. D. and W. E. Kerr. 1957. A genetical theory to explain sex determination by arrhenotokous parthenogenesis. Forma et Functio 1(4): 33–36.

Ebbert, M. A. 1991. The interaction phenotype in the Drosophila willistoni-spiroplasma symbiosis. Evolution 45: 971–988.

Ebbert, M. A. 1993. Endosymbiotic sex ratio distorters in insects and mites. Pp. 150–191. In: Evolution and Diversity of Sex Ratio in Insects and Mites. D. L. Wrensch and M. A. Ebbert, Eds. Chapman and Hall, New York.

Eickbush, D. G., T. H. Eickbush and J. H. Werren. 1992. Molecular characterization of repetitive DNA sequences from a B chromosome. Chromosoma 101: 575–583.

Ephrussi, A. and R. Lehmann. 1992. Induction of germ cell formation by oskar. Nature 358: 387–392.

Gherna, R. L., J. H. Werren, W. Weisburg, R. Cote, C. R. Woese, L. Mandelco and D. J. Brenner. 1991. Arsenophonus nasoniae gen. nov., sp. nov., the causative agent of the son-killer trait in the parasitic wasp Nasonia vitripennis. Int. J. Syst. Bact. 41: 563–565.

Gorman, M. M., I. Kuroda and B. S. Baker 1993. Regulation of the sex-specific binding of the

maleless dosage compensation protein to the male X chromosome in *Drosophila*. Cell 72: 39–
49.

Granadino, B. P. Santamaria and L. Sanchez. 1993. Sex determination in the germ line of *Drosophila melanogaster*: Activation of the gene *Sex-lethal*. Development 118: 813–816.

Hackett, K. J., D. E. Lynn, D. L. Williamson, A. S. Ginsberg and R. F. Whitcomb. 1986. Cultivation of the *Drosophila* sex-ratio spiroplasma. Science 232: 1253–1255.

Haig, D. 1993a. The evolution of unusual chromosomal systems in coccoids: Extraordinary sex ratios revisited. J. Evol. Biol. 6: 69–77.

Haig, D. 1993b. The evolution of unusual chromosomal systems in sciarid flies: Intragenomic conflict and the sex ratio. J. Evol. Biol. 6: 249–261.

Hamilton, W. D. 1967. Extraordinary sex ratios. Science 156: 477–488.

Hartl, D. L. and S. W. Brown. 1970. The origin of male haploid genetic systems and their expected sex ratio. *Theor. Popul. Biol.* 1: 165–190.

Hilfiker-Kleiner, D., A. Dubendorfer, A. Hilfiker and R. Nothiger. 1993. Developmental analysis of two sex-determining genes, M and F, in the housefly, *Musca domestica*. Genetics 134: 1187–1194.

Howard, R. S. and C. M. Lively. 1994. Parasitism, mutation accumulation and the maintenance of sex. Nature 367: 554–557.

Hunter, M. S., U. Nur and J. H. Werren. 1993. Origin of males by genome loss in an autoparasitoid wasp. Heredity 70: 162–171.

Hurst, L. D. 1993. The incidences, mechanisms and evolution of cytoplasmic sex ratio distorters in animals. Biol. Rev. 68: 121–193.

Hurst, G. D. D. and M. E. N. Majerus. 1993. Why do maternally inherited microorganisms kill males? Heredity 71: 81–95.

Hurst, L. D. and A. Pomiankowski. 1991. Causes of sex ratio bias may account for unisexual sterility in hybrids: A new explanation of Haldane's rule and related phenomena. Genetics 128: 841–858.

Hurst, G. D. D., L. D. Hurst and M. E. N. Majerus. 1993a. Altering sex ratios: The games microbes play. BioEssays 15: 695–697.

Hurst, G. D. D., M. E. N. Majerus and L. E. Walker. 1993b. The importance of cytoplasmic male killing elements in natural populations of the two spot ladybird, *Adalia bipunctata* (Linnaeus) (Coleoptera: Coccinellidae). Biol. J. Linn. Soc. 49: 195–202.

Inoue, H. and T. Hiroyoshi. 1986. A maternal-effect sex-transformation mutant of the housefly, *Musca domestica* L. Genetics 112: 469–482.

Jones, R. N. and H. Rees. 1982. B *Chromosomes*. Academic Press, New York.

Kajita, H. 1993. Induction of males in the thelytokous wasp *Encarsia formosa* Gahan (Hymenoptera: Aphelinidae). Appl. Entomol. Zool. 28: 115–117.

King, B. H. 1992. Sex ratio manipulation by parasitoid wasps. Pp. 418–441. In *Evolution and Diversity of Sex Ratio in Insects and Mites*. D. L. Wrensch and M. Ebbert, Eds. Chapman and Hall, New York.

Kraemer, C. and E. R. Schmidt. 1993. The sex determining region of *Chironomus thummi* is associated with highly repetitive DNA and transposable elements. Chromosoma 102: 553–562.

Krueger, C. M., M. E. Degrugillier and S. K. Narang. 1993. Size difference among 16S rRNA genes from endosymbiotic bacteria found in testes of *Heliothis virescens*, *H. subflexa* (Lepidoptera: Noctuidae), and backcross sterile male moths. Fla. Entomol. 76: 382–390.

Kuroda, M. I., M. J. Kernan, R. Kreber, B. Ganetzky, and B. S. Baker. 1991. The *maleless* protein associates with the X chromosome to regulate dosage compensation in *Drosophila*. Cell 66: 935–947.

LaChance, L. E. 1979. Genetic strategies affecting the success and economy of the sterile insect release method. Pp. 8–18. In: *Genetics in Relation to Insect Management*. M. A. Hoy and J. J. McKelvey, Jr., Eds. Rockefeller Found. Press, New York.

Lauge, G. 1985. Sex determination: Genetic and epigenetic factors. Pp. 298–317. In: *Comprehensive*

*Insect Physiology Biochemistry and Pharmacology.* Vol. 1. *Embryogenesis and Reproduction.* G. A. Kerkut and L. I. Gilbert, Eds. Pergamon Press, Oxford.

Leslie, J. F. 1984. A "sex-ratio" condition in *Oncopeltus fasciatus.* Heredity 75: 260–264.

Leventhal, E. 1968. The sex ratio condition in *Drosophila bifasciata:* Its experimental transmission to several species of *Drosophila.* J. Invertebr. Pathol. 11: 170–183.

Lyttle, T. W. 1993. Cheaters sometimes prosper: Distortion of Mendelian segregation by meiotic drive. Trends Genet. 9: 205–210.

Makela, M. E. and M. D. Huettel. 1979. Model for genetic control of *Heliothis virescens.* Theor. Appl. Genet. 54: 225–233.

Maynard Smith, J. 1978. *The Evolution of Sex.* Cambridge Univ. Press, Cambridge.

Michod, R. E. and B. R. Levin, Eds. 1988. *The Evolution of Sex.* Sinauer Assoc., Sunderland, MA.

Miller, S. G., M. D. Huettel, M. B. Davis, E. H. Weber and L. A. Weber. 1986. Male sterility in *Heliothis virescens* × *H. subflexa* backcross hybrids: Evidence for abnormal mitochondrial transcripts in testes. Mol. Gen. Genet. 203: 451–461.

Naito, T. and H. Suzuki. 1991. Sex determination in the sawfly, *Athalia rosae ruficornis* (Hymenoptera): Occurrence of triploid males. J. Hered. 82: 101–104.

Nothiger, R. and M. Steinmann-Zwicky. 1985. A single principle for sex determination in insects. Cold Spring Harbor Symp. Quant. Biol. 50: 615–621.

Nur, U., J. H. Werren, D. G. Eickbush, W. D. Burke and T. H. Eickbush. 1988. A "selfish" B chromosome that enhances its transmission by eliminating the paternal genome. Science 240: 512–514.

O'Neill, S. L., R. Giordano, A. M. E. Colbert, T. L. Karr and H. M. Robertson. 1992. 16S rRNA phylogenetic analysis of the bacterial symbiont associated with cytoplasmic incompatibility in insects. Proc. Natl. Acad. Sci. USA 89: 2699–2702.

Parkhurst, S. M. and D. Ish-Horowicz. Common denominators for sex. Current Biol. 2: 629–631.

Pauli, D. and A. P. Mahowald. 1990. Germ-line sex determination in *Drosophila melanogaster.* Trends Genet. 6: 259–264.

Rao, S. R. V. and M. Padmaja. 1992. Mammalian-type dosage compensation mechanism in an insect—*Gryllotalpa fossor* (Scudder)—Orthoptera. J. Biosci. 17: 253–273.

Reed, K. M. 1993. Cytogenetic analysis of the paternal sex ratio chromosome of *Nasonia vitripennis.* Genome 36: 157–161.

Retnakaran, A. and J. Percy. 1985. Fertilization and special modes of reproduction, Pp. 213–293. In: *Comprehensive Insect Physiology Biochemistry and Pharmacology,* Vol. 1, *Embryogenesis and Reproduction.* G. A. Kerkut and L. I. Gilbert, Eds. Pergamon Press, Oxford.

Rousset, F. and M. Raymond. 1991. Cytoplasmic incompatibility in insects: Why sterilize females? Trends Ecol. Evol. 6: 54–57.

Rousset, F., D. Vautrin and M. Solignac. 1992. Molecular identification of *Wolbachia,* the agent of cytoplasmic incompatibility in *Drosophila simulans,* and variability in relation to mitochondrial types. Proc. R. Soc. London B 247: 163–168.

Shirk, P. D., D. A. O'Brochta, P. E. Roberts and A. M. Handler. 1988. Sex-specific selection using chimeric genes. Pp. 135–146. In: *Biotechnology for Crop Protection.* P. A. Hedin, J. J. Menn and R. M. Hollingsworth, Eds. Am. Chem. Soc. Symp. Series No. 379., Am. Chem. Soc. Publ.

Slee, R. and M. Bownes. 1990. Sex determination in *Drosophila melanogaster.* Q. Rev. Biol. 65: 175–204.

Somerson, N. L., L. Ehrman, J. P. Kocka and F. J. Gottlieb. 1984. Streptococcal L-forms isolated from *Drosophila paulistorum* semispecies cause sterility in male progeny. Proc. Natl. Acad. Sci. USA 81: 282–285.

Steinmann-Zwicky, M. 1992. How do germ cells choose their sex? *Drosophila* as a paradigm. BioEssays 14: 513–518.

Steinmann-Zwicky, M. 1993. Sex determination in *Drosophila: sis-b,* a major numerator element of the X:A ratio in the soma, does not contribute to the X:A ratio in the germ line. Development 117: 763–767.

Stevens, L. 1993. Cytoplasmically inherited parasites and reproductive success in *Tribolium* flour beetles. Anim. Behav. 46: 305–310.

Stouthamer, R. and J. H. Werren. 1993. Microbes associated with parthenogenesis in wasps of the genus *Trichogramma*. J. Invertebr. Pathol. 61: 6–9.

Stouthamer, R. and R. F. Luck. 1993. Influence of microbe-associated parthenogenesis on the fecundity of *Trichogramma deion* and *T. pretiosum*. Entomol. Exp. Appl. 67: 183–192.

Stouthamer, R., R. F. Luck and W. D. Hamilton. 1990. Antibiotics cause parthenogenetic *Trichogramma* (Hymenoptera/Trichogrammatidae) to revert to sex. Proc. Natl. Acad. Sci. USA 87: 2424–2427.

Stouthamer, R., R. F. Luck and J. H. Werren. 1992. Genetics of sex determination and the improvement of biological control using parasitoids. Environ. Entomol. 21: 427–435.

Stouthamer, R. , J. A. J. Breeuwer, R. F. Luck and J. H. Werren. 1993. Molecular identification of microorganisms associated with parthenogenesis. Nature 361: 66–68.

Thornhill, R. and J. Alcock. 1983. *The Evolution of Insect Mating Systems*. Harvard Univ. Press, Cambridge, MA.

Traut, W. and U. Willhoeft. 1990. A jumping sex determining factor in the fly *Megaselia scalaris*. Chromosoma 99: 407–412.

Werren, J. H. and L. W. Beukeboom. 1993. Population genetics of a parasitic chromosome: Theoretical analysis of PSR in subdivided populations. Amer. Natur. 142: 224–241.

Werren, J. H., U. Nur and D. Eickbush. 1987. An extrachromosomal factor causing loss of paternal chromosomes. Nature 327: 75–76.

White, M. J. D. 1973. *Animal Cytology and Evolution*. Cambridge Univ. Press, Cambridge.

Whiting, P. W. 1943. Multiple alleles in complementary sex determination of *Habrobracon*. Genetics 28: 365–382.

Williams, G. C. 1975. *Sex and Evolution*. Princeton Univ. Press, Princeton, NJ.

Wood, R. J. and M. E. Newton. 1991. Sex-ratio distortion caused by meiotic drive in mosquitoes. Amer. Natur. 137: 379–391.

Wrensch, D. L. and M. A. Ebbert, Eds. 1993. *Evolution and Diversity of Sex Ratio in Insects and Mites*. Chapman and Hall, New York.

Younger-Shepherd, S., H. Vaessin, E. Bier, L. Y. Jan and Y. M. Jan. 1992. *dead-pan*, an essential panneural gene encoding an HLH protein, acts as a denominator in *Drosophila* sex determination. Cell 70: 911–922.

Zchori-Fein, E., R. T. Roush and M. S. Hunter. 1992. Male production induced by antibiotic treatment in *Encarsia formosa* (Hymenoptera: Aphelinidae), an asexual species. Experientia 48: 102–105.

# Chapter 12

# Molecular Genetics
# of Insect Behavior

## Overview

The study of insect behavior involves the analysis of all activities performed by an insect in relation to its surrounding environment. Behavior genetics is the study of the underlying hereditary basis of the behavior. Genetic analyses of insect behavior typically have employed Mendelian genetic analyses for traits determined by one or a few genes, or quantitative genetic methods for traits determined by 'many' genes. Fate mapping is a special technique that allows researchers to locate the anatomical site affected by mutations that alter behavior of *Drosophila*.

Several behavioral phenotypes in *Drosophila* have been dissected with the tools of molecular genetics, and inter- and intraspecific comparisons of the DNA sequences can be made. The *period* gene (*per*) is one of the best-studied genes. It is a component of the biological clock in *D. melanogaster* and other *Drosophila* species. It influences activity patterns and other circadian rhythms, as well as altering song cycles in courting males. The *per* locus has been cloned and sequenced in *D. melanogaster* and *D. simulans*. After the *per* gene of *D. simulans* was inserted by P-element-mediated transformation into a strain of *D. melanogaster* that is arrhythmic, *D. melanogaster* males were developed that produced song cycles like those of *D. simulans*. The region of DNA controlling the differences in song rhythm in the two species maps to a small segment of the *per* locus that may vary by as few as four amino acids. Learning mutants, such as *dunce* and *couch potato* also have been studied in some detail in *D. melanogaster* and could provide an understanding of the fundamental processes involved in learning in insects and other organisms.

The genetic analysis of behavior determined by many genes may be revolutionized by the use of molecular genetic techniques, for it now should be possible to map the number and location of genes affecting complex traits by correlating their inheritance with restriction fragment length polymorphisms or other DNA markers.

## Introduction

The study of insect behavior covers a very wide range of activities, including locomotion, grooming, feeding, communication, reproduction, dispersal, flight, learning, migration, host or prey selection, diapause, and various responses to environmental hazards such as temperature, humidity, parasites, and toxins (Dingle 1978, Beck 1980, Dingle and Hegmann 1982, Alcock 1984, Tauber et al. 1986, Gatehouse 1989). One definition of behavior is any action that an individual carries out in response to a stimulus or to its environment, especially an action that can be observed and described. However, insects also behave spontaneously, in the absence of any obvious stimulus. Thus, behavior includes studies to understand how an insect takes in information from its environment, processes that information, and acts. Processing information in the central nervous system may involve integrating information over time, including stimuli such as hormones coming from within the insect. Thus, the connection between stimulus and response can be delayed and indirect.

The genetic analysis of behavior rightfully has been perceived to be more complex than the analysis of morphological or anatomical traits. One of the complications in analysis of the genetic basis of behavior is the difficulty in defining the behavior in a clear manner. Behavior genetics began to grow as a field of study in the 1960s, but was limited to demonstrating that a behavioral trait was heritable, determining whether its mode of inheritance was either dominant or recessive, and resolving whether the variation was due to a single gene or multiple genes (Ehrman and Parsons 1973).

Genetic analyses of insect behavior require careful control of environmental conditions, because even subtle differences in test conditions can influence the assays. Objective measures of insect behavior often are difficult to make, and considerable efforts have been devoted to devising specific and appropriate assays. The possible influence of learning must be considered in many behavioral analyses and, to complicate matters further, learning rates no doubt vary among different strains of the same species, so that both heredity and environment must be considered in the genetic analysis of behavior. The field of insect behavior genetics involves, in many cases, analyses of the physiological or morphological changes that are associated with the change in behavior.

The genetic basis of insect behavior has been analyzed most extensively using *Drosophila melanogaster* and a few other species, such as grasshoppers and crickets (Benzer 1973, Matthews and Matthews 1978, Ehrman and Parsons 1981, Hall 1984, Kalmring and Elsner 1985, Huettel 1986, Huber et al. 1989). Traditional behavior–genetic analysis primarily employs two approaches: crossing experiments and selection experiments. A third approach, limited to analysis of behavior in *D. melanogaster*, involves analysis of **fate maps** in genetic mosaics to locate the anatomical site of abnormalities that affect behavior (Hotta and Benzer 1972).

While a specific behavior sometimes can be altered by the mutation of a single gene in a pathway leading to the behavior, an insect's behavior is influenced by many genes (Plomin 1990). Thus, analysis of behavior generally requires the use of **quantitative genetic** methods. Sometimes, as will be demonstrated in later examples, mutations in a single gene or a few major genes will alter a behavior, and the mode of inheritance can be assessed by traditional Mendelian techniques.

Analyses of insect behavior employ techniques from different disciplines, including anatomy, biochemistry, ecology, ethology (study of animal behavior in the natural environment), genetics, psychology, physiology, and statistics (Matthews and Matthews 1978, Hay 1985, Bell 1990, Holman et al. 1990, Via 1990, Barton Browne 1993). These disciplines are required because an insect perceives the environment through its sensory systems. The external sensory stimuli are transduced into electrical information, which is then processed and decoded, leading to a behavioral response. Behavior can be divided into several sequential steps: stimulus recognition, signal transduction, integration, and response or motor output.

Molecular genetic techniques provide powerful new tools to analyze insect behavior. It is now possible to clone specific genes and to determine their sequences and protein products. Because P-element-mediated transformation of *Drosophila* species is feasible, genes from one species of *Drosophila* can be inserted into the genome of another and their effect(s) on behavior can be determined. Molecular genetic analyses of learning and memory in *Drosophila* may provide a means to learn about one of the most challenging frontiers in biology— neurobiology. Molecular genetic methods may allow us to localize and identify some of the individual genes among the many genes involved in determining the interesting and complex behaviors exhibited by insects.

## The Insect Central Nervous System

The insect brain contains about $10^5$ to $10^6$ neurons. It consists of three main divisions—the protocerebrum, deutocerebrum, and tritocerebrum. In each of these divisions, different neuropil regions are located, with some highly structured and others lacking obvious structure.

In the protocerebrum, higher sensory centers (e.g., the corpora pedunculata or mushroom bodies and central complex) are present that are associated with vision and other sensory receptors. The superior protocerebrum, with the pars intercerebralis, contains different sets of neurosecretory cells that supply neurohemal organs in the corpora cardiaca and corpora allata, which are located in the head or prothorax in insects. The optic lobes flanking the protocerebrum consist of the most well-organized neuropils in the brain. The antennal centers are found in the deutocerebrum, and in the tritocerebrum, neurosecretory neurons and neurons associated with the control of feeding and foregut activity are found (Homberg et al. 1989). The brain is connected to the subesophageal ganglion via connectives and to the thoracic and abdominal ganglia, or ventral nerve cord (Strausfeld 1976). The structural organization of the adult and immature chemosensory system of D. melanogaster has been reviewed as the basis for developmental and molecular studies of chemosensory gene function (Stocker 1994).

Information is transmitted in the insect via nerves and by neuropeptides that coordinate development and behavior. Both neurosecretory cells and neurons use neuropeptides as messengers. More than 100 different types of neuropeptides have been identified, including proctolin and adipokinetic hormone, which serve as both hormones and neurotransmitters or neuromodulators (Scharrer 1987, Masler et al. 1993, Raina and Menn 1993). Neuropeptides range in size from three amino acid residues (thyrotropin-releasing hormone) to more than 50 amino acids (insulin). They are generated from larger precursor proteins, ranging from 90 to 250 amino acids. A number of the genes have been identified that code for neuropeptides, including bombyxin or prothoracicotropic hormone (PTTH), eclosion hormone (EH), FMRFamine-related (Phe–Met–Arg–Phe–$NH_2$) peptides, diapause hormone, and pheromone biosynthesis-activating neuropeptide (PBAN) (for a review, see Nassel 1993, Sato et al. 1993).

Neuropeptides are released as cotransmitters and modulate fast transmission at neuromuscular junctions. A given neuropeptide may occur at several different sites and at different levels, including central nervous system circuits, peripheral synapses, and at the peripheral targets—muscles and glands. Neuropeptides regulate behavior by coordinating temporal and spatial activity of many neuronal circuits. Each of the circuits controlling behavior employs sets of sensory neurons, interneurons, and motor neurons. Thus, multiple neural networks share neural elements. Molecular genetic analysis will provide rapid progress in understanding neuropeptide receptors and second messenger pathways. Some research on neuropeptides and their receptors indicates that they may have roles during embryonic development and as cytokines in the immune systems of insects (Nassel 1993).

## Analysis of Behavior by Traditional Methods

### Crossing Experiments

A crossing experiment involves mating individuals that differ in a particular kind of behavior and then examining the behavior of their $F_1$ and backcross progeny. Ideally, the environment is constant, or controlled so that all individuals experience the same conditions. It is easiest to interpret the results of the experiment if the individuals that are crossed differ only in a single behavioral attribute. The phenotype of the $F_1$ and backcross progeny indicates whether the behavior is determined by a single gene or by multiple genes, and whether there is dominance, sex linkage, or maternal influence. If the trait is determined by many genes, it is generally impossible to determine the number of loci, their relationship to each other, or their location on specific chromosomes because most insect species lack sufficient genetic markers. Sufficient genetic markers for each chromosome are available, however, to conduct detailed analyses of some complex behaviors with *Drosophila*. Honey bee behavior provides a rare example of a trait that appears to be determined by a few genes.

### Susceptibility to American foulbrood in Apis mellifera

Susceptibility of the haplodiploid honey bee, *Apis mellifera*, to the disease American foulbrood, which is caused by *Bacillus larvae*, has been analyzed by crossing two inbred strains with differing levels of resistance. The differences in resistance are attributed to differences in 'hygienic behavior' in worker (sterile female) bees (Rothenbuhler 1964). Resistant workers consistently remove dead larvae and pupae from the brood nest at a high rate, thus slowing the spread of the bacteria through the colony by reducing contamination (hygienic). Crosses between 'hygienic' and susceptible 'nonhygienic' bees yield $F_1$ worker progeny that are nonhygienic, indicating that the gene or genes conferring resistance are recessive, and wild alleles are dominant.

Progeny produced by backcrosses to the homozygous recessive hygienic strain yielded approximately 25% of the workers that were hygienic, which is consistent with the hypothesis that the hygienic behavior is determined by two recessive loci (Table 12.1). Hygienic worker bees are homozygous for two genes,

**Table 12.1**

Genetic Hypothesis of Differences in Nest-Cleaning Behavior among Inbred Lines, and $F_1$ and Backcross Progeny of the Haplodiploid Honey Bee *Apis mellifera*[a]

| Parental lines | Hygienic queen | $uu, rr \times U^+U^+, R^+R^+$ | | | Unhygienic male |
|---|---|---|---|---|---|
| $F_1$ progeny | | All workers $U^+u, R^+r$ | | | Unhygienic |
| Backcross progeny (from cross of $U^+u, R^+r$ × $u, r$) | 1: $uu, rr$ Hygienic | 1: $uu, R^+r$ Uncaps, doesn't remove | 1: $U^+u, rr$ Removes, doesn't uncap | 1 $U^+u, R^+r$ Unhygienic | |

[a]Data obtained from Rothenbuhler 1964.

$u$ and $r$, and the bees both uncap the cells containing dead brood and remove them. Analysis indicated that $uu$, $r^+r$ individuals will uncap the cells, but not remove dead brood. The $u^+u$, $rr$ individuals do not uncap brood, but will remove them if the cells are uncapped for them. Individuals that are $u^+u$, $r^+r$ are unhygienic, and will neither uncap nor remove brood.

### House-Entering Behavior in Aedes aegypti

House-entering behavior by the mosquito *Aedes aegypti* from East Africa has been analyzed by crosses and shown to be controlled by several genes with additive effects (Trpis and Hausermann 1978). Different populations of this mosquito behave differently. Some populations commonly enter houses (domesticated or D), while others rarely do so (either peridomestic, P, or feral, F). This behavior is important in determining whether the population transmits a disease such as yellow fever to humans.

Samples of three populations collected either inside houses (D), near a village (P), or from tree holes in a forest (F) were bred in insectaries and crossed to produce hybrid DP, PD, DF, FD, PF, FP populations. The original populations and the hybrid populations were then marked with different-colored fluorescent powders and released near houses. Marked mosquitoes were captured at two sites: inside houses and in the village area. Of the mosquitoes entering houses, 45% were from the domestic (D) population, 13.9 % from the hybrids between the domestic and the peridomestic population (DP and PD), 9.8% from the peridomestic population (P), and 5.7% from the hybrids between the domestic and feral populations (DF and FD). Only 1.5 and 0.6% of the hybrids PF and FP were collected in the house, and the feral population entered the house with a frequency of 0.6%. The recaptures in the village area were in the reverse order.

Domesticity in *A. aegypti* is a complex phenomenon that includes a variety of behaviors, including a preference for ovipositing in man-made containers, ability of larvae to develop in drinking water stored in clay pots with a low nutritional content, and preferences for feeding on man (rather than birds) inside houses, as well as resting and mating indoors. No doubt *A. aegypti* speciated long before man began to build houses, but this insect has adapted rapidly to human habitats, and the domestic form of *A. aegypti* is the only mosquito known that is entirely dependent on man (Trpis and Hausermann 1978).

### Selection Experiments

Selection experiments provide a way to determine the degree to which a given behavior is determined genetically. In a selection experiment, individuals with a specific behavioral attribute are allowed to reproduce and this process is repeated over succeeding generations. Eventually, the behavior of the selected population is altered *if* genetic variation for the attribute is present in the initial colony *and* the selection procedures have been appropriate. The response of the population to selection can be analyzed to estimate the heritability of the trait.

For example, migratory and nonmigratory behavior of the large milkweed bug, *Oncopeltus fasciatus*, is under genetic control. Palmer and Dingle (1989) selected *O. fasciatus* for wing length and for propensity to fly. Bidirectional selection on wing length was performed for 13 generations, and the flight behavior of individuals was monitored. Individuals also were selected for flight time, and those whose flight times totaled 30 minutes were considered 'fliers', while those with the shortest flight times were labeled 'nonfliers'.

Selection response on wing length was rapid. Flight tests of the long- and short-winged flies indicated there was a positive correlation between wing length and flight duration. Selection after two generations for flight or nonflight likewise resulted in divergent responses, indicating a large genetic component to flight behavior.

To estimate the degree of influence of genes on a specific behavior, two measures are used, the selection differential and the estimate of heritability. The **response to selection (R)** is the difference in mean phenotypic value between the offspring of the selected parents and the mean phenotypic value of the entire parental generation before selection (Falconer 1989),

$$R = h^2 S,$$

where $R$ is the improvement or response to selection, $h^2$ is the heritability of the characteristic under selection in the population, and $S$ is the selection differential. The **selection differential** is the average superiority of the selected parents expressed as a deviation from the population mean (Falconer 1989). The selection differential measures the difference between the average value of a quantitative character in the whole population and the average value of those selected to be parents of the next generation. It is measured in standard deviation units.

**Heritability** *in the broad sense* is the degree to which a trait is genetically determined. Because behavioral traits are influenced by both genes and environment, heritability is expressed as the ratio of the total genetic variance to the phenotypic variance ($V_G/V_P$). Heritability *in the narrow sense* is the degree to which a trait is transmitted from parents to offspring, and is expressed as the ratio of the additive genetic variance to the total phenotypic variance ($V_A/V_P$) (Falconer 1989).

Heritability could be estimated to be 'zero' if the specific population being selected had no variability for the behavioral attribute under study, perhaps because it was inbred. Heritability could be estimated to be 'one' if the trait was completely determined by genes, and environment had little impact on the phenotype, although this would be an unusual outcome.

Heritability estimates provide no information about the actual mode of inheritance of a quantitative trait because they represent the cumulative effect of all loci affecting the trait. The number of loci involved generally can be determined only with elaborate and specially designed experiments. A number of assumptions are made when estimating heritability: (1) all loci affecting the trait act *independently* of one another and, (2) the loci are *unlinked*. Another assumption (3) is that environment affects all genotypes in a similar fashion. These three assumptions are not always justified. Thus, heritability estimates are difficult to interpret, although they are useful for predicting response to selection under specific environmental conditions.

Heritability estimates usually are made by regression-correlation analyses of close relatives (parent-offspring, full sibs, half sibs), experiments involving response to selection, or analysis of variance components. Traits with high heritabilities respond readily to selection with an appropriate selection method. The magnitude of the response to the selection, the difference in mean values between parent and progeny generations, provides an estimate of heritability in the narrow sense ($h_n^2$). This estimate is valid only for the population being examined, under the test conditions employed, for the behavior observed, and for the method of measurement employed.

Heritability of most insect behaviors is relatively high, probably because many arthropod behaviors are highly stereotyped (Ehrman and Parsons 1981). For example, the heritability of locomotor activity of *D. melanogaster* has been estimated to be 0.51, and heritability of mating speed of male *D. melanogaster* has been estimated to be 0.33. Heritability for honey production from honey bees ranges from 0.23 to 0.75, depending upon the experimental conditions and colonies tested (Rinderer and Collins 1986). Italian honey bees are less able to remove the parasitic mite *Varroa jacobsoni* than Africanized bees, and the heritability of this ability was estimated to be 0.71 (Moretto et al. 1993). Heritability of the length of the prereproductive period, which is when migratory flight occurs in the noctuid moth *Helicoverpa armigera*, ranged from 0.54 to 0.16 (Colvin and Gatehouse 1993). Heritability of host selection behavior by *Asobara tabida*, a parasitoid of *Drosophila subobscura*, ranged from 0.03 to 1.0, depending upon the test method employed (Mollema 1991).

### Fate Mapping in Drosophila

A special approach to genetic analysis of behavior involves 'fate mapping' using *Drosophila melanogaster* (Hotta and Benzer 1972, Ashburner, 1989). One method for conducting fate mapping involves developing genetic mosaics in *D. melanogaster* by generating an abnormal **ring chromosome** from the X chromosomes. Ring chromosomes normally do not migrate successfully to the poles in mitosis or meiosis and are lost. Thus, a fly can be generated that is a genetic mosaic, consisting of both male (XO) and female (XX) tissues, which is determined by the presence of one or two X chromosomes, respectively. The XO sections of an embryo can identified by using X-linked mutants for eye color, body color, or bristle shape. These mutants allow the investigator to identify precise anatomical sites with abnormalities that affect a specific behavior (Hotta and Benzer 1972).

Fate mapping using flies with the *wings-up* gene provides an example of the method. Flies with this mutation (*wup* A) raise both wings vertically soon after emergence and maintain them permanently in that position. The mutant adults cannot fly but behave normally otherwise. The phenotype is expressed in all hemizygous males and 95% of homozygous females, indicating that the gene is located on the X chromosome and is recessive.

What is the basis of the abnormal behavior? The wings-up syndrome could result from a defect in the wing articulation, the wing muscles, or the neuro-muscular junctions of the central nervous system. Hotta and Benzer (1972) found that the defect was not associated with the cuticle of the thorax because some flies were found in which the wings were held up even though the entire external thorax, including the wings, was normal. Thus, the wing or its articulation is not the cause of the mutant phenotype.

The cause of the 'wings-up' behavior was found to be located close to the ventral midline. During embryonic development, this area invaginates to produce mesoderm, which suggested that a muscle defect is responsible. Subsequent dissection of mutant flies showed that flight muscles of *wup* A flies are atrophied. Fate mapping thus revealed lesions in specific structures, and indicated the primary focus of specific mutant genes. If a histological anomaly of flight muscles were due to abnormal neurogenesis rather than to abnormal muscle development, the focus would have mapped to the area of the blastoderm destined to produce the nervous system.

## Some Behaviors Are Influenced by One or a Few Genes

Crossing experiments have shown that a specific behavior is influenced by one or a few genes in a number of insects, including the flour moth *Ephestia kuhniella* (silk mat spinning by larvae prior to pupation), the mosquito *Aedes atropalpus* (egg maturation without an exogenous source of protein such as blood), and the parasitic wasp *Habrobracon juglandis* (flightlessness) (Ehrman and Parsons 1981). In the silk moth *Bombyx mori*, females with the *piled egg* gene deposit eggs in a peculiar manner. *B. mori* larvae with the *Non-preference* gene are unable to discriminate mulberry leaves from others (Tazima et al. 1975), and Huettel and Bush (1972) found that when two monophagous tephritid flies (*Procecidochares*) were crossed, the host preference behavior segregated in a manner consistent with control by a single locus. Diapause, usually considered to be a multigenic trait (Hoy 1978), is inherited as a simple autosomal recessive trait in *Drosophila melanogaster* populations from Canada (Williams and Sokolowski 1993).

A variety of mutants determined by major genes have been identified in *D. melanogaster* that affect behavior, as might be expected of such an intensively studied species (Grossfield 1975, Hall 1985). These include a group of sex-linked, incompletely dominant mutants (*Shaker, Hyperkinetic*, and *eag*) that are expressed when the flies are anesthetized with ether. The *Hyperkinetic* (*Hk*) locus causes a vigorous steady leg shaking, while mutations at the *Shaker* (*Sh*) locus cause vigorous and erratic shaking and a strong scissoring of wings and twitching of the abdomen. The *eag* mutant flies (*ether a-go-go*) are less vigorous in their shaking.

Other sex-linked genes include a temperature-sensitive recessive mutant (*para^{ts}*) that causes *D. melanogaster* to become immobile above 29°C. An autosomal recessive mutation, *fruity*, affects mating behavior of *D. melanogaster* males, so that homozygous males actively court each other, even in the presence of females. The *couch potato* (*cpo*) locus is unusually complex, spanning more than 100 kb and encoding three different messages (Bellen et al. 1992). *cpo* is differentially spliced and may encode three different proteins. Adult *cpo* flies are hypoactive and exhibit abnormal geotaxis, phototaxis, and flight behavior. Possibly, *cpo* is expressed in the fly's peripheral nervous system and controls the processing of RNA molecules required for the proper functioning of the peripheral nervous system (Bellen et al. 1992).

Many 'single-gene' mutants that affect the morphology of *D. melanogaster* also affect behavior. Some mutant flies exhibit abnormal behavior because they are unable to react to a stimulus due to altered effector structures. Other mutants exhibit altered behavior because perception of cues is impaired. For example, flies with *white* eyes may exhibit abnormal courtship behavior (Grossfield 1975).

Alterations in the brain affect behavior, and genes involved in the behavior can be analyzed by mutagenesis and mosaic analysis. For example, eight different genes affecting walking behavior in *Drosophila melanogaster* strains have been analyzed by chemical mutagenesis, histological analysis, and mosaic analysis (Strauss and Heisenberg 1993). These genes—*agnostic, central-body-defect, central-complex-broad, central-complex-deranged, central-brain-deranged, central-complex, ellipsoid-body-open,* and *no-bridge*—caused structural defects in the central complex of the brain and affected walking motivation, fast **phototaxis** (response to light) and negative **geotaxis** (orientation to gravity) (Strauss and Heisenberg 1993). Because the aberrant behaviors were associated with deficits in the central complex of the insect brain, they confirmed that this structure is involved in behavior. The central complex is suspected to regulate other behaviors, such as flight, singing, preening, and escape, as well. Thus, both the central complex and the mushroom bodies of the brain are involved in regulating behavior (Homberg 1987).

A portion of the genetic basis of pheromone communication in the European corn borer *Ostrinia nubialis* has been determined (Klun and Maini 1979, Klun and Huettel 1988, Lofstedt et al. 1989, Lofstedt 1990). Females of the E- and Z-strains of *O. nubialis* produce different **enantiomeric** ratios of sex pheromone. Hybrids between the two strains produce an intermediate pheromone blend. Analysis of the phenotypes of the $F_2$ and backcross progeny indicates the pheromones produced are controlled by two alleles at a single autosomal locus, although one or more modifier genes controls the precise ratio of the isomers in heterozygous females.

Males of the two O. *nubialis* strains are attracted to the appropriate pheromone blends in the field, and hybrid males respond preferentially to the pheromone produced by heterozygous females rather than to the pheromones produced by the two parental female types. The response of males to the pheromone is determined by a single sex-linked gene with two alleles. The olfactory sensillae of the two types of males are different and are controlled by an autosomal locus with two alleles. Hybrid males give intermediate results when tested for their electrophysiological responses, with E- and Z-cells yielding approximately equal spike amplitudes. The genes determining variation in pheromone production and organization of male olfactory sensillae are not closely linked and are probably on different chromosomes (Lofstedt 1990).

### Some Polygenically Determined Behaviors

Relatively few behaviors are determined by one or a few genes (Plomin 1990). More often, behavior is a continuous variable, with multiple genes contributing small additive effects. With such behaviors, the task of teasing apart the respective roles of genes and environment requires statistical analysis. Examples of *Drosophila* behavior determined by multiple genes include locomotor activity, **chemotaxis**, duration of copulation, **geotaxis**, host plant preference, mating speed, phototaxis, preening, and the level of sexual isolation within and between species.

Several genes influence host plant adaptation and host preference in insects, and learning also may affect host preference (for reviews, see Papaj and Prokopy 1989, Via 1990). The choice of a host plant usually is not a simple behavior, but is a hierarchy of several components. For example, attraction to a site from a distance and preference for an oviposition site (egg laying at the site) are genetically distinct in *Drosophila tripunctata* (Jaenike 1986).

The genetic basis of host-plant specialization in *Drosophila sechellia* and *D. simulans* is determined by a minimum of three or four different loci that affect egg production, survival, and host preference (R'Kha et al. 1991). *Drosophila sechellia* breeds in a single plant, *Morinda citrifolia*, which is toxic to other *Drosophila* species. Its sympatric relative, *Drosophila simulans*, breeds on a variety of plants. The two species can be crossed, and the $F_1$ hybrid embryos produced by *D. simulans* females are susceptible to *Morinda* fruit because susceptibility is maternally inherited and fully dominant.

*D. sechellia* is stimulated by *Morinda* to produce eggs, but *D. simulans* is inhibited by this plant. In hybrid progeny, the inhibition observed in *D. simulans* is dominant. *Morinda* is an oviposition attractant for *D. sechellia* but a repellent for *D. simulans*, and $F_1$ hybrids and backcross progeny exhibit intermediate, approximately additive, behavior. An analysis by Higa and Fuyama (1993) suggests that the species differences in these behaviors are controlled by genes located on the second chromosome. These differences result in isolation of the two species in nature, although they are geographically sympatric. Thus, the ecological niches of the two species are determined by tolerance to toxic products in the ripe *Morinda* fruit, strong preference for *Morinda* fruits by *D. sechellia*, ability to detect fragrant volatiles from *Morinda* over a long distance by *D. sechellia*, stimulation of egg production by *Morinda* in *D. sechellia*, and inhibition of egg production by *Morinda* in *D. simulans*.

Other insects in which specific behavioral attributes have been analyzed include *Musca domestica* (number of attempts to mate by males); *Phormia regina* (high and low ability to learn to extend the proboscis to a stimulus applied to the forelegs); hybrid crickets (call rhythm of males, female response to calling songs); *Anopheles albimanus* (ability to avoid pesticides); *Apis mellifera* (high and low collection of alfalfa pollen, and stinging behavior) (Ehrman and Parsons 1981, Hall 1985, Rinderer 1986). Likewise, Gould (1986) found that the propensity for cannibalism by larvae of the moth *Heliothis virescens* is polygenically determined. Most of these behaviors were analyzed by selection experiments.

Some behaviors have been dissected by mutational analysis. Both *Drosophila melanogaster* and *D. simulans* produce contact pheromones, which consist of cuticular hydrocarbons that elicit wing displays in males. The cuticle of *D. simulans* contains at least 13 hydrocarbons with 23, 25, 27, or 29 carbons that are either linear or 2-methyl-branched alkanes. The most important hydrocarbons involved in mate recognition or stimulation are 7-tricosene and 7-pentacosene. The cuticular hydrocarbons vary quantitatively by sex, with females showing higher levels of 7-tricosene than males. The variation in pheromone levels is under polygenic control, with some genes sex-linked and others autosomal. One mutation, *Ngbo*, influences the ratios of 7-tricosene and 7-pentacosene hydrocarbons in *D. simulans*. Another mutation, *kete*, reduces the amount of 7-tricosene and all other linear hydrocarbons produced but does not affect the ratio of 7-tricosene and 7-pentacosene (Ferveur and Jallon 1993). In most insects, linear and branched hydrocarbons seem to be derived by decarboxylation from long-chain fatty acids resulting from repeated elongations of medium-size fatty acids (Blomquist et al. 1987). It is possible that *kete* might act on the early biosynthetic steps, especially the fatty acid synthesis of linear compounds (Ferveur and Jallon 1993). Flies homozygous for both *kete* and *Ngbo* have reduced viability and fertility, perhaps because they have very little 7-tricosene. No doubt, other mutations will be identified that influence pheromonal communication and mating behavior in *D. simulans*.

## Molecular Genetic Analyses of Insect Behavior

The genetic analysis of behavior offers a 'new frontier' for molecular biology (Plomin 1990). Behavior is the most complex phenotype that can be studied because it involves the functioning of the whole organism, is dynamic, and changes in response to the environment. Insect molecular genetics is unlikely to replace traditional methods of behavior analysis. However, the ability to identify, clone, and sequence specific genes, particularly in D. melanogaster, makes it possible to analyze the molecular basis of several behaviors (including the periodicity of biological rhythms, mating behavior, locomotion, and memory) that are influenced by single genes. It is now possible to clone a gene from one Drosophila species, insert it into a P-element vector, and introduce the exogenous gene into mutant strains of D. melanogaster to confirm that the putative gene does, in fact, code for the behavior of interest. Cloned genes from Drosophila can be used as probes to clone genes from other arthropods, which then can be sequenced and compared.

## The period Gene of Drosophila

The potential for molecular genetic analysis of insect behavior influenced by a single major gene is best exemplified by the extensive analyses conducted using the period gene of D. melanogaster (Kyriacou 1990, Takahashi 1992, Young 1993, Chalmers and Kyriacou 1993). Many insects exhibit particular behavior at a specific time of the day. This periodicity is thought to be due to the action of a **circadian clock** that allows the insect to measure time.

Circadian rhythms have a number of characteristics:

1. The clocks usually are 'free running' in constant environments and are not simple responses to changes in light or temperature.
2. Although the rhythms are free running, an initial environmental signal is required to entrain the rhythmicity or start the clock. Among the cues that set the clock are alternating light and dark cycles, high- and low-temperature cycles, or short pulses of light.
3. The circadian rhythm is relatively insensitive to changes in temperature (temperature compensated).
4. The clock can be reset by altering the cues that entrain the clock.

D. melanogaster born and reared in constant darkness exhibit circadian locomotor activity rhythms as adults. However, the rhythms of the individual flies composing these populations are not synchronized with one another (Sehgal et al. 1992). Rhythms of flies can by synchronized if dark-reared flies are exposed to light treatments as first-instar larvae. Light treatments occurring at times preceding hatching of the first-instar larvae fail to synchronize adult locomotor activity rhythms but work subsequently, suggesting that the clock controlling circadian rhythms functions continuously from larval hatching to adulthood. The rhythm can be advanced, delayed, or unchanged, depending on the phase of the cycle at which the cue is given.

The circadian rhythm is presumed to oscillate within an approximate 24-

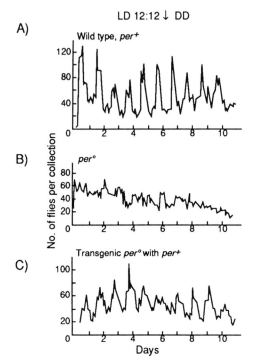

**Figure 12.1** Profiles of eclosion (emergence of adults from pupal cases) for populations of (A) *D. melanogaster* wild-type females, (B) *per⁰* males and females (C) and transgenic *per⁰* individuals that have received a wild-type *per⁺* gene by P-element-mediated transformation. (Modified from Bargiello et al. 1984.)

hour period. This oscillation somehow mediates the timing of specific behavior. Despite many years of study, the mechanisms of the clock are not fully understood (Hall 1990). It appears, however, that molecular genetic analyses of *Drosophila* clock mutants may provide an understanding of the mechanisms of the circadian clock in insects, as well as in other organisms (Takahashi 1992, Dunlap 1993).

The *per* locus is found on the X chromosome, and mutations influence several behaviors, including eclosion, locomotor activity, and the length of the interpulse interval of the courtship song. Eclosion of wild-type flies (emergence of adults from the pupal case) typically occurs approximately every 24 hours under normal light conditions, around dawn, when the presence of dew and high relative humidity increases their survival rate. Flies with the *per⁰* mutation eclose arrhythmically (Figure 12.1B), but the periodicity in eclosion can be restored by P-element-mediated transformation of arrhythmic flies using the

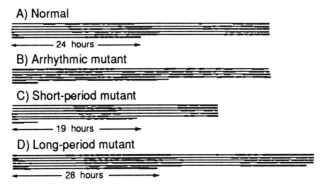

**Figure 12.2** Locomotor activity rhythms of flies with the wild-type (A), arrhythmic (B), short-period (C), and long-period alleles (D) of the *per* locus in *D. melanogaster*. Records read from left to right, with each new line representing the start of an interval.

wild-type allele (Bargiello et al. 1984) (Figure 12.1C). Locomotor activity is also greater around dawn, then decreases during midday and is followed by increased activity again in the evening (Figure 12.2). Three classes of mutant alleles exist; they either shorten (*per^S* mutants have 19-hour eclosion rhythms), lengthen (*per^L* mutants have 29-hour eclosion rhythms), or completely abolish circadian eclosion and locomotor activity rhythms (*per^O* mutants). Figure 12.2 illustrates the differences in locomotor activity rhythms for normal, arrhythmic, short-period, and long-period flies.

The *per* gene is approximately 7 kb long and encodes a 4.5-kb transcript with eight exons, the first of which is noncoding (Figure 12.3). At least three different RNA transcripts can be produced from this gene (Citri et al. 1987). The *per* locus has been sequenced and the polypeptide has a predicted molecular weight of 127,000. Two minor RNA transcripts can be produced by alternative splice sites near the 3' end so that at least three different proteins can be produced, but the significance of the three proteins is unknown.

One of the most striking features of the protein is a series of threonine-glycine (Thr-Gly) repeats in exon 5, which also have been observed in a *Neurospora* clock gene called *frequency* (McClung et al. 1989), and a gene expressed in the mammalian circadian pacemaker. It appears that these sequence similarities are due to coincidence rather than to conservation of a homologous core sequence (Chalmers and Kyriacou 1993). The region which encodes the Thr–Gly repeat is polymorphic in length within and between *Drosophila* species in Europe. For example, in *D. melanogaster* populations, either 17, 20, or 23 repeats are found. A clinal pattern has been found along a north–south axis in Europe and North Africa, with the shorter sequences predominantly found in southern Europe (Costa et al. 1992). Costa et al. (1992) suggested that the length poly-

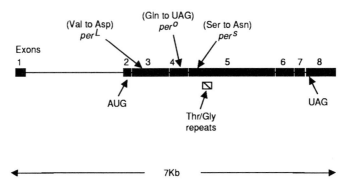

**Figure 12.3**   The exon/intron structure of the *D. melanogaster per* gene. The gene is approximately 7 kb long, with 7 exons. The locations of the *per^L*, *per^O*, and *per^S* mutations are indicated, as is the region which codes for the variable number (17, 20, or 23) of Thr/Gly repeats.

morphism cline may be maintained by selection under different temperature conditions.

A large number of tissues express the *per* product, including embryonic, pupal, and adult nervous systems, as well as the esophagous, gut, and ovaries. Liu et al. (1992) demonstrated that the period gene product is predominantly found in cell nuclei in adult *Drosophila*, and Hardin et al. (1992) showed that *per* mRNA levels undergo daily fluctuations in abundance. These fluctuations constitute a feedback loop in which the per protein affects the oscillations of its own mRNA. The fluctuations in *per* mRNA are due to fluctuations in *per* gene transcription, because the *per* mRNA has a relatively short half-life. Regular fluctuations in per protein were observed in cells of the visual system and central brain when anti-per antibody was used to label tissues of adult flies sectioned at different times of the day (Zerr et al. 1990). These results are consistent with the hypothesis that the per protein may act as a transcription factor.

How might the per protein(s) be used? Is the per product a component of the clock mechanism itself, or is it a protein that coordinates various physiological systems that express the clock's output? The protein encoded by *per* is similar to the proteins produced by at least three other genes: a gene encoding the human aryl hydrocarbon receptor nuclear transporter, the human arylhydrocarbon receptor, and the *single-minded* gene in *D. melanogaster*. These genes share a domain that encompasses approximately 270 amino acids. Per proteins may use this domain to dimerize and could interact with a partner that contains a DNA-binding region, thereby functioning as a transcriptional regulator (Takahashi 1992).

Several models for the biological clock involve membrane structures and functions such as ionic fluxes. The presence of three alternatively spliced *per* products implies a multifunctional role for this gene. Post-translational processing could yield an additional array of gene products with different functions, perhaps including small neuropeptides that could be related to humoral aspects of *per* function. A circadian clock model involves the synthesis of protein(s) cyclically over the course of a day. The *per* gene seems to be a major factor in building or maintaining a variety of rhythms in *Drosophila*. The *per* genes from *D. simulans*, *D. virilis*, *D. pseudoobscura*, and *D. yakuba* have been cloned and sequenced. Parts of the gene are conserved and parts are highly diverged in the different species. The results suggest that conserved regions of the protein may encode basic functions common to all the species (clock-type functions), but species-specific differences such as love songs, locomotor activity, and eclosion profiles may be encoded within the variable regions (Kyriacou 1990).

Other clock genes have been discovered, including the X-linked *Clock* and *Andante* loci (Kyriacou 1993). *Clock* and *Andante* cause slightly shortened and lengthened cycles, respectively. In addition, autosomal circadian rhythm mutations induce flies to eclose early in a light–dark cycle. *Phase-angle* causes flies to emerge in the predawn part of the cycle instead of just after dawn. Another mutation, *gate*, causes flies to fail to eclose during this usually narrow time window. While much remains to be learned about how *Drosophila* keeps time, molecular genetics provides a potent new approach to dissecting this behavior. Preliminary results suggest that the core mechanism of the circadian clock in *Drosophila* may be fundamentally similar in different species, including mammals (Takahashi 1992).

### Song Cycle Behavior in Transgenic Drosophila

The courtship song is produced when males vibrate their wings, and consists of two components: (1) courtship hums, and (2) a series of pulses with interpulse intervals, which can fluctuate between 15 and 85 milliseconds (ms) (Figure 12.4) (Kyriacou and Hall 1989). The variation in interpulse intervals within a courtship song is not random, but cycles with a period of 56 seconds in *D. melanogaster* and 35–40 seconds in *D. simulans*. *D. melanogaster* males with the *per*s mutation sing with 40-second periods, *per*L males sing with 76-second periods, and *per*O males are arrhythmic (Figure 12.4). The genetic basis of species-specific song instructions was confirmed by the transfer of the *per* gene cloned from *D. simulans* into *D. melanogaster* via P-element-mediated transformation (Wheeler et al. 1991). The *D. simulans* per gene restored a rhythm in *D. melanogaster*, and transgenic *D. melanogaster* males gave mean period lengths in their song cycles of approximately 35 seconds, which is characteristic of *D. simulans* males. Wheeler et al. (1991) concluded that substitutions in four or fewer amino acids in the *per* locus are responsible for the species-specific courtship behavior.

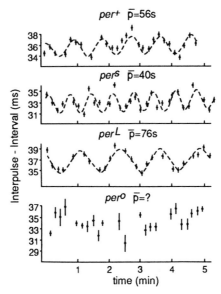

**Figure 12.4** The mean interpulse intervals (IPI) in courtship songs of *D. melanogaster* males varies. Each point is a mean IPI in milliseconds for 10 seconds of courtship. The IPIs in songs of wild-type *per*+, *per*S, and *per*L males average 56, 40, and 76 ms. *per*O mutants lack a periodicity in IPI of the song cycle. (Modified from Kyriacou and Hall 1980.)

### Other Effects of per

The three *per* alleles affect several additional behaviors, including locomotion, cellular rhythms, and development time. Flies with the *per*S mutation develop faster than wild type flies, and *per*L flies develop more slowly than the wild type under various environmental conditions. Thus, *per* mutants may be affecting another clock-like process that is used in timing development. Locomotor activity rhythms are also influenced because when individual fly rhythms are examined by observing the activity and rest cycles of single flies, these cycles have a 24-hour period in wild type flies but *per* mutants have altered cycles in the predicted direction. Whether *per* mutants also have changed heartbeat rates, modified mitotic cycles in *per*S embryonic cell lines, and changed cellular circadian cycles in salivary gland cells is controversial (Kyriacou 1990, 1993).

It has long been thought that circadian oscillations provided the clock both for daily rhythms, such as eclosion and locomotion, and for the critical daylength measurements inherent in photoperiodically induced diapause in insects. However, *per* appears to have no influence on the photoperiodic clock in *D. melanogaster* (Saunders 1990). Females of a wild type strain of *D. melanogaster*

(Canton-S) and strains with *per* mutations were able to discriminate between diapause-inducing short days and noninductive photoperiods. *D. melanogaster* adult females exhibit an ovarian diapause when reared and held under short days and low temperature (12°C). Females exposed to long days at the same temperature reproduce. The critical daylength (the photoperiod at which 50% of the individuals enter diapause) for Canton-S females at 12°C is approximately 14 hours of light per 24 hours. Photoperiodic response curves for the *per^S*, *per^L*, and Canton-S strains were almost identical, whereas the *per^O* mutants showed shortened critical daylengths. However, *per^O* mutants are able to discriminate between a long day and a short day, and it appears that the photoperiodic clock in *D. melanogaster*, although based on a system of circadian oscillators, does not depend on the presence or normal activity of the *per* gene.

Many behaviors, including learning, involve temporally patterned events. The timing interval between presentation of the **conditioned stimulus** and **reinforcement** is important in **associative learning**. The conditioned stimulus must be presented before the unconditioned stimulus, and the unconditioned stimulus must follow the conditioned within a relatively brief interval. It was thought that the *per* gene could be involved in learning, based on the observation that *per^L* males in one experiment did not exhibit normal experience-dependent courtship behavior. However, males with the wild type or *per^S* and *per^O* alleles could be conditioned normally. Gailey et al. (1991) repeated these experiments and concluded that *per* mutant flies showed normal learned courtship behavior.

### Est-6 and Est-5 in Drosophila Species

Evolutionary changes in gene regulation are considered to be a prerequisite for macroevolutionary change and species divergence. One case study involves an analysis of the esterase 6 enzyme in *Drosophila melanogaster* and its homolog (esterase 5) in *D. pseudoobscura* (Brady and Richmond 1990). *Est-6* influences behavior in *D. melanogaster* but *Est-5* has a very different function in *D. pseudoobscura*.

The *Esterase-6* (*Est-6*) gene in *D. melanogaster* influences male mating speed and rate of remating by females. Fast and slow variants of esterase 6 protein, as detected by electrophoresis, are produced in natural populations of *D. melanogaster*. More esterase 6 protein is produced in adult males than in females by a factor of two to ten. The enzyme is highly concentrated in the anterior ejaculatory duct and is transferred to females during the first two to three minutes of the 20-minute copulation. Enzyme activity in females can be detected up to two hours after mating and has an influence, along with the rate of sperm use in females, on the timing of remating by females. Males also transfer a substance in their seminal fluid which is converted in the females' reproductive tract by esterase 6 into a pheromone that serves as an antiaphrodisiac. The antiaphrodisiac reduces the sexual attractiveness and receptivity of the female and reduces the likelihood that she will remate. Since the sperm from the female's most recent mate takes precedence over that of a previous male in fertilizing her eggs, this appears to be a mechanism to encourage monogamy in D. melanogaster (Richmond et al. 1986).

Est-6 also influences the rate of mating of males in D. melanogaster. Males with the slow-moving variant of Est-6, as determined by electrophoresis, require 10.2 minutes to achieve copulation with females, while males with the faster-moving protein require only 5.7 minutes. However, females show no mating preferences for males with different alleles of Est-6 in 'choice' experiments.

Once the Est-6 gene was cloned, it could be used as a probe to identify homologous genes in related species, and these comparisons can provide clues to the evolution of behavior (Brady and Richmond 1990). Est-6 was used to isolate a homologous gene (called Est-5) from a genomic library of D. pseudoobscura. Est-5 apparently has a different function in D. pseudoobscura. Approximately 40% of Est-5 is expressed in the eyes of D. pseudoobscura, with the remainder being expressed in the hemolymph of both sexes. Despite these different patterns of expression, Est-6 and Est-5 have similar protein products, transcripts, and DNA sequences.

When the Est-5 gene was cloned into a P element and introduced into D. melanogaster, its activity and pattern of expression matched those of D. pseudoobscura. These results imply that the regulatory sequences in Est-5 have been conserved since the divergence of the two species 20–46 million years ago. Brady and Richmond (1990) speculate that the enzyme in the common ancestor of these two species had a more extensive expression pattern. After the divergence of these two species, regulatory mutations may have occurred that enhanced Est-5 expression in the eyes of D. pseudoobscura, while similar mutations in Est-6 led to increased expression in the male ejaculatory duct of D. melanogaster.

### Learning in Drosophila

It is difficult to produce a single definition of learning that is comprehensive. Learning has been defined as a change in behavior with experience, but this definition would not exclude responses such as growth and maturation, or other processes that are triggered by events such as mating or feeding. Another definition of learning is a reversible change in behavior with experience, but this excludes phenomena in which the modification caused by some experience is fixed and resistant to further change. Yet another definition is that learning is a more-or-less permanent change in behavior that occurs as a result of practice, but this definition is ambiguous (Papaj and Prokopy 1989).

Papaj and Prokopy (1989) suggest that the following properties are characteristic of learning in insects: (1) The individual's behavior changes in a repeatable way as a consequence of experience. (2) Behavior changes gradually with continued experience, often following a 'learning curve' to an asymptote. (3) The change in behavior accompanying experience declines in the absence of continued experience of the same type or as a consequence of a novel experience or trauma.

Insect populations vary in their ability to learn. The existing genetic variability within strains has been used to analyze the physiology of learning in *Drosophila* and *Phormia* flies (McGuire and Hirsch 1977, McGuire 1984, McGuire and Tully 1987). *Drosophila melanogaster* is able to be sensitized and habituated, learn associations with positive or negative reinforcement, and be classically conditioned (Davis and Dauwalder 1991).

The study of the genetic basis of learning in *Drosophila* has been difficult for two reasons: (1) it is difficult to locate the cells where purported memory-affecting genes are active, and (2) it is difficult to confirm that 'learning' genes actually modify learning rather than subtly affecting the ability of the flies to respond because of motor, sensory, or other physiological defects.

In the 1970s, scientists in Seymour Benzer's laboratory discovered that *D. melanogaster* could be trained to avoid an odor associated with a shock (reviewed by Tully 1991). The learned avoidance lasted only a few hours, but that odor avoidance test was used to screen mutagenized flies for strains that had normal olfaction and aversion to shock, but an abnormally low ability to associate odors with shocks. The mutants obtained were considered to be poor learners, but had different phenotypes. One, *amnesiac*, had a nearly normal learning ability, but forgot rapidly.

Flies with the learning mutant *dunce* have a shortened memory for several different conditioned behaviors (Davis and Dauwalder 1991). *dunce* flies have a defective gene for an enzyme (cyclic adenosine monophosphate (cAMP)-specific phosphodiesterase) that regulates levels of cAMP, and *dunce* flies have elevated cAMP levels (Zhong and Wu 1991). cAMP is known, from studies on the sea slug *Aplysia*, to be a part of a 'second messenger' signaling pathway in nerve cells that help form associative memories. It is thought that *dunce* flies have impaired synaptic transmission because the excess of cAMP leads to hyper-polarization of the synaptic terminals, resulting in a chronically lowered availability of neurotransmitter at the synaptic terminals (Delgado et al. 1992).

The *dunce* gene is one of the largest identified to date in *Drosophila*, extending over 140 kb. It is unusually complex, producing, by the use of multiple transcription start sites, alternative splicing of exons, and differential processing of 3' sequences, at least eight to ten RNA molecules ranging in size from 4.2 to 9.5 kb. One intron is unusually large, 79 kb in size, and contains at least two genes (*Sgs-4* and *Pig-1*) within it (Chen et al. 1987, Qiu et al. 1991). This arrangement is unusual because it is uncommon for introns to contain functional genes. Also, it is not known whether these two genes affect expression of *dunce*.

One contained gene is the glue protein gene *(Sgs-4)*, which is expressed in larval salivary glands and which provides the glue used by larvae to attach themselves to the surface for pupariation. *Sgs-4* is transcribed in the same direction as *dunce*. The second gene, *pig-1 (pre-intermolt gene)*, also is expressed in larval salivary glands, but is transcribed in the opposite direction.

This 'genes within genes' organization is not unique, but this is the most elaborate example known. Little is known about the biological significance of the interspersed organization or the mechanism by which the genes became interspersed. It is possible that the *Sgs-4* and *Pig-1* genes invaded *dunce* by transposition. Alternatively, *dunce* may have increased in size over evolutionary time, including adjacent genes as more upstream promoters were recruited. Whether there is any relationship between the expression of *dunce* and the two genes within *dunce* is not known. *dunce* is apparently a conserved gene, because genes homologous to *dunce* have been identified in mice, rats, and humans. The evidence to date suggests that the mammalian counterparts of *dunce* function in the biochemistry underlying behavior, specifically in regulating mood (Tully 1991a).

The *dunce* gene is expressed in the **mushroom body** of the brain of *D. melanogaster* (Figure 12.5). There are two mushroom bodies in the brain of *D. melanogaster*, each consisting of about 2,500 neurons that send dendrites into a neuropil just ventral to the perikarya, where inputs arrive from the antennal lobes and, in some insects, from visual and other sensory systems (Davis 1993). Mushroom bodies receive olfactory information from the antennal lobes through their dendrites located in the calyx, a region of the brain just ventral to the mushroom body. They are important for olfactory learning and memory and are in some ways analogous to the hippocampus in mammals, which is thought to integrate sensory input and help form memories.

Mushroom bodies in the Hymenoptera are particularly large, which may reflect the importance of the mushroom bodies for functions underlying social ance memory in normal *D. melanogaster* was produced by conditioning of larvae. Normal adults retained such memories, but flies with *dunce* and *amnesiac* mutations did not learn as larvae or retain memories as adults, suggesting that long-term odor memory formation is influenced by these mutations (Tully et al. 1994). Associative odor learning in *D. melanogaster* adults was abolished completely when mushroom bodies were chemically ablated by feeding hydroxyurea to newly hatched larvae (de Belle and Heisenberg 1994).

Mushroom bodies in the Hymenoptera are particularly large, which may reflect the importance of the mushroom bodies for functions underlying social behavior and learning and memory in the honey bee (Rybak and Menzel 1993). Insect learning probably requires other brain centers, including the antennal lobes, the central complex, and the lateral protocerebrum in insects (Davis 1993).

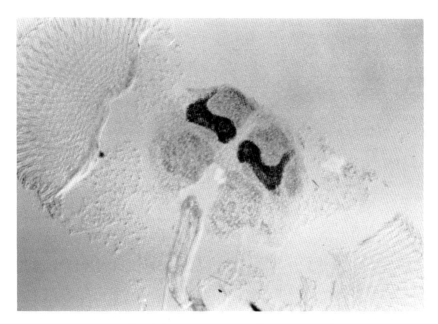

**Figure 12.5** Photograph of the brain structures known as mushroom bodies in *D. melanogaster*. The dark areas show the mushroom bodies stained with an antibody to the *dunce*-encoded enzyme. *dunce* encodes cAMP phosphodiesterase, which destroys cAMP, which is important in learning and memory. (Photograph provided by R. L. Davis.)

The mushroom bodies of the *Drosophila* brain can be stained with an antibody to the *dunce*-encoded protein (Figure 12.5). The activity of these genes has been identified by a technique called the **enhancer trap** (O'Kane and Gehring 1987). The technique involves placing a **reporter gene** (such as β-*galactosidase*, which turns the fly's brain tissues blue when the substrate is added) into the P element with the β-*galactosidase* gene under the control of a weak constitutive promoter. When this P element is brought in proximity to a tissue-specific enhancer element after insertion into the chromosome, the β-*galactosidase* gene will be regulated by the 'native' tissue-specific enhancer, resulting in the expression of the β-*galactosidase* reporter gene in a tissue- and stage-specific pattern. Ideally, the β-*galactosidase* gene will be expressed in a manner similar to that of the native gene normally controlled by the native enhancer. The enhancer trap technique is generally useful for identifying genes associated with behavior. For example, it has been used to identify genes expressed in both larval and adult olfactory organs (Riesgo-Escovar et al. 1992) and to identify cells that give rise to sensory structures of *D. melanogaster* (Blair et al. 1992).

To determine what genes are expressed in the mushroom body, the brains of flies are screened to see which have blue areas. Analysis of mutant flies produced in this manner identified some fifty mutants that affected learning, including several alleles of the *rutabaga* gene. Subsequently, *rutabaga* mutations were found to cause decreased expression of cAMP and the *rutabaga* protein product was identified as an adenylate cyclase. The disruption of conditioned behavior by *dunce* and *rutabaga* through their effects on the cAMP signaling system could be due to an alteration in the activity of molecules directly modulated by cAMP (i.e., cAMP-activated ion channels) or to alterations in the activity of protein kinase A, the major mediator of cAMP action (Davis 1993). Both *dunce* and *rutabaga* are expressed in the calyces and lobes of the mushroom body cells (Han et al. 1992). This evidence supports the hypothesis that cAMP is important in *Drosophila* olfactory memory, as well as the importance of the mushroom bodies as centers for olfactory learning.

The morphology of the brain also may be altered in learning mutants. Zhong et al. (1992) found that increased numbers of branches were found at larval motor axon terminals in *dunce* flies, but these branches were decreased in flies with both the *dunce* and *rutabaga* mutations. In both the rat and *Aplysia*, nerve terminal projections and synaptic connectivity are thought to be modified in association with learning processes. Thus, the fine tuning of morphological structures may be defective in *Drosophila* memory mutants.

*D. melanogaster* carrying the memory and learning mutant *turnip* have difficulty in olfactory discrimination, larval learning, conditioning of leg position, visual learning, and reward learning. The *turnip* gene is located on the X chromosome and is associated with reduced protein kinase C activity (Choi et al. 1991). Apparently, *turnip* is required for activating protein kinase C. Specifically, the *turnip* mutant is defective in phosphorylation of pp76, a membrane protein localized in head tissues. Protein phosphorylations have been implicated repeatedly in changes underlying learning and short-term memory.

Additional genes, including *radish, amnesiac, cabbage, latheo,* and *linotte,* that are involved in abnormal learning or memory are being discovered and characterized. For example, flies with the X-linked *radish* mutation initially learn in olfactory discrimination tests, but their subsequent memory decays rapidly at both early and late times after learning. The *radish* mutant flies show normal locomotor activity and sensitivity to odor cues and electric-shock reinforcements used in the learning tests. Anesthesia-resistant memory, or consolidated memory, is strongly reduced in flies with the *radish* phenotype, suggesting that this gene is involved in memory consolidation (Folkers et al. 1993).

The analysis of single gene mutations in *D. melanogaster* is providing a method for dissecting the physiology and genetics of behavior. While single-gene mutations usually have pleiotropic effects that complicate the study of whole animal behavior, they are providing powerful tools for discovering the fundamental mechanisms of learning and memory in insects. It is likely that the

enlightenment obtained from *Drosophila* learning mutants will provide an understanding of learning in higher organisms also (Tully 1991a,b).

# Neurobiochemistry of *Drosophila*

Molecular neurobiology is concerned about how the nervous system controls behavior at the molecular level (Glover and Hames 1989). What are the
biochemical substrates of behavior? A molecular genetic approach using *Drosophila* is providing interesting answers for both insects and mammals. For example, a potassium channel gene family was cloned first from *Drosophila* and subsequently from humans and mice using the *Drosophila* probes. The *Shaker*, *Shal*,
*Shab*, and *Shaw* subfamilies of the $K^+$ channel gene family have been found in
the Chordata, Arthropoda, and Mollusca, suggesting that the ancestral generic
$K^+$ channel gene had already given rise to these subfamilies by the time of the
Cambrian radiation (Salkoff et al. 1992).

A number of enzymes and receptors are involved in neurobiology, including
receptors for neurotransmitters and hormones, ion channel proteins and associated signal transduction components, brain-specific protein kinases, enzymes for
transmitter synthesis, neuropeptide processing enzymes, neuron-specific growth
or survival trophic factors and their receptors, inhibitors of neuronal growth,
glial-specific growth factors and their receptors, proteins associated with memory, neuronal cytoskeleton and axonal transport proteins, and others not listed
here or yet to be identified. A major endeavor in molecular neurobiology involves establishing the primary structure of all the categories of proteins involved in nerve signal reception and transmission (Barnard 1989). Based on
primary protein structure, families and superfamilies of proteins can be recognized, which provides possible insight into the roles of related molecules and
potential access to the genes that encode them.

## Electrical Signaling

The nervous system receives information about its internal and external
environment, processes this information, and produces an appropriate response.
The signaling of nerve cells depends on the electrical status of their outer
membranes. Nerve cells maintain a potential difference across the membrane
with the inside of the cell negative relative to the outside. The resting nerve cell
also maintains concentration gradients of sodium, calcium, and potassium ions.
Sodium and calcium ions are at a relatively high concentration outside the cell
while potassium ion levels are relatively high inside the cell. Signaling then
involves a change in the resting membrane potential which is brought about by
charge transfers carried by ionic fluxes through gated pores formed by transmembrane proteins called **channels**.

Ion channel proteins catalyze the transmembrane flow of ionic charge by forming narrow, hydrophilic pores through which ions can diffuse passively (Miller 1991). **Ion channels** must open or close rapidly in response to biological signals (gating). Furthermore, the open pore is generally selective and will determine which ions will permeate and which will not (**ionic selectivity**). Thus, a specific channel will permit $K^+$ but not $Na^+$ to pass, even though these ions are not geometrically elaborate structures and are thus not recognized specifically by enzymes.

Stimuli from the environment are perceived by specialized nerve cells (sensory cells). Each type of sensory cell responds to a particular stimulus such as light, sound, touch, heat, or chemicals such as pheromones. These sensory cells transform and amplify the energy provided by a stimulus into an electrical signal (sensory transduction). **Sensory transduction** is probably due to an alteration in the ionic permeability of the sensory cell membrane, which causes a depolarization of the membrane of the sensory cell from its resting level. The amplitude and duration of this departure generally increases logarithmically with the intensity of the stimulus. This signal is local and is not transmitted along the nerve cell; however, it acts as a stimulus to the axon and if the depolarization increases over a threshold level, the signal will trigger a change in **action potential** in the axon. Action potentials are all-or-nothing electrical impulses that propagate without distortion or attenuation along the entire length of an axon.

The generation and propagation of an action potential alters ionic conditions within the cell. When axonal membranes are depolarized, sodium channels open and allow sodium ions to flow down their gradient into the cell, producing the depolarizing phase of an action potential. Within milliseconds after the sodium channels are opened, they are inactivated, but at about the same time the membrane depolarization activates potassium channels, and the reciprocal $K^+$ flow repolarizes the cell and restores the membrane resting potential. During the course of an action potential, the sodium currents in one region of the axon membrane cause the depolarization and firing of an action potential in an adjacent region of the membrane so that the action potential is propagated along the full length of the axon.

The electrical signal is transmitted between cells at special sites called **synapses**, which occur between two nerve cells as well as between nerve cells and effectors such as muscle cells. The signal is relayed by a chemical neurotransmitter which is packaged in membrane-bound vesicles. When an action potential reaches the presynaptic terminal, the depolarization activates calcium channels in the presynaptic membrane and the subsequent influx of calcium ions leads to the release of neurotransmitter. The neurotransmitter diffuses to the postsynaptic cell and interacts with specific receptors on that cell surface. Receptors are activated in response to binding of the specific neurotransmitter molecules. Generally the size and duration of a synaptic potential reflects the amount of transmitter released by the presynaptic terminal. By depolarizing the postsynaptic cell above the threshold, the synaptic potential triggers the generation of an action potential, which continues the signaling one step further along the neural pathway (Ganetzky and Wu 1989).

## Neurotransmitters

Acetylcholine (ACh) is the major neurotransmitter in the central nervous system of *Drosophila* and other insects. Choline acetyltransferase (ChAT) is the biosynthetic enzyme, and acetylcholinesterase (AChE) is the degradative enzyme. AChE terminates synaptic transmission by rapidly hydrolyzing acetylcholine. Both enzymes are found in the *Drosophila* central nervous system and both have been isolated and cloned (Ganetzky and Wu 1989, Fournier et al. 1989, Wu et al. 1993).

The *acetylcholinesterase* gene, *Ace*, has been cloned in *Drosophila*. The transcription unit is 34 kb long and is split into ten exons, with the splicing sites of the two last exons precisely conserved among *Drosophila* and vertebrate cholinesterases (Fournier et al. 1989). The deduced mature *Ace* transcript is 4291 nucleotides long. The differential processing of transcripts may produce different subunit types. Fournier et al. (1989) suggest that the vertebrate cholinesterase organization is representative of the ancestral gene and that the introns found in *Drosophila* appeared later in evolution.

A gene for an acetylcholine receptor subunit has also been identified and cloned. Interestingly, the amino acid sequence of this AChR shares a high homology with vertebrate sequences.

## Sodium Channels

Two types of ion channels, permeable to sodium ($Na^+$) or potassium ions ($K^+$), are known to be responsible for most, if not all, membrane electrical phenomena. Sodium channels are critical in the generation and propagation of action potentials in nerve cells. They are being studied in *Drosophila* using two different approaches. In one, cDNA probes are used to clone genes that are homologous to the genes encoding the $\alpha$-subunit of vertebrate sodium channels. In the second, mutations affecting sodium channels are isolated and characterized and are subsequently cloned with these two strategies.

The sodium and potassium channels are encoded by multigene families, and the sodium channels are extensive and diverse. Several genes that affect sodium channels have been cloned (Salkoff et al. 1987), including *nap^ts* (*no action potential, temperature-sensitive*) and *para* (*paralytic*). The *nap^ts* gene affects the level of sodium channel activity and, at high temperatures, causes paralysis associated with a loss of action potentials (Kernan et al. 1991). The mutation *para^ts* is a temperature-sensitive mutation that causes instantaneous paralysis of adults at 29°C and of larvae at 37°C, and encodes sodium channels (Loughney et al. 1989). Molecular analyses indicate that the *para* locus encodes a protein that shares regions of extensive amino acid similarity with the α-subunit of vertebrate sodium channels.

## Potassium Channels

At least four distinct types of potassium channels have been found in *Drosophila* that mediate early and late voltage-dependent currents, and early and late calcium-dependent currents, respectively. Mutations of several different genes *(Shaker, Shal, Shab,* and *Shaw)* that alter behavior have been shown to alter potassium currents (Covarrubias et al. 1991). One of the best-studied is *Shaker (Sh)*.

As noted earlier in the chapter, *D. melanogaster* carrying the *Shaker* allele exhibit aberrant behavior, shaking their legs when anesthetized with ether. The underlying mechanism for this phenotype has been determined using molecular analyses. Voltage-clamp experiments in pupal and larval muscle showed that flies with the *Shaker* mutation have a fast, transient K$^+$ current. *Shaker* has been cloned and shown to code for a K$^+$ channel of the transient type that is present in both nerves and muscles (Papazian et al. 1987). Furthermore, at least four probable components of K$^+$ channels are encoded by *Shaker* by alternatively spliced transcripts.

Flies in which the *Shaker* gene is deleted still have K$^+$ currents similar to those coded by the *Shaker* gene, which suggested that K$^+$ channel proteins also are encoded by other genes. Butler et al. (1989) used a cDNA probe for *Shaker* and low stringency hybridization of a cDNA library to isolate three additional family members—*Shab, Shaw,* and *Shal*. These genes are organized similarly to *Shaker* in that only a single domain containing six presumed membrane-spanning segments is coded by each mRNA. This suggests that the diversity in K$^+$ channels found in different cells and tissues could be due to an extended gene family, as well to alternate splicing of the primary transcripts. These four genes appear to define four separate K$^+$ channel subfamilies in *Drosophila* and homologous genes isolated from vertebrates all appear to fall into one of these four subclasses.

Other potassium channel genes, including *ether-a-go-go (eag)* (Warmke et al. 1991, Bruggemann et al. 1993) and a calcium-activated potassium channel gene *(slo)* (Atkinson et al. 1991), have been isolated. The *eag* gene shares sequence similarities with several ionic channel proteins. It is likely that the eag protein represents a new type of ion channel that can mediate potassium-outward and calcium-inward currents, making it important in modulating synaptic efficiency in both central and peripheral nervous systems (Bruggemann et al. 1993). It is likely that potassium channels comprise a large and diverse group of integral membrane proteins. Another neurotransmitter, γ-aminobutyric acid (GABA), is a major inhibitory agent in the insect nervous system. The synthesis of GABA is controlled by the enzyme glutamic acid decarboxylase (GAD). A mammalian GAD cDNA probe was used to isolate cDNA from *Drosophila melanogaster* (Jackson et al. 1990).

*Drosophila* is perhaps unique among eukaryotes in the variety and sophistication of techniques that can be used to understand its neurobiology. Rapid progress is being made in understanding the genetic, molecular, and electrophysiological components of insect neurobiology.

## Sex Pheromone Communication in Moths

Pheromones are chemicals that are released by organisms into the environment, where they serve as messages to other individuals of the same species. Sex pheromones are widely used by insects to attract a mating partner or to stimulate that partner to copulate. Insect sex pheromones are used by humans either to detect the distribution and abundance of pest insects, or to disrupt their normal mating behavior in the hope that the insects will no longer be able to reproduce and damage plants (Shorey and McKelvey 1977). They also may be used to attract pest insects to poisoned baits, thus providing direct control.

Insects can detect extremely low concentrations of sex pheromones, sometimes from long distances. How they do so is beginning to be understood at the molecular level. For example, the antennae of a male *Antheraea polyphemus* moth have about 55,000 sensory hairs, with each sensillum containing the pheromone-sensitive dendrites of two to three sensory cells bathed in sensillum lymph—which is a proteinaceous fluid. Two soluble proteins are found in the lymph, a pheromone-degrading esterase and a pheromone-binding protein. The pheromone-binding proteins bind species-specific pheromones and are present in very high concentrations. Volatile hydrophobic odorant molecules have to enter an aqueous compartment and traverse a hydrophilic barrier before reaching olfactory neurons. The function of the pheromone binding proteins is not fully resolved (van den Berg and Ziegelberger 1991), although they are thought to be involved in carrying the hydrophobic odorant through the sensillum lymph toward the receptor proteins located in the dendrite membranes. Pheromone-binding proteins from a number of moths have been identified, including *Heliothis zea, Manduca sexta, Lymantria dispar, Antheraea pernyi,* and *A. polyphemus* (Raina et al. 1989, Raming et al. 1990).

Raming et al. (1989) isolated pheromone-binding protein from the antennae of male *A. polyphemus* moths and developed a synthetic probe to screen a cDNA library derived from antennae of these males. The complete sequence of the gene encoding this protein consists of 1186 nucleotides and the open reading frame encodes a 163-amino-acid protein. The mature protein consists of 142 amino acid residues. The primary structure of the pheromone-binding protein shows no similarity to odorant-binding proteins from vertebrates, suggesting that soluble odorant-binding proteins in insects and vertebrates evolved independently or that there are several classes of such proteins (Raming et al. 1990)

## Mapping Genes That Influence Complex Behaviors

Many behavioral attributes in insects are probably determined by many genes. Furthermore, behavior is determined, not only by genetic composition, but also by the environment. The effects of the environment may mask those of the genotype, making genetic analysis imprecise. While elaborate statistical techniques have been developed for quantitative genetic analyses (Falconer 1989), a thorough understanding of behaviors that are quantitatively determined has been slow to develop. Understanding the mechanisms of behavior determined by single genes is hard work, but relatively straight forward. Understanding the mechanisms of behavior determined by many genes requires ingenuity and extensive effort. At present, there are no molecular genetic analyses of polygenically determined insect behaviors.

Two molecular genetic techniques offer a new approach to studying behaviors that are polygenic and influenced by environmental forces. For example, **restriction fragment length polymorphism (RFLP)** mapping techniques are applicable to species for which little formal genetic information is available— which includes the majority of insects! RFLPs are short DNA sequences that vary in length from one individual to the next and can be mapped to specific chromosome locations. The variation in length is detected by cutting DNA with a specific restriction enzyme that recognizes a specific sequence. If different bands are produced in different individuals, these variations are due to mutations that create or abolish recognition sites for the restriction enzyme.

RFLPs may provide markers that can be used to identify the genetic bases of behavior determined by many genes if the behavioral traits segregate with the RFLP markers. If an RFLP marker and a behavior of interest are consistently inherited together, it is likely that they are located close together on the same chromosome. Mapping with RFLP markers requires that single-copy DNA be cloned from the species of interest and used as a probe to follow the segregation of homologous regions of the genome in individuals from segregating populations such as $F_2$ or backcross progeny. Many different clones of single-copy genes can be tested, and genetic linkage maps can be constructed that contain a large number of RFLP markers at close intervals. RFLP markers are typically codominant, allowing the genotype of a locus to be determined in individuals derived from any mating scheme. With RFLPs, it should be possible to develop markers for the entire genome.

RFLP markers have been used to identify polygenic traits useful in plant and animal breeding (Beckman and Soller 1987, Tanksley et al. 1989). For example,

RFLP markers have located genes in tomato, including a gene coding for fruit mass; four genes for soluble solids concentration; and five genes for acidity. RFLP probes from tomato were used to develop a molecular map of potato (Tanksley et al. 1989). RFLP analysis makes it possible to reduce complex or polygenic traits to their individual genetic components, and to study the individual and combined effects of these polygenes in different genetic backgrounds. Furthermore, RFLP analysis allows breeders to transfer traits into new crop varieties more rapidly because it is possible to monitor traits and closely linked RFLP markers and thus select individuals with the desired genes more efficiently.

The level of variation for RFLP markers in natural insect populations is large and the majority of RFLP markers are thought to be phenotypically neutral. RFLP markers that are found tightly linked to traits of interest could allow researchers to identify the number and locations of specific loci associated with a specific behavior in insects. Once located, it may be possible to clone these genes by chromosome walking.

A polymerase chain reaction (PCR) method called RAPD (random amplified polymorphic DNA) also may be useful for identifying genes that influence complex traits in insects (see Chapter 9 for a description of the technique). DNA bands produced by RAPD-PCR could also provide potential markers with which to identify the number and locations of specific loci associated with specific behaviors in populations with different behaviors. RAPD-PCR has been used, for example, to identify and localize specific disease resistance genes in plants (Michelmore et al. 1991). An advantage of RAPD-PCR is that no information about the sequence of specific genes is required to obtain DNA markers.

Analysis of insect behavior using insect molecular genetic approaches is still in its infancy. The disciplines of ethology, genetics, physiology, and molecular biology have come together and promise to revolutionize the study of insect behavior and neurobiology.

## References Cited

Alcock, J. 1984. *Animal Behavior: An Evolutionary Approach*. Sinauer Assoc., Sunderland, MA.

Ashburner, M. 1989. Drosophila: A *Laboratory Handbook*. Cold Spring Harbor Laboratory Press, Cold Spring Harbor, NY.

Atkinson, N. S., G. A. Robertson, and B. Ganetzky. 1991. A component of calcium-activated potassium channels encoded by the *Drosophila slo* locus. Science 253: 551–555.

Bargiello, T. A., F. R. Jackson and M. W. Young. 1984. Restoration of circadian behavioural rhythms by gene transfer in *Drosophila*. Nature 312: 752–754.

Barnard, E. A. 1989. Molecular neurobiology—An introduction. Pp. 1–7. In: *Molecular Neurobiology*. D. M. Glover and B. D. Hames, Eds. IRL Press, Oxford.

Barton Browne, L. 1993. Physiologically induced changes in resource-oriented behavior. Annu. Rev. Entomol. 38: 1–25.

Beck, S. D. 1980. *Insect Photoperiodism*. 2nd ed. Academic Press, New York.

Beckman, J. S. and M. Soller. 1987. Molecular markers in the genetic improvement of farm animals. Bio/Technology 5: 573–576.

Bell, W. J. 1990. Searching behavior patterns in insects. Annu. Rev. Entomol. 35: 447–467.

Bellen, H. J., S. Kooyer, D. D'Evelyn and J. Pearlman. 1992. The *Drosophila* couch potato protein is expressed in nuclei of peripheral neuronal precursors and shows homology to RNA-binding proteins. Genes Dev. 6: 2125–2136.

Benzer, S. 1973. Genetic dissection of behavior. Sci. Am. 229: 24–27.

Blair, S. S., A. Giangrande, J. B. Skeath and J. Palka. 1992. The development of normal and ectopic sensilla in the wings of *hairy* and *Hairy wing* mutants of *Drosophila*. Mech. Dev. 38: 3–16.

Blomquist, G. J., J. W. Dillwith and T. S. Adams. 1987. Biosynthesis and endocrine regulation of sex pheromone production in Diptera. Pp. 217–243. In: *Pheromone Biochemistry*. G. Prestwich and G. J. Blomquist, Eds. Academic Press, San Diego.

Brady, J. P. and R. C. Richmond. 1990. Molecular analysis of evolutionary changes in the expression of *Drosophila* esterases. Proc. Natl. Acad. Sci. USA 87: 8217–8221.

Bruggemann, A., L. A. Pardo, W. Stuhmer and Olaf Pongs. 1993. *Ether-a-go-go* encodes a voltage-gated channel permeable to $K^+$ and $Ca^{2+}$ and modulated by cAMP. Nature 365: 445–448.

Butler, A., A. Wei, K. Baker and L. Salkoff. 1989. A family of putative potassium channel genes in *Drosophila*. Science 243: 943–947.

Chalmers, D. E. and C. P. Kyriacou. 1993. Glowing reports on biorhythm research. BioEssays 15: 755–756.

Chen, C., T. Malone, S. K. Beckendorf and R. L. Davis. 1987. At least two genes reside within a large intron of the *dunce* gene of *Drosophila*. Nature 329: 721–724.

Choi, K. W., R. F. Smith, R. M. Buratowski and W. G. Quinn. 1991. Deficient protein kinase C activity in *turnip*, a *Drosophila* learning mutant. J. Biol. Chem. 266: 15999–16006.

Citri, Y., H. V. Colot, A. C. Jacquier, Z. Yu, J. C. Hall, D. Baltimore and M. Rosbash. 1987. A family of unusually spliced biologically active transcripts encoded by a *Drosophila* clock gene. Nature 326: 42–47.

Colvin, J. and A. G. Gatehouse. 1993. Migration and genetic regulation of the prereproductive period in the cotton-bollworm moth, *Helicoverpa armigera*. Heredity 70: 407–412.

Costa, R. A. A. Peixoto, G. Barbujani and C. P. Kyriacou. 1992. A latitudinal cline in a *Drosophila* clock gene. Proc. R. Soc. Lond. B 250: 43–49.

Covarrubias, M., A. Wei and L. Salkoff. 1991. *Shaker, Shal, Shab*, and *Shaw* express independent $K^+$ current systems. Neuron 7: 763–773.

Davis, R. L. 1993. Mushroom bodies and *Drosophila* learning. Neuron 11: 1–14.

Davis, R. L. and B. Dauwalder. 1991. The *Drosophila dunce* locus: Learning and memory genes in the fly. Trends Genet. 7: 224–229.

de Belle, J. S. and M. Heisenberg. 1994. Associative odor learning in *Drosophila* abolished by chemical ablation of mushroom bodies. Science 263: 692–695.

Delgado, R., R. Latorre and P. Labarca. 1992. $K^+$-channel blockers restore synaptic plasticity in the neuromuscular junction of *dunce*, a *Drosophila* learning and memory mutant. Proc. R. Soc. Lond. B 250: 181–185.

Dingle, H., Ed. 1978. *Evolution of Insect Migration and Diapause*. Springer-Verlag, New York.

Dingle, H. and J. P. Hegmann, Eds. 1982. *Evolution and Genetics of Life Histories*. Springer-Verlag, New York.

Dunlap, J. C. 1993. Genetic analysis of circadian clocks. Annu. Rev. Physiol. 55: 683–728.

Ehrman, L. and P. A. Parsons. 1973. *The Genetics of Behavior*. Sinauer Assoc., Sunderland, MA.

Ehrman, L. and P. A. Parsons. 1981. *Behavior Genetics and Evolution*. McGraw-Hill, New York.

Falconer, D. S. 1989. *Introduction to Quantitative Genetics*. 3rd Ed. Longman, New York.

Ferveur, J. F. and J. M. Jallon. 1993. Genetic control of pheromones in *Drosophila simulans*. II. *kete*, a locus on the X chromosome. Genetics 133: 561–567.

Folkers, E., P. Drain and W. G. Quinn. 1993. radish, a Drosophila mutant deficient in consolidated memory. Proc. Natl. Acad. Sci. USA 90: 8123–8127.

Fournier, D., F. Karch, J. M. Bride, L. M. C. Hall, J. B. Berge and P. Spierer. 1989. Drosophila melanogaster acetylcholinesterase gene structure, evolution and mutations. J. Mol. Biol. 210: 15–22.

Gailey, D. A., A. Villella and T. Tully. 1991. Reassessment of the effect of biological rhythm mutations on learning in Drosophila melanogaster. J. Comp. Physiol. A 169: 685–97.

Ganetzky, B. and C.F. Wu. 1989. Molecular approaches to neurophysiology in Drosophila. Pp. 9–61. In: Molecular Neurobiology. D. M. Glover and B. D. Hames, Eds. IRL Press, Oxford.

Gatehouse, A. G. 1989. Genes, environment and insect flight. Pp. 116–138. In: Insect Flight. G. J. Goldsworthy and C. Wheeler, Eds. CRC Press, Boca Raton, FL.

Glover, D. M. and B. D. Hames, Eds. 1989. Molecular Neurobiology. IRL Press, Oxford.

Gould, F. 1986. Genetic constraints on the evolution of cannibalism in Heliothis virescens. Pp. 55–62. In: Evolutionary Genetics of Invertebrate Behavior: Progress and Prospects. M. D. Huettel, Ed. Plenum, New York.

Grossfield, J. 1975. Behavioral mutants of Drosophila. Pp. 679–702. In: Handbook of Genetics. Vol. 3. Invertebrates of Genetic Interest. R. C. King, Ed. Plenum, New York.

Hall, J. C. 1984. Complex brain and behavioral functions disrupted by mutations in Drosophila. Dev. Genet. 4: 355–532.

Hall, J. C. 1985. Genetic analysis of behavior in insects. Pp. 287–373. In: Comprehensive Insect Physiology Biochemistry and Pharmacology. Vol. 9. G. A. Kerkut and L. I. Gilbert, Eds. Pergamon, New York.

Hall, J. C. 1990. Genetics of circadian rhythms. Annu. Rev. Genet. 24: 659–697.

Han, P. L., L. R. Levin, R. R. Reed and R. L. Davis. 1992. Preferential expression of the Drosophila rutabaga gene in mushroom bodies, neural centers for learning in insects. Neuron 9: 619–627.

Hardin, P. E., J. C. Hall and M. Rosbash. 1992. Circadian oscillations in period gene mRNA levels are transcriptionally regulated. Proc. Natl. Acad. Sci. USA 89: 11711–11715.

Hay, D. A. 1985. Essentials of Behaviour-Genetics. Blackwell, Palo Alto, CA.

Higa, I. and Y. Fuyama. 1993. Genetics of food preference in Drosophila sechellia. Genetica 88: 129–136.

Holman, G. M., R. J. Nachman and M. S. Wright. 1990. Insect neuropeptides. Annu. Rev. Entomol. 35: 201–217.

Homberg, U. 1987. Structure and functions of the central complex in insects. Pp. 347–367. In: Arthropod Brain: Its Evolution, Development, Structure and Functions. A. P. Gupta, Ed. Wiley, New York.

Homberg, U., T. A. Christensen and J. G. Hildebrand. 1989. Structure and function of the deutocerebrum in insects. Annu. Rev. Entomol. 34: 477–501.

Hotta, Y. and S. Benzer. 1972. Mapping of behaviour in Drosophila mosaics. Nature 240: 527–535.

Hoy, M. A. 1978. Variability in diapause attributes of insects and mites: Some evolutionary and practical implications. Pp. 101–126. In: Evolution of Insect Migration and Diapause. H. Dingle, Ed. Springer-Verlag, New York.

Huber, F., T. E. Moore and W. Loher, Eds. 1989. Cricket Behavior and Neurobiology. Cornell Univ. Press, Ithaca, NY.

Huettel, M. D., Ed. 1986. Evolutionary Genetics of Invertebrate Behavior: Progress and Prospects. Plenum, New York.

Huettel, M. D. and G. L. Bush. 1972. The genetics of host selection and its bearing on sympatric speciation in Procecidochares (Diptera: Tephritidae). Entomol. Exp. Appl. 15: 465.

Jackson, F. R., L. M. Newby and S. J. Kulkarni. 1990. Drosophila GABAergic systems: Sequence and expression of glutamic acid decarboxylase. J. Neurochem. 54: 1068–1078.

Jaenike, J. 1986. Genetic complexity of host selection behavior in Drosophila. Proc. Natl. Acad. Sci. USA 83: 2148–2151.

Kalmring, K. and N. Elsner, Eds. 1985. Acoustic and Vibrational Communication in Insects. Paul Parey, Hamburg.

Kernan, M. J., M. I. Kuroda, R. Kreber, B. S. Baker and B. Ganetzky. 1991. $nap^{ts}$, a mutation affecting sodium channel activity in Drosophila, is an allele of mle, a regulator of X chromosome transcription. Cell 66: 949–959.

Klun, J. A. and M. D. Huettel. 1988. Genetic regulation of sex pheromone production and response: Interaction of sympatric pheromonal types of European corn borer, Ostrinia nubilalis (Lepidoptera: Pyralidae). J. Chem. Ecol. 13: 2047–2061.

Klun, J. A. and S. Maini. 1979. Genetic basis of an insect chemical communication system. Environ. Entomol. 8: 423–426.

Kyriacou, C. P. 1990. The molecular ethology of the period gene in Drosophila. Behav. Genet. 20: 191–211.

Kyriacou, C. P. 1993. Molecular genetics of complex behavior: Biological rhythms in Drosophila. Pp. 329–356. In: Molecular Approaches to Fundamental and Applied Entomology. J. Oakeshott and M. J. Whitten, Eds. Springer-Verlag, New York.

Kyriacou, C. P. and J. C. Hall. 1980. Circadian rhythm mutations in Drosophila melanogaster affect short-term fluctuations in the male's courtship song. Proc. Natl. Acad. Sci. USA 77: 6729–6733.

Kyriacou, C. P. and J. C. Hall. 1989. Spectral analysis of Drosophila courtship song rhythms. Anim. Behav. 37: 850–859.

Liu, X., L. J. Zwiebel, D. Hinton, S. Benzer, J. C. Hall and M. Rosbash. 1992. The period gene encodes a predominantly nuclear protein in adult Drosophila. J. Neurosci. 12:2735–2744.

Lofstedt, C. 1990. Population variation and genetic control of pheromone communication systems in moths. Entomol. Exp. Appl. 54: 199–218.

Lofstedt, C., B. S. Hansson, W. Roelofs and B. O. Bengtsson. 1989. No linkage between genes controlling female pheromone production and male pheromone response in the European corn borer, Ostrinia nubilalis Hubner (Lepidoptera: Pyralidae). Genetics 123: 553–556.

Loughney, K., R. Kreber and B. Ganetzky. 1989. Molecular analysis of the para locus, a sodium channel gene in Drosophila. Cell 58: 1143–1154.

Masler, E. P., T. J. Kelly and J. J. Menn. 1993. Insect neuropeptides: Discovery and application in insect management. Arch. Insect Biochem. Physiol. 22: 87–111.

Matthews, R. W. and J. R. Matthews. 1978. Insect Behavior. Wiley Interscience, New York.

McClung, C. R., B. A. Fox and J. C. Dunlap. 1989. The Neurospora clock gene frequency shares a sequence element with the Drosophila clock gene period. Nature 339:558–562.

McGuire, T. R. and J. Hirsch. 1977. Behavior-genetic analysis of Phormia regina: Conditioning, reliable individual differences, and selection. Proc. Natl. Acad. Sci. USA 74: 5193–5197.

McGuire, T. R. 1984. Learning in three species of Diptera: The blow fly, Phormia regina, the fruit fly, Drosophila melanogaster, and the house fly, Musca domestica. Behav. Genet. 14: 479–526.

McGuire, T. R. and T. Tully. 1987. Characterization of genes involved with classical conditioning that produce differences between bidirectionally selected strains of the blow fly Phormia regina. Behav. Genet. 17: 97–107.

Michelmore, R. W., I. Paran and A. R. V. Kellsli. 1991. Identification of markers linked to disease-resistance genes by bulked segregant analysis: A rapid method to detect markers in specific genomic regions by using segregating populations. Proc. Natl. Acad. Sci. USA 88: 9828–9832.

Miller, C. 1991. 1990: Annus mirabilis of potassium channels. Science 251: 1092–1096.

Mollema, C. 1991. Heritability estimates of host selection behaviour by the parasitoid Asobara tabida. Neth. J. Zool. 41: 174–183.

Moretto, G., L. S. Goncalves and D. De Jong. 1993. Heritability of africanized and european honey bee defensive behavior against the mite Varroa jacobsoni. Rev. Brasil. Genet. 16: 71–77.

Nassel, D. R. 1993. Neuropeptides in the insect brain: A review. Cell Tissue Res. 273: 1–29.

O'Kane, C. J. and Gehring, W. J. 1987. Detection *in situ* of genomic regulatory elements in *Drosophila*. Proc. Natl. Acad. Sci. USA 84: 9123–9127.

Palmer, J. O. and H. Dingle. 1989. Responses to selection on flight behavior in a migratory population of milkweed bug (*Oncopeltus fasciatus*). Evolution 43: 1805–1808.

Papaj, D. R. and R. J. Prokopy. 1989. Ecological and evolutionary aspects of learning in phytophagous insects. Annu. Rev. Entomol. 34: 315–350.

Papazian, D. M., T. L. Schwarz, B. L. Tempel, Y. N. Jan and L. Y. Jan. 1987. Cloning of genomic and complementary DNA from *Shaker*, a putative potassium channel gene from *Drosophila*. Science 237: 749–753.

Plomin, R. 1990. The role of inheritance in behavior. Science 248: 183–188.

Qiu, Y., C. N. Chen, T. Malone, L. Richter, S. K. Beckendorf and R. L. Davis. 1991. Characterization of the memory gene dunce of *Drosophila melanogaster*. J. Mol. Biol. 222: 553–565.

Raina, A. K., H. Jaffe, T. G. Kempe, P. Keim, R. W. Blacher, H. M. Fales, C. T. Riley, J. A. Klun, R. L. Ridgway and D. K. Hayes. 1989. Identification of a neuropeptide hormone that regulates sex pheromone production in female moths. Science 244: 796–798.

Raina, A. K. and J. J. Menn. 1993. Pheromone biosynthesis activating neuropeptide: From discovery to current status. Arch. Insect Biochem. Physiol. 22: 141–151.

Raming, K., J. Krieger and H. Breer. 1989. Molecular cloning of an insect pheromone-binding protein. FEBS Lett. 256: 215–218.

Raming, K., J. Krieger and H. Breer. 1990. Primary structure of a pheromone-binding protein from *Antheraea pernyi*: Homologies with other ligand-carrying proteins. J. Comp. Physiol. B 160: 503–509.

Riesgo-Escovar, J., C. Woodard, P. Gaines and J. Carlson. 1992. Development and organization of the *Drosophila* olfactory system: An analysis using enhancer traps. J. Neurobiol. 23: 9947–964.

R'Kha, S., P. Capy and J. R. David. 1991. Host-plant specialization in the *Drosophila melanogaster* species complex: A physiological, behavioral, and genetical analysis. Proc. Natl. Acad. Sci. USA 88: 1835–1839.

Richmond, R. C., S. D. Mane, and L. Tompkins. 1986. The behavioral effects of a carboxylesterase in *Drosophila*. Pp. 223–236. In: *Evolutionary Genetics of Invertebrate Behavior: Progress and Prospects*. M. D. Huettel, Ed. Plenum, New York.

Rinderer, T. E., Ed. 1986. *Bee Genetics and Breeding*. Academic Press, Orlando.

Rinderer, T. E. and A. M. Collins 1986. Behavioral genetics. Pp. 155–176. In: *Bee Genetics and Breeding*. T. E. Rinderer, Ed. Academic Press, Orlando.

Rothenbuhler, W. C. 1964. Behavior genetics of nest cleaning in honey bees. IV. Responses of $F_1$ and backcross generations to disease-killed brood. Am. Zool. 4: 111–123.

Rybak, J. and R. Menzel. 1993. Anatomy of the mushroom bodies in the honey bee brain: The neuronal connections of the alpha-lobe. J. Comp. Neurol. 334: 444–465.

Salkoff, L., A. Butler, A. Wei, N. Scavarda, K. Giffen, C. Ifune, R. Goodman and G. Mandel. 1987. Genomic organization and deduced amino acid sequence of a putative sodium channel gene in *Drosophila*. Science 237: 744–753.

Salkoff, L., K. Baker, A. Butler, M. Covarrubias, M. D. Pak and A. Wei. 1992. An essential 'set' of $K^+$channels conserved in flies, mice and humans. Trends Neurosci. 15: 161–166.

Sato, Y., M. Oguchi, N. Menjo, K. Imai, H. Saito, M. Ideda, M. Isobe and O. Yamashita. 1993. Precursor polyprotein for multiple neuropeptides secreted from the suboesophageal ganglion of the silkworm *Bombyx mori*: Characterization of the cDNA encoding the diapause hormone precursor and identification of additional peptides. Proc. Natl. Acad. Sci. USA 90: 3251–3255.

Saunders, D. S. 1990. The circadian basis of ovarian diapause regulation in *Drosophila melanogaster*: Is the *period* gene causally involved in photoperiodic time measurement? J. Biol. Rhythms 5: 315–331.

Scharrer, B. 1987. Insects as models in neuroendocrine research. Annu. Rev. Entomol. 32: 1–16.

Sehgal, A., J. Price and M. W. Young. 1992. Ontogeny of a biological clock in Drosophila melanogaster. Proc. Natl. Acad. Sci. USA 89: 1423–27.

Shorey, H. H. and J. J. McKelvey, Jr. 1977. Chemical Control of Insect Behavior. Theory And Application. Wiley-Interscience, New York.

Stocker, R. F. 1994. The organization of the chemosensory system in Drosophila melanogaster: A review. Cell Tissue Res. 275: 3–26.

Strausfeld, N. J. 1976. Atlas of an Insect Brain. Springer-Verlag, Berlin.

Strauss, R. and M. Heisenberg. 1993. A higher control center of locomotor behavior in the Drosophila brain. J. Neurosci. 13: 1852–1861.

Takahashi, J. S. 1992. Circadian clock genes are ticking. Science 258: 238–239.

Tanksley, S. D., N. D. Young, A. H. Paterson, M. W. Bonierbale. 1989. RFLP mapping in plant breeding: New tools for an old science. Bio/Technology 7: 257–264.

Tauber, M. J., C. A. Tauber and S. Masaki. 1986. Seasonal Adaptations of Insects. Oxford Univ. Press, Oxford.

Tazima, Y., H. Doira and H. Akai. 1975. The domesticated silkmoth. Pp. 63–1124. In: Handbook of Genetics. Vol. 3. Invertebrates of Genetic Interest. R. C. King, Ed. Plenum, New York.

Trpis, M. and W. Hausermann. 1978. Genetics of house-entering behaviour in East African populations of Aedes aegypti (L.) (Diptera: Culicidae) and its relevance to speciation. Bull. Entomol. Res. 68: 521–532.

Tully, T. 1991a. Genetic dissection of learning and memory in Drosophila melanogaster. Pp. 29–66. In: Neurobiology of Learning, Emotion and Affect. J. Madden IV, Ed. Raven Press, New York.

Tully, T. 1991b. Physiology of mutations affecting learning and memory in Drosophila—the missing link between gene product and behavior. Trends Neurosci. 14: 163–164.

Tully, T., V. Cambiazo and L. Kruse. 1994. Memory through metamorphosis in normal and mutant Drosophila. J. Neurosci. 14: 68–74.

van den Berg, M. J. and G. Ziegelberger. 1991. On the function of the pheromone binding protein in the olfactory hairs of Antheraea polyphemus. J. Insect Physiol. 37: 79–85.

Via, S. 1990. Ecological genetics and host adaptation in herbivorous insects: The experimental study of evolution in natural and agricultural systems. Annu. Rev. Entomol. 35: 421–426.

Warmke, J., R. Drysdale and B. Ganetzky. 1991. A distinct potassium channel polypeptide encoded by the Drosophila eag locus. Science 252: 1560–1562.

Wheeler, D. A., C. P. Kyriacou, M. L. Greenacre, Q. Yu, J. E. Rutila, M. Rosbash and J. C. Hall. 1991. Molecular transfer of a species-specific behavior from Drosophila simulans to Drosophila melanogaster. Science 251: 1082–1085.

Williams, K. D. and M. B. Sokolowski. 1993. Diapause in Drosophila melanogaster females: A genetic analysis. Heredity 71: 312–317.

Wu, D., N. Schormann, W. Lian, J. Deisenhofer and L. B. Hersh. 1993. Expression, purification, and characterization of recombinant Drosophila choline acetyltransferase. J. Neurochem. 61: 1416–1422.

Young, M. W., Ed. 1993. Molecular Genetics of Biological Rhythms. Dekker, New York.

Zerr, D. M., J. C. Hall, M. Rosbash and K. K. Siwicki. 1990. Circadian fluctuations of period protein immunoreactivity in the CNS and the visual system of Drosophila. J. Neurosci. 10: 2749–2762.

Zhong, Y. and C. F. Wu. 1991. Altered synaptic plasticity in Drosophila memory mutants with a defective cyclic AMP cascade. Science 251: 198–201.

Zhong, Y., V. Budnik and C. F. Wu. 1992. Synaptic plasticity in Drosophila memory and hyperexcitable mutants: Role of cAMP cascade. J. Neurosci. 12: 644–651.

# Chapter 13

# Insect Molecular Systematics and Evolution

## Overview

Systematics is the study of phylogeny and taxonomy. Taxonomy can be divided into descriptive taxonomy and identification. DNA analysis is suitable for systematics studies because it is the most direct analysis of the genetic material possible and is unlikely to be confounded by life-stage- or environmentally induced variability. Molecular techniques used include analyses of isozymes, molecular cytogenetics, immunology, DNA–DNA hybridization, restriction analyses of DNA sequences, and DNA sequencing. Each technique has virtues and limitations in the amount and type of information provided, its technical difficulties, and its costs.

There are several significant controversies associated with using molecular techniques to study systematics and evolution. These include debates over the relative importance of molecular versus morphological data, the constancy of the molecular clock for evaluating time from divergence of taxa, the proper use of the terms "homology" and "similarity," and the neutrality of DNA sequence variation. The immense diversity of insects and their long evolutionary history provide a challenge for systematists. The use of molecular data provides new opportunities to discern the long and diverse evolutionary histories of arthropods. With the use of molecular methods, systematists and population geneticists are beginning to use common approaches to study both intraspecific and interspecific genetic diversity.

## Introduction

A formalized hierarchical system of nomenclature was established by Linnaeus in 1758, and the notion that a classification of organisms should be based on phylogenetic, or evolutionary, relationships developed after Darwin's publication of *The Origin of the Species* in 1859. The identification, description, and explanation of the diversity of organisms is known as **systematics**.

Modern systematics encompasses the study of both **phylogeny** and microevolutionary change. Systematists and population geneticists often use common molecular methods to test hypotheses. These include analyses of isozymes, molecular cytogenetics, immunology, DNA–DNA hybridization, restriction analyses, or sequencing of nuclear and mitochondrial DNA sequences. This chapter introduces these methods and describes their applications, limitations, and relative costs for systematic and evolutionary studies.

Systematics studies conducted prior to the 1960s primarily used morphological and behavioral attributes as characters, although cytogenetic characters were also employed in some cases (Mayr 1970, White 1973, 1978). In the 1960s, electrophoresis of proteins began to provide new characters for analysis. Lewontin and Hubby (1966) demonstrated that protein-coding genes often are **polymorphic** (have more than one allele) and that gel electrophoresis of proteins could reveal the presence of functionally similar forms of enzymes or **isozymes**. Protein electrophoresis provides a relatively inexpensive method for analyzing several genes from 20–30 individuals at the same time (Hames and Rickwood 1981, Pasteur et al. 1988). This technique is useful for analyzing mating systems, heterozygosity, relatedness, geographic variation, hybridization, species boundaries, and phylogenetic analyses of divergences within the past 50 million years (Table 13.1).

Immunological analyses also were employed in the 1960s, but are less often used for analyses of insect systematics or evolution today (Table 13.1). Immunological techniques can provide qualitative or quantitative estimates of amino acid sequence differences between homologous proteins (Maxson and Maxson 1990).

Cytogenetic analyses of variation in chromosome structure and number have been used for studies of hybridization, species boundaries, and phylogenies of organisms that diverged 5–50 million years ago (mya) (White 1973). Specific DNA sequences can be localized by *in situ* hybridization and new staining techniques can reveal the fine structure of chromosomes by revealing banding patterns (Table 13.1).

Differences in single-copy DNA sequences revealed by **DNA–DNA hybridization** have been available for analysis of phylogenies since the late 1960s (Werman et al. 1990). DNA–DNA hybridization is employed for analyses of species and higher taxa relationships, up to the family and order level (Table 13.1).

Restriction enzyme analyses are used to analyze differences in DNA sequences. **Restriction fragment length polymorphism (RFLP) analysis** is a very versatile technique, although it is more expensive and laborious than DNA–DNA hybridization (Table 13.1). RFLP analyses provide information on the

Table 13.1

Applications of Various Molecular Techniques to Systematics Problems[a]

| Problem | Isozymes[b] | Immunology[b] | Cytogenetics[b] | DNA–DNA hybridization[b] | Restriction analysis[b] | DNA/RNA sequencing[b] |
|---|---|---|---|---|---|---|
| Mating systems | A | I | M | I | M | $ |
| Clonal detection | A | I | M | I | A | $ |
| Heterozygosity | A | I | I | I | A | M |
| Paternity testing | M | I | I | I | A | A |
| Relatedness | M | I | I | I | A | $ |
| Geographic variation | A | I | M | I | A | A |
| Hybridization | A | I | A | I | A | M |
| Species boundaries | A | I | A | I | A | $ |
| Phylogeny (0–5 mya) | A | M | M | M | A | A |
| Phylogeny (5–50 mya) | A | A | A | A | A | A |
| Phylogeny (50–500 mya) | M | M | M | M | M | A |
| Phylogeny (500–3500 mya) | I | I | I | I | I | A |

[a]Modified from Hills and Moritz 1990.
[b]I, inappropriate use of the technique; M, marginally appropriate or appropriate under limited circumstances; $, appropriate use of technique but probably not cost effective; A, appropriate and effective method.

nature as well as the extent of differences between sequences in nuclear or mitochondrial DNA (Dowling et al. 1990). RFLPs are variations within a species in the length of DNA fragments generated by a specific restriction endonuclease. RFLP variations are caused by mutations that create or eliminate recognition sites for the restriction enzymes. RFLP analyses can be used effectively, and relatively economically, to analyze clonal populations, heterozygosity, relatedness, geographic variation, hybridization, species boundaries, and phylogenies ranging in age from 0 to 50 mya. It is possible to analyze more loci per individual by RFLP analysis than by DNA sequencing because RFLPs are less time consuming and expensive. Recently, RFLP analysis has been simplified by employing the polymerase chain reaction in a technique called PCR-RFLP, as described in Chapter 14, making RFLP analysis more rapid and inexpensive.

Analysis of nuclear or mitochondrial DNA sequences is one of the most demanding techniques available but provides very large amounts of detailed data (Table 13.1, Chapter 8). DNA sequencing allows specific DNA (or RNA) sequences to be used in phylogenetic analyses and the number of potential characters that can be examined theoretically is limited only by the number of nucleotides in the DNA of the organism. DNA sequencing can be used to produce molecular phylogenies of particular genes or gene families, trace genealogies within species, and allow construction of species phylogenies (Table 13.1). DNA sequences are used most often to reconstruct phylogenies over periods of time greater than 50 million years. The sequence data are appropriate for analysis of intraspecific variation, cryptic species, geographic variation, reproductive behavior, and heterozygosity estimates. Unfortunately, with the current technology, the time and cost required to obtain sequence data limits its applications. However, advances in DNA sequencing techniques over the past ten years have been phenomenal and, as costs decline, DNA sequencing may become more commonly employed.

## Molecular Evolution

**Molecular evolution** encompasses two areas of study: (1) analyzing the evolution of DNA and proteins and the mechanisms responsible for such changes, and (2) deciphering the evolutionary history of genes and organisms, or **molecular phylogeny**. The first area of study attempts to elucidate the causes and effects of evolutionary changes in molecules, while the second uses molecules as a tool to reconstruct the history of organisms. The two disciplines are intimately interrelated because phylogenetic knowledge is essential for determining the order of changes in the molecular characters being studied, while knowledge of the pattern and rate of change of a molecule is crucial in efforts to reconstruct the evolutionary history of a group of organisms (Li and Graur 1991).

Analyses of different genes provide information about different levels of phylogenetic analyses over a broad range of taxa. For example, ribosomal genes are widely used for phylogenetic analyses because they are highly conserved but have regions that change rapidly and regions that change slowly. Friedlander et al. (1992) evaluated 14 candidate nuclear genes as useful for higher level phylogenetic analysis of arthropods. The genes were identified by comparing known sequences from *Drosophila melanogaster* with their homologs in other animals in the GenBank and EMBL nucleic acid sequence databases. Criteria for choosing these genes included their degree of sequence conservation, which would make them useful for analyzing higher taxonomic levels, and gene structure, which would make them easy to analyze. The genes included *elongation factor-1α, elongation factor-2*, Na$^+$/K$^+$ATPs, *enolase, glucose-6-phosphate dehydrogenase, phosphoenolpyruvate carboxykinase*, and *dopa decarboxylase*. However, the level at which these fourteen genes are informative remains to be determined by comparing results of analyses using taxa of known relationships.

## Molecular Systematics

There is a broad overlap in the use of the terms systematics and taxonomy. Mayr and Ashlock (1991) define systematics as "the scientific study of the kinds and diversity of organisms and of any and all relationships among them" or the "science of the diversity of organisms" and taxonomy as "the theory and practice of classifying organisms." Taxonomy can be divided into descriptive taxonomy and identification (Post et al. 1992). Systematics deals with populations, species, and higher taxa. It is also concerned with variation within taxa. Thus, DNA analysis is particularly suitable for systematics studies because it is the most direct analysis of the genetic material possible and is unlikely to show life-stage- or environmentally induced variability. However, while DNA analyses offer abundant characters for analysis, they are not used without controversy.

## Controversies in Molecular Systematics and Evolution

Several significant controversies are associated with using molecular techniques to study systematics and evolution. These include debates over the relative importance of molecular versus morphological data, the constancy of evolutionary rates (the molecular clock), the use of the terms homology and similarity, and the neutrality of DNA sequence variation. These issues are described briefly.

### Molecular vs Morphological Traits

There is continuing debate over whether morphological or molecular characters are *better* for constructing phylogenies. When comparisons have been made, it appears that morphological changes and molecular changes often are independent, and either are responding to different evolutionary pressures or differently to evolutionary pressures. Hillis and Moritz (1990) conclude that the real issue in choosing a technique to answer a hypothesis should be whether: (1) the specific characters chosen exhibit the variation that is appropriate to the question posed, (2) the characters have a genetic basis, and (3) data are collected and analyzed in a way that makes it possible to utilize both morphological and molecular information. Molecular and morphological data each have advantages and disadvantages.

   DNA sequence data have the advantage of having a clear genetic basis and the amount of data is limited only by the genome size (and the time and funds of the scientist!). Morphological data have the advantage that they can be obtained from fossils (if available) and preserved collections, and can be interpreted in the context of ontogeny. (Only limited amounts of DNA data can be obtained from preserved fossils by the PCR because ancient DNA is usually highly degraded.) Moritz and Hillis (1990) conclude the debate is not an either/or situation and studies that incorporate both types of data should provide better results than those using just one approach. Furthermore, some problems can only be resolved with morphological data while other problems are better resolved with molecular data.

### The Molecular Clock

   Until the 1960s, the analysis of fossils was the only source of information about the *time* when ancestors of extant organisms lived. Molecular genetic studies in the 1960s provided a concept, called the **molecular clock**, that could be used to estimate the evolutionary history and time of divergence of organisms. The molecular clock was particularly useful for living species that have a poor fossil record, which is a very high proportion of extant species.

   Insects have a relatively extensive fossil record, with 1263 families of fossil insects known, compared with approximately 825 families of fossil tetrapod vertebrates (Labandeira and Sepkoski 1993). Labandeira and Sepkoski (1993) found 472 references on fossils covering 1263 insect families with all of the 30 commonly recognized extant orders of insects represented as fossils. While few fossil insects are known from the lower Devonian, a massive radiation began sometime during the early Carboniferous period, more than 325 million years ago. Insects continued to increase dramatically in diversity during the late Carboniferous and middle Permian periods (Table 13.2).

   The molecular clock is based on the assumption that basic processes such as DNA replication, transcription, protein synthesis, and metabolism are remarkably similar in all living organisms and that the proteins and RNAs that carry out these 'housekeeping functions' should be highly conserved. Of course, over time, nucleotide substitutions in housekeeping genes have occurred and DNA and protein sequences have changed. The changes have tended to preserve the *function* of the gene rather than modify or improve it. Thus, changes in the DNA

**Table 13.2**

Geological Time Scale in Millions of Years and Types of Fossil Insects Found[a]

| Era | Period | Epoch | Began (mya) | *Extinct* and extant fossil insect orders first found |
|-----|--------|-------|-------------|---------------------------------------------------------|
| Cenozoic | Quaternary | Recent | | Protura, Zoraptera, and Phthiraptera first appeared in fossil record. |
| | | Pleistocene | 1.6 | |
| | Tertiary | Pliocene | 5 | |
| | | Miocene | 25 | |
| | | Oligocene | 35 | |
| | | Eocene | 60 | Mantodea first appeared in fossil record. |
| | | Paleocene | 65 | |
| Mesozoic | Cretaceous | | 145 | Isoptera first appeared in fossil record. |
| | Jurassic | | 210 | Dermaptera first appeared in fossil record. |
| | Triassic | | 245 | Odonata, *Titanoptera*, Grylloblattodea, Tricoptera, Lepidoptera and Hymenoptera first appeared in the fossil record. |
| Paleozoic | Permian | | 285 | *Permothemistida*, Plecoptera, Embioptera, *Protelytroptera*, *Glosselytrodea*, Psocoptera, Thysanoptera, Hemiptera, *Antliophora*, Mecoptera, Diptera, *Amphiesmenoptera*, Neuroptera, Megaloptera, and Coleoptera first appeared in the fossil record. |
| | Carboniferous | | 360 | Pterygotes radiated into stem groups of all major lineages, with seven surviving to modern times (ephemeroids, odonatoids, plecopteroids, orthopteroids, blattoids, hemipteroids, and endopterygotes). Diplura, *Monura*, Thysanura, *Diaphanopterodea*, *Megasecoptera*, *Permothemistida*, *Protodonata*, *Paraplecoptera*, Orthoptera, Blattodea, *Caloneurodea*, *Blattinopsodea*, and *Miomoptera* were present. |
| | Devonian | | 400 | Collembola (*Rhyniella praecursor*) and Archaeognatha |
| | Silurian | | 440 | Mites, opilionids, scorpions, pseudoscorpions, centipedes, spiders found in pre-Devonian strata. |
| | Ordovician | | 500 | |
| | Cambrian | | 600 | |

[a]Modified from Kukalova-Peck (1991). Taxa in italics are extinct.

code should occur that have minimal or no effect on function. For example, changes in DNA sequences can occur but, because the genetic code is degenerate, the third base can be altered without affecting which amino acid is designated. Changes in the code also can occur without changing the function of the protein if amino acid changes occur in the region(s) that do not affect the function of the molecule, or one amino acid is replaced by a similar amino acid. The molecular clock hypothesis assumes that nucleotide substitutions in the housekeeping genes that constitute the clock occur at a *constant* rate, thus providing a reliable method for measuring time.

The molecular clock hypothesis was proposed after Zuckerkandl and Pauling examined amino acid substitutions in hemoglobin and cytochrome c proteins from different vertebrate species (Zuckerkandl and Pauling 1965). They found that the rate of molecular evolution was approximately constant over time *in all lineages*. Thus, they concluded that the sequences coding for these highly conserved macromolecules could be used to measure the evolutionary distance (time) between organisms by counting the number of accumulated changes. Phylogenetic history thus could be reconstructed, and the time when branches occurred among the lineages leading to modern species could be resolved. However, evidence from analyses of different protein sequences shows that the rates of change *do* vary among different proteins and lineages.

For example, cytochrome c has an acceptable clockwise behavior over the span of organisms studied. However, copper–zinc superoxide dismutase (SOD) behaves like an erratic clock (Ayala 1986). The average rate of amino acid substitutions in SOD per 100 residues per 100 million years is a minimum of 5.5 when fungi and animals are compared. The genetic code makes it possible to calculate the *minimum* number of nucleotide substitutions that would be required to change a codon for one amino acid to a codon for another. Some amino acid replacements can be made with the change of a single nucleotide, but others would require a minimum of two or three nucleotide substitutions (Wilson et al. 1977). The rate of substitutions in SOD is 9.1 amino acids per 100 residues per 100 million years when comparisons are made between insects and mammals, and 27.8 when mammals are compared with each other (Ayala 1986). Thus, the molecular clock should be calibrated with data that are independently derived, preferably with fossil evidence, if the absolute time of divergence is desired. Wilson et al. (1987) pointed out that morphological and molecular analyses of species with abundant fossil records have reduced the uncertainty in estimating the time of divergence by several orders of magnitude. The molecular clock is thought by some to be more useful in calculating *relative* times rather than absolute times of divergence.

Moran et al. (1993) compared sequences in 16S ribosomal DNA in endosymbiotic bacteria of aphids and their aphid hosts, and found the clock was approximately constant for these two organisms. These symbiotic bacteria live within specialized aphid cells, are maternally inherited, and are essential for growth and reproduction of their hosts, indicating a long association between them. Analyses of 16S ribosomal DNA sequences of the bacterial symbionts indicate that the bacteria *Buchnera aphidicola* are distinct within diverse aphid taxa. The phylogeny of *B. aphidicola* is concordant with the phylogeny of the aphid hosts, suggesting that the current distribution of endosymbionts among aphid species is due to vertical transfer from an ancestral aphid in which the original infection occurred. The data indicate that cospeciation may have occurred, with the aphids and their endosymbionts radiating synchronously. The aphid fossil record allowed Moran et al. (1993) to estimate when the aphid and bacterial radiations occurred. The "approximate constancy" of the 16S rDNA sequences, evolving at a relatively constant rate of 0.01 to 0.02 substitutions per site per 50 million years, suggested that they can be used as a molecular clock for the evolution of the endosymbionts. Using this rate, the association between the aphids and their endosymbionts is estimated to have begun about 160–280 million years ago.

## The Neutral Theory of Evolution

Another controversy involves the mechanism(s) of molecular evolution. At the core of the dispute is the **neutral theory of molecular evolution** (Kimura 1987). The neutrality theory recognizes that for any gene a large proportion of all possible mutations (alleles) are deleterious and that these are eliminated or maintained at a very low frequency by natural selection. The evolution of morphological, behavioral, and ecological traits is governed largely by natural selection, because it is determined by the selection of favorable alleles and works against deleterious ones. However, a number of mutations can result in alleles which are equivalent to each other. These mutations are *not* subject to selection relative to one another because they do not affect the fitness of the individual carrying them. Neither do they affect their morphology, physiology, or behavior. Thus, the neutrality theory states that the majority of nucleotide substitutions in the course of evolution are the result of the gradual, random fixation of neutral or nearly neutral changes, rather than the result of positive Darwinian selection. Neutral mutations can spread in a population because only a relatively small number of gametes are sampled each generation (random genetic drift). By chance, they can be transmitted to the next generation at a higher frequency (Kimura 1968, 1983).

The neutrality theory is a basic assumption of some methods of estimating phylogeny, and also affects the molecular clock hypothesis. The controversy has been raging since Kimura published his first paper in the 1960s and continues. Data indicate that many protein, chromosome, and DNA variations *are* under selection. Data are also available that support the hypothesis that much molecular variation is essentially neutral. The debate is thus over how *many* molecular variants are selectively neutral or nearly neutral. Moritz and Hillis (1990) suggest that each molecular marker should be tested for neutrality. They also note that because most departures from neutrality are locus specific, selection will have relatively minor effects on analyses if many different loci are studied.

### Homology and Similarity

A fourth issue concerns terminology. **Homology** is an important concept in biology and historically has had the precise meaning of "having a common evolutionary origin" (Reeck et al. 1987). However, homology often has been used in a more loose sense when discussing protein and nucleic acid sequence comparisons. Protein and nucleic acid sequences have been called homologous when they are *similar*. According to the traditional definition of homology, amino acid or nucleotide sequences are either homologous or they are not. They cannot exhibit a 'level of homology' or 'percentage homology'.

Reeck et al. (1987) point out that use of the term homology to mean similarity can cause three different problems: First, sequence similarities may be called homologies, but the sequences being compared are *not evolutionarily related*, which is certainly inconsistent with the definition of homology. Second, similarities in sequences (again called homologies) are discussed but evolutionary origins are not, which can lead the reader to believe that coancestry is involved when it is not. Third, the similarities (called homologies) are used to support a hypothesis of evolutionary homology. The problem is that while similarity is easy to document, a common evolutionary origin usually is more difficult to establish, especially if fossil evidence is lacking. When in doubt, it is better to talk about 'percentage similarity'.

Several evolutionary processes other than homology could account for sequence similarities, including **convergent evolution**, which is the independent evolution of the same characteristic in separate branches of a phylogenetic tree.

## Techniques for Molecular Systematics

The interest in molecular methods for systematics and evolutionary studies is reflected by the recent publication of detailed protocols for specific molecular methods and guidelines for data analysis. For example, Hillis and Moritz (1990) provide an introduction to molecular systematics that includes guidelines for sampling, collecting, and storing tissues; protocols for isozyme electrophoresis, immunological techniques, molecular cytogenetics, DNA–DNA hybridization, restriction site analysis, and nucleic acid sequencing; as well as analytical methods for intraspecific differentiation and phylogeny reconstruction. Pasteur et al. (1988) provide protocols and methods of isozyme genetic analysis, giving a number of examples from their studies of insects. Weir (1990) provides guidelines on analyzing population structure, phylogeny construction, and diversity using molecular and morphological data. Howe and Ward (1989) provide protocols and information on nucleic acid sequencing and data analysis, and Doolittle (1990) supplies information on computer programs for analysis of DNA and protein sequences. Gribskov and Devereaux (1991) provide a complete introduction to sequencing and analysis of nucleic acid sequence data. Li and Graur (1991) provide an introductory textbook on molecular evolution, and Nei (1987) gives an introduction to molecular evolutionary genetics.

This chapter provides the reader with a brief overview of the principles of molecular genetic techniques and some examples of their application. Detailed protocols and discussions of the relative merits of particular methods of data analysis are found in the references and books cited.

### Protein Electrophoresis

The term **isozyme** is a general designation for multiple forms of a single enzyme. Isozymes are complex proteins made up of paired polypeptide subunits; their subunits may be coded for by different loci. For example, protein Z could be a tetramer made up of two polypeptide units, A and B. Five isozymes of protein Z could exist and be symbolized as follows: AAAA, AAAB, AABB, ABBB, and BBBB. Isozymes will catalyze the same reaction, but may differ in properties such as the pH or substrate concentration at which they function best. They may have different isoelectric points and be separated by gel electrophoresis.

The term **allozyme** is used to refer to variant proteins produced by *allelic* forms of the *same locus*. Thus, A is now A'. A different mutation of A could produce A". Allozymes are a *subset* of isozymes; allozymes can differ by net charge or size so they can be separated by electrophoresis.

The process of analyzing isozymes or allozymes can be divided into five steps, including extraction of proteins, separation, staining, interpretation, and application. Proteins are more difficult to handle than DNA because they are more susceptible to degradation. Proteins must be frozen and stored at $-70°C$, but even at those temperatures some proteins can degrade within months. Proteins are separated in an electric field on a gel. In gels with a single pH, the proteins move through the gel at a continuous rate, but in gels with a pH gradient, they move until they reach their isoelectric point and then stop. The resultant electrophoretic bands can be visualized by appropriate staining (Murphy et al. 1990, May 1992). If a general protein detection system is used, only those proteins present in large quantities are detected, but more specific stains can also be used. Specific stains and buffer recipes are available for more than 50 enzymes (May 1992). The banding phenotypes observed on the gels can be interpreted in terms of genes and their alleles (Pasteur et al. 1988, May 1992).

Protein-coding genes are often **codominant**, with both alleles being expressed in heterozygous organisms. This makes it possible to relate a particular phenotype to a given genotype, if we assume that isozyme data reflect changes in the encoding DNA sequence. To interpret the banding patterns, the number of subunits in the enzyme and the distribution of enzymes in particular cells or tissues should be known (May 1992).

Analyses of isozymes remains a very cost-efficient and useful method for deciphering the systematics, population genetics, and evolution of insects. Protein electrophoresis can be conducted using starch (horizontal or vertical gel systems), polyacrylamide, agarose, and cellulose acetate gels as substrates (Hames and Rickwood 1981). Each has specific advantages and disadvantages (Moritz and Hillis 1990). However, isozyme or allozyme data are useful for estimating the evolution of only a portion of the genome: those genes coding for enzymes that have a different charge and size. The data also are most useful for analyzing relatively closely related taxa. Unfortunately, allozyme variation in some insects, for example, aphids and Hymenoptera, is very low and other molecular techniques are required.

Allozyme variation was used to analyze divergence and dispersal among populations and species of North American cave crickets (Caccone and Sbordoni 1987) and can serve as an example of the method. The authors examined 49 populations of nine species of cave crickets by analyzing variation at 41 loci. Cricket species inhabiting caves had lower genetic variation than species found outside of caves, as expected. Cave species had lower rates of gene exchange between populations in different geographic sites (caves) than noncave species. Within the cave species, the degree of genetic differentiation among populations of the same species was correlated with the degree of isolation conferred on the populations by the structure and location of the caves. Thus, where the limestone caves were continuous and highly fissured, populations were less differentiated because dispersal was enhanced. Populations occurring in regions where caves were farther apart from each other were more highly differentiated. The branching patterns produced by cluster analyses of the allozyme patterns resulted in a phylogeny that is similar to, but not completely concordant with, the phylogeny based on morphology.

A second example of allozyme analysis illustrates an economically important application of the technique. Twenty-four populations consisting of three subspecies of *Culicoides variipennis* (Diptera, Ceratopogonidae) from different geographic regions were examined for genetic differences using isozyme electrophoresis (Tabachnick 1992). Twenty-one loci were examined among the 24 populations of this vector of bluetongue virus, a disease that causes losses of U.S. $120 million annually to the U.S. livestock industry. The results were analyzed with a stepwise discriminant analysis.

Average heterozygosities for the three subspecies were estimated to be 0.134, 0.164, and 0.187. The complete analysis is consistent with the conclusion that there are three North American subspecies, and that the three subspecies may even be sufficiently differentiated to be considered different species. Furthermore, the geographic variation in bluetongue disease epidemiology is correlated with the distributions of the three subspecies. The data support the hypothesis that one subspecies is more effective as a vector of the virus. If confirmed, the results could have broad implications for pest management programs and significant economic impacts. If these results are supported by additional research, the areas inhabited only by the other two (nonvector) subspecies could be considered virus-free regions. Animals raised in such areas would not have to undergo extensive testing when livestock and livestock germ plasm from these areas are exported to regions without the disease.

### Molecular Cytology

Three breakthroughs in cytogenetic techniques have revived such approaches to molecular evolutionary studies. The first was the discovery that hypotonic treatment spreads metaphase chromosomes, allowing more accurate counts of chromosome numbers and details of chromosome morphology. The second was the development of chromosome banding techniques that allow the identification of specific types of DNA within homologous chromosomes. The third was the development of *in situ* hybridization techniques, which allow specific DNA sequences to be localized to particular segments of the chromosomes.

*In situ* hybridization involves annealing single-stranded probe molecules and target DNA to form DNA duplexes. It is effective in locating satellite DNA, ribosomal gene clusters, or duplicated genes of polytene chromosomes and can even locate single-copy DNA on mitotic chromosomes (Figure 13.1). Chromosomal DNA is denatured in such a way that it will anneal with high efficiency to complementary single-stranded nucleic acid probes to form hybrid duplexes. Because chromosomal DNA is complexed with proteins and RNA, the efficiency of *in situ* hybridization is determined by how well the chromosomal DNA can be denatured, how much DNA is lost during fixation and treatment, and whether chromosomal proteins are present in the region of interest (Sessions 1990). Sites where hybridization between a radioactive probe and its target DNA occur are visualized by autoradiography or by using nonradiographic labeling techniques such as biotinylation.

Chromosome morphology may also be used as a taxonomic character. In many cases, chromosomes can be identified by their relative size, centromere position, and secondary constrictions. Many chromosomes, particularly insect polytene chromosomes, have complex patterns of bands or other markers that can be used to identify specific populations or to discriminate between closely related species. Distinctive patterns can be obtained by Q-, G-, or C-banding that identify chromosomes in most species.

**Q-banding** is the simplest technique, and involves treating chromosome preparations with quinacrine mustard or quinacrine dihydrochloride, which pro-

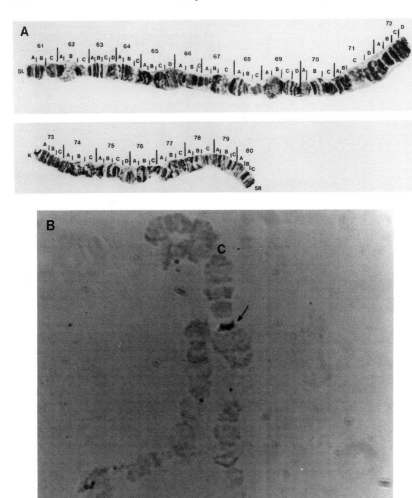

**Figure 13.1** (A) The left and right arms of chromosome 5 of the Mediterranean fruit fly, *Ceratitis capitata*. Banding regions are identified by numbers and letters. K, centromere. Chromosome 5L is homologous to the X chromosome of *Drosophila melanogaster*. (B) The arrow points to *in situ* hybridization signals for the *vitellogenin* gene at site 72A of chromosome 5. C refers to the centromere. (Photos provided by A. Zacharopoulou.)

duces fluorescent bands that are brightest in AT-rich regions of the chromosomes. Q-banding is visible only with UV optics and the bands fade rapidly. **G-banding** involves treating chromosome preparations with trypsin or NaOH and staining with Giemsa in a phosphate buffer, which yields alternating light and dark bands. The dark bands are primarily AT-rich regions and thus correspond to most Q-bands. **C-banding** requires a stringent extraction step that can result in loss of chromosomal DNA. During C-banding, chromosomes are treated with a strong base at a high temperature, incubated in a sodium citrate solution again at high temperature and stained in a concentrated Giemsa solution. C-banding extracts almost all of the non-C-band chromatin, leaving only constitutive heterochromatin, which usually contains rapidly reassociating repeated DNA sequences (Sessions 1990).

Cytogenetic data provide information independent from morphological, biochemical, or behavioral data for phylogenetic analyses. Cytogenetic data can reveal differences or similarities that may not be obvious at the morphological level. Chromosome size, shape, number, and ploidy levels can provide insights into the genetic architecture of taxa. Banding studies reveal aspects of the structural organization of chromatin on individual chromosomes, while probes of DNA sequences with *in situ* hybridization can reveal finer details of chromosome anatomy in terms of spatial arrangement, and the presence or absence of particular kinds of DNA sequences.

## DNA–DNA Hybridization

**DNA–DNA hybridization** allows an estimate of the degree of difference in single-copy DNA sequences from two different sources. This technique is based on the thermal stability of double-stranded DNA molecules. If a dsDNA molecule contains two strands from the same source (**homoduplex** molecules), the DNA will produce a characteristic melting curve if it is denatured by heating to 100°C. Subsequent cooling of the solution allows the complementary strands to reanneal. If a dsDNA molecule contains strands from *different* sources, the differences in nucleotide sequence will reduce how well they reanneal and will reduce the stability of the resultant **heteroduplex** molecule (Figure 13.2). Upon heating, the heteroduplex molecules will denature into single strands at lower temperatures than homoduplex DNA (Werman et al. 1990, Li and Graur 1991).

DNA–DNA hybridization involves creating hybrid DNA molecules by slowly cooling a mixture of denatured DNA from two different sources. Experimental conditions (salt concentration, temperature, fragment size) determine the amount of base pair mismatch that can occur in the hybrid molecules that form. Under high stringency conditions, base pairing between DNA from different species will occur only between well-matched sequences. At lower stringency conditions, more mismatches are tolerated.

The procedures are as follows: double-stranded DNA is isolated and purified

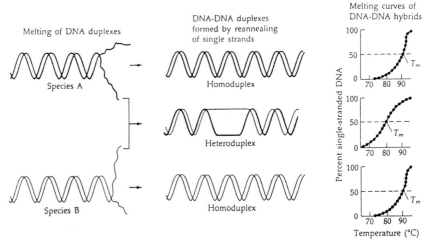

**Figure 13.2**   Sequence similarity is estimated from DNA–DNA hybridization analyses. DNA is melted to form single-stranded DNA. Homoduplexes and heteroduplexes are formed by reannealing single strands. The temperature at which 50% of the DNA melts into single strands, $T_m$, is determined for both hetero- and homoduplex DNA. $T_m$ values between the homoduplexes may differ, and the reciprocal heteroduplex DNA types will differ from both homoduplexes. (Modified from Avers 1989.)

to remove RNA and protein. The DNA, consisting of single-copy and repetitive DNA, is fragmented into short pieces. The amount of single-copy DNA is determined by a procedure called generating a $C_0t$ curve (Britten et al. 1974), which is used to facilitate separating the single-copy DNA from repetitive DNA by a hydroxyapatite chromatography column. It is necessary to remove the repetitive-copy DNA because this DNA is much more variable in quantity than single-copy DNA in insect genomes. Then, the fractionated single-copy DNA from one species is radioactively labeled (**tracer DNA**) and hybridized with unlabeled DNA (**driver**) from the same species (homoduplex reaction) and from different species (heteroduplex reactions). When hybridization is complete, the DNA is then denatured over a range of temperatures and the percentage of ssDNA in the solution is determined at each temperature by determining the melting profiles for the homoduplex and heteroduplex reactions. The thermal stability of the hybrid DNA is described by the term **median melting temperature**, or $T_m$, which is the temperature at which 50% of the heteroduplex dsDNA becomes single stranded. This $T_m$ can be compared with the temperatures at which 50% of the two different homoduplex dsDNAs becomes single-stranded, $\Delta T_m$, which provides an estimate of the genetic distance between the two DNAs.

Some disadvantages to DNA–DNA hybridization are that DNA–DNA hybridization produces data in the form of distance information rather than specific DNA sequences; comparisons are restricted primarily to the single-copy DNA in the genome; differences in the amount of single-copy DNA in the species pairs being examined could produce errors in estimating genetic distances; and DNA–DNA hybridization is relatively complex and expensive. It requires the use of radioisotopes and relatively large quantities of DNA (milligram amounts) so that adequate replicates can be run. DNA–DNA hybridization primarily measures the neutral drift of single-copy DNA (Werman et al. 1990).

Large-scale application of DNA hybridization to systematics problems was first accomplished by Sibley and Ahlquist (1981), who evaluated bird and mammal phylogenies. While the Sibley and Ahlquist DNA–DNA hybridization analyses were criticized because these workers adjusted the results to account for technical difficulties, DNA–DNA hybridization remains a useful technique if precision and accuracy are maintained during the hybridization process (Werman et al. 1990).

DNA–DNA hybridization studies have been conducted on a number of arthropods, including *Drosophila* species and cave crickets in two genera (*Euhadenoecus* and *Hadenoecus*) (Caccone and Powell 1987, 1990, Caccone et al. 1988a,b, 1992, Goddard et al. 1990, Powell and Caccone 1990). DNA–DNA hybridization studies of *Drosophila* species suggested that insect genomes: (1) exhibit more DNA heterogeneity than mammalian or bird genomes; (2) evolve more rapidly than mammalian and bird genomes; and (3) have less congruence between degree of DNA sequence divergence and morphological changes and/or ability to form interspecific hybrids than mammals and birds (Caccone and Powell 1990).

### Restriction Fragment Length Polymorphism Analysis

The analysis of **restriction fragment length polymorphisms (RFLPs)** allows comparison of DNA sequences and produces data that can be converted to estimates of sequence divergence. The number and size of fragments produced when DNA is digested with restriction nucleases provides sequence divergence information relatively rapidly and inexpensively (Dowling et al. 1990). This technique makes it possible to analyze *more loci* than DNA sequencing, although the *amount* of information provided for each locus is less complete. RFLPs can be used to provide data on variation in mitochondrial DNA, as well as unique- and repeated-sequence nuclear DNA (Lansman et al. 1981). They can provide information about variation within and among populations, levels of gene flow, effective population size, analyses of parentage and relatedness, and hybrid zones (Table 13.1). Higher-level systematics studies only rarely have used RFLPs.

Over 1400 restriction enzymes are known that cut DNA at a specific position within a specific recognition sequence (see examples in Table 6.3 and Brown 1991). Such recognition sequences are typically 4–6 bp long, although they can range up to 12 bp. The specificity of the restriction enzymes means that a complete digestion of DNA will yield a reproducible array of DNA fragments. Changes in the number and size of fragments can occur from changes in DNA sequence through rearrangements (inversions, tandem duplication, inverted duplication) or addition, deletion, or substitution of specific bases.

Once the DNA is digested with a restriction enzyme, the fragments produced are sorted by size using agarose or polyacrylamide gel electrophoresis. DNA fragments of known length are run on each gel to serve as an internal standard and to allow the size of the experimental fragments to be estimated. The DNA fragments in the gel are visualized by several methods, including staining with ethidium bromide (if the DNA was previously amplified by the PCR), or probing Southern blots with labeled probes. The detection technique employed depends on the amount of DNA present in the gel.

Staining with ethidium bromide is the simplest and least expensive, but also the least sensitive. The minimal amount of DNA in a band that can be detected by ethidium bromide is about 2 ng, so small fragments can be detected only if a large amount of DNA is present. DNA probes can be end labeled by adding $^{32}$P-labeled nucleotides to the ends of DNA probes produced by the restriction enzymes. Intensity of labeling is independent of fragment size and is more sensitive than EtBr, with 1–5 ng of DNA easily visualized. If primers are available, DNA can be first amplified by the PCR, cut with a restriction enzyme, and labeled by ethidium bromide.

If less DNA is available, radiolabeled DNA probes can used to visualize fragments. Southern blot hybridizations are highly sensitive and picogram quantities of DNA can be detected, although small fragments less than 50 bp are more difficult to detect. Southern blots require a suitable probe with sufficient sequence similarity to the target DNA that a stable hybrid can be formed at moderate to high stringency. The use of probes from other species (**heterologous** probes) makes interpretation of the results more difficult.

RFLP analyses can be conducted on mitochondrial DNA, single-copy nuclear DNA, and repeated sequences such as ribosomal DNA or hypervariable sequences. RFLP analyses of hypervariable sequences are becoming particularly popular for population studies (see Chapter 14). **Hypervariable sequences** are tandem repeats of a short minisatellite sequence and are dispersed throughout the genome. Minisatellite sequences from different species usually are different, but some hypervariable regions contain a sequence similar to one found in the phage M13, which means that a 'universal' probe is potentially available. Highly repetitive DNA sequences vary within and between insect subspecies. Tares et al. (1993) found that a family of highly repeated *Alu* I sequences (each about 176 nucleotides long) constitute about 2% of the nuclear genome of the honey bee, *Apis mellifera*. Approximately 23,000 copies are present per haploid genome. Sequence divergence was found, however, in four different honey bee subspecies (*Apis m. mellifera*, *A. m. ligustica*, *A. m. caucasica*, *A. m. scutellata*), and the bee species *A. cerana* and *A. dorsata*, suggesting that the *Alu* I sequences could be used for diagnosing species and subspecies of honey bees.

Examples of RFLP analyses in insects include the Colorado potato beetle *Leptinotarsa decemlineata* (Azeredo-Espin et al. 1991), 11 species of the *Aedes scutellaris* group (Kambhampati and Rai 1991b), *Pissodes* bark weevil species and individuals (Boyce et al. 1989), the Mediterranean fruit fly *Ceratitis capitata* (Sheppard et al. 1992), 17-year periodical cicadas *Magicicada* (Martin and Simon 1990), the screwworm *Cochliomyia hominivorax* (Roehrdanz 1989, Taylor et al. 1991), the mosquito *Anopheles quadrimaculatus* (Cockburn et al. 1990), the monarch butterfly *Danaus plexippus* (Brower and Boyce 1991), vespid wasps (Schmitz and Moritz 1990), the cricket *Gryllus firmus* (Rand and Harrison 1989), and the blowfly *Phormia regina* (Goldenthal et al. 1991). The analysis of RFLPs in mtDNA has revealed extensive intraspecific variation in most species studied, although variation within single populations is uncommon. mtDNA RFLPs can be used to estimate phylogenies of populations.

## DNA Sequencing

Information on sequences of protein, RNA, and DNA has only developed during the past 40 years. The first sequence information was obtained from proteins in the mid-1950s. RNA was sequenced in the mid-1960s and DNA sequences were obtained in 1975 when DNA sequencing methods were developed. Techniques for DNA sequencing were described in Chapters 8 and 9 and are available from many sources (Hillis et al. 1990, Innis et al. 1990). The use of the PCR reaction makes DNA sequencing of 300–600 base pairs from specific amplified regions less time-consuming and expensive for systematic studies (Simon et al. 1990, 1991). This section thus focuses on using DNA sequences for molecular systematics and evolution.

DNA sequences can be used to: (1) construct molecular phylogenies to evaluate the evolution of particular genes or gene families, (2) evaluate evolutionary changes within species, and (3) construct phylogenies of different species. DNA sequences can be obtained for single copy genes, mitochondrial DNA, and ribosomal DNA. Sequences can be used to study most systematics problems from intraspecific variability to the phylogeny of all organisms (Table 13.1). However, DNA sequencing is expensive and time consuming. For studies where it is important to examine many individuals, DNA sequencing may not be cost effective. Thus, studies of geographic variation, heterozygosity estimates, hybridization, cryptic species, and recent phylogenies are better analyzed by isozyme electrophoresis or RFLPs (Table 13.1).

The use of DNA sequences for systematics investigations depends upon the purity of the DNA used and the fidelity of the sequencing procedures. The DNA sequence data must then be analyzed. It is important to sequence both strands to avoid errors. Sequences can either be aligned with known sequences, or databases can be searched by matching the sequence to all other sequences in a database such as GenBank. When computer analyses of the sequences are conducted, an algorithm is used to align the nucleotide sequences. There are three major methods for comparing sequence similarity: matrix plots, global alignments, and local alignments (Hillis et al. 1990). Both alignment and phylogenetic inferences involve assumptions and subjective decisions (Hillis et al. 1990, Howe and Ward 1989, Gribskov and Devereaux 1991). The alignments are usually made based on the assumptions of parsimony. **Parsimony** dictates that an alignment of sequences be based on the minimal number of changes needed to transform one sequence into the other. Parsimony analysis is a powerful tool for the study of biological evolution and is used to construct phylogenetic trees, evaluate alternative hypotheses objectively, and to study evolutionary patterns and processes. However, there are some pitfalls in parsimony analysis (Stewart 1993).

The scientific community recognized that sequence information would be most useful if the data were available to everyone in an organized manner, and thus genetic sequences are being computerized and stored in three major databases: GenBank in the U.S.A., the EMBL Data Library in Europe, and the DNA Data Bank of Japan. The amount of sequence information available for DNA has grown exponentially over the past ten years and sequences continue to accumulate at an ever faster pace (Doolittle 1990). The importance of submitting DNA sequence data to GenBank is recognized by many journals that require that sequences be submitted prior to, or simultaneous with, publication. DNA sequencing for systematics and phylogenetic studies commonly examines sequences of single copy DNA, mitochondrial DNA, or the coding portions of both nuclear and mitochondrial ribosomal RNA (Hillis et al. 1990).

The three databases collaborate in collecting and distributing sequence data. By 1990, data on 1200 organisms (including animals, plants, microorganisms, and viruses) had been entered into the GenBank database, with about half of the sequences consisting of protein-coding regions and with over 1600 tRNA gene sequences. About 28% of the *Escherichia coli* genome is included, but only 0.2% of the human genome. About 150 'complete' genomes are available for viruses, plasmids, and organellar chromosomes. By 1991, more than 50 million nucleotides of DNA sequences were in the GenBank database (Gribskov and Devereux 1991).

### RAPD-PCR of Genomic DNA

The random amplified polymorphic DNA method of the PCR (described in Chapter 9) has been used to discriminate between cryptic sympatric species of sandfly (Psychodidae) in Venezuela that are thought to be vectors of leishmaniasis (Adamson et al. 1993). A diagnostic band was found for *Lutzomyia youngi* that was reproducible and reliable.

Species of *Culicoides* (Ceratopogonidae) were analyzed using RAPD-PCR (Raich et al. 1993). In this case RAPD-PCR did not allow differentiation of *Culicoides* species because no species-specific DNA bands were detected. However, screening was conducted with only four different primers, and it is likely that unique RAPD primers could be found that discriminate among the species if additional primers are tested. Alternatively, multiple markers may have to be employed to produce a banding pattern that can be analyzed by discriminant analysis, although the need to conduct multiple RAPD reactions would make RAPD-PCR more expensive and time-consuming.

The species status of the sweetpotato whitefly, *Bemisia tabaci*, is an economically significant systematic puzzle. A new 'biotype,' B, of this whitefly became a widespread and serious agricultural pest in the U.S.A. during 1992, but its species status and origin remain controversial. Gawel and Bartlett (1993) showed consistent DNA differences between the A and B biotypes using 20 different RAPD-PCR primers, suggesting that the two are separate species rather than biotypes of the same species.

### *t*DNA-PCR

Raich et al. (1993) analyzed species of *Culicoides* by the PCR using transfer DNA primers. **tDNA-PCR** involves using consensus primers to generate characteristic 'fingerprint' banding patterns that allow discrimination at the genus or species level. The bands generated by tDNA-PCR were suitable for discriminating among four species of *Culicoides*.

The virtues of both RAPD-PCR and tDNA-PCR are that prior sequence data are not needed to generate primers for either technique (Hadrys et al. 1992, Kambhampati et al. 1992). Both are rapid and relatively inexpensive and easier to carry out than traditional RFLP analysis. Both require only a small amount of DNA for amplification. Furthermore, only a fragment of the insect need be used, and the remainder can be saved for evaluating other taxonomic criteria (Raich et al. 1993).

## Subjects of Molecular Analysis

Nuclear DNA, mitochondrial DNA, and ribosomal DNA have been employed in systematic studies. The following discussion describes some of the unique attributes of mitochondrial, ribosomal, and satellite DNA.

### *Mitochondrial DNA*

mtDNA is a powerful subject for evolutionary studies, and has been used to study population structure and gene flow, hybridization, biogeography, and phylogenetic relationships (Avise et al. 1987). The small size, relatively rapid rate of evolutionary change, and maternal inheritance of mtDNA make it suitable for examining population history and evolution among closely related taxa (Gray 1989, Lansman et al. 1981, Simon et al. 1991). Molecular studies of mtDNA have employed both RFLPs (described above) and sequencing of specific regions of the mtDNA following cloning or amplification by the polymerase chain reaction (Satta and Takahata 1990, Pashley and Ke 1992).

mtDNA has a number of positive attributes for evolutionary and systematics studies, including (1) inheritance through the mother, (2) general conservation of gene order and composition, (3) a relatively rapid rate of sequence divergence, and (4) small size compared with the nuclear genome. The rate of evolution of insect mtDNA varies among lineages, among genes, and within genes, but averages one to two times that of single-copy nuclear DNA. However, at least one exception exists: the rates of mtDNA and single copy nuclear DNA (scnDNA) divergence within species of *Drosophila* are not greatly different (Satta et al. 1987, Caccone et al. 1988b, DeSalle and Grimaldi 1991). In fact, two classes of scnDNA appear to be present in eight species of the *Drosophila melanogaster* subgroup, with one changing rapidly and the second relatively slowly (Caccone et al. 1988b). Within mitochondria, there are regions that diverge rapidly, while other regions are highly conserved, making the different regions suitable for analysis of different taxonomic levels (Simon et al. 1991, Liu and Beckenbach 1992, Tamura 1992). Interestingly, insect mtDNA evolves more slowly than human mtDNA; human mtDNA evolves five to ten times faster than nuclear DNA (Powell et al. 1986).

There are difficulties in working with mtDNA. These include the lack of recombination in mitochondria, which makes mtDNA essentially a single heritable unit. This potentially produces gene diversity estimates that have larger standard errors than those determined using nuclear loci. There is more variation in copy number of certain tandemly repeated sequences (heteroplasmy) than formerly realized. This may be due to errors in DNA replication and may not be heritable. Biparental inheritance of mitochondria occurs occasionally in insects, which can complicate population studies (Lansman et al. 1983, Kondo et al. 1990, Matsuura et al. 1991). Mitochondrial introgression between *Drosophila* species has been suggested as an explanation of why individuals of some species have mtDNA from a related species (Aubert and Solignac 1990). Finally, mtDNA is less abundant in cells than nuclear DNA, so it is somewhat more difficult to isolate and purify (Tegelstrom 1992, White and Densmore 1992).

Animal mtDNA is a single circular molecule, ranging from 16 kb to about 20 kb in length, containing 12 or 13 protein-encoding genes, the genes for 22 tRNAs, and 2 ribosomal RNA subunits, as well as a noncoding region containing the origin of replication (Gray 1989, Moritz et al. 1987). mtDNA lacks introns, and intergenic sequences are usually small or absent. The control region containing the origin of replication is extremely rich in adenine and thymine in

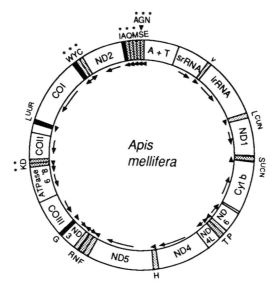

**Figure 13.3** Map of mitochondrial DNA of the honey bee *Apis mellifera*. Genes for tRNAs are denoted by the one-letter code for their corresponding amino acids. tRNA genes with asterisks are in a different position compared with the same genes in the genome of *Drosophila yakuba*. Protein-encoding genes are denoted COI, COII, COIII for the genes encoding subunits one, two, or three of *cytochrome c oxidase*, Cyt *b* for the *cytochrome b* gene and ND1-6 and ND4L for the genes encoding subunits 1–6 and 4L of the NADH dehydrogenase system. The AT-rich region believed to contain the origin of replication is denoted A+T. The direction of transcription for each coding region is shown by arrows. (Redrawn from Crozier and Crozier 1993.)

insects. The replication of mtDNA is unidirectional and highly asymmetric (Figure 13.3).

Within the insects, the tRNA genes in the mitochondria are known to vary in position between the orders Diptera and Hymenoptera, and within the order Diptera (Crozier and Crozier 1993). The complete sequence of the mitochondrial genome of *Drosophila yakuba* (Clary and Wolstenholme 1985a,b) and of the honey bee *Apis mellifera* (Crozier and Crozier 1993) are known. Complete sequences for the mosquitoes *Anopheles quadrimaculatus* and *Anopheles gambiae* are available (Mitchell et al. 1993, Beard et al. 1993). The noncoding control region, or D loop region, was sequenced for several species of Lepidoptera (Taylor et al. 1993). A map of *Drosophila yakuba* mitochondria was given in Chapter 4 (Figure 4.8), and a map of the mtDNA of the honey bee is shown in Figure 13.3. Honey bee mtDNA is 16,343 bp long and 11 of the tRNA genes are in altered positions compared with their positions in D. *yakuba*, but the other genes and regions are in the same relative locations (Cornuet and Garnery 1991). However, the length of the long intergenic sequence located between the COI and COII genes of A. *mellifera* varies between and within subspecies compared with D. *yakuba*, because different numbers of repeated sequences are found in this region in A. *mellifera* (Cornuet et al. 1991). This same region is even longer in several species of the bark weevil genus *Pissodes*, resulting in an unusually large mitochondrial genome size of 36 kb (Boyce et al. 1989).

Partial sequences of mtDNA are known for the locust *Locusta migratoria* (McCracken et al. 1987, Uhlenbusch et al. 1987), the mosquito *Aedes albopictus* (HsuChen and Dubin 1984, HsuChen et al. 1984), the lepidopteran *Spodoptera frugiperda* (Pashley and Ke 1992), the cicada *Magicicada* (Simon et al. 1990), 21 species of leafhoppers (Fang et al. 1993), seven species of black flies (Xiong and Kocher 1991), field crickets (Harrison et al. 1987), greenbug *Schizaphis graminum* (Powers et al. 1989), a number of hymenopteran species (Derr et al. 1992a,b), and several species of *Drosophila* (DeSalle et al. 1987, DeSalle 1992, Satta and Takahata 1990).

### Ribosomal RNA

Ribosomes are a major component of cells that are involved in translating messenger RNA into proteins. Ribosomes consist of ribosomal RNA (rRNA) plus proteins. All ribosomes can be dissociated into two subunits, a large and a small, each containing rRNA and protein molecules. The larger subunit (28S) may contain a smaller RNA molecule in addition to a larger rRNA. Ribosomal RNAs are frequently used to infer evolutionary relationships among species because they are universally present in all forms of life that have a protein-synthesizing system. Ribosomal RNAs contain regions that are 'conserved' and regions that are more variable, so that rRNAs can serve as both slow and fast clocks.

In eukaryotes, the genes encoding the 18S (small subunit) and 28S (large subunit) rRNAs are clustered as tandem repeats in the nucleolus-organizing regions of the nuclear chromosomes (Figure 13.4), but two ribosomal genes are found in mitochondria, including the large ribosomal subunit (16S rRNA) (Figure 13.3). In most animals, there are 100 to 500 copies of rDNA in the nuclear genome in repeated transcription units. The repeated transcription unit is composed of a leader promoter region known as the External Transcribed Spacer (ETS), an 18S rDNA coding region, an Internal noncoding Transcribed Spacer region (ITS), a 28S rRNA coding region, and an InterGenic nontranscribed Spacer segment (IGS).

Different portions of the repeated transcription unit evolve at different rates in the nuclear genome (Collins et al. 1989, Fallon et al. 1991, Kambhampati and

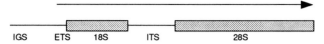

**Figure 13.4**  A simplified diagram of the ribosomal DNA repeat unit of eukaryotes. IGS is the intergenic spacer. ETS is the external transcribed spacer, 18S is the small subunit rRNA gene, ITS is the internal transcribed spacer, 28S is the large subunit rRNA gene. The arrow indicates the direction of transcription.

Rai 1991a, Baldridge et al. 1992). Thus, evolutionary studies employ analysis of different segments, depending on the taxonomic level being studied. In general, a high degree of polymorphism has been found in the *noncoding segments* of the repeat unit (ETS, ITS, IGS), and the most variable part of the repeated unit is the intergenic spacer (IGS), which typically contains reiterated subrepeats ranging from about 50 to several hundred bp in length (Cross and Dover 1987, Tautz et al. 1987). Baldridge and Fallon (1991, 1992) sequenced the rDNA intergenic spacer of the mosquito *Aedes albopictus* and found its structure to be similar to that of *D. melanogaster*.

PCR analysis can provide a relatively rapid and inexpensive method for identifying cryptic species. For example, species-specific variation in the IGS region was used to identify cryptic sibling species of the *Anopheles gambiae* complex (Collins et al. 1988, Scott et al. 1993). In a blind study of members of the *A. gambiae* complex, PCR analysis of ribosomal DNA was shown to be reliable for identifying these cryptic mosquito species (Paskewitz et al. 1993). The ITS region also was used to identify the cryptic mosquito species *Anopheles freeborni* and *A. hermsi* (Porter and Collins 1991), but when rDNA and mtDNA analyses were conducted, no species-specific differences were observed in mtDNA of five colonies of *A. freeborni* and three colonies of *A. hermsi* (Collins et al. 1990). Orrego and Agudelo-Silva (1993) sequenced a portion of the ITS region from parasitoid wasps in the genus *Trichogramma* after amplification by the PCR and concluded that the sequence variation found showed promise as a rapid and inexpensive method for identifying species that are notoriously difficult to identify.

Seventeen populations of the mosquito *Aedes albopictus* (Skuse) from around the world were found to vary among and within populations in the size and sequences of the nontranscribed intergenic spacer region (IGS) (Black et al. 1989). Black et al. (1989) showed that *A. albopictus* individuals within a population carried unique IGS regions. This high degree of intrapopulation variation is different from the variability in IGS regions in *Drosophila melanogaster*, where most individuals within a population share a common IGS region, but exhibit high variation among different populations (Coen et al. 1982).

The coding regions of the repeated unit change relatively little, and can be used for systematic studies of higher level taxa or for ancient lineages. Highly conserved regions are no doubt important for maintaining the characteristic secondary and tertiary structure of the rRNA molecules (Simon et al. 1991, van de Peer et al. 1993). Cano et al. (1992), for example, were able to isolate and amplify DNA by PCR from a 25- to 40-million-year-old bee (*Proplebeia domini-cana*) isolated from Dominican amber using primers based on sequences of the 18S rRNA genes from the yeast *Saccharomyces cerevisiae*. They sequenced portions of the 18S rRNA gene and were able to compare the sequences with those of two extant species of bees (*Plebeia jatiformis* and *P. frontalis*). The results indicate that approximately 7% of the base pairs were substituted over a period of 25 to 40 million years.

Two retrotransposable element types, called *R1* and *R2*, have been found in the 28S ribosomal RNA genes of many insect species. These elements were first found in the rDNA of *D. melanogaster*. *R1* insertions have been found in approximately 60% of the rDNA sequences in *D. melanogaster*, and rDNA containing these insertions has been shown to be inefficiently transcribed. The *R1* elements are approximately 5 kb long and most *R1* elements in the genome are precisely located at the same nucleotide position within the 28S RNA gene (Jakubczak et al. 1990, 1991). Most *R2* elements are located about 74 bp upstream from the site of the *R1* insertions. These elements have been studied in the mosquito *Anopheles gambiae*, the silk moth *Bombyx mori*, the sand fly *Calliphora erythrocephala*, the Japanese beetle *Popillia japonica*, the fungus gnat *Sciara coprophila*, two parasitoid wasps *Diadromus pulchellus* and *Eupelmus vuilleti*, and *Drosophila melanogaster* (Eickbush and Robins 1985, Jakubczak et al. 1990, Besansky et al. 1992, Bigot et al. 1992, Williams and Robbins 1992, Burke et al. 1993).

The *R1* and *R2* retrotransposable elements belong to a family of elements that lack long terminal repeats (LTRs) and are found in fungi, plants, and animals. Because *R1* and *R2* elements insert into the 28S ribosomal RNA genes in specific sites, they may encode their own endonuclease and thus be capable of initiating a specific cut at the target site in the 28S gene (Jakubczak et al. 1991, Burke et al. 1993). The 74-bp distance separating most *R1* and *R2* elements is highly conserved, except for those found in *Popillia japonica* and *Sciara coprophila*. Furthermore, the DNA sequences of the reading frames of the *R1* and *R2* elements from species in the orders Diptera, Lepidoptera, Hymenoptera, and Coleoptera have been conserved, and may be similar in all insects (Burke et al. 1993). Surprisingly, most elements have not accumulated mutations that would make them inactive. Two species examined, *Popillia japonica* and *Nasonia vitripennis*, had multiple families of either *R1* or *R2* elements, and sequence identity between the different families was low. The presence of multiple, highly divergent families of *R1* or *R2* elements within a species suggests either that each insertion family is able to maintain its copy number without eliminating other families, or that there has been extensive horizontal transfer of *R1* and *R2* elements between species.

Another unusual feature has been found in 28S rRNA of aphids. The 28S rRNA from most insects dissociates into two equal-sized subunits, α and β, under denaturing conditions (Ogino et al. 1990). The two subunits are hydrogen-bonded to each other at the regions close to the cleavage site, which is at a halfway point in the 28S rRNA molecule. This specific break, which has been called the hidden break, has also been found in the 23S rRNA of plant chloro-plasts. In three insects—*D. melanogaster, Sciara coprophila,* and *Bombyx mori*—the hidden break and its flanking sequences are AU-rich and may form a stem–loop structure. However, aphids appear to be exceptional, because they are the only group of insects studied to date that lack the hidden break. Ogino et al. (1990) sequenced DNA from the pea aphid *Acyrthosiphon pisum* and found that this region of the rDNA is GC-rich rather than AU-rich.

### Satellite DNA and VNTRs

Satellite DNA sequences are usually species specific, perhaps because this DNA evolves at a very high rate. Satellite DNA change may occur by sequence amplification, unequal crossing over, point mutation, and gene conversion. In the parasitoid wasps *Diadromus collaris* and *Eupelmus orientalis*, satellite DNA from each was different and also different from that of *D. pulchellus* and *E. vuilleti* (Rojas-Rousse et al. 1993). There are only a few cases in which the same satellite DNA sequences have been found throughout an entire genus. In many genera, satellite DNA can be used for species diagnoses (Bachmann et al. 1993, Heath et al. 1993, Jin and Chakraborty 1994).

Minisatellite DNA, also known as Variable Number of Tandem Repeats (VNTRs) can be amplified by the PCR. Heath et al. (1993) used published sequences of VNTRs to design primers for the PCR. They amplified fragments that showed little variation within a species but did show differences between species of fish and birds. Heath et al. (1993) named this procedure Directed Amplification of Minisatellite-region DNA (DAMD).

VNTR loci can be used to produce a DNA fingerprint for evolutionary studies of closely related populations or species. However, DNA fingerprinting does not provide information about allele-frequency distributions, because nei-ther the number of loci nor the identification of specific alleles is directly observable. Jin and Chakraborty (1994) provided a method for calculating bias-corrected estimators of Nei's standard and minimum genetic distances and showed that VNTR data produce values similar to those obtained with locus-specific allele frequencies.

## Molecular Systematics Methodology

The immense diversity of insects and their long evolutionary history provides a challenge for systematists. Due to mutation, high reproductive rates, natural selection and stochastic events, populations change through time. A process of gradual change through thousands of years can result in a different species; a change within a single lineage is called **phyletic speciation**. Another form of speciation is called **cladogenic speciation**. In this form, two populations of a species become isolated and diverge genetically as a result of independent mutation, natural selection, and genetic drift. Other models for speciation include speciation through hybridization and polyploidy, or by modification of regulatory genes. The attributes of an organism used by systematists to establish their relation to other organisms are called **characters**. Characters can be based on morphology, physiology, ecology, behavior, biochemistry, or genetics.

There are at least three main approaches to the process of developing classifications (see, for example, Hennig 1966, Mayr and Ashlock 1991, Sneath and Sokal 1973, Forey 1993). Unfortunately, the debates over which approach is more objective or appropriate or practical have not been resolved. **Phenetic systematics** focuses on classifications based on overall similarities among organisms, involves all possible characters, and a calculation of average similarities with all characters assumed to be equally useful. In some cases classifications based on phenetic similarities may also reflect the phylogeny of the taxa because those that are most similar may well have shared a most recent ancestor, but in other cases this is not so because of convergent evolution.

**Cladistic (phylogenetic) systematics** uses only cladistic relationships as a basis for constructing classifications. The rate or amount of change is not considered and only monophyletic taxa are allowed. **Evolutionary systematics** focuses on the order of origin of lineages and also takes into account the amount and nature of evolutionary change which occurs after cladogenesis. Not all characters are assumed to be of equal value, and are weighted accordingly. One of the major difficulties in any reconstruction of phylogeny is to determine which character state is primitive or ancestral (**plesiomorphic**), and which is derived (**apomorphic**).

Classifications are often represented in graphic forms as tree-like dichotomous branching graphs or **dendrograms**. A dendrogram produced from phenetic information is called a **phenogram**. A phenogram shows how similar the group is, but does not provide information about probable lines of descent.

When a dendrogram is produced from cladistic information, it is called a **cladogram**. A cladogram shows the sequence of origin of clades and indicates the times at which the various cladogenic events have taken place. If the dendrogram includes both phenetic and phylogenetic data, it is a phylogram, or **phylogenetic tree**, and indicates not only the cladistic branching but the relative amount of change that has occurred. Those species which show the closest relationship are grouped together into larger, more inclusive groups or genera. Genera are grouped into families, and families into orders, classes, and phyla.

### Intraspecific Differentiation

For most species there is substantial intraspecific variation due to environmental influences, genetics, differences in age, sex, stage of life cycle, disease, and parasitism. Identification of intraspecific genetic variation with molecular genetic methods is discussed in Chapter 14.

### Sequence Analysis Methods

Once DNA sequence information for nuclear DNA, mtDNA, or rDNA has been obtained, analysis can provide several types of information, including possible structure, function, and characteristics of the protein. The similarity of the sequences to sequences obtained from other organisms can be compared. The tasks of collating, assembling, and correcting the sequence data are usually performed with the help of a variety of computer programs (Gribskov and Devereux 1991). Because great volumes of DNA are being sequenced in large projects, including the *Drosophila* and human genome projects, automated sequencing methods are being improved rapidly, as are the computer hardware and software necessary to handle the sequences (Howe and Ward 1989, Weir 1990).

### Constructing Phylogenies

Swofford and Olsen (1990), Felsenstein (1982, 1988), and Miyamoto and Cracraft (1991) review the techniques used to infer phylogenetic relationships from molecular data. There are many techniques for inferring phylogenetic relationships from molecular data, and nearly as many controversies about which are most appropriate. Only a brief outline of the different approaches is provided here. Details about phylogenetic techniques should be obtained from the reviews just cited.

Inferring a phylogeny is an estimation procedure and is based on incomplete information. The selection of one or more trees from among the set of possible phylogenies is based on one of two approaches: (1) defining a specific sequence of steps, an algorithm, for constructing the best tree, or (2) defining a criterion for comparing alternative phylogenies with one another and deciding whether they are equally good, or if one is better. Some methods of phylogeny construction are based on different explicit evolutionary assumptions, while others are not.

Phylogenetic trees represent evolutionary pathways and there is a difference between species trees and gene trees. Branches in a species tree join extant species to an ancestral species and represent the times since those species diverged. The data used to construct the tree often represent a single region of the genome of those species. A gene tree constructed from a short region of the genome may not be the same as the species tree. Two species may carry genes that diverged prior to the species split, or introgression or transposition may have resulted in genes having diverged after the species split.

Phylogenies are presented as rooted or unrooted trees. A **rooted tree** conveys the temporal ordering of the species or genes on a tree, but an **unrooted tree** reflects the distances between units with no indication of which was ancestral to which. Most of the analytical techniques result in an unrooted tree or unrooted phylogeny, one in which the earliest point in time is unidentified.

Molecular data used to construct trees are either discrete characters or similarities (or **distances**). Examples of discrete molecular characters include DNA sequences, allozyme frequencies, or restriction map data. Most methods assume independence and homology among discrete characters. Distance data specify a relationship between pairs of taxa or molecules. DNA–DNA hybridization data provide distances directly. Immunological and nucleic acid hybridization methods directly produce data in the form of pairwise similarities, but sequence, restriction map, and allozyme data must be transformed to produce distance data. Once the data have been gathered and transformed into appropriate values, there are three broad categories of methods to estimate phylogeny. These include distance matrix methods, maximum parsimony methods, and maximum likelihood methods, which are discussed in detail in Swofford and Olson (1990) and Weir (1990). The following descriptions provide only a brief overview of the methods.

**Distance matrix methods** are based on the set of distances calculated between each pair of species. The computations are relatively simple and the quality of the resulting tree depends on the quality of the distance measure. Clustering usually is used to group the taxonomic units into a phenetic unit. Several methods of clustering can be used, but the most widely used is called **UPGMA** (Unweighted Pair—Group Method using an Arithmetic average). It defines the intercluster distance as the average of all the pairwise distances for members of two clusters. The results of the clustering can be presented in a dendrogram, in which the branch points are placed midway between two sequences or clusters. The distance between a pair of sequences is the sum of the branch lengths. The UPGMA method is often used for distance matrices, and it generally performs well when the nucleotide substitution rates are the same along all branches of the tree. However, the assumption of nearly equal nucleotide substitution rates (or that a molecular clock is operating) is crucial for the UPGMA method.

For situations in which the assumptions of a molecular clock are inappropriate, the Fitch–Margoliash algorithm can be used (Weir 1990). If information for an outgroup is available, the resultant tree can be rooted. The Fitch and Margoliash method allows for the possibility that the tree found is incorrect and recommends that other trees be compared based on a measure-of-goodness of fit. The best tree will have the smallest percentage in standard deviation. The Fitch–Margoliash and UPGMA methods should result in very similar trees if a molecular clock is operating.

**Maximum parsimony methods** focus on the character values observed for each species, rather than working with the distances between sequences that summarize differences between character values. These methods minimize the numbers of changes in sequences between species over the tree, usually making the assumption that there have been approximately constant rates of change. Branch lengths are usually not obtained. For each possible tree, the sequences at each node are inferred to be those that require the least number of changes to give each of the two sequences of the immediate descendants. The total number of changes required over the whole tree is found, and the tree with the minimum number of changes is the most parsimonious.

Parsimony methods assume that the best tree is the shortest or has the smallest number of character state changes, or branching points. However, if there are large amounts of change, parsimony methods can give estimated trees that are inaccurate (Swofford and Olson 1990). Stewart (1993) pointed out that parsimony analysis can be problematic for two general reasons: (1) failure to find the shortest tree and (2) the shortest tree is not the correct phylogeny. Failure to find the shortest tree can occur if too many taxa or too few phylogenetically informative data are used.

**Likelihood methods** of analyzing DNA sequence data rely on genetic models, and can provide a basis for statistical inference. They are more difficult to compute than the methods described in the preceding paragraphs (Weir 1990). Maximum likelihood methods of tree construction assume the form of the tree and then choose the branch length to maximize the likelihood of the data given that tree. These likelihoods are then compared over different possible trees and the tree with the greatest likelihood is considered to be the best estimate.

Unfortunately, the number of possible trees increases very rapidly as the number of taxa under consideration increases, and this is especially true when using maximum likelihood methods. Thus, if three species are being compared, the number of possible unrooted trees is one; with four species, it is three trees; with six species, it is 105 trees; and with eight species, it is 10,395 trees. Maximum likelihood methods provide consistent estimates of branch lengths, indicating that the estimates approach the true values as the amount of data increases. To estimate the likelihood that a particular tree estimate is the true tree, **bootstrapping** techniques can be employed. Bootstrapping involves repeated sampling, with replacement, of the data to produce an estimate of the variance.

Methods for analyzing molecular data are still undergoing development, because none of the techniques currently available is fully satisfactory. The immense amount of DNA sequence data that is becoming available makes it difficult to use maximum likelihood methods. Maximum likelihood algorithms have been developed to build trees from pairwise distances, but they employ only a summary of the data and information is thus lost. Parsimony methods are fast, but may be appropriate only for very slow rates of evolutionary change. Furthermore, phylogenetic analyses of molecular data are complex and best handled with one of a variety of computer software packages. These, too, will change as new methods of analysis become available.

### Software Packages

Four major computer software packages for phylogenetic analyses are widely available, supported, and in widespread use (Swofford and Olsen 1990). Other packages are also available. Because software evolves rapidly, it is impossible to describe the available programs in detail. However, some information is provided so that sources can be contacted for recent information.

PHYLIP is a Phylogeny Inference Package available from Joseph Felsenstein (Department of Genetics SK-50, University of Washington, Seattle, WA 98195). It is a collection of about 30 independent programs implementing maximum likelihood, parsimony, compatibility, distance, and invariants methods. Some of the programs provide bootstrap methods for estimating confidence limits.

PAUP is Phylogenetic Analysis Using Parsimony, available from D. L. Swofford (Illinois Natural History Survey, 607 East Peabody Drive, Champaign, IL 61820). PAUP performs parsimony analysis under a variety of models, and bootstrapping routines are available. Also available is BIOSYS-2, which includes cluster analysis and distance Wagner routines for gene frequency data.

Hennig86 is a small, fast, and effective program for parsimony analysis under the Wagner and Fitch models.

MacClade, written by W. P. Maddison and D. R. Maddison, and distributed by Sinauer Associates (Sunderland, MA 01375), is useful in the analysis of character evolution and the testing of phylogenetic hypotheses under the same parsimony models described for PAUP plus some additional ones. MacClade is available only for Macintosh computers.

## The Phylogeny of Insects

Insects are highly diverse and ancient organisms. The Crustacea are considered the sister group of the Tracheata (= Myriapoda + Hexapoda or Insecta). Relatively advanced Crustacea are found in the Cambrian period [600 million years ago (mya)], so it is assumed that tracheates were present by this time as well (Kukalova-Peck 1991). Labandeira et al. (1988) showed that a bristletail (Archaeognatha) from the Early Devonian resembles modern archaeognathans. Arthropods have apparently been found on land since Devonian times (Table 13.2). Two collembola species, including *Rhyniella praecursor*, also were found in the Lower Devonian (400 mya). These collembola resemble recent extant Isotomidae and Neanuridae, suggesting that terrestrial arthropods had already radiated in the Ordovician period (ca. 500 mya).

A number of extinct and extant orders of primitive insects have been found in diverse Late Paleozoic fauna (Table 13.2). During the Carboniferous (which began 360 mya) a diverse array of extinct and extant (underlined) insects were present, including the: Diplura, Monura, Diaphanopterodea, Palaeodictyoptera, Megasecoptera, Permothemistida, Ephemoptera (mayflies), Protodonata, Paraplecoptera, Plecoptera (stoneflies), Orthoptera (grasshoppers and crickets), Blattodea (cockroaches), Caloneurodea, Blattinopsodea, and Miomoptera. During the Permian (which began 285 mya), additional extinct and extant insect groups are found in the fossil record, including Plecoptera (stone flies), Embioptera (web spinners), Protelytroptera, Glosselytrodea, Thysanoptera (thrips), Hemiptera (bugs and leafhoppers), Antliophora, Mecoptera (scorpion flies), Diptera (true flies), Asmphiesmenoptera, Neuroptera (lacewings, ant lions), Megaloptera (dobsonflies), and Coleoptera (beetles). By the Triassic (245 mya), nearly all modern orders of insects are found in the fossil record, including the Lepidoptera (butterflies and moths), Trichoptera (caddisflies), and Hymenoptera (bees and wasps). By the Jurassic (210 mya), many recent families are present. Tertiary insects (65 mya) are essentially modern and include genera nearly indistinguishable from living fauna.

Labandeira and Sepkoski (1993) suggest that the great diversity of insects was achieved by low extinction rates rather than by high origination rates. The great radiation of modern insects began 245 million years ago and was not accelerated by the expansion of the angiosperm plants during the Cretaceous period.

Living insects are classified in at least 29 orders and more than 750 families, with the orders Coleoptera (>300,000 named species), Lepidoptera (>120,000 species), Hymenoptera (>120,000 species), and Diptera (>150,000 species) containing the most species. Insects are diverse, numerous, and ancient (Daly et al. 1978). An understanding of their systematics and phylogeny will require the combined use of the fossil record, traditional morphological data, and molecular methods.

### Molecular Analyses of Arthropod Phylogeny

The origin and phylogeny of the Arthropoda have been analyzed using ribosomal sequence data (Field et al. 1988, Turbeville 1991). The slowly evolving core segments of ribosomal RNA allow comparisons to be made which permit the reconstruction of phylogenies of phyla and kingdoms. To study the phylogenetic relationship at the family level, domains that change more rapidly must be analyzed, such as a 324-bp sequence from the second expansion segment of the 28S gene.

The molecular phylogeny of the animal kingdom was investigated by se-

quencing the 18S ribosomal RNA gene and using a distance-matrix method of
analysis (Field et al. 1988). Ribosomal RNA sequences from individuals from 22
classes in ten animal phyla were compared. The branching patterns obtained
were compared with branching patterns derived from morphological, embryological, and fossil data. Sequence data from the 18S ribosomal RNA gene
indicated that arthropods are not close relatives of the annelids, suggesting an
early divergence of arthropods from other metameric lineages.

One hypothesis treats the Arthropoda as a monophyletic group, with the
Uniramia and Crustacea combined as Mandibulata. According to this viewpoint, the Uniramia (which includes the insects), Crustacea, and Chelicerata
each descend from a single ancestral arthropod taxon, and all arthropods are
more closely related to each other than to any other organisms (Briggs and
Fortey 1989, Grosberg 1990). A differing view is that arthropods are polyphyletic, because embryological evidence suggests that crustaceans may have arisen
from polychaete-like annelids while Uniramia may have evolved from
oligochaete-like annelids.

In the tree developed for arthropods based on the 18S ribosomal RNA
sequences, chelicerates are represented by the horseshoe crab *Limulus*, crustaceans by the brine shrimp *Artemia*, Uniramia by *Drosophila*, and the millipede by
*Spirobolus*. Field et al. (1988) note that, while the number of arthropod species
they sampled is limited, the results support the hypothesis that arthropods represent a coherent phylum of single origin, rather than a polyphyletic group with
several distinct annelid ancestors. However, Field et al. (1988) sampled relatively few arthropods, and were unable to resolve the relationships within the
Arthropoda.

A second study to resolve the relationships within the Arthropoda using
more groups was conducted by Turbeville et al. (1991). Partial 18S ribosomal
RNA sequences of five chelicerate arthropods, plus a crustacean, myriapod,
insect, chordate, echinoderm, annelid, and platyhelminth were compared. The
sequence data were used to infer phylogeny using three methods—a maximum-
parsimony method, an evolutionary-distance method, and the evolutionary-
parsimony method. The results generated by maximum-parsimony and distance
methods support both monophyly of the Arthropoda and monophyly of the
Chelicerata within the Arthropoda (Figure 13.5). These results are congruent
with phylogenies based on cladistic analyses of morphological characters. The
most parsimonious tree suggests that relationships between the nonchelicerate
arthropods and relationships within the chelicerate clade cannot be reliably
inferred with the partial 18S rRNA sequence data. However, support for monophyly of the Arthropoda is found in the majority of the combinations analyzed if
the coelomates are used as outgroups; monophyly of the Chelicerata is supported
in most combinations. The use of additional sequence data should resolve the
issues surrounding the evolution of the Arthropoda that remain undetermined
by both studies.

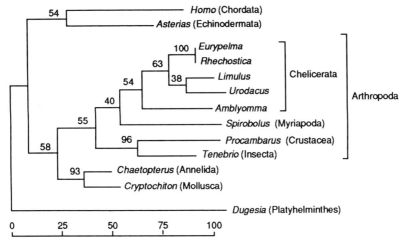

**Figure 13.5** A minimal-length phylogenetic tree derived from maximum parsimony analysis using PAUP with sequence DNA from a partial 18S rRNA sequence of five chelicerate arthropods plus a crustacean, myriapod, insect, chordate, echinoderm, annelid, and platyhelminth. Branch lengths are proportional to the number of nucleotide substitutions. The tree length is 576 nucleotide substitutions, with an overall consistency index of 0.688. Scale bars indicate number of nucleotide substitutions. (Redrawn from Turbeville et al. 1991.)

## The Congruence between Morphological and Molecular Methods

A central aim of evolutionary biology is to understand the relationship between organisms and their evolutionary history. Unfortunately, the fossil record is often inadequate for this purpose, so inferences made about lineages of organisms are based on what the scientist can observe and measure. Many scientists are concerned about using any single method, such as isozyme data or sequence data or morphological traits, to deduce evolutionary patterns.

The possibility exists that inferences concerning phylogenetic relationships that are based on molecular data may not reflect accurately the historical relationship of the taxa from which the data were obtained, producing a 'gene tree/species tree' problem. For example, Powell (1991) pointed out that molecular studies on members of the *Drosophila pseudoobscura* group can lead to conclusions of monophyly, paraphyly, and polyphyly, depending upon which set of data is used to construct the trees. Extensive information is available on the morphology, mating behavior, hybrid formation, chromosomes, and allozymes of the four taxa and all lead to the same conclusions regarding the relationships of these species. DNA–DNA hybridization data of single-copy DNA among the taxa

also yielded results consistent with all previous data and the expected phylogenetic tree. However, a molecular phylogeny of these same taxa based on
RFLP data for a segment of the third chromosome resulted in different conclusions. The different sets of data used to construct the two trees are presumed
accurate and thus neither tree is 'wrong', but reflects different aspects of the
history of the same taxa when different data are considered. DNA–DNA hybridization data are used to evaluate the average divergence between taxa of the
total single-copy nuclear DNA, which consists of about 70% of the *Drosophila*
genome. Probably less than 10% of the genome codes for amino acids of proteins, so the single-copy nuclear DNA analysis of DNA-DNA hybridization
represents virtually all coding DNA and provides an average over nearly the
whole genome, leading to a 'correct' species tree (Powell 1991). However, the
genetic divergence of a single gene or gene region may not accurately reflect
the phylogeny of species. For example, in this case, the third chromosome
represents only about 20% of the total genome, and much of the DNA in this
chromosome will yield the wrong species tree because it contains an inversion
polymorphism that occurred prior to the species split. Furthermore, the inversion polymorphism reduced gene exchange in that region of the chromosome
and was under selection, and therefore was not neutral, which could lead to its
longer persistence in the lineage.

Relatively few studies have examined the degree of consensus obtained
when the same organism is studied using both morphological and molecular
approaches. The following examples show that morphological and molecular
data can lead to different conclusions in some cases, but are congruent in others.

### 12S Ribosomal RNA Sequences of Onychophorans and Arthropods

The relationship of arthropods to other invertebrates has generated considerable debate. One group of particular interest is the Onychophora, or velvet
worms. Onychophorans resemble slugs with legs, and have been involved in the
argument as to whether arthropods are mono- or polyphyletic. Onychophorans
have been described both as mollusks and as the missing link between arthropods and annelids. A polyphyletic hypothesis of arthropod relationships proposes that either (1) onychophorans are more closely related to certain annelids
than to arthropods, or (2) 'arthropodization' has occurred independently at least
three times. A monophyletic hypothesis proposes that (1) onychophorans are
primitive, or (2) closely allied to myriapods and hexapods.

The resolution to the debate appears to be provided by analyses of 329-bp
sequences of the 12S ribosomal RNA from a wide range of arthropods, including
six onychophoran species and organisms from other phyla. Ballard et al. (1993)
concluded that the 12S rRNA sequence data can resolve arthropod relationships
over a broad taxonomic range and that arthropods are monophyletic and include

the onychophorans. This suggests that the characters of the onychophorans, such as a poorly defined head region, represent a degenerate feature as a response to their specialized life style. The conclusions reached by Ballard et al. (1993), however, are based on sequences from a single gene and their strength will be increased if the phylogeny of onychophorans can be confirmed by sequence data from additional genes.

### DNA–DNA Hybridization and Dolichopoda Cave Crickets

Allegrucci et al. (1992) compared the genetic divergence in nuclear genes as determined by DNA–DNA hybridization and the degree of morphological differentiation for different sets of characters of *Dolichopoda* cave crickets. Six populations consisting of five species from central and southern Italy were studied. The overall genetic divergence at nuclear genes was estimated by single-copy DNA–DNA hybridization, and by analyzing allozyme frequencies at 26 loci.

Multivariate analyses were conducted on two groups of morphological characters—(1) epiphallus morphology, and (2) body and appendage morphology. Seven measurements were made on epiphallus shapes and five measurements were used for body and appendage morphology, including pronotum length and width, and metafemur, metatibia, and metatarsus length of adult males. Genetic distance was calculated using Nei's allozyme distance values (Figures 13.6A and 13.6B). Nei's genetic distance method is the most commonly used technique to transform allelic and genotypic frequency data to genetic distances (Nei 1972, 1978, 1987). Nei's method measures the number of codon substitutions per locus that have occurred after a pair of taxa have diverged. However, Nei's estimate is valid only if the rate of substitution per locus is uniform across both loci and lineages.

The results obtained by Allegrucci et al. (1992) showed a close agreement between the branching patterns obtained using the $\Delta T_m$ values from DNA hybridization studies and Nei's distance values obtained using isozyme data (Figures 13.6C and 13.6D). In both cases, the clustering analyses produced trees with similar topologies although they had different branch lengths. Isozyme electrophoresis allowed discrimination between species within each cluster, and populations within species. DNA–DNA hybridization of single-copy DNA separated the two clusters, but was less efficient than isozyme electrophoresis at the intra-cluster level.

However, the two morphological matrices are not congruent with each other or with any of the two molecular matrices (Figures 13.6A and 13.6B), although the branching pattern based on morphology of one character matched to some extent the phylogenies inferred from molecular data. The authors concluded that the species similarities obtained from the morphometric study are more likely to be due to *phenotypic plasticity* and/or short-term adaptation to local

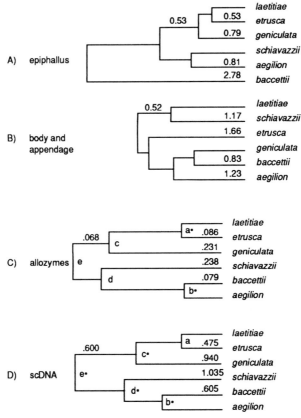

**Figure 13.6** Genetic divergence in nuclear genes as determined by DNA–DNA hybridization and the degree of morphological differentiation for different sets of characters of *Dolichopoda* cave crickets. Six populations consisting of five species from central and southern Italy were studied. The overall genetic divergence at nuclear genes was estimated by single-copy DNA–DNA hybridization and allozyme frequencies at 26 loci. UPGMA trees are based on (A) allozymes, (B) single-copy DNA–DNA hybridization, (C) seven morphometric measurements of the epiphallus, and (D) five morphometric measurements which describe body and appendage shape. The numbers indicate branch length. (Redrawn from Allegrucci et al. 1992.)

environmental conditions than actual phylogenetic relationships (Allegrucci et al. 1992). Unfortunately, the amount of influence nongenetic variation has on the morphological traits chosen for this analysis is not known, and the authors conclude that morphological characters should be chosen carefully to reduce the confusion from such variation.

The absolute rate of molecular change in these *Dolichopoda* cave cricket species, based on single-copy DNA–DNA hybridization data, was estimated to be at least 0.98% divergence per million years per lineage, which is in agreement with calibrations attempted using other insects (Allegrucci et al. 1992). The estimates of time of divergence based on allozyme data (Nei's D) were highly consistent with the estimate obtained from geological data (Allegrucci et al. 1992).

## Adh DNA Sequences and Drosophila Evolution

An examination of the relationship between molecular and morphological data in the Drosophilidae was made by DeSalle and Grimaldi (1991) and by Thomas and Hunt (1993). In an analysis of phylogenetic relationships in the genus *Drosophila*, Thomas and Hunt (1993) compared the results of using DNA sequences for the *alcohol dehydrogenase* (*Adh*) gene with phylogenies produced by independent molecular data (on larval hemolymph protein immunological distances and mitochondrial DNA), by internal morphology characters, or by adult morphology characters. Times of divergence of lineages estimated from the *Adh* sequence data are consistent with available biogeographic and fossil data. Congruence was found for phylogenies produced by the other molecular data and by internal morphology. Phylogenies produced by *Adh* DNA sequences and adult morphology characters were not congruent, but the reasons for the disparities remain unresolved. Likewise, Pelandakis and Solignac (1993) compared nucleotide sequences of 72 species of Drosophilidae from large ribosomal RNA. The molecular phylogenetic trees produced did not correspond to the taxonomical hierarchy. However, so much information (molecular, genetic, and developmental data) is available for this genus that the discrepancies may be resolved soon.

## 16S rDNA Sequences and the Phylogeny of the Hymenoptera

The Hymenoptera is a large and diverse insect order, with two suborders, the Symphyta (sawflies, horntails, and parasitic wood wasps) and Aprocrita (parasitic wasps, ants, solitary and social wasps and bees). Hymenopteran phylogenies have been developed using several characters, but no consensus exists regarding the relationships among them. The Aprocrita are generally interpreted as monophyletic and divided into two groups, the Aculeata and the Terebrantes (Parasitica). The aculeate Hymenoptera include the ants, solitary and social wasps and bees, and are considered monophyletic. The monophyly of the terebrants (parasitic wasps or parasitoids) is controversial, and one classification would consider the Aculeata and the terebrant superfamily Ichneumonoidea (Ichneumonidae + Braconidae) to be sister groups (reviewed in Derr et al. 1992a).

The PCR was used to amplify DNA from the large ribosomal subunit (16S rRNA) of the mitochondrial genome from species in six superfamilies within the suborders Symphyta and Apocrita of the Hymenoptera, and a phylogenetic analysis was conducted using the sequences obtained from the PCR-amplified DNA (Derr et al. 1992a,b). Parsimony analysis provided strong evidence for monophyly of both the Aculeata and terebrants, including the Ichneumonoidea.

The phylogeny of eusociality in honey bees and stingless bees (Apidae) was evaluated using 16S ribosomal RNA gene sequences from 14 species in four tribes (Cameron 1993). The results are not congruent with previous phylogenies based on morphology. For example, the molecular data indicate a close phylogenetic relationship between the primitively eusocial bumble bees and the stingless bees. The results also suggest that advanced eusocial behavior evolved twice independently within the species evaluated (Cameron 1993).

### 16S rDNA Sequences and Deltocephalus-like Leafhoppers

Phylogenies were constructed using mitochondrial 16S rDNA sequences by Fang et al. (1993) for 21 species of *Deltocephalus*-like leafhoppers and compared with the phylogenies produced earlier based on morphological characters. PCR-amplified DNA from field-collected and museum-preserved dried specimens was sequenced. A previous phylogeny based on the morphology of the male genitalia (aedeagus) placed genera into four groups, with the first group having a linear, fused, connective articulated with the aedeagus; the second group having a linear connective articulated with the aedeagus; the third group having a Y-type connective articulated with the aedeagus; and the fourth group having a Y-type connective fused with the aedeagus.

*Deltocephalus*-like leafhoppers (group one) were considered to be monophyletic, based on morphology, distribution, and feeding habits. Phylogenies derived from the rDNA sequences using either parsimony or distance analyses indicated that 18 of the 19 genera in group one belonged to a monophyletic group, thus agreeing with the original classification (Fang et al. 1993). The assignment of the single genus that did not appear to belong to the group one genera has long been controversial and the molecular data support placing this genus in another group. Based on the sequences of 16S rDNA sequences, these leafhopper genera may have diverged more than 200 mya, assuming that substitution rates in 16S rDNA sequences from *Drosophila* and leafhoppers are equal. This suggests that the *Deltocephalus*-like genera of leafhoppers began exploiting grasses as hosts during the very early evolution of grasses.

### RFLP Analyses of mtDNA in Aedes and Anopheles Mosquitoes

Kambhampati and Rai (1991) analyzed seven species in the *Aedes scutellaris* subgroup and four in the *A. albopictus* subgroup for RFLP variation in mtDNA and found extensive variation. Analyses using allozyme data, morphology, biogeography, and ecology were not completely congruent with the phylogenies drawn from the mtDNA data. Despite a lack of significant morphological divergence among the closely related species in the A. *scutellaris* and the A. *albopictus* subgroups, the mtDNA had diverged extensively. Possibly this is because different parts of the genome (nuclear and mitochondrial) evolved at different rates and because population bottlenecks have influenced the mtDNA diversity. The lack of congruence between the phylogenies based on mtDNA and other characters indicates that more than one approach should be employed in analyzing systematics problems with these mosquitoes.

DNA analyses often provide more detailed information on systematic status than isozyme analyses, but this is not always the case. Mitchell et al. (1992) compared the ease with which four morphologically identical sibling species of the mosquito *Anopheles quadrimaculatus* could be identified using allozymes and RFLP analyses of mtDNA and rDNA. They found few intraspecific differences in RFLP patterns, but did find consistent species-specific differences in RFLP patterns among the four species for both mtDNA and rDNA, and no evidence for hybridization among the four species. However, RFLP analysis of mtDNA was less appropriate than isozyme analysis for studying genetic structure of populations within these mosquito species because little variability was found within or between populations. It is likely that mosquitoes exhibit less variation in mtDNA than *Drosophila*, because Mitchell et al. (1992) found only 2–3% sequence divergence between the mitochondrial genomes of these four species of A. *quadrimaculatus* compared with species of *Drosophila*, which exhibit 5–10% nucleotide divergence.

### mtDNA Phylogeny of Papilio Species Groups

An analysis of the mitochondrial DNA variation in several swallowtail butterfly (Papilionidae) species groups has provided information on their phylogeny and evolution (Sperling 1993a,b). The phylogeny constructed from restriction analysis of mtDNA in the *Papilio machaon* species group, which exhibits a confusing set of color morphs, ecological races, and incomplete reproductive barriers, is largely congruent with one based on allozymes and color pattern (Sperling 1993a). Analysis of mtDNA in the *Papilio glaucus* and P. *troilus* groups generally confirmed traditional species limits, except for P. *rutulus* and P. *eurymedon*, which share their mtDNA yet maintain morphological and ecological differences in sympatry (Sperling 1993b). Viable hybrid adults of both sexes of P. *rutulus* and P. *eurymedon* can be produced in the laboratory, although few natural hybrids have been found.

## Molecular Genetics and Speciation

One of the central questions of biology is how a continuous process of evolution can produce morphologically discontinuous 'species' (Coyne 1992). A genetical theory of speciation proposes that species are groups of interbreeding populations that are reproductively isolated from other such groups, with reproductive barriers divided into prezygotic isolating factors (mating discrimination, different habitat preferences, etc.) and postzygotic isolating factors (hybrid inviability and sterility). Reproductive isolation, in concert with selection and genetic drift, creates and expands the morphological differences among species living in the same area (Dobzhansky 1937, Mayr 1942). Physical isolation (allopatry) leads inevitably to evolutionary change through natural selection or drift, and reproductive isolation evolves as a by-product of the genetic changes. Any resultant hybrid inviability could be due to the development of divergent developmental systems. Reproductive isolation could be increased if incompletely isolated populations become sympatric (live together in the same area) so that selection would fix the alleles that reduce interspecific mating. The process of increasing isolation is called 'reinforcement', but how often this process occurs is now debated (Coyne 1992). Likewise, the extent of sympatric speciation, in which reproductive isolation occurs without geographic isolation, remains controversial; the sympatric host races of the tephritid *Rhagoletis pomonella* represent one well-documented example (Feder et al. 1988).

How many genes are involved in speciation? The genetic basis of speciation is assumed to be due to changes in more than one gene, but the number is unknown in most cases. Changes in more than two genes are considered the minimum required for reproductive isolation. Changes in segments of the genome, such as inversions or translocations, can also result in reproductive isolation. Most often, reproductive isolation is considered to be polygenically determined. However, few genetic studies have been conducted on the genetics of reproductive isolation, or temporal or habitat isolation, so that few generalizations can be made about the number of genes responsible for speciation (Coyne 1992).

Many questions about speciation remain unanswered. How important are conventional gene mutations compared with novel genetic elements, such as repeated sequences, microorganisms, or transposable elements? How often are 'speciation' genes altered in their coding sequence and in noncoding regulatory regions? How often is reproductive isolation based on polyploidy, or on chromosomal rearrangements of chromosomes? How often is reproductive isolation and speciation influenced by hybrid inviability caused by infectious bacteria such as *Wolbachia*? If transposable elements, polyploidy, or infectious microorganisms cause speciation events, would they produce a rapid change without significant genetic change in their host organism? Finally, what is the relative importance of selection and drift in speciation?

Molecular genetic techniques may be able to provide answers to some of these questions (Coyne 1992). Molecular mapping could determine how often reproductive isolation has a simple or complex genetic basis. Cloning and char-

acterizing genes important in speciation may provide information on how repro-
ductive isolation occurs at the molecular level. The sequencing of the *Drosophila melanogaster* genome may provide new methods for analyzing genome evolution in other arthropods.

## Research Needed in Molecular Systematics and Evolution

Perhaps the greatest need in molecular systematics and evolution is to resolve which genes are informative for which questions and to determine how to analyze molecular data appropriately (Crozier 1993). The wealth of informa-tion obtained from DNA sequences can provide insights into evolution and speciation, but we don't know how to interpret the data yet. For example, how does one choose appropriate genes for a specific problem? How can estimates of genetic distance (D) be used to make judgments about species status or date of speciation events? There is little doubt that molecular markers can document patterns of variation with a high degree of resolution, but their interpretation depends upon whether the speciation event was sympatric or allopatric, with the value of genetic distance lower for sympatric species pairs than for allopatric species pairs of *Drosophila*. Harrison (1991) suggests that extreme caution should be employed when using D as an indicator of species status because different species have different genetic architectures, barriers to genetic exchange, and different evolutionary histories.

More molecular data are being produced than we know how to analyze. Within the next few years, much of the genome of *Drosophila melanogaster* will be sequenced, which will provide opportunities to examine questions about genome evolution, rather than just the evolution of a few genes or portions of a gene. Ideally, the *Drosophila* genome analysis will provide information on how best to analyze insect phylogenies, and on additional genes suitable for answering spe-cific systematic and evolutionary questions.

## Some Relevant Journals

*Biochemical Systematics and Ecology*, Pergamon Press.
*Evolution*, Haworth Press, and sponsored by the Society for the Study of Evolution.
*Molecular Biology and Evolution*, University of Chicago Press, and sponsored by the Society for Molecular Biology and Evolution.
*Molecular Phylogenetics and Evolution*, Academic Press.
*Journal of Molecular Evolution*, Springer-Verlag and sponsored by the International Society of Mo-lecular Evolution.

## References Cited

Adamson, R. E., R. D. Ward, M. D. Feliciangeli and R. Maingon. 1993. The application of random amplified polymorphic DNA for sandfly species identification. Med. Vet. Entomol. 7: 203–207.

Allegrucci, G., A. Caccone, D. Cesaroni and V. Sbordoni. 1992. Evolutionary divergence in *Dolichopoda* cave crickets: A comparison of single copy DNA hybridization data with allozymes and morphometric distances. J. Evol. Biol. 5: 121–148.

Aubert, J. and M. Solignac. 1990. Experimental evidence for mitochondrial DNA introgression between *Drosophila* species. Evolution 44: 1272–1282.

Avers. C. J. 1989. *Process and Pattern in Evolution*. Oxford Univ. Press, New York.

Avise, J. C., J. Arnold, R. M. Ball, E. Bermingham, T. Lamb, J. E. Neigel, C. A. Reeb and N. C. Saunders. 1987. Intraspecific phylogeography: The mitochondrial DNA bridge between population genetics and systematics. Annu. Rev. Ecol. Syst. 18: 489–522.

Ayala, F. J. 1986. On the virtues and pitfalls of the molecular evolutionary clock. Heredity 77: 226–235.

Azeredo-Espin, A. M. L., R. F. W. Schroder, M. D. Huettel and W. S. Sheppard. 1991. Mitochondrial DNA variation in geographic populations of Colorado potato beetle, *Leptinotarsa decemlineata* (Coleoptera: Chrysomelidae). Experientia 47: 483–485.

Bachmann, L., J. M. Schibel, M. Raab and D. Sperlich. 1993. Satellite DNA as a taxonomic marker. Biochem. System. Ecol. 21: 3–11.

Baldridge, G. D. and A. M. Fallon. 1991. Nucleotide sequence of a mosquito 18S ribosomal RNA gene. Biochim. Biophys. Acta 1089: 396–400.

Baldridge, G. D. and A. M. Fallon. 1992. Primary structure of the ribosomal DNA intergenic spacer from the mosquito, *Aedes albopictus*. DNA Cell Biol. 11: 51–59.

Baldridge, G. D., M. W. Dalton and A. M. Fallon. 1992. Is higher-order structure conserved in eukaryotic ribosomal DNA intergenic spacers? J. Mol. Evol. 335: 514–523.

Ballard, J. W. O., G. J. Olsen, D. P. Faith, W. A. Odgers, D. M. Rowell and P. W. Atkinson. 1993. Evidence from 12S ribosomal RNA sequences that onychophorans are modified arthropods. Science 258: 1345–1348.

Beard, C. B., D. M. Hamm and F. S. Collins. 1993. The mitochondrial genome of the mosquito *Anopheles gambiae*: DNA sequence, genome organization and comparison with the mitochondrial sequences of other insects. Insect Mol. Biol., 2: 100–124.

Besansky, N. J., S. M. Paskewitz, D. M. Hamm and F. H. Collins. 1992. Distinct families of site-specific retrotransposons occupy identical positions in the rRNA genes of *Anopheles gambiae*. Mol. Cell. Biol. 12: 5102–5110.

Bigot, Y., F. Lutcher, M. H. Hamelin and G. Periquet. 1992. The 28S ribosomal RNA-encoding gene of Hymenoptera: Inserted sequences in the retrotransposon-rich regions. Gene 121: 347–352.

Black, W. C., D. K. McLain and K. S. Rai. 1989. Patterns of variation in the rDNA cistron within and among world populations of a mosquito, *Aedes albopictus* (Skuse). Genetics 121: 539–550.

Boyce, T. M., M. E. Zwick and C. F. Aquadro. 1989. Mitochondrial DNA in the bark weevils: Size, structure and heteroplasmy. Genetics 123: 825–836.

Briggs, D. E. G. and R. A. Fortey. 1989. The early radiation and relationships of the major arthropod groups. Science 246: 241–243.

Britten, R. J., D. E. Graham and B. R. Neufeld. 1974. Analysis of repeating DNA sequences by reassociation. Methods Enzymol. 29: 363–418.

Brower, A. V. Z. and T. M. Boyce. 1991. Mitochondrial DNA variation in monarch butterflies. Evolution 45: 1281–1286.

Brown, T. A., Ed. 1991. *Molecular Biology LabFax*. Bios Sci. Publ., Oxford.

Burke, W. D., D. G. Eickbush, Y. Xiong, J. Jakubczak and T. H. Eickbush. 1993. Sequence relationship of retrotransposable elements R1 and R2 within and between divergent insect species. Mol. Biol. Evol. 10: 163–185.

Caccone, A. and V. Sbordoni. 1987. Molecular evolutionary divergence among North American cave crickets. I. Allozyme variation. Evolution 41: 1198–1214.

Caccone, A. and J. R. Powell. 1987. Molecular evolutionary divergence among North American cave crickets. II. DNA–DNA hybridization. Evolution 41: 1215–1238.

Caccone, A. and J. R. Powell. 1990. Extreme rates and heterogeneity in insect DNA evolution. J. Mol. Evol. 30: 273–280.

Caccone, A., R. DeSalle and J. R. Powell. 1988a. Calibration of the change in thermal stability of DNA duplexes and degree of base pair mismatch. J. Mol. Evol. 27: 212–216.

Caccone, A., G. D. Amato and J. R. Powell. 1988b. Rates and patterns of scnDNA and mtDNA divergence within the *Drosophila melanogaster* subgroup. Genetics 118: 671–683.

Caccone, A., J. M. Gleason and J. R. Powell. 1992. Complementary DNA-DNA hybridization in *Drosophila*. J. Mol. Evol. 34: 130–140.

Cameron, S. A. 1993. Multiple origins of advanced eusociality in bees inferred from mitochondrial DNA sequences. Proc. Natl. Acad. Sci. USA 90: 8687–8691.

Cano, R. J., H. N. Poinar, D. W. Roubik and G. O. Poinar, Jr. 1992. Enzymatic amplification and nucleotide sequencing of portions of the 18s rRNA gene of the bee *Proplebeia dominicana* (Apidae: Hymenoptera) isolated from 25–40 million year old Dominican amber. Med. Sci. Res. 20: 619–622.

Clary, D. O. and D. R. Wolstenholme. 1985a. The mitochondrial DNA molecule of *Drosophila yakuba*: Nucleotide sequence, gene organization, and genetic code. J. Mol. Evol. 22: 252–271.

Clary, D. O. and D. R. Wolstenholme. 1985b. The ribosomal RNA genes of *Drosophila* mitochondria DNA. Nucleic Acids Res. 13: 4029–4045.

Cockburn, A. F., S. E. Mitchell and J. A. Seawright. 1990. Cloning of the mitochondrial genome of *Anopheles quadrimaculatus*. Arch. Insect Biochem. Physiol. 12: 31–36.

Coen, E. S., J. M. Thoday and G. Dover. 1982. Rate of turnover of structural variants in the rDNA gene family of *Drosophila melanogaster*. Nature 295: 564–568.

Collins, F. H., V. Finnerty and V. Petrarca. 1988. Ribosomal DNA-probes differentiate five cryptic species in the *Anopheles gambiae* complex. Parasitologia 30: 231–240.

Collins, F. H., C. H. Porter and S. E. Cope. 1990. Comparison of rDNA and mtDNA in the sibling species *Anopheles freeborni* and *A. hermsi*. 1990. Am. J. Trop. Med. Hyg. 42: 417–423.

Cornuet, J. M. and L. Garnery. 1991. Mitochondrial DNA variability in honeybees and its phylogeographic implications. Apidologie 22: 627–642.

Cornuet, J. M., L. Garnery and M. Solignac. 1991. Putative origin and function of the intergenic region between COI and COII of *Apis mellifera* L. mitochondrial DNA. Genetics 1128: 393–403.

Coyne, J. A. 1992. Genetics and speciation. Nature 355: 511–515.

Cross, N. C. P. and G. A. Dover. 1987. Tsetse fly rDNA: An analysis of structure and sequence. Nucleic Acids Res. 15: 15–30.

Crozier, R. H. 1993. Molecular methods for insect phylogenies. Pp. 164–221. In: *Molecular Approaches to Fundamental and Applied Entomology*. J. Oakeshott and M. J. Whitten, Eds. Springer-Verlag, New York.

Crozier, R. H. and Y. C. Crozier. 1993. The mitochondrial genome of the honeybee *Apis mellifera*: Complete sequence and genome organization. Genetics 133: 97–117.

Daly, H. V., J. T. Doyen, and P. R. Ehrlich. 1978. *Introduction to Insect Biology and Diversity*. McGraw-Hill, New York.

Derr, J. N., S. K. Davis, J. B. Woolley and R. A. Wharton. 1992a. Variation and the phylogenetic utility of the large ribosomal subunit of mitochondrial DNA from the insect order Hymenoptera. Mol. Phylogenet. Evol. 1: 136–147.

Derr, J. N., S. K. Davis, J. B. Woolley and R. A. Wharton. 1992b. Reassessment of the 16S rRNA nucleotide sequence from members of the parasitic Hymenoptera. Mol. Phylogenet. Evol. 1: 338–341.

DeSalle, R. 1992. The phylogenetic relationships of flies in the family Drosophilidae deduced from mtDNA sequences. Mol. Phylogenet. Evol. 1: 31–40.

DeSalle, R. and D. A. Grimaldi. 1991. Morphological and molecular systematics of the Drosophilidae. Annu. Rev. Ecol. Syst. 22: 447–475.

DeSalle, R., T. Friedman, E. M. Prager and A. C. Wilson. 1987. Tempo and mode of sequence evolution in mitochondrial DNA of Hawaiian *Drosophila*. J. Mol. Evol. 26: 157–164.

Dobzhansky, T. 1937. *Genetics and the Origin of Species*. Columbia Univ. Press, New York.

Doolittle, R. F., Ed. 1990. *Molecular Evolution: Computer Analysis of Protein and Nucleic Acid Sequences* (Methods Enzymol., vol. 183). Academic Press, New York.

Dowling, T. E., C. Moritz and J. D. Palmer. 1990. Nucleic acids II: Restriction site analysis. Pp. 250–317. In: *Molecular Systematics*. D. M. Hillis and C. Moritz, Eds. Sinauer Assoc., Sunderland, MA.

Eickbush, T. H. and B. Robins. 1985. *Bombyx mori* 28S ribosomal genes contain insertion elements similar to the Type I and II elements of *Drosophila melanogaster*. EMBO J. 4: 2281–2285.

Fallon, A. M., R. J. Blahnik, G. D. Baldridge and Y. J. Park. 1991. Ribosomal DNA structure in *Aedes* mosquitoes (Diptera: Culicidae) and their cell lines. J. Med. Entomol. 28: 637–644.

Fang, Q. W. C. Black IV, H. D. Blocker and R. F. Whitcomb. 1993. A phylogeny of new world *Deltocephalus*-like leafhopper genera based on mitochondrial 16S ribosomal DNA sequences. Mol. Phylogenet. Evol. 2: 119–131.

Feder, J. L., C. A. Chilcote and G. L. Bush. 2988. Genetic differentiation between sympatric host races of the apple maggot fly *Rhagoletis pomonella*. Nature 336: 61–64.

Felsenstein, J. 1982. Numerical methods for inferring evolutionary trees. Q. Rev. Biol. 57: 379–404.

Felsenstein, J. 1988. Phylogenies from molecular sequences: Inference and reliability. Annu. Rev. Genet. 22: 521–565.

Felsenstein, J. 1993. PHYLIP: Phylogeny Inference Package, Version 3.5, Univ. Washington, Seattle, WA.

Field, K. G., G. J. Olsen, D. J. Lane, S. J. Giovannoni, M. T. Ghiselin, E. C. Raff, N. R. Pace and R. A. Raff. 1988. Molecular phylogeny of the animal kingdom. Science 239: 748–753.

Forey, P. L. 1992. *Cladistics: A Practical Course in Systematics*. Oxford Univ. Press, Oxford.

Friedlander, T. P., J. C. Regier and C. Mitter. 1992. Nuclear gene sequences for higher level phylogenetic analysis: 14 promising candidates. Syst. Biol. 41: 483–490.

Gawal, N. J. and A. C. Bartlett. 1993. Characterization of differences between whiteflies using RAPD-PCR. Insect Mol. Biol. 2: 33–38.

Goddard, K. A. Caccone and J. R. Powell. 1990. Evolutionary implications of DNA divergence in the *Drosophila obscura* group. Evolution 44: 1656–1670.

Goldenthal, M. J., K. A. McKenna and D. J. Joslyn. 1991. Mitochondrial DNA of the blowfly *Phormia regina*: Restriction analysis and gene localization. Biochem. Genet. 29: 1–11.

Gray, M. W. 1989. Origin and evolution of mitochondrial DNA. Annu. Rev. Cell Biol. 5: 25–50.

Gribskov, M. and J. Devereux, Eds. 1991. *Sequence Analysis Primer*. Stockton Press, New York.

Grosberg, R. K. 1990. Out on a limb: Arthropod origins. Science 250: 632–669.

Hadrys, H., M. Balick and B. Schierwater. 1992. Applications of random amplified polymorphic DNA (RAPD) in molecular ecology. Mol. Ecol. 1: 55–63.

Hames, B. D. and D. Rickwood, Eds. 1981. *Gel Electrophoresis of Proteins: A Practical Approach*. IRL Press, Oxford.

Harrison, R. G. 1991. Molecular changes at speciation. Annu. Rev. Ecol. Syst. 22: 281–308.

Harrison, R. G., D. M. Rand and W. C. Wheeler. 1987. mtDNA variation in field crickets across a narrow hybrid zone. Mol. Biol. Evol. 4: 144–158.

Heath, D. D., G. K. Iwama and R. H. Devlin. 1993. PCR primed with VNTR core sequences yields species specific patterns and hypervariable probes. Nucleic Acids Res. 21: 5782–5785.

Hennig, W. 1966. *Phylogenetic Systematics*. Univ. Illinois Press, Urbana.

Hillis, D. M. and C. Moritz. 1990. An overview of applications of molecular systematics, Pp. 502–515. In: *Molecular Systematics*. D. M. Hillis and C. Moritz, Eds. Sinauer Assoc., Sunderland, MA.

Hillis, D. M., A. Larson, S. K. Davis and E. A. Zimmer. 1990. Nucleic Acids III: Sequencing. Pp. 318–370. In: *Molecular Systematics*. D. M. Hillis and C. Moritz, Eds. Sinauer Assoc., Sunderland, MA.

Howe, C. J. and E. S. Ward, Eds. 1989. *Nucleic Acids Sequencing: A Practical Approach.* IRL Press, New York.

HsuChen, C. C. and D. T. Dubin. 1984. A cluster of four transfer RNA genes in mosquito mitochondrial DNA. Biochem. Int. 8: 385–391.

HsuChen, C. C., R. M. Kotin and D. T. Dubin. 1984. Sequences of the coding and flanking regions of the large ribosomal subunit RNA gene of mosquito mitochondria. Nucleic Acids Res. 12: 7771–7785.

Innis, M. A., D. H. Gelfand, J. J. Sninsky and T. J. White, Eds. 1990. *PCR Protocols: A Guide to Methods and Applications.* Academic, San Diego, CA.

Jakubczak, J. L., Y. Xiong and T. H. Eickbush. 1990. Type I (R1) and Type II (R2) ribosomal DNA insertions of *Drosophila melanogaster* are retrotransposable elements closely related to those of *Bombyx mori.* J. Mol. Biol. 212: 37–52.

Jakubczak, J. L., W. D. Burke and T. H. Eickbush. 1991. Retrotransposable elements R1 and R2 interrupt the rRNA genes of most insects. Proc. Natl. Acad. Sci. USA 88: 3295–3299.

Juan, C., P. Vazquez, J. M. Rubio, E. Petitpierre and G. M. Hewitt. 1993. Presence of highly repetitive DNA sequences in *Tribolium* flour-beetles. Heredity 70: 1–8.

Kambhampati, S. and K. S. Rai. 1991a. Temporal variation in the ribosomal DNA nontranscribed spacer of *Aedes albopictus* (Diptera: Culicidae). Genome 34: 293–297.

Kambhampati, S. and K. S. Rai. 1991b. Variation in mitochondrial DNA of *Aedes* species (Diptera: Culicidae). Evolution 45: 120–129.

Kambhampati, S., W. C. Black IV and K. S. Rai. 1992. Random amplified polymorphic DNA of mosquito species and populations (Diptera: Culicidae): Techniques, statistical analysis and applications. J. Med. Entomol. 29: 939–945.

Kimura, M. 1968. Evolutionary rate at the molecular level. Nature 217: 624–626.

Kimura, M. 1983. *The Neutral Theory of Molecular Evolution.* Cambridge Univ. Press, Cambridge.

Kimura, M. 1987. Molecular evolutionary clock and the neutral theory. J. Mol. Evol. 26: 24–33.

Kondo, R., Y. Satta, E. T. Matsuura, H. Ishiwa, N. Takahata and S. I. Chigusa. 1990. Incomplete maternal transmission of mitochondrial DNA in *Drosophila.* Genetics 126: 657–663.

Kukalova-Peck, J. 1991. Fossil history and the evolution of hexapod structures. Pp. 141–179. In: *The Insects of Australia.* Vol. I. 2nd Ed. Melbourne Univ. Press, Melbourne.

Labandeira, C. C. and J. J. Sepkoski, Jr. 1993. Insect diversity in the fossil record. Science 261: 310–315.

Labandeira, C. C., B. S. Beall and F. M. Hueber 1988. Early insect diversification: Evidence from a lower Devonian bristletail from Quebec. Science 242: 913–916.

Lansman, R. A., J. C. Avise and M. D. Huettel. 1983. Critical experimental test of the possibility of "paternal leakage" of mitochondrial DNA. Proc. Natl. Acad. Sci. USA 80: 1969–1971.

Lansman, R. A. R. O. Shade, J. F. Shapiro and J. C. Avise. 1981. The use of restriction endonucleases to measure mitochondrial sequence relatedness in natural populations. III. Techniques and potential applications. J. Mol. Evol. 17: 214–226.

Lewontin, R. C. and J. Hubby. 1966. A molecular approach to the study of genic heterozygosity in natural populations. II. Amounts of variation and degree of heterozygosity in natural populations of *Drosophila pseudoobscura.* Genetics 54: 595–609.

Li, W. H. and D. Graur. 1991. *Fundamentals of Molecular Evolution.* Sinauer Publ., Sunderland, MA.

Li, J. and R. Chakraborty. 1994. Estimation of genetic distance and coefficient of gene diversity from single-probe multilocus DNA fingerprinting data. Mol. Biol. Evol. 11: 120–127.

Liu, H. and A. T. Beckenbach. 1992. Evolution of the mitochondrial cytochrome oxidase II gene among 10 orders of insects. Mol. Phylogenet. Evol. 1: 41–52.

Martin, A. and C. Simon. 1990. Differing levels of among-population divergence in the mitochondrial DNA of periodical cicadas related to historical biogeography. Evolution 44: 1066–1080.

Matsuura, E. T., H. Fukuda and S. I. Chigusa. 1991. Mitochondrial DNA heteroplasmy maintained in natural populations of *Drosophila simulans* in Reunion. Genet. Res., Camb. 57: 123–126.

Maxon, L. R. and R. D. Maxon. 1990. Proteins II: Immunological techniques. Pp. 127–155. In: *Molecular Systematics*. D. M. Hillis and C. Moritz, Eds. Sinauer Assoc., Sunderland, MA.

May, B. 1992. Starch gel electrophoresis of allozymes. Pp. 1–27. In: *Molecular Genetic Analysis of Populations: A Practical Approach*. A. R. Hoelzel, Ed. IRL Press, Oxford.

Mayr, E. 1942. *Systematics and the Origin of Species*. Columbia Univ. Press, New York.

Mayr, E. 1970. *Populations, Species, and Evolution*. Harvard Univ. Press, Cambridge, MA.

Mayr, E. and P. D. Ashlock. 1991. *Principles of Systematic Zoology*. 2nd Ed. MGraw-Hill, New York.

McCracken, A., I. Uhlenbusch and G. Gellissen. 1987. Structure of cloned *Locusta migratoria* mitochondrial genome: Restriction mapping and sequence of its ND-1 (URF-I) gene. Curr. Genet. 11: 625–630.

Mitchell, S. E., S. K. Narang, A. F. Cockburn, J. A. Seawright and M. Goldenthal. 1992. Mitochondrial and ribosomal DNA variation among members of the *Anopheles quadrimaculatus* (Diptera: Culicidae) species complex. Genome 35: 939–950.

Mitchell, S. E., A. F. Cockburn and J. A. Seawright. 1993. The mitochondrial genome of *Anopheles quadrimaculatus*, species A: Complete nucleotide sequence and gene organization. Genome 36: 1058–1073.

Miyamoto, M. M. and J. Cracraft, Eds. 1991. *Phylogenetic Analysis of DNA Sequences*. Oxford Univ. Press, Oxford.

Moran, N. A., M. A. Munson, P. Baumann and H. Ishikawa. 1993. A molecular clock in endosymbiotic bacteria is calibrated using the insect hosts. Proc. R. Soc. Lond. B. 253: 167–171.

Moritz, C. and D. M. Hillis. 1990. Molecular systematics: Context and controversies. Pp. 1–12. In: *Molecular Systematics*. D. M. Hillis and C. Moritz, Eds. Sinauer Assoc., Sunderland, MA.

Moritz, C., T. E. Dowling and W. M. Brown. 1987. Evolution of animal mitochondrial DNA: relevance for population biology and systematics. Annu. Rev. Evol. Syst. 18: 269–92.

Murphy, R. W., J. W. Sites, Jr., D. G. Buth and C. H. Haufler. 1990. Proteins I: Isozyme electrophoresis. Pp. 45–126. In: *Molecular Systematics*. D. M. Hillis and C. Moritz, Eds. Sinauer Assoc., Sunderland, MA.

Nei, M. 1972. Genetic distance between populations. Amer. Natur. 106: 283–292.

Nei, M. 1978. Estimation of average heterozygosity and genetic distance from a small number of individuals. Genetics 89: 583–590.

Nei, M. 1987. *Molecular Evolutionary Genetics*. Columbia Univ. Press, New York.

Ogino, K., H. Eda-Fujiwara, H. Fujiwara and H. Ishikawa. 1990. What causes the aphid 28S rRNA to lack the hidden break? J. Mol. Evol. 30: 509–513.

Orrego, C. and F. Agudelo-Silva. 1993. Genetic variation in the parasitoid wasp *Trichogramma* (Hymenoptera: Trichogrammatidae) revealed by DNA amplification of a section of the nuclear ribosomal repeat. Fla. Entomol. 76: 519–524.

Pashley, D. P. and L. D. Ke. 1992. Sequence evolution in mitochondrial ribosomal and ND-1 genes in Lepidoptera: Implications for phylogenetic analyses. Mol. Biol. Evol. 9: 1061–1075.

Paskewitz, S. M., K. Ng, M. Coetzee and R. H. Hunt. 1993. Evaluation of the polymerase chain reaction method for identifying members of the *Anopheles gambiae* (Diptera: Culicidae) complex in southern Africa. J. Med. Entomol. 30: 953–957.

Pasteur, N., G. Pasteur, F. Bonhomme, J. Catalan and J. Britton-Davidian. 1988. *Practical Isozyme Genetics*. Ellis Horwood, Chichester.

Pelandakas, M. and M. Solignac. 1993. Molecular phylogeny of *Drosophila* based on ribosomal RNA sequences. J. Mol. Evol. 37: 525–543.

Porter, C. H. and F. H. Collins. 1991. Species-diagnostic differences in a ribosomal DNA internal transcribed spacer from the sibling species *Anopheles freeborni* and *Anopheles hermsi* (Diptera: Culicidae). Am. J. Trop. Med. Hyg. 45: 271–279.

Post, R. J., P. K. Flook and M. D. Wilson. 1992. DNA analysis in relation to insect taxonomy, evolution and identification. Pp. 21–34. In: *Insect Molecular Science*. J. M. Crampton and P. Eggleston, Eds. Academic Press, London.

Powell. J. R. 1991. Monophyly/paraphyly/polyphyly and gene/species trees: An example from *Drosophila*. Mol. Biol. Evol. 8: 892–896.

Powell, J. R. and A. Caccone. 1990. The TEACL method of DNA-DNA hybridization: Technical considerations. J. Mol. Evol. 30:267–272.

Powell, J. R., A. Caccone, G. D. Amato and C. Yoon. 1986. Rates of nucleotide substitution in *Drosophila* mitochondrial DNA and nuclear DNA are similar. Proc. Natl. Acad. Sci. USA 83: 9090–9093.

Powers, T. O., S. G. Jensen, S. D. Kindler, C. J. Stryker and L. J. Sandall. 1989. Mitochondrial DNA divergence among greenbug (Homoptera: Aphididae) biotypes. Ann. Entomol. Soc. Am. 82: 298–302.

Raich, T. J., J. L. Archer, M. A. Robertson, W. J. Tabachnick and B. J. Beaty. 1993. Polymerase chain reaction approaches to *Culicoides* (Diptera: Ceratopogonidae) identification. J. Med. Entomol. 30: 228–232.

Rand, D. M. and R. G. Harrison. 1989. Molecular population genetics of mtDNA size variation in crickets. Genetics 121: 551–569.

Reeck, G. R., C. deHaen, D. C. Teller, R. F. Doolittle, W. M. Fitch, R. E. Dickerson, P. Chambon, A. D. McLachlan, E. Margoliash, T. H. Jukes and E. Zuckerkandl. 1987. "Homology" in proteins and nucleic acids: A terminology muddle and a way out of it. Cell 50: 667.

Roehrdanz, R. L. 1989. Intraspecific genetic variability in mitochondrial DNA of the screworm fly (*Cochliomyia hominivorax*). Biochem. Genet. 27: 551–569.

Rojas-Rousse, D., Y. Bigot and G. Periquet. 1993. DNA insertions as a component of the evolution of unique satellite DNA families in two genera of parasitoid wasps: *Diadromus* and *Eupelmus* (Hymenoptera). Mol. Biol. Evol. 10: 383–396.

Satta, Y., H. Ishiwa and S. I. Chigusa. 1987. Analysis of nucleotide substitutions of mitochondrial DNAs in *Drosophila melanogaster* and its sibling species. Mol. Biol. Evol. 4: 638–650.

Satta, Y. and N. Takahata. 1990. Evolution of *Drosophila* mitochondrial DNA and the history of the *melanogaster* subgroup. Proc. Natl. Acad. Sci. USA 87: 9558–9562.

Schmitz, J. and R. F. A. Moritz. 1990. Mitochondrial DNA variation in social wasps (Hymenoptera, Vespidae). Experientia 46: 1068–1072.

Scott, J. A., W. G. Brogdon and F. H. Collins. 1993. Identification of single specimens of the *Anopheles gambiae* complex by the polymerase chain reaction. Am. J. Trop. Hyg. 49: 520–529.

Sessions, S. K. 1990. Chromosomes: molecular cytogenetics. Pp. 156–203. In: *Molecular Systematics*. D. M. Hillis and C. Moritz, Eds. Sinauer Assoc., Sunderland, MA.

Sheppard, W. S., G. J. Steck and B. A. McPheron. 1992. Geographic populations of the medfly may be differentiated by mitochondrial DNA variation. Experientia 48: 1010–1013.

Sibley, C. G. and J. E. Ahlquist. 1981. The phylogeny and relationships of the ratite birds as indicated by DNA-DNA hybridization. Pp. 301–335. In: *Evolution Today*. G. G. E. Scudder and J. L. Reveal, Eds. Carnegie-Mellon Univ., Pittsburgh, PA.

Simon, C. 1991. Molecular systematics at the species boundary: Exploiting conserved and variable regions of the mitochondrial genome of animals via direct sequencing from amplified DNA. Pp. 35–71. In: *Molecular Techniques in Taxonomy*. G. M. Hewitt, A. W. B. Johnston and J. P. W. Young, Eds. Springer-Verlag, Berlin.

Simon, C., S. Paabo, T. D. Kocher and A. C. Wilson. 1990. Evolution of mitochondrial ribosomal RNA in insects as shown by the polymerase chain reaction. Pp. 235–244. In: *Molecular Evolution*. M. T. Clegg and S. J. O'Brien, Eds. UCLA Symp. Molec. Cell. Biol. New Series, Vol. 122, A. R. Liss, New York.

Simon, C., A. Franke and A. Martin. 1991. The polymerase chain reaction: DNA extraction and amplification. Pp. 329–355. In: *Molecular Techniques in Taxonomy.* G. M. Hewitt, W. Johnston and J. P. W. Young, Eds. Springer-Verlag, Berlin.

Sneath, P. H. A. and R. R. Sokal. 1973. *Numerical Taxonomy.* 2nd Ed. Freeman, New York.

Sperling, F. A. H. 1993a. Mitochondrial DNA phylogeny of the *Papilio machaon* species group (Lepidoptera: Papilionidae). Mem. Entomol. Soc. Can. 165: 233–242.

Sperling, F. A. H. 1993b. Mitochondrial DNA variation and Haldane's rule in the *Papilio glaucus* and *P. troilus* species groups. Heredity 71: 227–233.

Stewart, C. B. 1993. The powers and pitfalls of parsimony. Nature 361: 603–607.

Swofford, D. L. and G. J. Olsen. 1990. Phylogeny reconstruction. Pp. 411–501. In: *Molecular Systematics.* D. M. Hillis and C. Moritz, Eds. Sinauer Assoc., Sunderland, MA.

Tabachnick, W. J. 1992. Genetic differentiation among populations of *Culicoides variipennis* (Diptera: Ceratopogonidae), the North American vector of bluetongue virus. Ann. Entomol. Soc. Am. 85: 140–147.

Tamura, K. 1992. The rate and pattern of nucleotide substitution in *Drosophila* mitochondrial DNA. Mol. Biol. Evol. 9: 814–825.

Tares, S. J. M. Cornuet and P. Abad. 1993. Characterization of an unusually conserved *Alu* I highly reiterated DNA sequence family from the honeybee, *Apis mellifera.* Genetics 134: 1195–1204.

Tautz, D., C. Tautz, D. Webb and G. A. Dover. 1987. Evolutionary divergence of promoters and spacers in the rDNA family of four *Drosophila* species. J. Mol. Biol. 195: 525–542.

Taylor, D. B., L. Hammack and R. L. Roehrdanz. 1991. Reproductive compatibility and mitochondrial DNA restriction site analysis of New World screwworm, *Cochliomyia hominivorax,* from North Africa and Central America. Med. Vet. Entomol. 5: 145–151.

Taylor, M. F. J., S. W. McKechnie, N. Pierce and M. Kreitman. 1993. The Lepidopteran mitochondrial control region: Structure and evolution. Mol. Biol. Evol. 10: 1259–1272.

Tegelstrom, H. 1992. Detection of mitochondrial DNA fragments. Pp. 89–113. In: *Molecular Genetic Analysis of Populations: A Practical Approach.* A. R. Hoelzel, Ed. IRL Press, Oxford.

Thomas, R. H. and J. A. Hunt. 1993. Phylogenetic relationships in *Drosophila:* A conflict between molecular and morphological data. Mol. Biol. Evol. 10: 362–374.

Turbeville, J. M., D. M. Pfeifer, K. G. Field and R. A. Raff. 1991. The phylogenetic status of arthropods, as inferred from 18S rRNA sequences. Mol. Biol. Evol. 8: 669–686.

Uhlenbusch, I., A. McCracken and G. Gellissen. 1987. The gene for the large (16S) ribosomal RNA from the *Locusta migratoria* mitochondrial genome. Curr. Genet. 11: 631–650.

van de Peer, Y., J. M. Neefs, P. De Rijk and R. De Wachter. 1993. Reconstructing evolution from eukaryotic small-ribosomal-subunit RNA sequences: Calibration of the molecular clock. J. Mol. Evol. 37: 221–232.

Weir, B. S. 1990. *Genetic Data Analysis.* Sinauer Assoc., Sunderland, MA.

Werman, S. D., M. S. Springer and R. J. Briteen. 1990. Nucleic acids I: DNA–DNA hybridization. Pp. 204–249. In: *Molecular Systematics.* D. M. Hillis and C. Moritz, Eds. Sinauer Assoc., Sunderland, MA.

White, M. J. D. 1973. *Animal Cytology and Evolution.* 3rd Ed. Cambridge Univ. Press, Cambridge.

White, M. J. D. 1978. *Modes of Speciation.* Freeman, San Francisco.

White, P. S. and L. D. Densmore III. 1992. Mitochondrial DNA isolation. Pp. 29–58. In: *Molecular Genetic Analysis of Populations: A Practical Approach.* A. R. Hoelzel, Ed. IRL Press, Oxford.

Williams, S. M. and L. G. Robbins. 1992. Molecular genetic analysis of *Drosophila* rDNA arrays. Trends Genet. 8: 335–340.

Wilson, A. C., S. S. Carlson and T. J. White. 1977. Biochemical evolution. Annu. Rev. Biochem. 46: 573–639.

Wilson, A. C., H. Ochman and E. M. Prager. 1987. Molecular time scale for evolution. Trends Genet. 3: 241–247.

Xiong, B. and T. D. Kocher. 1991. Comparison of mitochondrial DNA sequences of seven morphospecies of black flies (Diptera: Simuliidae). Genome 34: 306–311.

Zacharopoulou, A. M. Frisardi, C. Savakis, A. S. Robinson, P. Tolias, M. Konsolaki, K. Komitopoulou and F. C. Kafatos. 1992. The genome of the Mediterranean fruitfly *Ceratitis capitata*: Localization of molecular markers by in situ hybridization to salivary gland polytene chromosomes. Chromosoma 101: 448–455.

Zuckerkandl, E. and L. Pauling. 1965. Evolutionary divergence and convergence in proteins. Pp. 97–166. In: *Evolving Genes and Proteins*. V. Bryson and H. J. Vogel, Eds. Academic Press, New York.

# Chapter 14

# Insect Population Ecology
# and Molecular Genetics

## Overview

Molecular genetic techniques provide powerful new tools for the study of insect biology, ecology, and population genetics both in natural populations and in the laboratory. Analysis of proteins by electrophoresis has been useful with many insects, but some groups of insects with low levels of detectable genetic variation cannot be studied unless more sensitive DNA markers are used. DNA analyses can identify biotypes, determine whether hybridization or introgression occurs, and provide information on founder effects, inbreeding, genetic bottlenecks, dispersal, and selection intensity. Use of DNA markers in population ecology in the past has been limited by technical difficulties. RFLPs require significant amounts of DNA for analysis, and traditional PCR methods require information about DNA sequences, which is difficult to obtain for most arthropod species. Nevertheless, technical advances now make it possible to sample large amounts of genetic variation rapidly and inexpensively in large numbers of individuals by the RAPD method of the polymerase chain reaction (PCR-RAPD), by DNA fingerprinting of minisatellite or microsatellite DNA, or by restriction enzyme digests of DNA amplified by PCR (PCR-RFLP). The application of these molecular techniques to population biology and ecology will provide new opportunities to resolve both fundamental and applied questions in insect population ecology, population genetics, and pest management.

## Introduction

The fields of ecology and population genetics generally employ synthetic, rather than reductionist, research approaches. It is intriguing that one of the reductionist approaches to biology (molecular genetics) may provide a valuable new tool for resolving problems in population genetics and ecology.

388

Insect ecology is the study of insects in their environment. Insect ecologists are concerned especially with the biology of groups of insects and with the pattern of relationships between insects and their environment. Ecology thus is concerned with organisms, populations, and communities in ecosystems. Ecology sometimes is divided into autecology and synecology. **Autecology** deals with the study of the individual organism or an individual species, its life history, behavior, and adaptations to the environment. **Synecology** investigates groups of organisms associated as a unit (Odum 1971, Price 1975, Southwood 1978, Huffaker and Rabb 1984). At present, most ecological research employing molecular genetic techniques is autecological.

Population geneticists study how genetic principles apply to entire populations (Hartl 1981). One of the most striking features of insect populations is their phenotypic diversity. An underlying assumption that population geneticists make is that this phenotypic diversity is matched by genetic diversity. They then attempt to deal with both phenotypic and genotypic differences among individuals. Extensive protein (allozyme) variation has been found in many natural insect populations studied by electrophoresis. Exceptions include some haplodiploid Hymenoptera and clonal organisms such as aphids (Crozier 1977, Lester and Selander 1979).

Allozymes have been used to analyze mating systems (random versus assortative mating), inbreeding, genetic drift, hybridization, effective population size, degree of genetic differentiation among populations, and migration. Yet, there is evidence that allozyme studies greatly underestimate the amount of variation in insect populations, detecting only about 30% of the genetic diversity. Molecular genetic methods provide a way to examine DNA directly and to uncover greater amounts of variation. Which DNA technique should be employed in a particular project depends on the goals of the research and the level of DNA variation in the species of interest (Moritz et al. 1987).

Population genetics and population ecology have been distinct disciplines, although interest in combining them has emerged in the past few years (Slatkin 1987). The study of genes and genetic systems may provide insights for both autecological and synecological studies because an insect's heredity determines its behavior and ability to survive in specific environments and communities. Changes in genes and gene frequencies in populations over evolutionary time are important for both speciation and community structure.

The application of molecular genetic techniques to the study of insect population ecology is in its infancy, but should play a more significant role as soon as insect ecologists discover the power (and limitations) of these new tools. Analyses of DNA could provide more refined evaluations of biotypes, colonization, dispersal, intraspecific and interspecific variability, kinship, insect–plant interactions, and hybridization or introgression, although analysis of DNA data typically is more complex than for allozymes. This chapter provides a survey of methods and some examples in which molecular genetic methods have been used to answer different kinds of population biology and ecology problems, with the focus on insect species other than *Drosophila* (Table 14.1).

## Molecular Techniques Available

Proteins, nuclear DNA, and mitochondrial DNA (mtDNA) can be analyzed, and each has its particular advantages and disadvantages. They vary in ease of study, amount of variation that can be detected, and cost. Both single-copy and multiple-copy DNA sequences can be analyzed in nuclear DNA, which allows analysis of genetic variation at the individual, population, or species level. mtDNA analysis provides sufficient variation for studies of individuals, populations, or species, depending upon the region of the mtDNA molecule that is analyzed.

**Table 14.1**

Selected Examples of Problems in Insect Population Biology and Ecology Evaluated by Molecular Techniques

| Research problem | Techniques employed | Selected examples and references |
|---|---|---|
| Biotype identification | RAPD-PCR | *Trioxys pallidus* (Edwards and Hoy 1993) |
| | | Several aphid species (Black et al. 1992) |
| | | Several aphid parasitoids (Roehrdanz et al. 1993) |
| | | *Anaphes* and *Trichogramma* species (Landry et al. 1993) |
| | RAPD-PCR and allozymes | *Diuraphis noxia* worldwide (Puterka et al. 1993) |
| | RFLP analysis of rDNA | *Schizaphis graminum* (Black 1993) |
| Colonization and origin | RFLP | *Drosophila subobscura* (Rozas and Aguade 1991) |
| | RAPD-PCR | *Trioxys pallidus* (Edwards and Hoy, submitted) |
| | Allozymes | *Aedes albopictus* (Kambhampati et al. 1991) |
| Distribution | DNA fingerprinting | *Schizaphis graminum* (Shufran et al. 1991) |
| Genetic diversity and population structure | RFLPs of mtDNA | *Melanoplus sanguinipes* (Chapco et al. 1992b) |

*(continues)*

**Table 14.1** (*Continued*)

| Research problem | Techniques employed | Selected examples and references |
|---|---|---|
| | RAPD-PCR | Fourteen species of grasshoppers (Chapco et al. 1992a) |
| | Allozymes | *Ceratitis capitata* (Gasperi et al. 1991) |
| | mtDNA variation | *Ceratitis capitata* (Sheppard et al. 1992) |
| | | *Rhagoletis pomonella* (Feder et al. 1988) |
| | DNA fingerprinting | *C. capitata* (Haymer et al. 1992) |
| | Allozymes | *Magicicada* (Archie et al. 1985) |
| | mtDNA and allozymes | (Martin and Simon 1988) |
| | mtDNA RFLPs | *Leptinotarsa decemlineata* (Azeredo-Espin et al. 1991) |
| | DNA fingerprinting | *Bicyclus anynana* (Saccheri and Bruford 1993) |
| | Nuclear RFLPs, mtDNA, RAPD-PCR, allozymes, DNA fingerprinting | *Apis mellifera* (Hall 1986, 1990, 1992b, Hall and Muralidharan 1989, Smith et al. 1989, Hall and Smith 1991, Muralidharan and Hall 1990, Sheppard et al. 1991a,b, Hunt and Page 1992, Tares et al. 1993) |
| | RFLPs of rDNA | *Rhopalosiphum maidis* (Lupoli et al. 1990) |
| | Allozymes and rDNA | *Polistes* species (Davis et al. 1990) |
| | Allozymes | *Spodoptera frugiperda* (Pashley 1986, 1989) |
| | RFLPs | (Lu et al. 1992) |
| | mtDNA RFLPs | *Cochliomyia hominivorax* (Roehrdanz 1989) |
| | RAPD-PCR | *Trichogramma* and *Anaphes* species (Landry et al. 1993) |
| Host–parasite interactions | PCR | *Borrelia burgdorferi* in *Ixodes* ticks (Persing et al. 1990) |
| | | St. Louis encephalitis in mosquitoes (Howe et al. 1992) |
| Hybrid zone analysis | Allozymes and mtDNA | *Gryllus pennsylvanicus* and *G. firmus* (Rand and Harrison 1989) |
| | mtDNA | *Caledia captiva* (Marchant 1988) |
| | rRNA, mtDNA | (Marchant et al. 1988) |

(*continues*)

**Table 14.1**   (*Continued*)

| Research problem | Techniques employed | Selected examples and references |
|---|---|---|
|  | mtDNA and karyotype | (Marchant and Shaw 1993) |
|  | mtDNA | *Apis mellifera iberica* (Smith et al. 1991) |
| Identifying clones | DNA fingerprinting | *Myzus persicae* (Carvalho et al. 1991) |
| Insect–plant interactions | RFLPs | *Pemphigus betae* (Paige and Capman 1993) |
| Kinship patterns | DNA fingerprinting | *Parachartergus colobopterus* (Choudhary et al. 1993) |
|  | RAPD-PCR | *Aedes aegypti* (Apostol et al. 1993) |
|  | DNA fingerprinting | *Specius speciosus* (Pfennig and Reeve 1993) |
| Monitoring natural enemies of pests | DNA hybridization | Gypsy moth NPV (Keating et al. 1989) |
|  | RAPD-PCR | *Diaeretiella rapae* and *Lysiphlebus testaceipes* in aphid hosts (Black et al. 1992) |
|  | Allozymes | *Coccinella septempunctata* (Krafsur et al. 1992) |
| Paternity analysis | RAPD-PCR | *Anax parthenope* (Hadrys et al. 1993) |
|  |  | *Apis mellifera* (Fondrk et al. 1993) |
| Pesticide resistance | RFLPs of esterases | *Culex pipiens* (Raymond et al. 1991) |
|  | DNA sequencing | *Aedes aegypti* (Thompson et al. 1993) |
|  | Dot hybridization | *Myzus persicae* (Devonshire 1989, Field et al. 1989) |
| Species identification | Dot blots and squashes | *Anopheles gambiae* (Hill et al. 1991) |
|  | PCR of rDNA fragments | (Paskewitz and Collins 1990) |
|  | Allozymes, PCR | *Bemisia tabaci*/species? (Perring et al. 1992, 1993, Campbell et al. 1993, Bartlett and Gawal 1993) |
|  | Sequencing of ITS1 and ITS2 regions of rDNA | *Ixodes dammini* (Wesson et al. 1993) |
|  | 16S rRNA sequence | *Simulium* sibling species (Xiong and Kocher 1993) |

(*continues*)

**Table 14.1** (*Continued*)

| Research problem | Techniques employed | Selected examples and references |
|---|---|---|
| | Allozymes and mtDNA RFLPs | *Anopheles albitarsis* complex (Narang et al. 1993) |
| Sperm precedence | DNA fingerprinting | *Poecilimon veluchianus* (Achmann et al. 1992) |
| Subspecies identification | RAPD-PCR | *Aedes aegypti* (Ballinger-Crabtree et al. 1992) |

Potential methods for analyzing proteins, nuclear DNA, and mtDNA include allozyme electrophoresis, restriction fragment length polymorphisms (RFLPs) of mitochondrial or nuclear DNA, DNA fingerprinting, sequencing of both nuclear and mtDNA, standard PCR, and random amplified polymorphic (RAPD)-PCR analysis of nuclear DNA. These techniques were described in Chapters 6, 8, and 9. Each method varies in the time required, ease of execution, cost, and level of genetic variability that can be detected (Table 14.2).

Protein electrophoresis is useful for analyzing mating systems, discriminating among clones, evaluating degree of genetic variation (heterozygosity), testing for paternity or relatedness, and monitoring geographic variation, hybridization, sibling species or species boundaries. It provides data on genotype frequency distributions in populations. It is one of the most cost-effective techniques available and is relatively easy to perform. It has been used to identify Japan as the likely origin of the mosquito *Aedes albopictus* populations that recently colonized both the U.S.A. and Brazil (Kambhampati et al. 1991). It has also been used to demonstrate genetic differentiation between sympatrically occurring hawthorn and apple populations of the apple maggot fly *Rhagoletis pomonella* (Feder et al. 1988). Unfortunately, it may not detect sufficient variation to answer some questions, and the number of analyses that can be performed with very small insects may be limited because of inadequate amounts of proteins (Table 14.2).

RFLP analysis, like protein electrophoresis, is one of the most versatile techniques available for analyzing variation in both mitochondrial and nuclear DNA. Depending on which restriction enzymes and target sequences are analyzed, extensive variation can be discerned. However, RFLP analyses require relatively large amounts of DNA (which may not be obtainable from single individuals of small insect species), and the DNA must be electrophoresed, blotted, and probed to detect the variation. Probes must be developed, either as consensus sequences from other species or after cloning and sequencing species-specific DNA. Working with large numbers of individual insects is time-consuming and expensive (Table 14.2).

**Table 14.2.**

Techniques Most Useful for Insect Population Biology and Ecology Studies That Require Assays of Large Numbers of Insects Employ the PCR Reaction

| Technique | Level of discrimination and type of data obtained (gene frequencies or base pair changes) | Advantages (+) and disadvantages (−) | Selected references |
|---|---|---|---|
| Allele-specific PCR | Detect single nucleotide differences in individuals and populations | (+) Small amounts of DNA required; relatively rapid and inexpensive; results can be visualized by staining with ethidium bromide and other labels. | See references in Chapter 9; Innis et al. 1990, Erlich 1989 |
| | Gene frequency and base pair changes | (−) DNA sequence information needed for primers, or consensus primers; each reaction provides information for only one locus. | |
| DNA fingerprinting | | | Bruford et al. 1992, Kirby 1990 |
| Multilocus | Detect differences in individuals and populations in tandemly repeated units in nuclear DNA<br>Neither gene frequency nor base pair changes | (+) High levels of variation present.<br><br>(−) Comigrating bands may not be identical alleles at a locus unless single-locus probes are used; relatively large amounts of clean, high-molecular-weight DNA required; labeled probes required; relatively expensive and labor intensive; time and effort are required to identify repeated units. | |
| Single locus | Detect differences in individuals and populations in nuclear DNA | (+) Specific loci and alleles can be identified; easier to analyze than multiple locus data; high level of specificity; more rapid and | |

| Technique | What it detects | Advantages (+) / Disadvantages (−) | References |
|---|---|---|---|
| | Neither gene frequency nor base pair changes | easy than multilocus probes; useful for complex mating systems. (−) See above, plus expensive and slow compared with PCR-based procedures. | |
| DNA fingerprinting with the PCR | See above. | (+) See above, plus faster and less expensive (−) Reliable PCR products can be difficult to obtain. | Hadrys et al. 1992 |
| RAPD-PCR | Differences in single nucleotides in nuclear DNA; Gene frequency data | (+) Useful for species with limited genetic information; efficient; relatively inexpensive; requires little DNA. (−) Sensitive to DNA concentration; no genetic information on PCR products; can yield nonreproducible products; markers are dominant and heterozygotes may be difficult to identify; incorrect scoring can occur if two different fragments comigrate. | Edwards and Hoy 1993, MacPherson et al. 1993, Landry et al. 1993 |
| RFLPs | Differences in single nucleotides detected by sequences recognized by restriction endonucleases in nuclear and mtDNA; Gene frequency and changes in base pairs | (+) mtDNA most often analyzed; standard probes are available. (−) Requires large amounts of DNA; usually requires radiolabeled probes; single locus or several loci only analyzed; relatively expensive and technically demanding. | Aquadro et al. 1992, Dowling et al. 1990, Tegelstrom, 1992, White and Densmore 1992 |
| PCR-RFLPs | Differences in single nucleotide sequences in nuclear and | (+) Requires only a small amount of DNA; DNA can be visualized with EtBr; less ex- | Karl and Avise 1993, Simon et al. 1993 |

*(continues)*

**Table 14.2** (*Continued*)

| Technique | Level of discrimination and type of data obtained (gene frequencies or base pair changes) | Advantages (+) and disadvantages (−) | Selected references |
|---|---|---|---|
| | mtDNA recognized by the specific restriction enzyme used<br><br>Gene frequency data | pensive and more sensitive than standard RFLPs.<br><br>(−) Specific primers required. | |
| Protein electrophoresis | Detect changes in charged amino acids<br><br>Gene frequency data | (+) Inexpensive; many protocols available; produces codominant Mendelian characters of enzymes important in physiology.<br><br>(−) Less sensitive than DNA tests; number of tests that can be performed may be limited in small insects; proteins subject to environmental influences. | May 1992, Pasteur et al. 1988, Murphy et al. 1990 |
| Sequencing PCR-amplified DNA[a] | Differences in single nucleotides of nuclear and mtDNA, including coding and noncoding regions<br><br>Gene frequency and changes in base pairs | (+) Relatively small amounts of DNA needed; high resolution possible; some universal PCR primers available.<br><br>(−) Time-consuming and expensive; relatively small portion of genome can be sampled; technically more demanding than analysis of PCR-RFLPs or RAPD-PCR. | Hoelzel and Green 1992 |

[a]Sequencing DNA that is not amplified by the PCR is so time-consuming and expensive, it is not listed as a useful option for population genetics or biology studies.

A modification of this method, called **PCR-RFLP**, eliminates many of the disadvantages of traditional RFLP analysis (Karl and Avise 1993). If no probe is available, a genomic DNA library is constructed and clones are isolated. Clones with inserts of 500 to 2000 bp are chosen and sequences of the first 100–150 nucleotides from both ends are determined so that PCR primers can be derived. Nuclear DNA is amplified by the PCR using these primers and digested with appropriate restriction enzymes. The cut DNA is visualized after electrophoresis by staining with ethidium bromide. The advantage of PCR-RFLP is that DNA extracted from a single individual is sufficient (after PCR amplification) to provide bands that can be visualized without using hybridization with labeled probes. PCR-RFLP makes RFLP analysis suitable for studying individual specimens of very small species, requires no labeled probes, and is faster and less expensive than standard RFLP analysis. If consensus primers are available, then the cloning step is not required. For example, Simon et al. (1993) analyzed mtDNA in 13- and 17-year periodical cicadas using standard primers for the COII-A6-A8-COIII segment of mtDNA (Simon et al. 1991).

**DNA fingerprinting** involves the use of 'microsatellite' or 'minisatellite' sequences which consist of arrays of up to several hundred 15–60 bp units, often widely scattered throughout the chromosomes of many organisms (Bruford et al. 1992, Blanchetot 1991, Blanchetot and Gooding 1993, Estoup et al. 1993). The tandem repeat units contain a common 'core' sequence in different species. DNA fingerprinting results in a pattern of DNA bands that resemble the 'bar codes' used by computers to identify items in stores. It is used to evaluate DNA variability at the individual and population level and was first conducted on humans and other vertebrates (Jeffreys 1987). The banding patterns produced often are specific to a particular individual (except for monozygotic twins), are inherited in a Mendelian manner, and are generally stable within an individual's germ line and somatic tissues.

Satellite regions consist of short tandemly repeated sequences that differ in the number of times they are repeated in the genome. Thus, DNA fingerprinting detects polymorphisms at many sites in the genome (Table 14.2). Polymorphisms are visualized by hybridizing a labeled probe to genomic DNA that has been cut with a restriction enzyme and separated into bands on a gel by electrophoresis.

DNA fingerprinting has been used to identify individual insects or their progeny and can resolve whether a mating has been successful (Burke 1989). It could be useful in monitoring the establishment and dispersal of specific biotypes of arthropods, including those with low levels of protein variation, such as parthenogenetic aphids or parasitic Hymenoptera (Table 14.1).

The need to isolate and clone probe sequences for each species has limited the use of DNA fingerprinting by entomologists. However, recent studies suggest that commercially available hypervariable minisatellite loci isolated from humans (for example, 33.15, 33.6) or from vectors (M13) may provide a probe

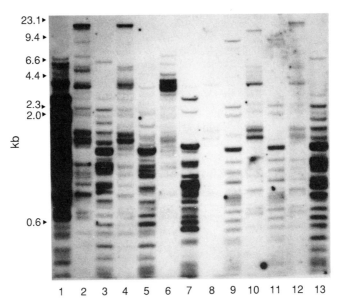

**Figure 14.1** Minisatellite DNA fingerprints for individual aphids from six partheno-genic lines of the grain aphid *Sitobion avenae* using a derivative of Jeffrey's 33.15 probe to show differences with two restriction enzymes. Genomic DNA was extracted and cut with *Sau* 3A and *Rsa* I. Lane 1 contains DNA from a control aphid *A. pisum* cut with *Hin* fl. Lanes 2 and 3 contain DNA from the Norwegian green clone cut with *Sau* 3A and *Rsa* I, respectively. Lanes 4 and 5 are from the Norwegian brown clone; lanes 6 and 7 are from the Rothamsted UK clone; lanes 8 and 9 are from the Dundee UK 3 clone; lanes 10 and 11 are from the Dundee 9 clone; and lanes 12 and 13 are from the Green Basingstoke UK clone. Lanes 6 and 8 are too faint to interpret. (Photograph provided by G. R. Carvalho.)

suitable for differentiating among closely related aphid individuals (Carvalho et al. 1991). DNA extracted from individuals of two different aphid species, *Myzus persicae* and *Sitobion avenae*, produced complex DNA fingerprints in Southern blot analyses when probed with a sequence (called 33.15) from human mini-satellite DNA (Figure 14.1). DNA profiles from individuals of six different *S. avenae* clones produced different DNA fingerprints. This is notable because no variation in the aphids was shown when 11 different enzymes were examined by electrophoresis. The proportion of fragments shared among different clones of *S. avenae* cut with the *Rsa* I restriction enzyme was used to estimate genetic related-ness by using a similarity coefficient (*D*).

$$D = 2N_{AB}/(N_A + N_B),$$

where $N_{AB}$ is the number of fragments shared by samples A and B and $N_A$ and $N_B$ are the fragments in A and B, respectively (Carvalho et al. 1991). DNA fingerprint differences appeared to be stable over nine parthenogenetic generations in M. *persicae*, which suggests that this technique could be used to monitor these clonal organisms in natural populations over several generations.

The 'standard' PCR is rapid, easy, and appropriate for many population biology or ecology studies (Arnheim et al. 1990). One modification of the PCR, RAPD-PCR, may be especially useful for population biologists and ecologists (see Chapter 9 for a description of RAPD). Hadrys et al. (1992) noted that RAPD-PCR is one of the most versatile population analysis methods available because it can be used to determine species identity, to assess kinship, and to analyze paternity. It estimates genetic variation within populations and can be used to monitor colonization. RAPD-PCR is suitable for studying insects for which very little genetic information is available, requires only very small amounts of DNA, and can be used with very small insects. It is rapid and relatively inexpensive compared with RFLP analysis or DNA sequencing (Table 14.2). Because RAPD-PCR uses short primers of arbitrary sequence (10-mers), it does not require the investigator to know the sequences of specific genes in order to develop primers for PCR. It is thought that RAPD-PCR primers sample both single copy and repetitive DNA. While the repeatability and reliability of RAPD-PCR has been questioned, it can provide repeatable and useful data if certain precautions are taken (Penner et al. 1993, Edwards and Hoy 1993, MacPherson et al. 1993) (see also Chapter 9).

RAPD-PCR was used to analyze the amount of genetic variation in the parasitic wasps *Trioxys pallidus* and *Diglyphus begini* (Figure 14.2). DNA from individual *T. pallidus* was amplified using 120 different primers (Edwards and

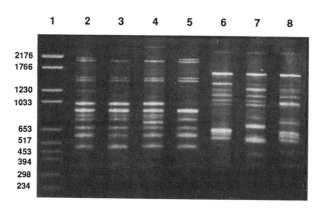

**Figure 14.2** RAPD-PCR banding patterns obtained from individual *Trioxys pallidus* male wasps (lanes 2–5) and individual *Diglyphus begini* males (lanes 6 through 8) using primer 5'-GCTGGACATC3'. Size standards are in lane 1 for reference. (Photograph provided by O. R. Edwards.)

Hoy 1993). Of the 120 primers tested, 92 produced a total of 342 scorable bands, of which 118 exhibited presence/absence polymorphisms. *D. begini* was evaluated with 25 primers, and 17 produced a total of 51 scorable bands. The level of genetic variation detected was greater than any found in Hymenoptera using allozymes (Menken 1991, Packer and Owen 1992) and comparable to the amount of variation detectable with RFLPs. The bands considered 'reliable' were inherited as dominant Mendelian traits (Figure 14.2). Because only small amounts of DNA are used for each RAPD-PCR reaction, multiple reactions can be conducted using the DNA from a single individual. In theory, up to 150 reactions could be conducted with DNA from a single *T. pallidus* male, providing an average of 1.5 informative bands per primer employed or a total of 225 presumptive loci (150 reactions × 1.5 bands per primer) that could be analyzed (Edwards and Hoy 1993). RAPD-PCR can be used to analyze population structure and gene flow, and monitor the establishment and dispersal of particular biotypes (Hadrys et al. 1993, Black et al. 1992, Kambhampati et al. 1992, Edwards and Hoy, submitted for publication).

RAPD-PCR has drawbacks. RAPD-PCR bands are inherited as dominant traits, and thus heterozygotes normally cannot be identified readily in diplodiploid organisms. In the haplodiploid Hymenoptera, this difficulty is overcome by performing the analysis on the haploid males only or by testing the genotype of females by testing their male progeny (Figure 14.2). Other problems with RAPD-PCR include the need to optimize reaction conditions for each species (Hadrys et al. 1992). 'Reliable' bands must be identified by repeating reactions to determine which bands are consistent. RAPD-PCR is very sensitive to the concentration of DNA in the reaction, so reaction conditions must be optimized carefully and DNA extraction techniques must be consistent (Edwards and Hoy 1993).

Sequencing of DNA can provide information about changes in the DNA at the nucleotide level in both mitochondrial and nuclear DNA. However, DNA sequencing samples only a tiny fraction of the genome because it is expensive and time consuming to conduct. Traditional sequencing protocols require cloning and identifying specific DNA segments to be sequenced. While sequencing of DNA amplified by the PCR is less time-consuming, information about sequences is needed in the PCR in order to develop appropriate primers. DNA sequencing by the PCR has been used infrequently for population studies (Table 14.2). It is unlikely that sequencing will be used often for large scale population studies unless the goals of the project demand that the DNA sequence be determined.

## Analysis of Molecular Data

A number of authors discuss methods of analyzing molecular and population data (for example, Weir and Cockerham 1984, Cockerham and Weir 1987,

Slatkin and Barton 1989, Lynch and Crease 1990, Weir 1990, Hoelzel and Dover 1991, Hoelzel and Bancroft 1992). A variety of analyses can be conducted on molecular data to estimate parameters such as genetic diversity (heterozygosity and proportion of polymorphic loci), interpopulation diversity, genetic distance, effective population size, kinship, paternity, and the effect of migration on population diversity, as described below. Fundamentally, the techniques described above produce one of two types of data: sequence data or allele frequency data (Table 14.2). Details of the statistical methods for each type of data are discussed in the references cited.

### Allozymes

The visualization and interpretation of allozyme data is described in Pasteur et al. (1988) and May (1992). Allozyme data can be used to obtain gene frequencies (Hoelzel and Bancroft 1992).

$$p = (2N_{AA} + N_{Aa})/2N \text{ and } q = (2N_{aa} + N_{Aa})/2N,$$

where $p$ is the frequency of the A allele, $q$ is the frequency of the $a$ allele, $N$ is the total number of individuals in the sample, and $N_{AA}$, $N_{aa}$, and $N_{Aa}$ are the number of individuals with AA, $aa$, and Aa genotypes, respectively. According to the Hardy–Weinberg rule, the proportion of AA individuals should be $p^2$, the proportion of $aa$ individuals should be $q^2$, and the proportion of heterozygotes should be $2pq$ in an ideal population (an infinitely large random-mating population) in which there is no selection, migration, or mutation. Such a population is in **Hardy–Weinberg equilibrium.**

**Polymorphism (P)** and **heterozygosity (H)** can be calculated for allozyme data (Hoelzel and Bancroft 1992). $P$ is the proportion of polymorphic loci and $H$ is the proportion of heterozygous loci. When the population is in Hardy-Weinberg equilibrium, heterozygosity can be calculated from allele frequencies at a given locus by:

$$h = 1 - \Sigma x_i^2,$$

where $x_i^2$ is the frequency of the $i$th allele at a given locus. The proportion of heterozygous individuals ($H$) is the average heterozygosity for all loci studied, so it is calculated as the mean of $h$ over all loci.

**Genetic distance** or gene diversity between populations can be calculated using allozyme data. Most analyses of genetic distance assume that molecular genetic changes are accumulating gradually at a constant rate and that most are selectively neutral. This suggests that the genetic changes can be used to estimate the time of genetic differentiation within and between populations. When DNA variation is measured directly, the statistical analyses assume that: (1) nucleotides are randomly distributed in the genome, (2) variation arises by base substitution, (3) substitution rates are the same for all nucleotides, and (4) all relevant bands or fragments can be detected and bands that comigrate but are different are not scored as identical (Hoelzel and Bancroft 1992). While the first three assumptions usually are not valid, it is thought that small deviations from them will not alter the results significantly.

The most common method for using protein polymorphisms to analyze genetic distance in populations is that of Nei (1972). In two populations, X and Y, the probability that two randomly chosen genes at a single locus ($j_k$) are identical is:

$$j_x = \Sigma x_i^2 \text{ and } j_y = \Sigma y_i^2,$$

where $x_i$ and $y_i$ are the frequencies of the $i$th alleles at a given locus in populations X and Y, respectively. If there are two alleles at this locus with frequencies $p$ and $q$, then

$$j = p^2 + q^2.$$

The probability that a gene is identical at the same locus in populations X and Y is

$$j_{xy} = \Sigma x_i y_i.$$

The normalized identity ($I$) between populations X and Y for all loci is

$$I = J_{XY}/(J_x J_y)^{1/2},$$

where $J_{XY}$, $J_x$, and $J_y$ are the arithmetic means of $j_{xy}$, $j_x$ and $j_y$, respectively, over all loci. **Nei's standard genetic distance (D)** between populations X and Y is then

$$D = -\ln(I),$$

where $I$ is multiplied by the natural logarithm (ln) to give a value that is 0.0 for genotypes that are completely dissimilar. The relationship between $D$ and time ($t$) is

$$t = 0.5aD,$$

where $a$ is the average rate of detectable change per locus per year.

Interpopulation diversity using allozyme data is usually measured using the **coefficient of gene differentiation** ($G_{ST}$). $G_{ST}$ is derived by estimating the average similarity within and between populations. It is an extension of Wright's correlation ($F_{ST}$) between two gametes drawn at random from each subpopulation. The coefficient of differentiation is

$$G_{ST} = (H_T - H_S)/H_T,$$

where $H_S$ is the average gene diversity within populations, and $H_T$ is the interpopulation gene diversity.

## DNA *Fingerprints*

DNA fingerprinting can identify multiple loci or single loci in individuals. Multiple-locus DNA fingerprinting uses satellite sequences scattered throughout the chromosomes of many insects to produce a series of bands that are often specific to an individual insect. Multiple-locus DNA fingerprinting may detect so much variation within populations that it is difficult to use DNA fingerprints in population biology unless inbreeding has occurred. DNA fingerprinting can also be done with the PCR using specific or consensus primers (Kirby 1990).

Population estimates of allele and genotype frequencies can be tested for correspondence with Hardy–Weinberg equilibrium conditions (Bruford et al. 1992). The high level of variation detected by satellite data makes it feasible to test for paternity and to conduct studies of variability within both sexual and clonal populations (Brookfield 1992). Variation and genetic distance can also be calculated (Hoelzel and Bancroft 1992). Single-locus DNA fingerprinting is easier to analyze because there are fewer bands.

## DNA *Sequences*

DNA sequence data are analyzed by computer programs to determine the best alignment (Doolittle 1990, Gribskov and Devereux 1991). The identity of two sequences is compared on the percentage of shared bases. Deletions and insertions are usually scored as a single change regardless of length (Hoelzel and Bancroft 1992). As with proteins or RFLP data, nucleotide diversity, genetic distance, and interpopulation diversity can be estimated (Hoelzel and Bancroft 1992).

## *RAPD-PCR*

RAPD-PCR DNA bands are considered as dominant loci, and scored as present or absent (Hadrys et al. 1992). Kambhampati et al. (1992) discussed appropriate methods of data analysis. It appears that RAPD-PCR loci can be used to determine paternity, kinship, and hybridization, as well as to estimate population heterozygosity, effective population size, genetic distance between populations, and interpopulation diversity (Table 14.1).

## *RFLPs*

Visualization and interpretation of RFLP data is described by Aquadro et al. (1992) and Dowling et al. (1990). Restriction patterns can be compared by using either the lengths of the fragments or actual restriction sites. Restriction patterns can be classified as haplotypes and a measure of diversity can be derived as a function of the frequency of the different haplotypes (Hoelzel and Bancroft 1992). (The term haplotype is a contraction of **haploid** and **genotype** and describes the combination of linked alleles in a cluster of related genes.) Likewise, genetic distance is measured as an estimate of the number of base substitutions per nucleotide separating the two populations. Interpopulation diversity $(G_{ST})$ can also be estimated in a manner similar to that for allozyme data, but gene identities must be estimated from RFLP patterns. RFLP data also can be analyzed as changes in base pairs if the assumption is made that each change in restriction pattern is caused by a change in a single base pair.

## Case Studies in Molecular Ecology and Population Biology

Extensive studies on the population biology and genetics of *Drosophila* species have been conducted using molecular techniques (Singh and Long 1992). The following examples illustrate a few of the ways in which molecular techniques have been applied to the study of insects, with a focus on species other than *Drosophila*.

The first case study involves a unique situation in which a species normally considered to be beneficial, the honey bee, has become a pest. The problem arose because geneticists attempted to improve the characteristics of the honey bee for use in the Neotropics. The results of this genetic experiment were spectacular and unpredicted, raising vexing genetic and ethical questions. Molecular genetic techniques may yield data of both theoretical and applied significance for the 'Africanized' honey bee problem in the New World.

### The 'Africanized Honey Bee' in the New World

The European honey bee *Apis mellifera* has numerous subspecies or races. These can be identified by morphological, behavioral, physiological, and ecological differences. While North and South America have many native bee species that pollinate plants and some produce honey, none are as productive and easy to manage as the European honey bee. Thus, a number of races of the European honey bee were transported to North and South America beginning in the 1500s. Early introductions into the Americas included the subspecies *A. m. iberica* (from Spain, 1500s), *A. m. mellifera* (black bee of Northern Europe, 1600s), *A. m. ligustica* (from Italy, mid-1800s), *A. m. carnica* (Carniolan, mid-1800s), and *A. m. caucasica* (Caucasian, mid-1800s). Recent analyses of mtDNA from feral honey bee colonies suggest that African honey bee mtDNA has persisted at a low frequency (1%) in U.S. populations since the introduction of Egyptian colonies in 1866 (Schiff and Sheppard 1993). The Carniolan, Caucasian, and Italian bees dominated commercial beekeeping throughout North and South America until the 1950s, but were able to establish large feral populations only in temperate North America.

In 1956, a small number of honey bees from Africa were introduced into Brazil as part of a breeding program designed to provide a hybrid honey bee more adapted to the Neotropics. At least 47 queens of the African honey bee, *A. m. scutellata* were introduced, and some of them (26 or 27) accidentally escaped (or were purposely distributed because of their superior honey-collecting ability). In any case, relatively few African queens provided the initial gene pool.

The nature of the African bee problem is disputed. The first question is whether the bees that have spread throughout South and Central America are relatively 'pure' African bees (A. m. scutellata) or whether the migrants are descendants of crosses between the African bees and European bees. If inter-breeding has occurred, to what degree do the bees retain their African characteristics? How serious a problem are the bees? How best can the bees be managed as they move north in the U.S.A.?

In this discussion, the term **'Africanized bees'** is adopted to describe bees that are descendants of the original African bees. This term is employed as a label without implying any specific degree or type of genetic relationship between the released African and European bees, which remains controversial. The colonization of the Western Hemisphere is still under way, and part of the controversy exists because the colonization process is dynamic. Different phases in the colonization process occur as the bees invade a new area, encounter European bees in different densities in different regions, and establish new populations in both managed apiaries and feral habitats.

What is clear is that after the African bees escaped, their descendants became unexpectedly good colonizers. They have established large feral populations in the Neotropics, and have also invaded managed apiaries. The 'Africanized bees' colonized South and Central America incredibly rapidly, moving at a rate of nearly 300–500 km per year. During this continent-wide colonization process, the bees appear to have retained many of their African behavioral characteristics. In South and Central America, more than two million apiaries have been taken over by Africanized bees in an area greater than 17 million square kilometers. In these regions, European bee colonies cannot be maintained unless beekeepers regularly requeen with European queens.

Considerable debate has ensued over the benefits and risks of these bees. The news media in the U.S.A. have labeled them 'killer bees' and created great alarm and even hysteria (Zuckerman 1985). How much is fact and how much is myth? Goncalves et al. (1988) reviewed the history of beekeeping in Brazil, and considered the Africanized bees to be polyhybrids which resulted from matings between African and European bees. According to Goncalves et al. (1988), Africanized bees have dramatically altered, but not devastated, Brazilian bee-keeping. Because Africanized bees are more aggressive and defensive, they are more difficult to manage and beekeeping now is less often practiced by amateurs in Brazil. Professional beekeepers had to change their management practices by wearing more effective protective clothing during hive maintenance and by moving hives away from people and domestic animals. This required more intensive management, which increased labor costs, but Brazilian beekeepers now produce large quantities of honey and some "actually prefer Africanized bees because it is easy to get bees and because they are good producers" (Gon-calves et al. 1988). Comparisons in Brazil also have shown that both Africanized and pure African bees are far superior in producing honey. Despite this relatively optimistic picture in Brazil, beekeeping has not fared as well in other South and Central American countries.

Is there something unusual about the genetics of Africanized bees that has enabled their populations to grow so quickly and to colonize such a large area so rapidly? Why are Africanized bees able to colonize such a diverse geographic range and displace existing populations of European bees? What are the evolu-tionary implications for feral European honey bee populations in the Americas? Will the aggressive behavior of the 'Africanized' bees be diluted by outcrossing? Background information on bee biology, ecology, and management practices necessary to understand these issues is described in Table 14.3 for those who are unfamiliar with honey bees.

The issue of how often Africanized and European bees interbreed and whether interbreeding will reduce the aggressive behavior of the hybrids is based on a number of assumptions that may not be valid. For example, Collins (1988) proposed a model of colony defense behavior with four sequential steps: alerta-tion, search, orientation, and attack. Within each step, several possible re-sponses are possible, each with varying intensities. Africanized and European bees exhibit distinct differences in these behaviors, but the differences are not inherited in a simple manner. Even among different true African bee colonies, differences are found in defensive behavior. In one experiment, 10% of the assayed African colonies were considered 'gentle', 70% intermediate, and 20 to 25% highly defensive. When the defensive behavior was divided into some of its component parts, several genes appeared to be involved.

Mendelian analyses have suggested that defensiveness is dominant, with at least two genes involved in determining the number of stings. Yet, the mode of inheritance of the various behaviors depended upon the specific crosses made. For example, progeny produced by crossing African bees with A. m. caucasica were more defensive than progeny produced by crossing African bees with A. m. ligustica. More responsive behavior to alarm pheromone is dominant to less responsive behavior, with two or three loci controlling the trait. During an artificial selection program, African bees selected for both greater and lesser defensive behavior were significantly different after only two generations of selection. Unfortunately, selection for less defensive behavior was less successful than selection for more defensive behavior, and Collins (1988) concluded that it would be better to prevent Africanization of commercial honey bee stocks rather than to select for a more manageable type of bee after apiaries have been invaded.

Reciprocal crosses of African with European bees appear to yield different outcomes in the field. Thus, Rinderer et al. (1985) concluded that Africanized

**Table 14.3**

Biology, Ecology, and Management Issues Relevant to the 'Africanized Bee Problem' in the U.S.A.

---

An understanding of the genetic questions surrounding the 'Africanized bee problem' is impossible without a thorough knowledge of honey bee biology. *Apis mellifera* colonies are essentially a family, usually with a single queen producing all the progeny. Each diploid queen mates outside the hive on several nuptial flights with 6–17 different haploid males (drones) and then she returns to the hive. Females use stored sperm to fertilize eggs that develop into diploid daughters; unfertilized eggs produce sons. Because queens mate with more than one male, workers may be either full or half sisters. Most females develop into nonreproductive, sterile, workers, but new queens develop if they are provided appropriate nutrients by the workers.

*Paternal gene flow* occurs when drones mate with queens from distant colonies. *Maternal gene flow* occurs when a queen and a contingent of workers fly to a new location to establish a new colony (swarm), leaving behind a portion of the colony with a queen. European bees move about 1.6 to 5 km when they swarm, but African swarms may move as far as 160 km. African bees tend to abscond often (the entire colony moves to a new site), swarm more often, and forage differently than European bees. African bees are better adapted to surviving with poor and unpredictable resources of pollen and nectar than European bees (Rinderer 1988).

The 'Africanized bees' in the New World are thought to be different from the populations from which they were derived for several reasons (Fletcher 1988). First, the gene pool of African bees (*A. m. scutellata*) imported into Brazil was small (about 26 queens), which probably resulted in founder effects. The colonies introduced into Brazil were only a limited portion of the range of *A. m. scutellata* in Africa (one queen was from Tanzania and 25 were from the Transvaal of South Africa). Some (strongly debated) level of hybridization with European bees has occurred. Furthermore, during their spread throughout South and Central America, the Africanized bees may have been strongly selected for increased absconding and/or swarming behavior (Fletcher 1988).

Honeybees are a significant component of U.S. agriculture, with pollination services provided by more than 1,600 commercial beekeepers managing more than four million colonies. Honeybees are responsible for at least 80% of the pollination of more than 100 crops.

Africanized bees are more easily provoked than European bees and sting readily, posing a public health threat. Because Africanized bees tend to swarm and abscond frequently, they are more difficult to manage than European bees. Thus management costs associated with the invasion of the Africanized bee will make the indispensable commercial pollination industry in the U.S.A. more expensive (Camazine and Morse 1988).

Feral populations of the Africanized bee will be a serious problem in the U.S.A. as they have been elsewhere. Feral Africanized bees can invade hives managed by amateur beekeepers and make it difficult for people to keep bees in their backyards. Only well-trained and well-equipped beekeepers will be able to manage bees in infested areas. A

---

*(continues)*

**Table 14.3**   (*Continued*)

large queen bee-rearing business in the southern U.S.A. probably will no longer be viable because these queens could mate with Africanized drones on their nuptial flights.

*How* to control feral Africanized bee populations in the U.S.A. in order to protect the public health and the beekeeping industry is unresolved. Proposals for managing apiaries include (1) quarantining and exterminating Africanized bees, and (2) stock certification and requeening of contaminated hives with European bees. These techniques are expensive and rely on the availability of pure European queens from uninfested regions. They also require rapid and inexpensive methods of identifying Africanized bees. The development of rapid, inexpensive, and reliable molecular techniques for identifying Africanized and European bees would aid in managing the Africanized bees in apiaries.

Several methods have been proposed for managing *feral* Africanized bees in the southern U.S.A., but they are even more difficult and expensive than managing apiaries, and rely on a number of assumptions about the genetics of Africanized bees that may not be valid. Proposals were made to control the Africanized bees before they entered the U.S.A. by (1) breeding for a gentle hybrid and releasing it to outcompete (or hybridize with) feral Africanized bees, or (2) producing gentle hybrids by making large numbers of European drones available to mate with feral Africanized queens. These proposals were highly controversial and the second plan, tested in Mexico, apparently was unsuccessful (Booth 1988). There was debate over whether control programs were even needed.

One group of researchers contended that as the Africanized bees moved north they would interbreed with European bees and their unpleasant characteristics would be diluted. These researchers believed, based on morphometric analyses, that the bees are a blend of African and European types and that additional dilution of the African traits will occur as the bees interbreed with European bees in the U.S.A. Others concluded that the Africanized bees arrived at the Texas border with African mitochondria and a high proportion of nuclear African genes. They predicted that the interbreeding of 'Africanized bees' with European bees is an unlikely solution to the problem of managing feral populations in the southern U.S.A. However, as the 'Africanized' bees move north into regions with colder climates, it is likely that a hybrid zone will produced.

---

male (drone) bees successfully migrated into European bee apiaries in Venezuela in large numbers, but Africanized colonies only rarely allowed European drones to persist. Because Africanized colonies exported half or more of their drones to European colonies and accepted almost no foreign drones, Africanized males had a strong mating advantage in Venezuela. This behavior could mean that a few Africanized colonies would produce all, or nearly all, of the drones in a mating area, and 'Africanization' of apiary bees in a region could be achieved within two to three generations.

Sheppard et al. (1991a,b) analyzed populations of bees in Argentina and Brazil using mitochondrial DNA, morphology, and enzymes. They concluded that considerable hybridization has occurred between European and Africanized bees. Hybridization between Africanized and European bees in the Yucatan peninsula of Mexico was analyzed by Rinderer et al. (1991) using morphological data and mitochondrial RFLP patterns. The results suggested that extensive introgression between European and Africanized honey bees had occurred in the apiaries and that genetic modification of Africanized honey bees could occur if sufficiently large numbers of European drones were present. These conclusions have been questioned because they were not based on studies of hybridization in feral populations in the tropics.

Difficulties in distinguishing between Africanized and European bees, by either their genotype or phenotype, have limited the ability to resolve these issues. Rapid and reliable identification methods are necessary if quarantines or eradication programs are to work. Likewise, control programs based on European queen stock certification and selective breeding for European characteristics require effective identification methods. Identification of Africanized bees by morphology is unreliable because morphological traits are determined by both genetic and environmental influences such as climate and diet; identifying bees derived from matings between European and Africanized bees is particularly difficult. The use of enzyme markers to discriminate between European and Africanized bees has been hindered by the scarcity of differences in allozymes, with only five having significant frequency differences between African and European populations. Nuclear and mtDNA polymorphisms have provided more precise identifications, but are not yet sufficiently rapid and inexpensive for widespread use (Hall and Smith 1991).

Analyses of both nuclear and mtDNA from Neotropical Africanized and European honey bees support the hypothesis that the bees moving into the U.S.A. are primarily African in genotype (Hall 1986, 1990, Hall and Muralidharan 1989, Smith et al. 1989, Hall and Smith 1991). mtDNA studies are particularly useful for determining the maternal ancestry of hybrid individuals, and thus the direction of gene flow in hybrid zones or hybrid populations, because mtDNA does not undergo recombination during reproduction. As a result, genetic markers on mtDNA are passed intact through the maternal lineage, making it possible to identify whether African queens have contributed to a hybrid population even after many generations of hybridization with European drones.

The polymerase chain reaction was used to identify mtDNA from European and African bees (Hall and Smith 1991). The primers used were from a large ribosomal subunit. A total of 129 colonies from Africa and Europe were tested, as were 41 managed and feral colonies in the U.S.A. and northern Mexico. As expected, European colonies had European mtDNA and South African colonies had African mtDNA. Of 76 colonies from central and southern Mexico, Honduras, Costa Rica, and Venezuela, 72 had African mtDNA and 4 had European mtDNA. These results led Hall and Smith (1991) to conclude that 'Africanized

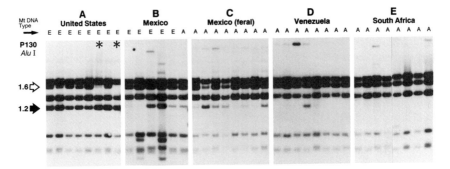

**Figure 14.3**   Nuclear DNA from bees from the U.S.A. (A), Mexico [B (domestic) and C (feral)], Venezuela (D), and South Africa (E) was extracted, and cut with *Alu* I and probed with P130. Each lane represents DNA from a mixture of siblings in a single colony. The solid arrow points to a 1.2-kb fragment found in European bees, which is allelic to the 1.6-kb fragment indicated by the open arrow. The 1.2-kb fragment is present in U.S. colonies (A) and absent in South African colonies (E). Most of the Mexican colonies (B) had reduced frequencies of the 1.2-kb fragment, which is consistent with the 'Africanization' of the colonies as a result of European queens mating with African drones. Some feral Mexican bees (C) showed varying faint 1.2-kb fragments owing to hybridization with European drones. Almost all managed Venezuelan colonies (D) lacked the 1.2-kb fragment typical of the European bees. The asterisks (lanes 6 and 8 in A) indicate that these U.S. bees were homozygous for the 1.2-kb allele typical of European bees. (Photograph provided by H. G. Hall.)

bees' have spread as nearly undiluted African maternal lineages and that there has been little European maternal gene flow into the feral Africanized bee populations. The low rate of European maternal gene flow perhaps is due to **negative heterosis** (reduced vigor or fitness in hybrids), which would reduce the survival of hybrid bees in feral populations.

Nuclear DNA of Africanized and European honey bees from a total of 126 colonies from Europe, the U.S.A., Mexico, Costa Rica, Venezuela, and South Africa was analyzed by Hall (1990) using RFLP analysis of random nuclear DNA probes. An example of the results is shown in Figure 14.3. Genomic DNA was extracted, cut with *Alu* I and probed. Each lane represents DNA from a mixture of siblings in a single colony. The solid arrow points to a 1.2-kb fragment found in 70% of European bees that is absent in South African colonies (Hall 1990). Two colonies from the United States that were sampled carried the European band (Figure 14.3A). Fewer managed Mexican colonies had the 1.2-kb band, which is consistent with the 'Africanization' of these colonies; some feral Mexican swarms showed a few faint 1.2-kb bands, probably because European drones had mated with Africanized queens. Almost all managed Venezuelan colonies lacked the 1.2-kb band typical of European bees. This analysis suggests there has been a low level of introgression of European nuclear DNA into feral Africanized colonies.

In a separate study, Muralidharan and Hall (1990) identified nuclear DNA markers that appeared to be specific to African honey bees. In a survey of RFLPs found in 16 U.S. colonies, 20 South African colonies, 26 feral Venezuelan colonies, 19 feral Mexican colonies, and 5 managed Mexican colonies, Muralidharan and Hall (1990) found that Neotropical Africanized bees are similar to African bees. Four African-specific RFLP markers were present in all Venezuelan and Mexican samples. A fifth African marker was present in more than half of these samples. European markers were absent in all Venezuelan samples and in half of the Mexican colonies. These results support the hypothesis that the 'Africanized' bees have retained a large amount of their mitochondrial and nuclear genome from the original African colonies.

Two genetic hypotheses have been proposed to explain how 'African bees' were able to colonize the Americas so incredibly rapidly. Evidence to support each hypothesis has been obtained using a variety of methods, but molecular genetic approaches have been most compelling. The first hypothesis suggests that the dramatic spread occurred because African drones mated with European queens, producing hybrid progeny that retained many of the 'African' traits. This is called the 'paternal gene flow' hypothesis. The alternative hypothesis suggests that Africanized bees colonized the Americas primarily as swarms of African matrilines, which resulted in 'maternal gene flow' (Hall and Muralidharan 1989, Smith et al. 1989).

mtDNA studies strongly support the second hypothesis (Hall and Muralidharan 1989). However, the RFLP analyses of nuclear genes suggest that some European paternal gene flow has occurred (Muralidharan and Hall 1990). The data suggest that introgressive hybridization, or gene flow, has been asymmetrical in feral populations, perhaps due to some type of negative heterosis or ecological selection. Negative heterosis (or reduced fitness in hybrid progeny) might be produced when European queens mate with African drones.

To test the hypothesis that there is a low rate of European maternal gene flow due to negative heterosis, metabolic capacities of African and European bees and their reciprocal $F_1$ and backcross progeny were analyzed as a measure of relative fitness (Harrison and Hall 1993). Measurements of $CO_2$ production during flight indicated that progeny of African queens $\times$ European drones and European queens $\times$ African drones have low, nonintermediate metabolic rates. The different reciprocal hybrid progenies had different metabolic capacities; progeny from European queens mated with African drones had a lower metabolic activity than the reciprocal cross. These results suggest that hybrids might not persist in feral populations in the Neotropics, perhaps due to differences in flight capacity or metabolism. These data could explain how the African mito-

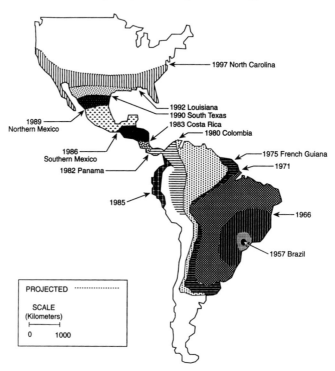

1997 North Carolina

1992 Louisiana
1990 South Texas
1983 Costa Rica
1980 Colombia

1989
Northern Mexico

1986
Southern Mexico

1982 Panama

1975 French Guiana
1971

1985

1966

1957 Brazil

PROJECTED  ·······················

SCALE
(Kilometers)

0       1000

**Figure 14.4**   The actual and projected rate of spread of Africanized bees in the Americas. The northward extent of the Africanized bees is controversial. (From Winston 1992.)

chondrial genotype has persisted in feral 'Africanized' bees during colonization of tropical Central and South America.

The outcome of the meeting between feral European and Africanized honey bees in the southern U.S.A. will be resolved soon. The 'killer bees' entered Texas in 1990 and have not slowed their colonization rate (Hunter et al. 1993). Africanized bees are expected to establish feral populations in the southern U.S.A., but feral populations are not expected to persist in more northern regions (Michener 1972, Camazine and Morse 1988, Taylor 1985, Winston 1992). Figure 14.4 illustrates only one of the projected distributions of the Africanized bee in the U.S.A. (Winston 1992).

Another prediction is that there will be a zone of hybridization between feral 'Africanized' and European bees in the U.S.A. How narrow or wide such a hybrid zone will be is disputed. Such a hybrid zone appears to exist in Spain where two other honeybee subspecies, A. m. iberica and A. m. intermissa, are found (Smith et al. 1991). The iberica subspecies is found throughout the Iberian peninsula, but the *intermissa* subspecies is found in North Africa. Two types of mtDNA were found in Spanish *A. m. iberica*. One is a west European type and the other is an African type like that of *A. m. intermissa*. The European type of mtDNA predominated in northern populations of *A. m. iberica* and the *intermissa*-like mtDNA predominated in southern populations of *A. m. iberica*. The data indicate that African and west European lineages of honey bees came into contact and formed a narrow hybrid zone, but the factors that generated a boundary between these populations in Spain remain unknown.

Careful sampling of both nuclear and mtDNA in feral and apiary populations of bees in the temperate U.S.A. before, during, and after the arrival of 'Africanized' bees will provide documentation of the pattern and time course of gene flow between Africanized and European populations and in the hybrid zone (Smith 1991, Moritz and Meusel 1992, Hall 1992a). It will be interesting to compare the results of these molecular genetic analyses with genetic analyses of the behavior of the new hybrid bee populations.

### Genetic Variability in the Fall Armyworm

The fall armyworm, *Spodoptera frugiperda*, is a polyphagous 'species' that attacks over 60 plants, particularly corn and bermuda grass. This migratory pest overwinters in southern Florida, southern Texas, and the Caribbean regions, but disperses north during the late spring and early summer. The 'fall armyworm' has exhibited anomalous differences in tolerance to pesticides and other life history traits important in pest management practices.

Molecular genetic data are sufficiently sensitive that cryptic sibling species (species that are similar phenotypically) can be detected readily. The undetected presence of cryptic species can have practical significance for pest management programs. Pashley (1986) showed that there may be two sibling species, each associated with different host plants. One is associated with rice and bermuda grass and the other with corn. The two types occur sympatrically, with a high level of reproductive isolation, and exhibit physiological adaptations to the different host plants and genetically based differences in development on them (Pashley 1988). Allozyme and RFLP analysis of mtDNA indicated the two types could be distinguished and that there is a near absence of gene flow (Pashley 1989). Lu et al. (1992) also reported RFLP differences in genomic DNA of the two types after analyzing 6 colonies with 22 different markers.

### Origin and Migration of Insecticide Resistance in Culex pipiens

The development of pesticide resistance by more than 450 species of arthropods is a significant applied and theoretical problem (Brown and Brogdon 1987, Roush and McKenzie 1987, Roush and Tabashnik 1990). Resistance to pesticides costs over one billion dollars annually, and resistance to virtually all pesticides has been found in many pest arthropods. For example, resistance to pesticides in mosquitoes has resulted in a worldwide resurgence of malaria, despite an extensive and expensive campaign by the World Health Organization to eradicate the vectors of this devastating disease.

Methods for limiting the development of resistance are essential for reducing agricultural production costs and avoiding environmental damage due to excessive pesticide use. Molecular genetic analyses may provide methods for reconstructing the evolutionary history of specific resistance genes or alleles within insect populations. Studies of DNA sequences may increase the precision with which the evolution of genes within populations can be reconstructed.

Raymond et al. (1992) argue that migration may be more important than mutation as a source of alleles for resistance to organophosphorus (OP) pesticides in the mosquito *Culex pipiens*. OP resistance in *C. pipiens* on three continents is due to gene amplification of the B2 esterase allele (Mouches et al. 1986). Raymond et al. (1991) analyzed the RFLP patterns of esterases B1 and B2 from mosquitoes collected from Texas, California, Pakistan, the Ivory Coast, the Congo, and Egypt. The amplified unit or amplicon, which includes the B1 esterase gene, contains a highly conserved 28-kb core sequence. The B1 and B2 esterase genes show a 96% similarity in sequence in a *Bam* HI-*Hind* III fragment (Raymond et al. 1991). Raymond et al. (1991) found that all amplified esterase B2 genes in mosquitoes from different geographic regions were similar in all 24 restriction sites examined. The flanking regions of the B2 esterase gene were also identical.

Raymond et al. (1991) concluded that amplified B2 esterase genes in these different mosquito populations originated from a recent single mutation that subsequently spread to diverse geographic sites. They suggested that the resistant mosquitoes could have moved around the world by way of international airlines or boats within the past 30 years. If the evolution of OP resistance in *C. pipiens* is influenced more by the extent of *dispersal* than by the rate of mutation, it has significant ramifications for managing this important disease vector. Quarantines and inspections of international transportation may be more critical than previously realized.

An alternative explanation is that the OP resistance is due to selection on a conserved gene in different mosquito populations in the different geographic regions. Selection for gene amplification in different human individuals or human cell cultures occurs repeatedly and independently after treatment with various drugs such as methotrexate (Schimke 1988) or organophosphorous insecticides (Prody et al. 1989). Likewise, cells of the mosquito *Aedes albopictus* also can be selected for resistance to methotrexate, indicating that amplification of other resistance genes can occur in insects. Gene amplification of the *dihydrofolate reductase* gene in *A. albopictus* results in approximately 1200 copies of a 140-kb-long region (Shotkoski and Fallon 1990, 1991, 1993). Gene amplification of an esterase gene in the aphid *Myzus persicae* results in resistance to insecticides which can be lost (Field et al. 1988). Such revertant aphids, however, retain the ability to respond rapidly to selection with pesticides (Field et al. 1989).

DNA amplification increases the relative amount of a region of a genome within a cell and has been observed in many organisms, from bacteria to humans. Two types of amplification have been identified. (1) Amplification can occur in all individuals in a species as part of normal development. An example is the amplification of the chorion genes of *Drosophila melanogaster* by repeated bidirectional initiation of DNA replication (see Chapter 5). (2) Amplification also can develop unexpectedly in a fraction of individuals within a population. The causes of such sporadic amplification are unknown. In the cases studied, the amplifications often involve arrays of repeated elements that may be inverted. These amplified regions may have been produced by localized overreplication of DNA or by unequal crossing over.

The migration of resistance alleles has been neglected as a significant phenomenon by scientists working on pesticide resistance (Denholm and Rowland 1992). In fact, relatively few genetic studies have been conducted on movements of resistant insects (Unruh 1990), and additional molecular analyses will provide information on the degree to which such large-scale migration occurs. If dispersal of resistance alleles is a common mechanism for development of resistance, an understanding of how often and when dispersal occurs may be critical to those designing methods for preserving the effectiveness of pesticides in pest management programs. If large-scale dispersal of resistance alleles is common, misuse of pesticides, even on a local scale, could significantly affect the evolution of pesticide resistance throughout the pest species' geographic range.

## Population Variability and Breeding for Resistance to Greenbug

The greenbug *Schizaphis graminum* is a serious pest of small grains and grain sorghum in the U.S.A. and other parts of the world. This aphid is parthenogenetic, although a sexual phase occurs naturally and can be induced in the laboratory by manipulating temperature and photoperiod. Control of *S. graminum* is achieved by combining pesticide applications and resistant crop varieties. Unfortunately, *S. graminum* can develop resistance to both pesticides and resistant crop cultivars. A knowledge of the amount and types of genetic variability in greenbug populations would provide useful information for both plant breeders and entomologists planning management practices (Shufran et al. 1991).

The amount of variation in the intergenic spacer region (IGS) of the ribosomal RNA (rDNA) genes of greenbug populations collected from wheat and sorghum in Kansas during two seasons was determined using DNA fingerprinting on 536 individuals (Shufran et al. 1991). Many IGS-length classes were found within and among aphids collected on both wheat and sorghum. Almost every individual collected in any particular field on any one date had a different

IGS size, and 13 major IGS size classes were present. No differences were found in populations collected from sorghum and wheat, and little relationship was found between IGS size patterns and collection sites.

The data suggest that those field populations of greenbug are made up of many unique clones rather than a proliferation of one or a very few. Thus, most of the genetic diversity in greenbug populations in Kansas could be sampled in individuals collected from a single sorghum leaf. Collecting from additional plants, fields, or counties did not increase the amount of genetic diversity in the greenbug samples. Shufran et al. (1991) concluded that when plant breeders are testing wheat and sorghum cultivars for resistance to greenbug in a single field, they are probably sampling most of the variability in greenbug populations in Kansas at that time. Because crop breeding is a continuous process, the data suggest that it will be important to sample a relatively few plants *over time* in order to detect changes in the response of greenbug populations to specific cultivars.

### Population Isolation and Introgression in Periodical Cicadas

Mitochondrial DNA markers have been used to decipher the evolutionary origins of three species of periodical cicadas (*Magicicada*). These cicadas feed underground on roots in the deciduous forests of the eastern United States. In the Mississippi Valley and southern U.S.A., their life cycle is 13 years, while it is 17 years in the north and west. Adult cicadas emerge from the ground synchronously in huge numbers to mate and deposit eggs nearly every year in some part of the range. The immense populations, sometimes as large as 1.5 million individuals per acre, that emerge in the same year are called broods. This synchronized emergence may be advantageous because the large numbers allow most of the individuals to escape predation. Likewise, the long life cycles may prevent predator populations from synchronizing with the emergences.

Thirteen of the possible 17 broods of the 17-year cicadas (M. *septemdecim*) and three of the possible 13 broods of the 13-year cicada (M. *tredecim*) have been identified and their emergence patterns have been identified and monitored. While most broods emerge as scheduled, small numbers of cicadas have emerged 'out of step' with their cohort. Some 17-year broods appear to have accelerated their emergence by four years in certain sites, and both 13- and 17-year broods emerge in the same geographic region. The life cycle of the 17-year cicada includes a 4-year inhibition of early nymphal growth. If this inhibition were eliminated, the brood could emerge after 13 years. It has been suggested that a single gene controls the length of the life cycle. The question thus is whether the cicadas emerging after 13 years are derived from some type of genetic change in a 17-year cicada brood. There is evidence in Illinois that a 17-year brood and a 13-year brood emerged together in 1868 and hybridized, which resulted in a population that subsequently emerged as adults every 13 years.

To answer this intriguing evolutionary and ecological question, Martin and Simon (1988) analyzed the abdominal sternite color, the frequency of allozyme polymorphisms (phosphoglucomutase, PGM), and mtDNA of the 'hybrid' Illinois population. The data indicated that mtDNA in the 13-year Illinois brood is like that in the adjacent 17-year brood and distinct from that of the neighboring 13-year brood. The new brood is like the 17-year brood in abdominal color and frequency of PGM polymorphism. These results are consistent with the hypothesis that the two populations hybridized and that the 13-year life-cycle trait is dominant.

Nevertheless, Martin and Simon (1988) rejected the hybridization hypothesis. They pointed out that most cases of hybridization involve narrow zones in which the species come into contact, yet this new periodical cicada population occurs over a large area. They also noted that the complete elimination of one of the mtDNA genotypes throughout the entire region would require *extremely* strong selection because there have been only eight generations since 1868 upon which selection could have occurred. Furthermore, if hybridization had occurred, intermediate phenotypes for abdominal sternite color and PGM frequencies should have occurred, but did not.

As a result, the findings are interpreted as evidence for a widespread life-cycle switch, *without hybridization*, in which a large number of 17-year cicadas underwent a four-year acceleration in development to become 13-year cicadas in northern Arkansas, Missouri, Illinois, and southeastern Iowa (Martin and Simon 1988). After 1868, a portion of the progeny of brood X in these areas had a permanent four-year acceleration in development rate. Some cicadas in the region did not switch life-cycle length after 1868 and continued to emerge after 17 years. Martin and Simon (1988) suggest that the switch in life cycle-length could have been triggered by environmental causes, perhaps by high-density populations.

This change in life-cycle length can disrupt gene flow and initiate genetic divergence in the formerly unified 17-year cicada population. The formation of a new temporally isolated brood creates new genetic opportunities because it can also bring together two genetically distinct populations. Thus, the individuals that switched their life cycle from 17 to 13 years now emerge synchronously with a previously isolated brood that emerges every 13 years. Gene flow between these previously isolated broods is now possible if they can mate and produce viable progeny. Obtaining the answers to this large-scale and lengthy genetic experiment requires considerable patience on the part of the scientists involved.

## Genetic Variation among Populations of the Russian Wheat Aphid

The Russian wheat aphid *Diuraphis noxia* is a serious pest of small grains, including wheat and barley. It originated in the Middle East, Asia Minor, and southern Russia, but has been accidentally spread to South Africa, Mexico, and the U.S.A. This species exhibits both anholocycly (continuous parthenogenetic reproduction by apomictic thelytoky, which involves fusion of nuclei or cells from the same parent to yield homozygous progeny) and holocyclic reproduction (parthenogenetic reproduction interrupted by an annual sexual generation in the fall, with the resultant eggs overwintering). Sexual reproduction is important in host race or biotype development, which is commonly observed in aphids. Biotypes are important because one approach to managing this pest would employ classic biological controls (the introduction and establishment of parasites, predators, or pathogens to control the introduced pest). When possible, biological control specialists attempt to collect the parasites, predators, or pathogens from the geographic origin of the pest so that the natural enemies are adapted to the particular pest biotype or race.

An extensive analysis of genetic variation and relationship among collections of *D. noxia* was conducted using allozyme and RAPD-PCR markers (Puterka et al. 1993). Thirty-six populations of *D. noxia* were collected from wheat, barley, or other grasses from the U.S.A., Canada, France, Jordan, Mexico, South Africa, Syria, Turkey, the Ukraine, or Moldavia. They included seven clonal populations that appeared to be distinct biotypes on different cereal cultivars. Twenty-three enzyme systems were tested and 17 provided useful bands. Using seven primers, 69 bands were produced by the RAPD-PCR in evaluations of 18 populations.

The analyses by either allozyme or RAPD-PCR markers were consistent within aphid clones. This suggests that parthenogenesis in *D. noxia* produces progeny that do not differ genetically from their stem mother, as expected. The populations examined with enzyme markers produced virtually no intrapopulation genetic variation. There was also little intrapopulation variation observed in RAPD-PCR markers, but more genetic polymorphisms were detected using RAPD-PCR and thus this technique was used to discriminate among populations. The results support previous evidence that allozyme analyses of aphids do not adequately reveal the full level of genetic variation. The greatest amount of allozyme and RAPD-PCR diversity was present among populations collected in the eastern and northeastern Mediterranean regions and the least was present among the introduced populations.

There was good agreement in the phylogenies derived by the allozyme and RAPD-PCR analysis. Both suggest that the French, South African, Mexican, and U.S. populations shared a common origin and that they may have originated from Turkey. The data suggest that the aphids have dispersed by human transport rather than through migration.

### Population Structure and Kinship in Polistes

The order Hymenoptera includes more than 250,000 species of ants, bees, and wasps, most of which are arrhenotokous. Fertilized eggs (2n) develop into females while unfertilized eggs (n) develop into haploid males, in which all the loci are sex linked. The relationship between the haplodiploid genetic system and sociality has been debated. Because sisters are more closely related to each other than they are to their mother, Hamilton (1964) proposed that **kin selection** has been important in the evolution of sociality, which appears to have developed as many as 11 times in the order Hymenoptera. The mechanisms by which sociality could have evolved in the Hymenoptera all involve assumptions about population genetics. Electrophoretic analyses of proteins in diverse hymenopterans suggest that they have significantly lower levels of variability than diplodiploid insects, and this has been attributed to haplodiploidy, sociality, inbreeding, and small population sizes.

Molecular genetic techniques provide a new tool for analyzing population structure in haplodiploid and social species. For example, Davis et al. (1990) analyzed RFLPs of ribosomal DNA and allozyme polymorphisms in four species of the primitively eusocial wasp *Polistes* to estimate $F_{ST}$, which measures the extent of genetic differentiation among subpopulations; $F_{IS}$, which measures the extent of local inbreeding; and theta ($\theta$), which is the coefficient of kinship between individuals from the same nest relative to the subpopulation (Cockerham and Weir 1987).

Davis et al. (1990) found that heterozygosity as estimated by evaluation of eight allozymes was low in the *Polistes* field populations examined. Likewise, no heterogeneity in length or restriction sites was found among ribosomal DNA within each species. Estimates of $F_{ST}$ values for three species were not significantly different from zero, but were significant for *Polistes exclamans*. The data suggest that dispersal in *Polistes exclamans* occurs through mated individuals.

Estimates of $F_{IS}$ did not differ significantly from zero except for one species, but all values of q were significantly different from zero. These results indicated that inbreeding can make a significant contribution to the coefficient of kinship within a nest. Davis et al. (1990) concluded that theories of the evolution of social behavior based on coefficents of kinship should include consideration of population structure, particularly the impact of local inbreeding.

### Paternal Investment by the Bushcricket

Many insect females mate with several males. Because sperm is stored within specialized organs in the female, a male's chance to fertilize eggs is reduced if sperm from another male is present. A variety of reproductive strategies have evolved to increase the probability that a male will fertilize eggs when multiple matings occur, including mechanically removing sperm from a previous mating. Most strategies, however, are less obvious, and obtaining an understanding of what is happening requires more detailed methods to determine which male's sperm were successful.

Achmann et al. (1992) analyzed sperm precedence in the bushcricket *Poecilimon veluchianus* (Tettigonioidea) using the minisatellite probe $(GATA)_4$.

Very complex banding patterns were produced, with an average of 15 scoreable bands per DNA fingerprint. Even siblings could be distinguished by their individual DNA fingerprints. Paternity was studied in five families. In four of the families, the second (and last) male to mate with the female fertilized 90% of the eggs. In one family, all progeny were descendants of the first male.

The authors concluded that DNA fingerprinting was ideal for studies on sperm competition because: (1) Many insect species lack visible markers or isozyme polymorphisms. (2) DNA fingerprinting provides a considerable number of highly polymorphic genetic markers, which allows individuals and their genetic relationship to be determined. (3) Probes are available for nearly every species (Epplen et al. 1991). (4) The high level of genetic discrimination by DNA fingerprinting makes it possible to establish the sperm precedence pattern even in multiply mated females without conducting a repeated series of mating experiments.

### Gypsy Moth Populations and Virus Infections

The gypsy moth, *Lymantria dispar*, is a serious pest of deciduous forests in Europe and the U.S.A. Populations undergo periodic outbreaks in very high densities that result in defoliation. Outbreaks of gypsy moth are often followed by population crashes caused by a nuclear polyhedrosis virus (NPV). A better understanding of the role of the gypsy moth NPV in the population dynamics could be useful for pest managers. Current monitoring methods involve collecting large numbers of larvae, rearing them until death, and then conducting a microscopic examination to determine the cause of death. An accurate, inexpensive assay suitable for large-scale population studies would significantly improve analyses of the impact of gypsy moth NPV in field populations.

Keating et al. (1989) used gypsy moth NPV DNA probes and compared slot blot and larval squash hybridization techniques. A **slot blot** allows multiple samples of DNA to be applied to nitrocellulose filters in specific sites (slots) using a vacuum. The larval squash technique simply involved squashing larvae between nitrocellulose filters. The resultant blots were probed with radiolabeled virus to measure virus incidence in larvae in a laboratory assay. The squash technique is less expensive and labor intensive than the current method. Larvae analyzed by slot or squash methods can be frozen and processed after the field season is over, when more time is available.

## Potential Future Applications

Intraspecific variation and genetic change in both pest and beneficial arthropods influences pest management strategies and tactics in different ways. The fact that pest and natural enemy populations change genetically in their composition often has been discussed, although comparatively little information has been available to document these important changes (Roush and Tabashnik 1990, Gould 1993, Kennedy 1993, Steiner 1993, Morton 1993). Molecular genetic techniques allow for rapid assessment of intraspecific genetic variation, changes in host preferences, insecticide resistance levels, and pest and natural enemy biotypes that are significant in biological control of pest arthropods and weeds.

Such methodologies are of more than academic interest. Monitoring resistance to pesticides by molecular methods remains relatively undeveloped at this time. An ideal assay for detecting resistance would be rapid and sufficiently inexpensive to screen large numbers of insects, sensitive enough to identify all resistant individuals, and able to resolve heterozygous and homozygous genotypes (Brown and Brogdon 1987, Brewer 1991). Current biochemical methods for identifying resistant individuals have numerous problems and drawbacks. Likewise, molecular genetic techniques at this time do not meet all of the desired criteria.

Potential methods to identify resistant individual insects by molecular genetic techniques include direct or indirect methods for detecting mutations. Linkage analysis, an indirect method, relies on markers that are coinherited with a resistance allele at a reliable frequency. Another indirect method could employ DNA sequences obtained by PCR-RFLPs or RAPD-PCR that are closely associated with the resistance locus to identify resistance genes that have not been identified and sequenced (Michelmore et al. 1991). These linked markers also could be used to determine the number of loci involved in resistance.

Direct methods would allow resistance genes to be detected by Southern hybridization, if a probe is available. Resistance caused by gene amplification, chromosome rearrangements, deletions, or insertions could be detected most readily in Southern blots, but resistances caused by single nucleotide substitutions would be more difficult to identify. For example, the amplified esterase DNA sequences of insecticide-resistant aphids, *Myzus persicae*, were detected in individual insects using a dot hybridization assay with an esterase cDNA as the radioactive probe (Field et al. 1989). This technique was sensitive enough to identify 'revertant' aphids that were susceptible but still retained their amplification of the esterase genes. PCR amplification of DNA followed by RFLP analysis could be sufficiently sensitive to detect changes in single nucleotides. The PCR-RFLP method produces enough DNA from an individual insect that banding patterns can be evaluated by staining with ethidium bromide so that radiolabeled probes are not necessary. An alternative PCR method would involve amplification of the mRNA to produce cDNA. The amount of mRNA could be quantified and would allow detection of resistant individuals that produce more enzyme, such as esterases, than susceptible individuals.

Another direct method for identifying resistant insect populations involves cloning and sequencing resistance genes using low stringency screening with a probe from *Drosophila melanogaster*. Thompson et al. (1993) used a γ-amino-butyric acid receptor gene that provides cyclodiene resistance in *D. melanogaster* as a probe to identify a cyclodiene resistance gene in *Aedes aegypti* with 87% sequence similarity. Sequencing of the gene from *A. aegypti* indicated that a single nucleotide change (G to T) is associated with resistance and causes a substitution of an alanine with a serine in the predicted lining of the ion channel; this same mutation occurs in resistant *D. melanogaster*. Thus, cyclodiene resistance is conserved between *Drosophila* and mosquitoes, and *Drosophila* probes could be used to identify resistant mosquito populations. If other resistance genes are conserved between *Drosophila* and other insects, then it may be possible to screen insect populations for specific mutations by sequencing DNA amplified by the PCR.

In addition to an analysis of resistance, molecular technologies will prove useful for the evaluation of other components of pest management. Insects collected and reared in the laboratory undergo genetic bottlenecks, inadvertent selection, and genetic drift (Bush et al. 1976, Chambers 1977, Hoy 1985, Roush 1990, Roderick 1992). Maintaining quality in insects reared for classic biological control, natural enemies mass reared for augmentative releases, or pest insects mass reared for genetic control programs is difficult (Bartlett 1985, Hoy 1976, 1985, Crossa et al. 1993, Delpuech et al. 1993). The analysis of population genetic architecture by allozymes often is less sensitive than desired. Molecular techniques such as DNA fingerprinting, RAPD-PCR, or PCR-RFLP might provide abundant genetic information from individual insects relatively rapidly and inexpensively. Information on effective population size and founder effects, as well as the impact of inbreeding and drift, could lead to improved rearing schemes.

Other potential ways in which molecular genetics could improve pest management methods are discussed in Chapter 15.

## Some Relevant Journals

*Biochemical Systematics and Ecology*, Pergamon Press.
*Molecular Ecology*, Blackwell Scientific Publ., Oxford.

## References Cited

Achmann, R., K. G. Heller and J. T. Epplen. 1992. Last-male sperm precedence in the bushcricket *Poecilimon veluchianus* (Orthoptera, Tettigonioidea) demonstrated by DNA fingerprinting. Mol. Ecol. 1: 47–54.
Apostol, B. L., W. C. Black IV, B. R. Miler, P. Reiter and B. J. Beaty. 1993. Estimation of the number of full sibling families at an oviposition site using RAPD-PCR markers: Applications to the mosquito *Aedes aegypti*. Theor. Appl. Genet. 86: 991–1000.
Aquadro, C. F., W. A. Noon and D. J. Begun. 1992. RFLP analysis using heterologous probes. Pp.

115–158. In: *Molecular Genetic Analysis of Populations: A Practical Approach*. A. R. Hoelzel, Ed. IRL Press, Oxford.

Archie, J., C. Simon and D. Wartenberg. 1985. Geographical patterns and population structure in periodical cicadas based on spatial analysis of allozyme frequencies. Evolution 39: 1261–1274.

Arnheim, N., T. White and W. E. Rainey. 1990. Application of PCR: Organismal and population biology. BioScience 40: 174–182.

Azeredo-Espin, A. M. L., R. F. W. Schroder, M. D. Huettel and W. S. Sheppard. 1991. Mitochondrial DNA variation in geographic populations of Colorado potato beetle, *Leptinotarsa decemlineata* (Coleoptera: Chrysomelidae). Experientia 47: 483–485.

Ballinger-Crabtree, M. E., W. C. Black IV and B. R. Miller. 1992. Use of genetic polymorphisms detected by the random-amplified polymorphic DNA polymerase chain reaction (RAPD-PCR) for differentiation and identification of *Aedes aegypti* subspecies and populations. Am. J. Trop. Med. Hyg. 47: 893–901.

Bartlett, A. C. 1985. Guidelines for genetic diversity in laboratory colony establishment and mainte-nance. Pp. 7–17. In: *Handbook of Insect Rearing*. Vol. 1. P. Singh and R. F. Moore, Eds. Elsevier-Science, Amsterdam.

Bartlett, A. C. and N. J. Grawel. 1993. Determining whitefly species. Science 261: 1334.

Black, W. C., IV. 1993. Variation in the ribosomal RNA cistron among host-adapted races of an aphid *(Schizaphis graminum)*. Insect Mol. Biol. 2: 59–69.

Black, W. C., IV, N. M. DuTeau, G. J. Puterka, J. R. Nechols and J. M. Pettorini. 1992. Use of the random amplified polymorphic DNA polymerase chain reaction (RAPD-PCR) to detect DNA polymorphisms in aphids (Homoptera: Aphididae). Bull. Entomol. Res. 82: 151–159.

Blanchetot, A. 1991. A *Musca domestica* satellite sequence detects individual polymorphic regions in insect genomes. Nucleic Acids Res. 19: 929–932.

Blanchetot, A. and R. H. Gooding. 1993. Genetic analysis by DNA fingerprinting in tsetse fly genomes. Insect Biochem. Mol. Biol. 23: 937–944.

Booth, W. 1988. USDA fights to repel African bees' invasion. Science 242: 368–369.

Brewer, K. K. 1991. Application of molecular techniques to improved detection of insecticide resistance. Am. Entomol. Summer: 96–103.

Brookfield, J. F. Y. 1992. DNA fingerprinting in clonal organisms. Mol. Ecol. 1: 21–26.

Brown, T. M. and W. G. Brogdon. 1987. Improved detection of insecticide resistance through conventional and molecular techniques. Annu. Rev. Entomol. 32: 145–162.

Bruford, M. W., O. Hanotte, J. F. Y. Brookfield and T. Burke. 1992. Single-locus and multilocus DNA fingerprinting. Pp. 225–269. In: *Molecular Genetic Analysis of Populations: A Practical Approach*. A. R. Hoelzel, Ed. IRL Press, Oxford.

Burke, T. 1989. DNA fingerprinting and other methods for the study of mating success. Trends Ecol. Evol. 4: 139–144.

Bush, G. L., R. Neck and G. B. Kitto. 1976. Screwworm eradication: Inadvertent selection for non-competitive ecotypes during mass rearing. Science 193: 491–493.

Camazine, S. and R. A. Morse. 1988. The Africanized honeybee. Am. Scientist 76: 465–471.

Campbell, B. C., J. E. Duffus and P. Baumann. 1993. Determining whitefly species. Science 261: 1333.

Carvalho, G. R., N. Maclean, S. D. Wratten, R. E. Carter and U. P. Thurston. 1991. Differentiation of aphid clones using DNA fingerprints from individual aphids. Proc. Roy. Soc. Lond. B 143: 109–114.

Chambers, D. L. 1977. Quality control in mass rearing. Annu. Rev. Entomol. 22:289–308.

Chapco, W., N. W. Ashton, R. K. B. Martel, N. Antonishyn and W. L. Crosby. 1992a. A feasibility study of the use of random amplified polymorphic DNA in the population genetics and systemat-ics of grasshoppers. Genome 35: 569–574.

Chapco, W., R. A Kelln and D. A. McFadyen. 1992b. Intraspecific mitochondrial DNA variation in the migratory grasshopper, *Melanoplus sanguinipes*. Heredity 69: 547–557.

Choudhary, M., J. E. Strassmann, C. R. Solfs and D. C. Queller. 1993. Microsatellite variation in a social insect. Biochem. Genet. 31: 87–96.

Cockerham, C. C. and B. S. Weir. 1987. Correlations, descent measures: Drift with migration and mutation. Proc. Natl. Acad. Sci. USA 84: 8512–8514.

Collins, A. M. 1988. Genetics of honey-bee colony defense. Pp. 110–117. In: *Africanized Honey Bees and Bee Mites*. G. R. Needham, R. E. Page, Jr., M. Delfinado-Baker and C. E. Bowman, Eds. Ellis Horwood, Chichester.

Crossa, J., C. M. Hernandez, P. Bretting, S. A. Eberhart and S. Taba. 1993. Statistical genetic considerations for maintaining germ plasm collections. Theor. Appl. Genet. 86: 673–678.

Crozier, R. H. 1977. Evolutionary genetics of the Hymenoptera. Annu. Rev. Entomol. 22: 263–288.

Davis, S. K., J. E. Strassmann, C. Hughes, L. S. Pletscher and A. R. Templeton. 1990. Population structure and kinship in *Polistes* (Hymenoptera, Vespidae): An analysis using ribosomal DNA and protein electrophoresis. Evolution 44: 1242–53.

Delpuech, J. M., Y. Carton and R. T. Roush. 1993. Conserving genetic variability of a wild insect population under laboratory conditions. Entomol. Exp. Appl. 67: 233–239.

Denholm, I. and M. W. Rowland. 1992. Tactics for managing pesticide resistance in arthropods: Theory and practice. Annu. Rev. Entomol. 37: 91–112.

Devonshire, A. L. 1989. Insecticide resistance in *Myzus persicae*: From field to gene and back again. Pestic. Sci. 26: 375–382.

Doolittle, R. F., Ed. 1990. *Molecular Evolution: Computer Analysis of Protein and Nucleic Acid Sequences*. (Methods Enzymol., vol. 183). Academic Press, New York.

Dowling, T. E., C. Moritz and J. D. Palmer. 1990. Nucleic acids II: Restriction site analysis. Pp. 250–317. In: *Molecular Systematics*. D. M. Hillis and C. Moritz, Eds. Sinauer Assoc., Sunderland, MA.

Edwards, O. R. and M. A. Hoy. 1993. Polymorphism in two parasitoids detected using random amplified polymorphic DNA (RAPD) PCR. Biol. Contr. Theor. Appl. Pest Managt. 3: 243–257.

Edwards, O. R. and M. A. Hoy (submitted). RAPD-PCR DNA markers used to monitor genetically-improved pesticide-resistant *Trioxys pallidus* (Hymenoptera: Aphidiidae) after releases into three California walnut orchards.

Epplen, J. T., H. Ammer and C. Epplen. 1991. Oligonucleotide fingerprinting using simple repeat motifs: A convenient, ubiquitously applicable method to detect hypervariability for multiple purposes. Pp. 50–69. In: *DNA Fingerprinting: Approaches and Applications*. T. Burke, G. Dolf, A. J. Jeffreys and R. Wolff, Eds. Birkhauser, Basel.

Erlich, H. A., Ed. 1989. *PCR Technology: Principles and Applications of DNA Amplification*. Stockton Press, New York.

Estoup, A., M. Solignac, M. Harry and J. M. Cornuet. 1993. Characterization of $(GT)_n$ and $(CT)_n$ microsatellites in two insect species: *Apis mellifera* and *Bombus terrestris*. Nucleic Acids Res. 21: 1427–1431.

Fang, Q., W. C. Black, IV, H. D. Blocker and R. F. Whitcomb. 1993. A phylogeny of new world *Deltocephalus*-like leafhopper genera based on mitochondrial 16S ribosomal DNA sequences. Mol. Phylogenet. Evol. 2: 119–131.

Feder, J. L., C. A. Chilcote and G. L. Bush. 2988. Genetic differentiation between sympatric host races of the apple maggot fly *Rhagoletis pomonella*. Nature 336: 61–64.

Field, L. M., A. L. Devonshire and B. G. Forde. 1988. Molecular evidence that insecticide resistance in peach-potato aphids (*Myzus persicae* Sulz.) results from amplification of an esterase gene. Biochem. J. 251: 309–312.

Field, L. M., A. L. Devonshire, R. H. ffrench-Constant and B. G. Forde. 1989. The combined use of immunoassay and a DNA diagnostic technique to identify insecticide-resistant genotypes in the peach-potato aphid, *Myzus persicae* (Sulz.). Pestic. Biochem. Physiol. 34: 174–178.

Fletcher, D. J. C. 1988. Relevance of the behavioral ecology of African bees to a solution to the Africanized-bee problem. Pp. 55–61. In: *Africanized Honey Bees and Bee Mites*. G. R. Needham, R. E. Page, Jr., M. Delfinado-Baker and C. E. Bowman, Eds. Ellis Horwood, Chichester.

Fondrk, M. K., R. E. Page, Jr. and G. J. Hunt. 1993. Paternity analysis of worker honeybees using random amplified polymorphic DNA. Naturwissensch. 80: 226–231.

Gasperi, G., C. R. Guglielmino, A. R. Malacrida and R. Milani. 1991. Genetic variability and gene flow in geographical populations of *Ceratitis capitata* (Wied.) (medfly). Heredity 67: 347–356.

Goncalves, L. S., A. C. Stort and D. DeJong. 1988. Beekeeping in Brazil. Pp. 359–372. In: *Africanized Honey Bees and Bee Mites*. G. R. Needham, R. E. Page, Jr., M. Delfinado-Baker and C. E. Bowman, Eds. Ellis Horwood, Chichester.

Gould, F. 1993. The spatial scale of genetic variation in insect populations. Pp. 67–85. In: *Evolution of Insect Pests/Patterns of Variation*. K. C. Kim and B. A. McPheron Eds. Wiley, New York.

Gribskov, M. and J. Devereux, Eds. 1991. *Sequence Analysis Primer*. Stockton Press, New York.

Hadrys, H., M. Balick and B. Schierwater. 1992. Applications of random amplified polymorphic DNA (RAPD) in molecular ecology. Mol. Ecol. 1: 55–63.

Hadrys, H., B. Schierwater, S. L. Dellaporta, R. DeSalle and L. W. Buss. 1993. Determination of paternity in dragonflies by random amplified polymorphic DNA fingerprinting. Mol. Ecol. 2: 79–87.

Hall, H. G. 1986. DNA differences found between Africanized and European honeybees. Proc. Natl. Acad. Sci. USA. 83: 4874–77.

Hall, H. G. 1990. Parental analysis of introgressive hybridization between African and European honeybees using nuclear DNA RFLPs. Genetics 125: 611–621.

Hall, H. G. 1992a. DNA studies reveal processes involved in the spread of new world African honeybees. Fla. Entomol. 75: 51–59.

Hall, H. G. 1992b. Further characterization of nuclear DNA RFLP markers that distinguish African and European honeybees. Arch. Insect Bioch. Physiol. 19: 163–175.

Hall, H. G. and K. Muralidharan. 1989. Evidence from mitochondrial DNA that African honey bees spread as continuous maternal lineages. Nature 339: 211–213.

Hall, H. G. and D. R. Smith. 1991. Distinguishing African and European honeybee matrilines using amplified mitochondrial DNA. Proc. Natl. Acad. Sci. USA 88: 4548–4552.

Hamilton, W. D. 1964. The genetical evolution of social behavior. J. Theor. Biol. 7: 1–16 and 17–52.

Harrison, J. F. and H. G. Hall. 1993. African-European honeybee hybrids have low nonintermediate metabolic capacities. Nature: 363: 258–260.

Hartl, D. L. 1981. *A Primer of Population Genetics*. Sinauer Assoc., Sunderland, MA.

Haymer, D. S., D. O. McInnis and L. Arcangeli. 1992. Genetic variation between strains of the Mediterranean fruit fly, *Ceratitis capitata*, detected by DNA fingerprinting. Genome 35: 528–533.

Hill, S. M., R. Urwin and J. M. Crampton. 1991. A comparison of non-radioactive labeling and detection systems with synthetic oligonucleotide probes for the species identification of mosquitoes in the *Anopheles gambiae* complex. Am. J. Trop. Med. Hyg. 44: 609–622.

Hoelzel, A. R. and G. A. Dover. 1991. Statistical interpretation of variation and genetic distance. Pp. 31–46 In: *Molecular Genetic Ecology*. IRL Press, Oxford.

Hoelzel, A. R. and D. R. Bancroft. 1992. Statistical analysis of variation. Pp. 297–305. In: *Molecular Genetic Analysis of Populations: A Practical Approach*. A. R. Hoelzel, Ed. IRL Press, Oxford.

Hoelzel, A. R. and A. Green. 1992. Analysis of population-level variation by sequencing PCR-amplified DNA. Pp. 159–187. In: *Molecular Genetic Analysis of Populations: A Practical Approach*. A. R. Hoelzel, Ed. IRL Press, Oxford.

Howe, D. K., M. H. Vodkin, R. J. Novak, R. E. Shope and G. L. McLaughlin. 1992. Use of the polymerase chain reaction for the sensitive detection of St. Louis encephalitis viral RNA. J. Virol. Methods 36: 101–110.

Hoy, M. A. 1976. Genetic improvement of insects: Fact or fantasy. Environ. Entomol. 5: 833–839.

Hoy, M. A. 1985. Improving establishment of arthropod natural enemies. Pp. 151–166. In: *Biological Control in Agricultural IPM Systems.* M. A. Hoy and D. C. Herzog, Eds. Academic Press, Orlando.

Huffaker, C. B. and R. L. Rabb, Eds. 1984. *Ecological Entomology.* Wiley, New York.

Hunt, G. J. and R. E. Page, Jr. 1992. Patterns of inheritance with RAPD molecular markers reveal novel types of polymorphism in the honey bee. Theor. Appl. Genet. 85: 15–20.

Hunter, L. A., J. A. Jackman and E. A. Sugden. 1993. Detection records of africanized honey bees in Texas during 1990, 1991 and 1992. Southwest. Entomol. 18: 79–89.

Innis, M. A., D. H. Gelfand, J. J. Ninsky and T. J. White, Eds. 1990. *PCR Protocols: A Guide to Methods and Applications.* Academic Press, San Diego.

Jeffreys, A. J. 1987. Highly variable minisatellites and DNA fingerprints. Biochem. Soc. Trans. 15: 309–317.

Kambhampati, S., W. C. Black IV and K. S. Rai. 1991. Geographic origin of the US and Brazilian *Aedes albopictus* inferred from allozyme analysis. Heredity 67: 85–94.

Kambhampati, S., W. C. Black and K. S. Rai. 1992. A RAPD-PCR based method for the identification and differentiation of mosquito populations: Techniques and statistical analyses. J. Med. Entomol. 29: 939–945.

Karl, S. A. and J. C. Avise. 1993. PCR-based assays of Mendelian polymorphisms from anonymous single-copy nuclear DNA: Techniques and applications for population genetics. Mol. Biol. Evol. 10: 342–361.

Keating, S. T., J. P. Burand and J. S. Elkinton. 1989. DNA hybridization assay for detection of gypsy moth nuclear polyhedrosis virus in infected gypsy moth (*Lymantria dispar* L.) larvae. Appl. Environ. Microbiol. 55: 1749–1754.

Kennedy, G. G. 1993. Impact of intraspecific variation on insect pest management. Pp. 425–451. In: *Evolution of Insect Pests/Patterns of Variation.* K. C. Kim and B. A. McPheron, Eds. Wiley, New York.

Kirby, L. T. 1990. *DNA Fingerprinting: An Introduction.* Stockton Press, NY.

Krafsur, E. S., J. J. Obrycki and R. V. Flanders. 1992. Gene flow in populations of the seven-spotted lady beetle, *Coccinella septempunctata.* J. Hered. 83: 440–444.

Landry, B. S., L. Dextraze and G. Boivin. 1993. Random amplified polymorphic DNA markers for DNA fingerprinting and genetic variability assessment of minute parasitic wasp species (Hymenoptera: Mymaridae and Trichogrammatidae) used in biological control programs of phytophagous insects. Genome 36: 580–587.

Lester, L. J. and R. K. Selander. 1979. Population genetics of haplodiploid insects. Genetics 92: 1329–1345.

Lu, Y. J., M. J. Adang, D. J. Isenhour and G. D. Kochert. 1992. RFLP analysis of genetic variation in North American populations of the fall armyworm moth *Spodoptera frugiperda* (Lepidoptera: Noctuidae). Mol. Ecol. 1: 199–208.

Lupoli, R. M. E. Irwin and C R. Vossbrinck. 1990. A ribosomal DNA probe to distinguish populations of *Rhopalosiphum maidis* (Homoptera: Aphididae). Ann. Appl. Biol. 117: 3–8.

Lynch, M. and T. J. Crease. 1990. The analysis of population survey data on DNA sequence variation. Mol. Biol. Evol. 7: 377–394.

MacPherson, J. M., P. E. Eckstein, G. J. Scoles and A. A. Gajadhar. 1993. Variability of the random amplified polymorphic DNA assay among thermal cyclers, and effects of primer and DNA concentration. Mol. Cell. Probes 7: 293–299.

Marchant, A. D. 1988. Apparent introgression of mitochondrial DNA across a narrow hybrid zone in the *Caledia captiva* species-complex. Heredity 60: 39–46.

Marchant, A. D., M. L. Arnold and P. Wilkinson. 1988. Gene flow across a chromosomal tension zone. I. Relicts of ancient hybridization. Heredity 61: 321–328.

Marchant, A. D. and D. D. Shaw. 1993. Contrasting patterns of geographic variation shown by mtDNA and karyotype organization in two subspecies of *Caledia captiva* (Orthoptera). Mol. Biol. Evol. 10: 855–872.

Martin, A. P. and C. Simon. 1988. Anomalous distribution of nuclear and mitochondrial DNA markers in periodical cicadas. Nature 336: 6196.

May, B. 1992. Starch gel electrophoresis of allozymes. Pp. 1–27. In: *Molecular Genetic Analysis of Populations: A Practical Approach*. A. R. Hoelzel, Ed. IRL Press, Oxford.

Menken, S. B. J. 1991. Does haplodiploidy explain reduced levels of genetic variability in Hymenoptera. Proc. Exp. Appl. Entomol. N.E.V. Amsterdam 2: 172–178.

Michelmore, R. W., I. Paran and R. V. Kesseli. 1991. Identification of markers linked to disease-resistance genes by bulked segregant analysis: A rapid method to detect markers in specific genomic regions by using segregating populations. Proc. Natl. Acad. Sci. USA 88: 9828–9833.

Michener, C. D. 1972. Committee on the African Honeybee. Final Report. National Academy of Sciences, Washington, DC.

Moritz, C., T. E. Dowling and W. M. Brown. 1987. Evolution of animal mitochondrial DNA: Relevance for population biology and systematics. Annu. Rev. Evol. Syst. 18: 269–92.

Moritz, R. F. A. and M. S. Meusel. 1992. Mitochondrial gene frequencies in Africanized honeybees (*Apis mellifera* L.): Theoretical model and empirical evidence. J. Evol. Biol. 5: 71–81.

Morton, R. A. 1993. Evolution of *Drosophila* insecticide resistance. Genome 36: 1–7.

Mouches, C. N. Pasteur, J. B. Berge, O. Hyrien, M. Raymond, B. Robert deSaint Vincent, M. de Silvestri and G. P. Georghiou. 1986. Amplification of an esterase gene is responsible for insecticide resistance in a California *Culex* mosquito. Science 133: 778–780.

Muralidharan, K. and H. G. Hall. 1990. Prevalence of Africa DNA RFLP alleles in Neotropical African honeybees. Arch. Insect Bioch. Physiol. 15: 229–236.

Murphy, R. W., J. W. Sites, Jr., D. G. Buth and C. H. Haufler. 1990. Proteins I: Isozyme electrophoresis. Pp. 45–126. In: *Molecular Systematics*. D. M. Hillis and C. Moritz, Eds. Sinauer Assoc., Sunderland, MA.

Narang, S. K., T. A. Klein, O. P. Perera, J. B. Lima and A. T. Tang. 1993. Genetic evidence for the existence of cryptic species in the *Anopheles albitarsis* complex in Brazil: Allozymes and mitochondrial DNA restriction fragment length polymorphisms. Biochem. Genet. 31: 97–111.

Nei, M. 1972. Genetic distance between populations. Amer. Natur. 106: 283–292.

Odum, E. P. 1971. *Fundamentals of Ecology*. 3rd Ed. Saunders, Philadelphia.

Packer, L. and R. E. Owen. 1992. Variable enzyme systems in the Hymenoptera. Biochem. Syst. Ecol. 20: 1–7.

Paige, K. N. and W. C. Capman. 1993. The effects of host-plant genotype, hybridization, and environment on gall-aphid attack and survival in cottonwood: The importance of genetic studies and the utility of RFLPS. Evolution 47: 36–45.

Pashley, D. P. 1986. Host-associated genetic differentiation in fall armyworm (Lepidoptera: Noctuidae): A sibling species complex? Ann. Entomol. Soc. Amer. 79: 898–904.

Pashley, D. P. 1988. Quantitative genetics, development, and physiological adaptation in host strains of fall armyworm. Evolution 42: 93–102.

Pashley, D. P. 1989. Host-associated differentiation in armyworms (Lepidoptera: Noctuidae): An allozymic and mitochondrial DNA perspective. Pp. 103–114. In: *Electrophoretic Studies on Agricultural Pests*. H. D. Loscale and J. den Hollander, Eds. Clarendon Press, Oxford.

Paskewitz, S. M. and F. H. Collins. 1990. Use of the polymerase chain reaction to identify mosquito species of the *Anopheles gambiae* complex. Med. Vet. Entomol. 4: 367–373.

Pasteur, N., G. Pasteur, F. Bonhomme, J. Catalan and J. Britton-Davidian. 1988. *Practical Isozyme Genetics*. Ellis Horwood, Chichester.

Penner, G. A., A. Bush, R. Wise, W. Kim, L. Domier, K. Kasha, A. Laroche, G. Scoles, S. J. Molar and G. Fedak. 1993. Reproducibility of random amplified polymorphic DNA (RAPD) analysis among laboratories. PCR Methods Appl. 2: 341–345.

Perring, T. M., A. Cooper and D. J. Kazmer. 1992. Identification of the Poinsettia strain of *Bemisia tabaci* (Homoptera: Aleyrodidae) on broccoli by electrophoresis. J. Econ. Entomol. 85: 1278–1284.

Perring, T. M., A. D. Cooper, R. J. Rodriguez, C. A. Farrar and T. S. Bellows, Jr. 1993. Identification of a whitefly species by genomic and behavior studies. Science 259: 74–77.

Persing, D. H., S. R. Telford III, P. N. Rys, D. E. Dodge, T. J. White, S. E. Malawista and A. Spielman. 1990. Detection of *Borrelia burgdorferi* DNA in museum specimens of *Ixodes dammini* ticks. Science 249: 1420–1423.

Pfennig, D. W. and H. K. Reeve. 1993. Nepotism in a solitary wasp as revealed by DNA fingerprinting. Evolution 47: 700–704.

Price, P. W. 1975. *Insect Ecology*. Wiley, New York.

Prody, C. A., P. Dreyfus, R. Zamir, H. Zakut and H. Soreq. 1989. *De novo* amplification within a "silent" human cholinesterase gene in a family subjected to prolonged exposure to organophosphorous insecticides. Proc. Natl. Acad. Sci. USA 86: 690–694.

Puterka, G. J., W. C. Black IV, W. M. Steiner and R. L. Burton. 1993. Genetic variation and phylogenetic relationships among worldwide collections of the Russian wheat aphid, *Diuraphis noxia* (Mordvilko), inferred from allozyme and RAPD-PCR markers. Heredity 70: 604–618.

Rand, D. M. and R. G. Harrison. 1989. Ecological genetics of a mosaic hybrid zone: Mitochondrial, nuclear and reproductive differentiation of crickets by soil type. Evolution 43: 432–449.

Raymond, M., A. Callaghan, P. Fort and N. Pasteur. 1991. Worldwide migration of amplified insecticide resistance genes in mosquitoes. Nature 350: 151–153.

Raymond, M., M. Marquine and N. Pasteur. 1992. Role of mutation and migration in the evolution of insecticide resistance in the mosquito *Culex pipiens*. Pp. 19–27. In: *Resistance 91: Achievements and Developments in Combating Pesticide Resistance*. I. Denholm, A. L. Devonshire and D. W. Hollomon, Eds. Elsevier Appl. Sci., London.

Rinderer, T. E. 1988. Evolutionary aspects of the Africanization of honey-bee populations in the Americas. Pp. 13–28. In: *Africanized Honey Bees and Bee Mites*. G. R. Needham, R. E. Page, Jr., M. Delfinado-Baker and C. E. Bowman, Eds. Ellis Horwood, Chichester.

Rinderer, T. E., R. L. Hellmich, R. G. Danka and A. M. Collins. 1985. Male reproductive parasitism: a factor in the Africanization of European honey-bee populations. Science 228: 1119–21.

Rinderer, T. E., J. A. Stelzer, B. P. Oldroyd, S. M. Buco and W. L. Rubink. 1991. Hybridization between European and Africanized honey bees in the neotropical Yucatan peninsula. Science 253: 309–11.

Roderick, G. K. 1992. Postcolonization evolution of natural enemies. Proc. Thomas Say Publ. Entomol. 1: 71–86.

Roehrdanz, R. L. 1989. Intraspecific genetic variability in mitochondrial DNA of the screwworm fly (*Cochliomyia hominivorax*). Biochem. Genet. 27: 551–569.

Roehrdanz, R. L., D. K. Reed and R. L. Burton. 1993. Use of polymerase chain reaction and arbitrary primers to distinguish laboratory-raised colonies of parasitic Hymenoptera. Biol. Contr. Theor. Appl. Pest Mgt. 3: 199–206.

Roush, R. T. 1990. Genetic variation in natural enemies: critical issues for colonization in biological control. Pp. 263–288. In: *Critical Issues in Biological Control*. M. Mackauer, L. E. Ehler and J. Roland, Eds. Andover, Intercept.

Roush, R. T. and J. A. McKenzie. 1987. Ecological genetics of insecticide and acaricide resistance. Annu. Rev. Entomol. 32: 361–380.

Roush, R. T. and B. E. Tabashnik, Eds. 1990. *Pesticide Resistance in Arthropods*. Chapman and Hall, New York.

Rozas, J. and M. Aguade. 1991. Using restriction-map analysis to characterize the colonization process of *Drosophila subobscura* on the American continent. I. *rp49* region. Mol. Biol. Evol. 8: 447–457.

Saccheri, I. J. and M. W. Bruford. 1993. DNA fingerprinting in a butterfly, *Bicyclus anynana* (Satyridae). J. Hered. 84: 195–200.

Schiff, N. M. and W. S. Sheppard. 1993. Mitochondrial DNA evidence for the 19th century introduction of African honey bees into the United States. Experientia 49: 530–532.

Schimke, R. T. 1988. Gene amplification in cultured cells. J. Biol. Chem. 263: 5989.

Sheppard, W. S., T. E. Rinderer, J. A. Mazzoli, J. A. Stelzer and H. Shimanuki. 1991a. Gene flow between African- and European-derived honey bee populations in Argentina. Nature 349: 782–784.

Sheppard, W. S., A. E. Soares, D. DeJong and H. Shimanuki. 1991b. Hybrid status of honey bee populations near the historic origin of Africanization in Brazil. Apidologie 22: 643–652.

Sheppard, W. S., G. J. Steck and B. A. McPheron. 1992. Geographic populations of the medfly may be differentiated by mitochondrial DNA variation. Experientia 48: 1010–1013.

Shotkoski, F. A. and A. M. Fallon. 1990. Genetic changes in methotrexate-resistant mosquito cells. Arch. Insect Bioch. Physiol. 15: 79–92.

Shotkoski, F. A. and A. M. Fallon. 1991. An amplified insect dihydrofolate reductase gene contains a single intron. Eur. J. Biochem. 201: 157–160.

Shotkoski, F. A. and A. M. Fallon. 1993. An amplified mosquito dihydrofolate reductase gene: Amplicon size and chromosomal distribution. Insect Molec. Biol. 2: 155–161.

Shufran, K. A., W. C. Black IV and D. C. Margolies. 1991. DNA fingerprinting to study spatial and temporal distributions of an aphid, *Schizaphis graminum* (Homoptera: Aphididae). Bull. Entomol. Res. 81: 303–313.

Simon, C., A. Franke and A. Martin. 1991. The polymerase chain reaction: DNA extraction and amplification. Pp. 329–355. In: *Molecular Techniques in Taxonomy*. G. M. Hewitt, A. W. B. Johnston and J. P. W. Young, Eds. Springer-Verlag, Berlin.

Simon, C., C. McIntosh and J. Deniega. 1993. Standard restriction fragment length analysis of the mitochondrial genome is not sensitive enough for phylogenetic analysis or identification of 17-year periodical cicada broods (Hemiptera: Cicadidae): The potential for a new technique. Ann. Entomol. Soc. Am. 86: 228–238.

Singh, R. S. and A. D. Long. 1992. Geographic variation in *Drosophila*: From molecules to morphology and back. Trends Ecol. Evol. 7: 340–345.

Slatkin, M. 1987. Gene flow and the geographic structure of a natural population. Science 236: 787–792.

Slatkin, M. and N. H. Barton. 1989. A comparison of three indirect methods for estimating average levels of gene flow. Evolution 43: 1349–1368.

Smith, D. R. 1991. African bees in the Americas: Insights from biogeography and genetics. Trends Ecol. Evol. 6: 17–21.

Smith, D. R., O. R. Taylor and W. M. Brown. 1989. Neotropical Africanized bees have African mitochondrial DNA. Nature 339: 213–15.

Smith, D. R., M. F. Palopoli, B. R. Taylor, L. Garnery, J. M. Cornuet, M. Solignac and W. M. Brown. 1991. Geographical overlap of two mitochondrial genomes in Spanish honeybees (*Apis mellifera iberica*). J. Hered. 82: 96–100.

Southwood, T. R. E. 1978. *Ecological Methods*. 2nd Ed. Halsted, New York.

Steiner, W. W. M. 1993. Genetic differentiation and geographic distribution in insects: patterns and inferences. Pp. 27–65. In: *Evolution of Insect Pests/Patterns of Variation*. K. C. Kim and B. A. McPheron Eds. Wiley, New York.

Tares, S., J. M. Cornuet and P. Abad. 1993. Characterization of an unusually conserved *Alu* I highly reiterated DNA sequence family from the honeybee, *Apis mellifera*. Genetics 134: 1195–1204.

Taylor, O. R., Jr. 1985. African bees: Potential impact in the United States. Bull. Entomol. Soc. Amer. 31: 15–24.

Tegelstrom, H. 1992. Detection of mitochondrial DNA fragments. Pp. 89–113. In: *Molecular Genetic Analysis of Populations: A Practical Approach*. A. R. Hoelzel, Ed. IRL Press, Oxford.

Thompson, M., F. Shotkoski and R. ffrench-Constant. 1993. Cloning and sequencing of the cyclodiene insecticide resistance gene from the yellow fever mosquito *Aedes aegypti*. FEBS 325: 187–190.

Unruh, T. R. 1990. Genetic structure among 18 west coast pear psylla populations: implications for the evolution of resistance. Amer. Entomol. Spring: 37–43.

Weir, B. S. 1990. *Genetic Data Analysis: Methods for Discrete Population Genetic Data*. Sinauer Assoc., Sunderland, MA.

Weir, B. S. and C. C. Cockerham. 1984. Estimating F-statistics for the analysis of population structure. Evolution 38: 1358–1370.

Wesson, D. M., D. K. McLain, J. H. Oliver, J. Piesman and F. H. Collins. 1993. Investigation of the validity of species status of *Ixodes dammini* (Acari: Ixodidae) using rDNA. Proc. Natl. Acad. Sci. USA 90: 10221–10225.

White, P. S. and L. D. Densmore III. 1992. Mitochondrial DNA isolation. Pp. 29–58. In: *Molecular Genetic Analysis of Populations: A Practical Approach*. A. R. Hoelzel, Ed. IRL Press, Oxford.

Winston, M. L. 1992. The biology and management of africanized honey bees. Annu. Rev. Entomol. 37: 173–193.

Xiong, B. and T. D. Kocher. 1993. Intraspecific variation in sibling species of *Simulium venustum* and *Simulium verecundum* complexes (Diptera: Simuliidae) revealed by the sequence of the mitochondrial 16S rRNA gene. Can. J. Zool. 71: 122–1206.

Zuckerman, E. 1985. They're called killer bees. New York Times Magazine, 15 September.

# Chapter 15

# Transgenic Pest and Beneficial Arthropods for Pest Management Programs

## Overview

Genetic manipulations of pest and beneficial arthropods for use in pest management programs share many problems and issues. They require methods for efficient and stable transformation, and knowledge of appropriate promoters and other regulatory elements needed to obtain an effective expression of the inserted gene in both space and time. The number of genes that are cloned and of potential value for pest management programs remains limited at this time; they are primarily single genes for resistance to pesticides or other toxins. Eventually, however, we will have cloned genes that code for other desired traits such as high or low tolerances to abiotic factors such as temperature and relative humidity, or the ability to modify sex ratios or developmental rates and fecundities. Before transgenic arthropods with alterations in these complex traits can be developed, we must understand the underlying mechanisms and identify the critical genes involved in these processes.

One factor hindering progress in the genetic manipulation of beneficial and pest arthropods is the lack of a 'universal' transformation system. We need a transformation method that will provide a rapid and general system for introducing exogenous DNA into species for which little genetic information is available. If a specific vector for each target species must be developed, then transformation may be limited to a few species of great economic importance because of the high cost and lengthy time needed to develop each vector system.

The use of recombinant DNA technology to modify pest and beneficial arthropods is in its infancy and has not yet yielded either an effective beneficial arthropod for a pest management program or the control of a pest arthropod population. Releases of transgenic arthropods into the environment will require an analysis of risk, which adds a significant cost in both time and resources to genetic manipulation projects, as it has with transgenic crops and microorganisms. It has taken years for companies to get to the point where transgenic crops could become commercially available, and it remains unclear how successful they will be in the marketplace. Once a genetically manipulated strain has been produced and tested in small plot trials, it can be deployed in a pest management program. Deployment of genetically manipulated pest and beneficial arthropods requires extensive information about their behavior, as well as their population biology, ecology, and genetics.

## Introduction

Thirty-two years ago, Sailer (1961) suggested that interspecific hybridization might provide useful genes for genetic improvement of beneficial arthropods. In so doing, he anticipated the possibilities offered by recombinant DNA techniques when he speculated that "there is no reason why a wide variety of useful characters cannot be similarly moved from one species to another and combined to form strains of insect parasites or plant pollinators that are superiorly adapted to the environments where they are needed." It is likely that advances in molecular genetics now will provide the opportunities for employing genes from a wide array of species to modify both pest and beneficial arthropod species for use in pest management programs (Beckendorf and Hoy 1985).

Molecular genetic manipulations of pest or beneficial arthropods share many of the same techniques and goals, but one major difference is critical. The goal is to *reduce* the reproductive or vector capability of pests. With arthropod natural enemies, the goal is to *enhance* the natural enemy so that improved biological pest control is achieved. In both situations, competitiveness and ability to function effectively under field conditions usually are required.

Genetic manipulation of domesticated or semidomesticated arthropods, such as silkworms and honey bees, may result in improved disease resistance, silk production or pollination (Rothenbuhler 1979, Yokoyama 1979). Because honey bees and silk moths are managed extensively by humans, their competitiveness in the field may not be critical (Hoy 1976). However, a significant constraint on genetic manipulation of most pest or beneficial arthropods is the anticipated difficulty of maintaining quality in mass-reared populations. One of the significant benefits of recombinant DNA techniques may be that it will be easier to maintain 'quality' in transgenic arthropods.

Arthropod natural enemies have been modified by artificial selection and hybridization of different strains to achieve heterosis or the use of mutagens to obtain a specific trait has been proposed, but these approaches have been employed only rarely (Hoy 1990a). Recombinant DNA techniques could make genetic improvement in arthropod natural enemies more efficient and less expensive because, once a useful gene has been cloned, it could be inserted into a number of beneficial species in a relatively short time. Recombinant DNA methods broaden the number and type of genes available for use, because a gene from a prokaryote or other organism can be used if provided with an appropriate promoter. Because long selection programs are unnecessary, there is less likelihood that laboratory adaptation will occur during the manipulation.

Pest arthropods have been modified by radiation or chemosterilization, induction of chromosomal aberrations, hybrid sterility, genetic incompatibility, and meiotic drive (Curtis 1979, LaChance 1979, Pal and Whitten 1974, Whitten 1979, Wright and Pal 1967). Only the sterile insect release method (SIRM), in which males are sterilized by irradiation, has been used extensively to control fruit flies, mosquitoes, and the screwworm *Cochliomyia hominivorax*. Control of pest arthropods could perhaps be improved with recombinant DNA techniques, again because they could reduce the negative impacts of laboratory adaptation or irradiation and increase the availability of useful characters.

The SIRM, in which sterile males are released to mate with fertile native females, requires rearing large numbers of males for release into the environment after they have been sterilized by radiation or chemicals. This approach was so successful that it resulted in the eradication of the screwworm *Cochliomyia hominivorax* from the U.S.A. and the program has since been expanded to eliminate the screwworm from all of Central America to the Isthmus of Panama. The success of the SIRM relies on careful monitoring of the flies' fitness and ability to mate with wild type females during their mass production to reduce loss of competitiveness (Bush 1979, Mangan 1991). The irradiation used to induce sterility causes somatic damage, which can reduce vigor and competitiveness of the released males. As a result, a very high ratio of released to native males is required, which increases rearing costs. Although the SIRM also has been successfully employed with the Mediterranean fruit fly *Ceratitis capitata*, a number of species are not suitable for the SIRM because of logistical, economical, or biological reasons, and alternative approaches to genetic control are necessary (LaChance 1979).

Traditional genetic manipulation projects involving artificial selection or mutagenesis by irradiation typically have three phases: conceiving and identifying the problem, developing the genetically manipulated strain, and evaluating and implementing the new biotype (Hoy 1990a). This chapter reviews the state of the art of genetic manipulation of pest and beneficial arthropods by recombinant DNA techniques from the point of view of improving pest management programs and discusses the issues surrounding the assessment of risks in releasing transgenic arthropods into the environment.

## Which Genetic Manipulations of Pests Might Be Useful?

Many speculations have been made regarding the role molecular genetics could play in the genetic control of pest arthropods that serve as the vectors of human and animal diseases or pests of agricultural crops (Besansky and Collins 1992, Cockburn et al. 1984, Cockburn and Seawright 1990, Crampton et al. 1990, Eggleston 1991, Fallon 1991, Handler and O'Brochta 1991, IAEA 1993, Kidwell and Ribeiro 1992, Meredith and James 1990, Mouches 1989, Whitten 1985). For example, it might be possible to develop insects that can be sterilized by the presence of lethal genes in the germ line; these lethal genes could be activated by stimuli such as heat, light, hormones, diet, or exposure to metals. The colony could be reared normally, but a promoter could be stimulated to activate a gene that would sterilize males before their release. Genes that can be mutated to cause male sterility are known in *Drosophila* and might be useful in the genetic control of agricultural pests such as the Mediterranean or Caribbean fruit flies. It might also be possible to release tropical populations of mosquitoes into temperate populations so that the hybrid progeny produced are unable to overwinter in diapause. Hanson et al. (1993) demonstrated reduced overwintering ability in hybrid progeny produced by temperate and tropical *Aedes albopictus* populations due to a lack of cold hardiness in overwintering eggs. However, it is possible that resistance to mating with the tropical population could develop in the temperate populations.

Yet another proposal is to alter a pest arthropod population so that it is unable to vector a disease. Mosquitoes or tsetse flies may become unable to vector diseases because their salivary glands or gut membranes are impermeable to the parasites or because they have toxic factors in their hemolymph. It may also be possible to alter the genome of the arthropod vector so that it is unable to feed on more than one host; even if the vector picks up the pathogen, it will not transmit it.

Proposed mechanisms of controlling a pest population include swamping a wild population with some kind of genetic load by releasing a few individuals with a trait that will spread and exert a delayed impact on the fitness and vectoring capacity of the wild population. Thus, transposons or meiotic drive mechanisms (mechanisms that result in the unequal recovery of the two types of gametes produced during meiosis by a heterozygote) might be introduced into wild populations (Kidwell and Ribeiro 1992). If transposons are introduced into a population, they can cause chromosomal mutations, shut off genes, or cause sterility. If a transposon could be engineered so that it could not be repressed by its host, then perhaps it could increase in number until it caused so much damage that the population crashed.

Possibly, specific arboviruses could be controlled by introducing plasmid constructs containing arbovirus sequences into the arthropod vector genome to interfere with virus infection and replication. McGrane et al. (1988) noted that *Aedes triseriatus*, an effective vector of LaCrosse virus, might be genetically manipulated to be refractory to virus infection because mosquitoes infected with the LaCrosse virus become resistant to superinfection, or to infection with related viruses. The molecular basis of the resistance is unknown, but is associated with virus replication in midgut cells. Of course, such transgenic mosquitoes would have to be reared, released, and established in the field, where they would replace the normal vector populations or introduce the gene into the vector population. This tactic assumes, once again, that the genetically manipulated population is competitive in the field.

Production of large numbers of competitive irradiated or chemosterilized males in the sterile insect release method (SIRM) requires careful quality control. In the near future it may be possible to alter the sex ratio of pest arthropods by genetically manipulating their sex-determining genes. The genetic basis of sex determination in *D. melanogaster* is being resolved (Slee and Bownes 1990) (Chapter 11). If this knowledge can be applied to other arthropods, it may allow the manipulation of the sex ratio of economically important species and make it possible to rear large numbers of sterile males by altering the developmental pathway with environmental cues.

A number of steps are involved in a program designed to control pest arthropods through molecular genetic manipulation (Figure 15.1). First, the target species must be identified as a significant pest for which conventional control tactics are ineffective, because genetic control is usually more expensive and difficult. Furthermore, genetic manipulation with recombinant DNA techniques may generate concerns about risk.

Second, the trait amenable to manipulation should be clearly identified. Starting a genetic manipulation project without a clear understanding of the pest species' weak point(s) is unlikely to yield an effective control program. How best might our knowledge about the pest species' physiology, ecology, or behavior be used against it? Once a target trait has been identified, it must be genetically altered.

After a modified strain is developed, it must then be evaluated in the laboratory for fitness and stability. Eventually the manipulated strain must be released into small plots in the field for evaluation. Permission to release a transgenic arthropod pest will have to be obtained from the appropriate regulatory agencies. Releases initially will be made into small plots and eventually into larger plots. If the strain(s) perform well, implementation and evaluation of efficacy will be needed.

## Which Genetic Manipulations of Beneficial Arthropods Might Be Useful?

Several species of silk moths, including *Bombyx mori*, *Philosamia* species and *Antheraea* species, are mass reared to produce silk in Japan, China, and India (Gopinathan 1992). Work has begun to develop a stable transformation system for *B. mori* (Tamura et al. 1990, Okano et al. 1992). Analysis and cloning of the

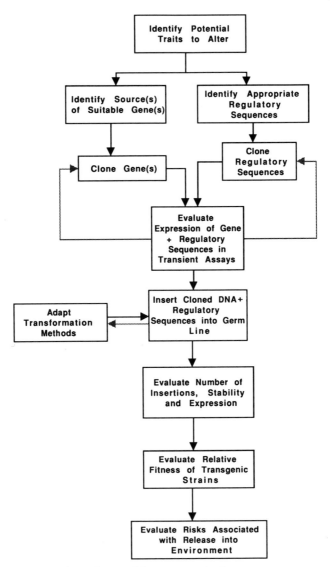

**Figure 15.1** An outline of some of the steps involved in genetic engineering of pest or beneficial arthropods up to the point when the transgenic arthropods are ready to be released into the environment.

light and heavy-chain fibroin genes from B. *mori* has been achieved, and it may be useful to transfer fibroin genes from a strain of silk moth that produces high quality silk into the more sturdy native races grown in India. This tactic also may be useful in B. *mori* because it appears that expression of fibroin genes is directly proportional to the gene dosage, which is normally increased through chromosomal endoduplication. Increasing the fibroin gene copy number to two or more per haploid genome through genetic manipulation might result in increased production of fibroin in silk moths.

Silk moths are susceptible to a variety of viral, bacterial, fungal, and protozoan infections. Increasing their immunity to such diseases through recombinant DNA techniques would be useful. Other attributes that might be modified include pupal diapause and fecundity.

Traditional genetic methods for improving arthropod natural enemies for biological control of pest arthropods have involved selecting for resistance to pesticides, lack of diapause, and increased tolerance to temperature extremes. Development of a flightless strain of the parasite *Microplitis croceipes* has been proposed for releases against cotton bollworms, *Helicoverpa zea*, because the wasps would be unable to leave the cotton fields where they are released (USDA 1993). Most genetic manipulation projects have involved selection of predatory mites (Acari: Phytoseiidae) for resistance to pesticides (Hoy 1990b). Pesticide-resistant predatory mites have been evaluated in the field and are being implemented in integrated pest management (IPM) programs for apples, pears, almonds, grapes, greenhouses, and strawberries (Hoy 1990b). Genetic manipulation has proven to be practical and cost effective when the trait(s) limiting efficacy can be identified, the improved strain retains its fitness, and methods for implementation have been developed (Headley and Hoy 1987).

Traits primarily determined by single major genes, such as pesticide resistance, are most appropriate for manipulation at this time because methods for manipulating and stabilizing traits that are determined by complex genetic mechanisms are not yet available. Genetic improvement can be useful when the natural enemy is known to be a potentially effective biological control agent except for a limiting factor; the limiting trait primarily is influenced by a single major gene; the gene can be obtained by selection, mutagenesis, or cloning; the manipulated strain is fit and effective; and the released strain can be maintained by some form of reproductive isolation.

The outcome of releasing genetically modified natural enemies is determined by the goals of the pest management program. Genetically manipulated natural enemies can be released in large numbers in order to have a significant impact on the pest population over a short interval, perhaps only one growing season (augmentative releases). Alternatively, smaller numbers of genetically manipulated natural enemies can be released, with the goal of permanently establishing them in the environment so that their effect can be maintained over several growing seasons in orchards and vineyards. The outcome of the releases will be different depending on whether native populations of the genetically manipulated natural enemy are present.

To date, most releases have involved pesticide-resistant natural enemy populations. If susceptible native populations are present, pesticides are applied to reduce these populations, and the resistant populations could become established in two different ways: (1) the resistant strain could completely replace the susceptible native population, especially if few or no natives remained after pesticide applications (the replacement model). (2) Alternatively, the resistant and susceptible populations could interbreed and the resistance gene could be selected for, with the resultant hybrid population becoming resistant (the introgression model). If no susceptible population is present, as is the case when genetically manipulated strains are released into greenhouses or in new geographic regions, the population can be maintained with the desired trait intact (Hoy 1990a, 1993).

## Steps in Genetic Manipulation by Recombinant DNA Methods

Genetic improvement by recombinant DNA techniques involves several steps (Figure 15.1). As discussed in the previous two sections, the goals of genetic manipulation of pests and of beneficial arthropods are different. The goal is to *reduce* or eliminate the pest population in the field, or its impact. By contrast, the goal is to *enhance* or increase the population of beneficial anthropods, or their impact, in the field. A successful outcome generally requires that we have a thorough knowledge of the biology, ecology and behavior of the target species. A critically important first step is to identify one or more specific traits that, if altered, potentially would achieve the goals of the project.

Next, suitable genes must be identified and cloned. Appropriate regulatory sequences must be identified so that the inserted gene will be expressed at appropriate levels in the correct tissues and at a relevant time.

It is likely that the construct containing the gene of interest and the regulatory sequences initially will be evaluated by means of **transient transformation**. This involves transferring a gene or genes of interest into an organism and assessing whether it is transcribed and translated. If the cloned gene and its regulatory sequences function well in a transient system, they may function well when inserted into the germ line.

Stable transformation involves incorporating the genetic information into the germ line so that the information is transmitted to succeeding generations. Usually several transformed lines are developed and evaluated to determine which lines are most fit and stable. It is likely that transgenic lines will next have to be evaluated in a contained greenhouse before they can be released into the environment. If the laboratory and greenhouse tests indicate that the transgenic strain is relatively fit and the trait is stable and appropriately expressed, the transgenic strain(s) should be evaluated in small field plots to confirm its efficacy and fitness.

Before any field tests can occur, however, permission to release the transgenic strain should be obtained (Hoy 1990a, 1992a,b). At least initially, it will be easier to obtain permission from regulatory agencies to release transgenic arthropod natural enemies than to release transgenic *pest* species.

Research is essential in all aspects in this sequence (Figure 15.1). First, we need to identify and clone useful genes. We need to ensure that the inserted genes will be expressed in specific tissues and at the appropriate time. Identifying appropriate regulatory sequences is nearly as important as identifying functional coding sequences. Effective germ line transformation methods are needed so that stable transformation of organisms can be achieved in an efficient and predictable manner.

Inserting cloned DNA into pest or beneficial arthropods could be accomplished by several different techniques (Table 15.1). The effects of the inserted DNA could be transient and short term, or stable and long term. If the inserted DNA is incorporated into the chromosomes in the cells that give rise to the germ line, the foreign genetic material should be transmitted faithfully and indefinitely to successive generations. Cloned DNA can be isolated from the same or other species, and it is technically feasible to insert genes from microorganisms into arthropods and have the DNA transcribed and translated. However, DNA coding sequences isolated from microorganisms must be attached to promoters (controlling elements) and other regulatory DNA sequences derived from a higher organism so that the gene will be expressed in arthropods. These regulatory sequences determine when a gene will be transcribed, at what level, in what tissues, and how long the messenger RNA can be used for translation. The following discussion reviews the status of these components of genetic manipulation of both pest and beneficial arthropods.

## Potential Germ Line Transformation Methods

Most research on stable (germ line) transformation methods has been accomplished with *Drosophila melanogaster* (Table 15.1). Initial efforts to genetically engineer *D. melanogaster* were rarely successful until Spradling and Rubin discovered that a transposable element called the P element could be genetically manipulated to serve as a vector to carry exogenous genes into the chromosomes of germ-line cells (Rubin and Spradling 1982, Spradling and Rubin 1982). The DNA carried by the P-element vector became stably integrated into the chromosomes of the fly and was expressed. This pioneering work has elicited immense amounts of research on fundamental analyses of gene structure, function,

**Table 15.1**

Some Potential Methods for Stably Transforming Arthropods

| Technique | Example(s) available | References |
|---|---|---|
| Artificial chromosomes | None in arthropods; feasible with yeast and mice | Schedl et al. 1992, Peterson et al. 1993 |
| Baculovirus vectors *Autographa californica* *Bombyx mori* | Primarily for protein expression in larvae or cell cultures; lethal to infected host unless addtional genetic modifications are conducted | Iatrou and Meidinger 1990, Miller 1988 |
| DNA delivered by microprojectiles | Transient expression in *Drosophila* embryos | Baldarelli and Lengyel 1990 |
| | Transient expression in *Anopheles gambiae* | Mialhe et al. 1993 |
| Electroporation | Transient transformation of *Drosophila* only | Kamdar et al. 1992 |
| Maternal microinjection | *Metaseiulus occidentalis* (Acari: Phytoseiidae) | Presnail and Hoy 1992 |
| Microinjection of eggs | Three mosquito species; P element apparently not functional | Miller et al. 1987, McGrane et al. 1988, Morris et al. 1989 |
| P-element vectors | *Drosophila* species only | Handler and O'Brochta 1991 |
| Soaking dechorionated eggs in DNA solution | *Drosophila* | Walker 1989 |
| Sperm as vectors of DNA | *Lucilia cuprina; Apis mellifera* | Atkinson et al. 1991, Milne et al. 1989 |
| Transfection of culured cells | *Aedes albopictus* | Fallon 1991 |
| Transformation of insect symbionts | Bacterial symbiont of *Rhodnius prolixus* engineered; symbionts inserted into symbiont-free insects were transmitted to successive generations and *Rhodnius* survived antibiotic treatment | Beard et al. 1992 |
| Transplant nuclei and cells | *Drosophila* | Zalokar 1981, Lawrence 1993 |

*(continues)*

**Table 15.1** (*Continued*)

| Technique | Example(s) available | References |
|---|---|---|
| Transposable element (TE) vectors from target species | None at this time, but TEs are known from several, including *Anopheles gambiae* and *Bombyx mori* | Michaille et al. 1990, Besansky 1990 |
| Yeast recombinase (FLP) mediated recombination on specific target DNA sequences (FRT) | *Aedes aegypti* | Morris et al. 1991, Kilby et al. 1993 |

and regulation in *Drosophila* and has given us a broad understanding of how flies develop (Lawrence 1992). Many genes have been identified, isolated, and cloned from *Drosophila*. At present, only a few of these genes appear to be potentially useful in genetic manipulations of either pest or beneficial arthropods.

### P-Element Vectors

In molecular genetics, vectors are self-replicating DNA molecules that transfer a DNA segment between host cells. The P element was genetically engineered to serve as a vector for exogenous DNA in order to develop transgenic *D. melanogaster*. P elements were discovered in *D. melanogaster* because they cause a syndrome called hybrid dysgenesis. An intact P element consists of 2907 bp of DNA, with a coding sequence that produces a single polypeptide with transposase activity (see Chapter 10). Intact P elements have 31-bp inverted repeats flanking the coding region, and these intact inverted repeats are required if the P element is to transpose. Thirty to fifty P elements are normally dispersed throughout the genome of P strains of *D. melanogaster* but do not produce hybrid dysgenesis because transposition is suppressed by unknown factors in the P cytotype. Hybrid dysgenesis occurs when male flies from a strain with chromosomes that contain P elements (P males) are mated with females from a strain lacking P elements (an M strain). Their $F_1$ progenies exhibit a high rate of mutation, chromosomal aberrations, and even complete sterility, caused by transposition of P elements in their germ line chromosomes. The reciprocal cross does not generate hybrid dysgenesis because the female's P cytotype suppresses transposition of the P elements. Transposition of P elements does not occur in somatic tissues, because required splicing of the transposase mRNA is germ line specific (Laski et al. 1986).

Since the pioneering research of Rubin and Spradling (1982), in which they reported the first use of P elements that were genetically manipulated to serve as vectors to carry DNA into the genome of D. *melanogaster*, P-element vectors have been investigated as possible carriers of exogenous DNA into the chromosomes of other arthropods (Handler et al. 1993). P-element vectors have been used effectively with other *Drosophila* species such as D. *similans* and D. *hawaiiensis*. Other insects, including three mosquitoes and the Mediterranean fruit fly, *Ceratitis capitata*, have received microinjected DNA cloned into P-element vectors. *Aedes aegypti* (Morris et al. 1989), *Anopheles gambiae* (Miller et al. 1987), and *Aedes triseriatus* (McGrane et al. 1988) have been transformed in a stable manner. Unfortunately, the rate of transformation was very low (less than 0.1% of the microinjected embryos) and there is no evidence that the process of transformation was P-element mediated. It appears that P-element-mediated transposition may be limited to *Drosophila* species (Handler and O'Brochta 1991, Handler et al. 1993). To date, there is no firm evidence that integration of any exogenous DNA in an insect outside the genus *Drosophila* has been P-element-mediated. As a result, a variety of other methods for achieving transformation have been considered and evaluated (Table 15.1).

### Other Transposable Element Vectors

P elements are only one family of transposable elements that are found in *Drosophila melanogaster* and other arthropods. Transposable elements are commonly found in all organisms (including bacteria, yeast, plants, nematodes, mice, humans, *Bombyx mori*, mosquitoes) whenever they have been sought, but they have been less well studied in other arthropods (Berg and Howe 1989).

It is possible that species-specific transposable elements, such as *mariner* and *hobo*, could be isolated and genetically modified for use as vectors in specific insects. The *mariner* element is found in a wide variety of insects (Robertson 1993a, Atkinson et al. 1993), but it is unclear whether an active element can be engineered for use as a vector. However, engineering vectors is neither a rapid nor inexpensive process and this approach may be limited to those arthropod species that are of major economic importance. Furthermore, because transgenic arthropods being released into the environment should be stably transformed, such transposable element vectors must be incapable of additional subsequent movement after the first, targeted transformation. Thus, issues of risk assessment should be considered in designing a genetic manipulation project involving 'native' transposable element vectors (Hoy 1992a,b).

### Microinjection

Microinjecting exogenous DNA carried in P-element vectors into *Drosophila* eggs is a well developed technique (Santamaria 1986). *Drosophila* eggs must be injected when freshly deposited, prior to the formation of pole cells (which will give rise to the germ line), when the nuclei are still present as a syncytium. The chorion is removed from *Drosophila* eggs, either chemically by soaking them in a sodium hypochlorite solution or mechanically by rubbing them on a piece of sticky tape. The eggs are then dehydrated to relieve internal pressure that would cause them to burst when the needle is inserted. The eggs are placed on double-sided sticky tape, covered with paraffin oil, and injected at the posterior pole with a fine glass needle (1- to 3-μm tip diameter). A micromanipulator with the capacity to move in three directions is desirable for the manipulations and injections. Eggs are usually kept in a cool chamber and injected immediately in a cool room (18–20°C) to slow development. After injection, the eggs are held for two days at 18–23°C under oil, after which the hatched larvae will be in the medium or moving about in the oil. Survival after injection varies with the experience and skill of the operator but averages 20–30%. Total mortality may equal 40–60% before the injected embryos become adults.

These microinjection methods had to be modified for mosquito eggs, with slightly different injection methods required for different genera (McGrane et al. 1988, Miller et al. 1987, Morris et al. 1989). The mosquito chorion is rigid and very difficult to remove, yet the chorion makes the eggs relatively impenetrable. However, if eggs are collected very soon after deposition (within 1–2 hours), the chorions are still flexible and penetrable, and microinjection is successful using tips 3–10 μm in diameter. Such new eggs are also very susceptible to desiccation and must be held in halocarbon oil saturated with water to maintain viability. Survival of mosquito embryos that have been microinjected is equivalent to survival rates obtained with *Drosophila*.

Milne et al. (1988) developed a method for microinjecting early honey bee (*Apis mellifera*) embryos. Honey bee eggs are more fragile and susceptible to damage than *Drosophila* eggs and contain about 14 times more mass. Thus, chorion removal and dehydration were not critical, but a smaller needle tip was. Also, the slower developmental time to formation of the blastoderm (about 10 hours) provided a longer interval for injections.

Presnail and Hoy (1992) found that eggs of the phytoseiid predator *Metaseiulus occidentalis* were extremely difficult to dechorionate and dehydrate and that the needle tip had to be modified for this species. Based on results to date, it appears that the *Drosophila* microinjection methodology will have to be adapted empirically to each insect species and will not be feasible with all. Variables to consider include whether to dechorionate, whether to dehydrate and for how long or under what conditions, at what age or stage to inject, what holding conditions to implement after injection, and what size and shape of needle to use.

### Injecting DNA without a Transposable Element Vector

It may be feasible to microinject exogenous DNA into insect embryos without using any transposable element vector. It is known that transformation of a number of organisms, including *Drosophila*, can be achieved by such a method, although at a relatively low rate (Walker 1989). Evaluation of the fate of injected DNA in *Drosophila* might offer an explanation of how this could happen (Steller and Pirrota 1985). Eggs of *Drosophila* contain about 10 nanoliters and are surrounded by a thin, stiff, vitelline membrane covered by a rigid proteinaceous chorion. Sperm enter through the micropyle, and after fertilization the pronucleus undergoes rapid, synchronous divisions without partitioning the nuclei by cell membranes. The resulting syncytium is maintained until after nine divisions (90 minutes at 25°C), when the majority of the 512 nuclei migrate to the periphery of the egg. A few nuclei which enter the posterior of the egg are enveloped in cell membranes (pole cells) and give rise to the germ line. After four more divisions, the nuclei begin to be cellularized and the embryo reaches the cellular blastoderm stage. Because cell membranes are absent, DNA injected into the early cleavage embryo might be relatively free to diffuse and find its way into nuclei when the nuclear membrane dissociates at each division.

Steller and Pirotta (1985) found that plasmid DNA injected into early *Drosophila* embryos becomes enclosed in nuclei or nucleus-like structures, where it remained throughout embryonic development. A portion of the exogenous DNA in these nuclei is converted to a high-molecular-weight form consisting largely of tandem oligomers, some of which may be derived from homologous recombination between plasmids. Some exogenous DNA also may engage in nonhomologous recombination between plasmid and genomic DNA and be inserted into the chromosomes in the following manner: the exogenous DNA could become a substrate for endogenous enzymes, with nicks in the DNA being repaired and topoisomerases converting the circles into supercoils that become integrated into the genome, although with a "rather low efficiency" and apparently without a specific transposition mechanism. Steller and Pirrota (1985) suggested that the frequency of genomic integration might be stimulated by the use of nicked plasmid DNA. Others have suggested that integration might be enhanced if single-stranded DNA were used.

### Maternal Microinjection

Another potential target for DNA-mediated transformation consists of the developing eggs of gravid female insects or mites. For example, early preblastoderm eggs present within adult females of the predatory mite *Metaseiulus occidentalis* were microinjected by inserting a needle through the cuticle of gravid females (Figure 15.2). This technique, called 'maternal microinjection', resulted in relatively high levels of survival and stable transformation without the aid of a transposase-producing helper plasmid (Presnail and Hoy 1992). In this example, the bacterial *lacZ* reporter gene regulated by the *Drosophila hsp70* promoter was expressed in larvae developing from the injected eggs and in subsequent generations of mites arising from the transformants. Stable transfor-

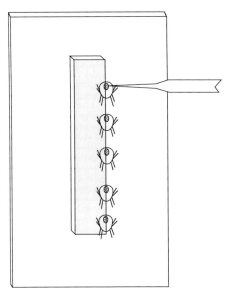

**Figure 15.2** Maternal microinjection has been used to obtain transgenic strains of the phytoseiid predator *Metaseiulus occidentalis*. During microinjection, females are confined to a glass microslide using two layers of double-stick tape cut to produce an overhanging ledge. Microinjection into the posterior region of the female may be directly into individual eggs or the ovary. (From Presnail and Hoy 1992.)

mation was confirmed in the sixth generation by polymerase chain reaction (PCR) amplification of a region spanning the *Drosophila/E. coli* DNA sequences inserted into the mite. Microinjection of another colony (COS) of M. *occidentalis* resulted in four transformed lines, and stable transformation was confirmed after about 16 generations both by PCR and by Southern blot analysis of genomic DNA (J. K. Presnail, A. Jeyaprakash and M. Hoy, unpublished observations).

Maternal microinjection of M. *occidentalis* is less laborious than microinjection of eggs that have been laid by females because the eggs do not need to be dechorionated or dehydrated prior to the injection. Survival rates of injected females were comparable to survival of microinjected *Drosophila* eggs. The transformation rate was approximately one-tenth the efficiency of P-element-mediated transformation of *Drosophila*, but comparable to techniques employed for species in which transformation is achieved without a P-element vector (Presnail and Hoy 1992).

Maternal microinjection may provide a DNA delivery system for other arthropod species. Presnail and Hoy (unpublished observations) found that

DNA could be injected into multiple eggs of Metaseiulus occidentalis and a different predatory mite species, Amblyseius finlandicus, as well as the parasitic wasp Cardiochiles diaphaniae. It should be possible to adapt needle diameter and tip structure so that a needle can be inserted into the region of the ovary(ies) of many arthropods. Injection may be facilitated by inserting the needle into membranous regions between sclerotized segments after preliminary dissections have been made to determine the locations of the ovaries (Presnail and Hoy 1992).

## Soaking Dechorionated Eggs in DNA
## May Result in Transformation

Before the development of P-element vectors, exogenous DNA was introduced into Drosophila embryos by soaking them in DNA solutions after dechorionation (reviewed by Walker 1989). Some of the adult flies produced exhibited somatic mosaicism, which appeared to be due to incorporation and expression of exogenous DNA in somatic cells. However, the method has not been pursued because of low uptake (<2%), variable phenotypes, and the difficulty in establishing stably transformed lines of flies. Most of these experiments used total genomic DNA, and Walker (1989) speculated that soaking embryos in specific cloned gene sequences rather than total DNA could result in higher rates of stable transformation. However, the high success rates achieved with P-element-mediated transformation eliminated the motivation for Drosophila workers to investigate these techniques. Whether this method can be used for other arthropod species has not been determined.

## Sperm as Vectors of DNA

The use of sperm as carriers of exogenous DNA has been evaluated for honey bees and the Australian sheep blowfly Lucilia cuprina (Atkinson et al. 1991). If sperm are used to transfer genes, the exogenous DNA should bind to the sperm so that it be carried into the nucleus of the egg. This approach to gene transfer would probably be limited to species such as the honey bee, for which semen can be collected and used for artificial insemination.

There are no examples of successful stable transformations using this technique, and considerable controversy surrounds this approach (Milne et al. 1989). Atkinson et al. (1991) indicate that sperm from both L. cuprina and A. mellifera are capable of binding labeled DNA, but the DNA is associated entirely with the outside sperm membrane. There is no evidence that the DNA is taken into the sperm and these authors concluded that internalization of DNA is important if the sperm is to be a vector. They concluded that "the use of sperm as vectors for DNA-mediated transformation does not appear to be a viable option". In contrast, Milne et al. (1989) argue that transformation of A. mellifera by DNA-mediated sperm is feasible.

## DNA Delivered by Microprojectiles

A novel method for delivering DNA into living plant cells involves coating microprojectiles with DNA or RNA, then shooting them into a plant cell with a gun. The so-called 'gene gun' has been used successfully to transform major crop plants, yeast, and cultured cells. Its use with arthropod eggs is, however, limited.

Baldarelli and Lengyel (1990) obtained transient expression of DNA in *Drosophila* embryos after ballistic introduction with DNA-carrying tungsten particles. The embryos were incubated overnight and tested the next day. The authors suggest this method may, with some modification, be suitable for stable germ line transformation. This technique would be especially useful with arthropods whose eggs may be difficult to dechorionate. Also, it may be particularly useful for species that deposit large numbers of eggs. It would not be advantageous for species, such as parasitic Hymenoptera, that deposit their eggs into the body of their insect host because obtaining large numbers of eggs by dissection would be extremely tedious. However, if an artificial ovipositional medium were available for the parasitic wasps, it might be possible to obtain reasonable numbers of eggs at the appropriate developmental stage.

### Electroporation

Introducing DNA into insect embryos using an electric field pulse allows many more embryos to be transformed at one time than does microinjection into individual dechorionated eggs. With short electric impulses above a certain strength, membranes are made more permeable for a short interval. This allows material to cross the perturbed membrane. Nondechorionated *Drosophila* embryos, incubated in a solution of DNA, can take up and transiently express this DNA (Kamdar et al. 1992). Efforts to use electroporation for germ line transformation of insects has been investigated, but stable transformation has not yet been reported.

### Engineered Chromosomes

Engineered artificial chromosomes have been constructed in yeast which behave much like the natural ones do. Apparently, the essential components needed for chromosomes include genes, centromeres, autonomously replicating sequences that serve as origins of chromosome replication, and telomeres. In yeast, artificially constructed chromosomes that were only 10 to 15 kb long did not behave like natural chromosomes (full-sized yeast chromosomes are about 150 kb long), but artificial chromosomes about 50 kb long were more stably inherited. The larger artificial chromosomes passed through meiosis and were present in a few copies per cell. Although the longer chromosomes performed better, they were lost about twice as often as normal chromosomes, perhaps because the spacing between different DNA segments is incorrect or the chromosomes still are not long enough.

It may be possible to use yeast artificial chromosomes (YACs) to transform arthropods. Transgenic mice have been obtained by injecting a yeast artificial chromosome into fertilized mouse oocytes (Schedl et al. 1992). These mice carried the YAC DNA and expressed the YAC-encoded tyrosinase gene, which caused the albino mice to become pigmented. The YAC integrated into the mouse genome and the presence of yeast telomeric sequences apparently did not reduce the efficiency of integration.

These results suggest that it may be possible to construct artificial chromosomes for arthropods that can confer useful traits in a stable manner. Such artificial chromosomes may be particularly useful for situations where it is desirable to insert a number of genes that are linked.

## Transplanting Nuclei and Cells

Zalokar (1981) reported methods for injecting and transplanting nuclei and pole cells into eggs of Drosophila, and Lawrence et al. (1993) made hybrid Drosophila embryos by transplanting pole cells from two species into sterile D. melanogaster hosts. Thus, it might be possible to genetically transform insect cells in cell culture, isolate the nuclei, and transplant them into the region where the germ line cells (pole cells) will develop in embryos.

## Transformation of an Insect Symbiont

Richards (1993) suggests that the insects that transmit the infectious agents of malaria, trypanosomiasis, filariasis, dengue fever, and viral encephalitis to humans, as well as the vectors that transmit hundreds of viral, fungal, and bacterial diseases to plants could be managed by genetically altering the symbionts of the insect vectors.

As discussed in Chapter 5, many symbionts supply nutrients that are essential for their hosts and are carried as inclusions within their hosts' cells. Many bacterial symbionts resemble Escherichia coli, which is relatively easy to genetically engineer. After the symbiont has been transformed, it could be inserted back into a symbiont-free arthropod host, which could be reared and released for a genetic control program. The symbionts would be vehicles for expressing foreign genes that would reduce or enhance the fitness of their arthropod hosts or reduce their ability to vector a disease (Richards 1993).

Beard et al. (1992, 1993) demonstrated that genetic engineering of insect symbionts is feasible by transforming a bacterial symbiont, Rhodococcus rhodniia, of the Chagas disease vector Rhodnius prolixus. The symbiont lives extracellularly in the insect gut lumen and is transmitted from adult to progeny by egg shell contamination or by contamination of food with infected feces. The symbiont was genetically engineered to be resistant to an antibiotic and the resistant symbionts were transmitted to insects lacking any symbionts. The insects containing the resistant symbionts subsequently were treated with the antibiotic and survived, transmitting the transformed symbiont to successive generations of insects. The protozoan that causes Chagas disease in humans lives in the insect's gut. Beard et al. (1992) speculate that the symbiont might be used to express a gene product within the insect gut that would negatively affect attachment or development of the Chagas disease agent, *T. cruzi*.

The symbionts of *Rhodnius* also have been transformed with a shuttle vector containing *ampicillin* and *thiostrepton* resistance genes and with genes coding for cecropin A and related pore-forming molecules (Richards 1993). The antibiotic resistance genes provide a selective advantage to the transgenic symbionts so they can survive antibiotics in the blood meal. The cecropin A and related molecules insert into certain membranes and make holes. The symbionts and *T. cruzi* are in close proximity in the gut of their host, and it is hoped that release of the pore-forming molecules from the symbiont will lead to lysis of the pathogen *T. cruzi*. A number of questions remain unanswered, including whether the cecropin molecules will be so toxic that the transgenic symbiont will be unable to survive.

Experiments also are under way to transform the symbionts of tsetse flies (*Glossina* species), which are vectors of both animal and human African trypanosomiasis (Richards 1993). One proposal is to transform the symbiont of tsetse flies with one of two different genes, the AF-2 toxin of *Pseudomonas fluorescens*, which is very active against African trypanosomes, or genes for antibody fragments directed against the procyclin surface molecules of the pathogen *Trypanosoma brucei*. If the antibodies are expressed by the symbiont and transported to where the trypanosomes are, they could inactivate these pathogens.

Another project is under way to modify the ability of the green peach aphid *Myzus persicae* to transmit the disease-causing pea ennation mosaic virus (PEMV) (Richards 1993). PEMV is ingested by the aphid and enters the hemolymph, where it attaches to a salivary gland of the aphid. When the aphid feeds on another plant, it secretes saliva carrying PEMV into the plant before it begins feeding. Each aphid carries between $10^6$ and $10^8$ of several species of symbiotic bacteria. These symbionts secrete proteins into the hemolymph that are used by the aphids as storage proteins (symbionin). PEMV has a protein coat that is vital to transmission of the PEMV because it interacts with a virus receptor in the insect salivary gland. If the symbionts could be engineered to express antibody fragments against the PEMV protein, in a manner similar to their expression of symbionin, it might be possible to block the interaction of the PEMV protein domain and its receptor in the aphid's salivary gland and thus convert the virus into a nontransmissible virus.

Much research must be conducted to learn how to isolate and transform these symbiont species. Likewise, extensive research is needed on how to obtain effective expression and export of proteins, as well as identification of appropriate genes for transformation. A method for deploying the transgenic symbionts or transgenic symbiont-containing insects must be developed. Methods for releasing insects containing transgenic symbionts will have to be developed and tested to determine whether the arthropod-borne diseases can be eliminated or whether resistance to the transgenic symbionts can develop.

### Transfer of Exogenous DNA into Cultured Insect Cells

Cultured insect cells can be induced to take up exogenous DNA by several methods and the transformed cells can be used to evaluate the expression of genes and promoters. One method involves the use of polycations, including polyornithine or polybrene, to enhance the adsorption of negatively charged DNA molecules onto *Aedes albopictus* cells (Fallon 1991). Calcium phosphate coprecipitation has been used successfully with *Drosophila* cells (Walker 1989). Other methods, including electroporation, liposomes, laser micropuncture, and several types of microinjection, have been used to transfer exogenous DNA into insect cells (Walker 1989). However, at present, adapting these methods appears to be an empirical process for each species.

Expression of the introduced DNA may occur only for a short period (transient expression) after introduction into insect cells, although stable expression of the introduced genes can sometimes be obtained (Fallon 1991). A popular sensitive assay gene is the *chloramphenicol acetyltransferase* (CAT) gene. Because insect cell lines are unable to acetylate chloramphenicol, expression of CAT is due to expression of the exogenous DNA in the transformed cell based on fusion with a eukaryotic promoter such as the *Drosophila heat shock protein 70* (*hsp70*) (Okano et al. 1992), the *Bombyx mori sericin-1* gene, or the *Drosophila copia* LTR promoters (Tamura et al. 1990).

Exogenous plasmid DNA molecules introduced into cells may join by end-to-end ligation, probably catalyzed by cellular repair enzymes. These end-to-end molecules, called **concatemers**, can occasionally be integrated into host chromosomes by unknown recombinational events and result in stable transformation (Monroe et al. 1992).

Stable transformation of other insect cell lines apparently has not yet been achieved, although the tools are available to attempt it. Several selectable markers are available, including resistance to neomycin or G418, methotrexate, or heavy-metal ions. There are several possible promoters available for testing, including the *copia* and *Drosophila hsp70* promoters. Actin, metallothionein, or baculoviral promoters may also be useful.

### Baculovirus Vectors

Nuclear polyhedrosis viruses, or baculoviruses, have double-stranded, circular DNA genomes contained within a rod-shaped protein coat. Baculoviruses infect a number of pest insects and several have been used as biological pesticides, including *Autographa californica* nuclear polyhedrosis virus (AcNPV) and *Lymantria dispar* NPV (LdNPV). These NPV and the *Bombyx mori* nuclear polyhedrosis virus (BmNPV) also have been exploited as vectors to carry exogenous DNA into insect cells (Miller 1988, Iatrou and Meidinger 1990, Yu et al. 1992). However, because insect cells or larvae die from their infection, baculovirus vectors are not suitable for producing stably transformed insects.

Baculoviruses have been engineered to be superb expression vectors, and it is possible to produce a very wide array of proteins from different species in insect cells or larvae. Genetic manipulation of baculoviruses has been directed toward two major goals: improved commercial production of proteins of biomedical or agricultural importance (Miller 1988), and improved efficacy of baculoviruses as biological pesticides (Wood and Granados 1991). If baculovirus vectors that are nonlethal to their hosts could be developed, they could also serve as vectors for stable transformation.

### Yeast (FLP)-Mediated Recombination

The ability to introduce cloned and modified DNA into the germ line at a predictable chromosomal site is especially desirable because it reduces the likelihood of position effects on gene expression. One method for accomplishing this is based on a system found in the yeast *Saccharomyces cerevisiae*. A gene for yeast recombinase, **FLP recombinase**, was found on a plasmid in *S. cerevisiae*. This plasmid also carries two inverted recombination target sites (FRTs) that are specifically recognized by the FLP recombinase. In yeast, FLP recombinase catalyzes the recombination of DNA between the FRT sites in the plasmid, inverting the sequences between them (intramolecular recombination). FLP will also catalyze intermolecular recombination between homologous sites on two different chromosomes (intermolecular recombination).

The FLP-FRT system has been modified to insert exogenous DNA into a specific site in a *Drosophila* chromosome. The FLP and FRT sequences were cloned from yeast and the FRT sites were inserted into the *Drosophila* chromosome through P-element-mediated transformation (Konsolaki et al. 1992, Simpson 1993). When yeast recombinase is added to the transformed lines, exogenous DNA can be inserted into a specific location in the genome, for example, between the FRT sites.

If the FRT sites can be inserted into other insects, the system may be useful for site-directed modification. Morris et al. (1991) showed that FLP-mediated, site-specific intermolecular recombination occurred in microinjected embryos of the mosquito *Aedes aegypti*. The results of these experiments did not allow them to determine whether the mosquitoes were stably transformed. However, this technique may be useful for assessing the effect of specific gene constructs on competence to vector disease or development.

The FRT-FLP system could provide a rapid method of inserting different

DNA sequences into a specific chromosomal site (where the FRT site is). However, because a stable FRT site must be present in the genome, different lines carrying FRT sites in different chromosomal locations will have to be evaluated to determine which site permits better expression of the foreign genes. Thus, the FLP system may be best suited for those species undergoing intensive and long term genetic analysis and manipulation.

## Which Genes Are Available?

Genes theoretically can be isolated from either closely or distantly related organisms for insertion into arthropods by recombinant methods. It may also be possible to isolate a gene from the species being manipulated, alter it, and reinsert it into the germ line. Assuming that a transformation method is available so that either transient or stable transformation can be achieved, the major issue then becomes whether the exogenous gene is expressed appropriately and effectively. Expression requires an appropriate promoter and other regulatory elements.

Many genes have been cloned from *Drosophila* and other species and inserted into *Drosophila* by P-element-mediated transformation. Most of the projects were directed at understanding gene regulation or developing a selectable marker for identifying transformants. These cloned genes may not be particularly useful for genetic manipulation of beneficial or pest arthropods. Genes used to identify transformants include microbial genes (*neomycin* or G418 resistance, *chloramphenicol acetyltransferase*, and β-*galactosidase*) and *Drosophila* eye color genes such as *rosy* or *white*. Additional genes cloned from *Drosophila* could be used directly for transforming pest or beneficial insects, or they could serve as probes for homologous sequences in other insect species. Cloned genes also could be modified by *in vitro* mutation to achieve a desired phenotype.

For the foreseeable future, resistance genes will probably be the most available and useful for transformation of arthropods. Potentially useful resistance genes have been identified in *Drosophila* (Morton 1993). A number of genes have been cloned from insects and other organisms, including a *parathion hydrolase* gene (*opd*) from *Pseudomonas*, a cyclodiene resistance gene (GABA$_A$) from *Drosophila*, β-tubulin genes from *Neurospora crassa* and *Septoria nodorum* that confer resistance to benomyl, an *acetylcholinesterase* gene (*Ace*) from *D. melanogaster* and the mosquito *Anopheles stephensi*, a glutathione transferase gene (*GST1*) from *Musca domestica*, a cytochrome P450-B1 gene (*CYP6A2*) associated with DDT resistance in *Drosophila*, the knockdown resistance gene associated with resistance to DDT and pyrethroids in *Musca domestica*, the amplification core and esterase B1 gene isolated from *Culex* mosquitoes, which are responsible for organophosphorus insecticide resistance, as well as the esterase genes conferring resistance in the aphid *Myzus persicae* (Table 15.2).

**Table 15.2**

Some Cloned Resistance Genes Possibly Useful for Genetic Manipulation of Pest and Beneficial Arthropods

| Gene, abbreviation, and resistance conferred | Sources | References |
|---|---|---|
| Acetylcholinesterase (Ace); pesticide resistances | D. melanogaster Anopheles stephensi | Hall and Spierer 1986, Hall and Malcolm 1991, Hoffmann et al. 1992, Fournier et al. 1989, 1992a |
| β-Tubulin; benomyl resistance | Neurospora crassa Septoria nodorum | Orbach et al. 1986, Cooley et al. 1991 |
| Catalase; $H_2O_2$ resistance | D. melanogaster | Orr and Sohal 1992 |
| γ-Aminobutyric acid $A_A$ receptor gene; cyclodiene resistance (Rdl) | D. melanogaster Aedes aegypti | ffrench-Constant et al. 1991, 1993a,b, ffrench-Constant and Rocheleau 1993, Thompson et al. 1993 |
| Cytochrome P450-B1; DDT resistance | D. melanogaster | Waters et al. 1992 |
| Cytochrome P450 | Musca domestica | Feyereisen et al. 1989 |
| Esterase B1 amplification core; organophosphate resistance | Culex species | Mouches et al. 1986, 1990 |
| Esterases E4 and PE4 | Myzus persicae | Field et al. 1993 |
| Glutathione S-transferase; DmGST 1-1, DMGST-2, DDT resistance | D. melanogaster | Toung et al. 1990, 1993 Beall et al. 1992 |
| Glutathione S-transferase; MdGST1; organophosphate resistance | Musca domestica | Wang et al. 1991, Fournier et al. 1992b |
| Knockdown resistance; kdr; DDT and pyrethroids | Musca domestica | Williamson et al. 1993 |
| Metallothionein genes; Mtn; copper resistance | D. melanogaster | Theodore et al. 1991 |
| Multidrug resistance; Mdr49, Mdr50, and Mdr65; colchicine resistance | D. melanogaster | Wu et al. 1991, Gerrard et al. 1993 |

(continues)

**Table 15.2**   (*Continued*)

| Gene, abbreviation, and resistance conferred | Sources | References |
|---|---|---|
| *neomycin phospho-transferase*; *neo*; resistance to kanamycin, neomycin, G418 | Transposon Tn5 | Beck et al. 1982 |
| Parathion hydrolase, *opd*; parathion, paraoxon resistance | *Pseudomonas diminuta* *Flavobacterium* sp. | Serdar et al. 1989, Dumas et al. 1990, Phillips et al. 1990, Mulbry and Karns 1989, Dave et al. 1993 |

Metallothionein genes have been cloned from *Drosophila* and other organisms that appear to function in homeostasis of copper and cadmium and in their detoxification (Theodore et al. 1991). Perhaps these genes could provide resistance to fungicides containing copper in arthropod natural enemies. In many crops, fungicides may have serious negative impacts on beneficial arthropods such as phytoseiid predators.

A family of genes of potential importance for pest management consists of the multidrug resistance genes in mammals. These genes, *mdr* or *pgp*, become amplified and overexpressed in multidrug-resistant cell lines, resulting in cross-resistances to a broad spectrum of compounds, including those used in cancer chemotherapy. The multidrug resistance genes code for a family of membrane glycoproteins that appear to function as an energy-dependent transport pump. Recently, three members of this multigene family were isolated from *D. melanogaster* and these genes (*Mdr49* and *Mdr65*) could provide resistance to a number of exogenous chemicals (Wu et al. 1991, Gerrard et al. 1993). For example, *D. melanogaster* strains that were made deficient for *Mdr49* were viable and fertile, but had an increased sensitivity to colchicine during development. Whether the insertion of multidrug resistance genes would provide a useful increase in tolerance to chemicals that arthropods might encounter in the environment remains to be determined.

Preliminary results suggest that microbial genes conferring resistance to pesticides can function in arthropods. The *opd* gene isolated from *Pseudomonas* and conferring resistance to organophosphorus pesticides has been inserted, using a baculovirus expression vector, into cultured fall armyworm *Spodoptera frugiperda* cells and larvae (Dumas et al. 1990). Phillips et al. (1990) also transferred the *opd* gene into *D. melanogaster*. The *opd* gene was put under control of the *Drosophila heat shock 70* promoter, *hsp70*, and stable active enzyme was produced and accumulated with repeated induction. It is likely that this gene could be used to confer resistance to organophosphorus pesticides in beneficial arthropod species, as well as to serve as a selectable marker for detecting transformation of pest species. If the *opd* gene were linked to the sex-determining system of pest species being reared for genetic control programs, and could be induced by a specific environmental stimulus such as heat shock, it is possible that the unwanted females could be eliminated, thereby reducing mass rearing costs. Whether a pest species containing a pesticide resistance gene should be released into the environment is subject to debate, unless they are fully sterile males.

Freeze resistance in frost-susceptible hosts may be increased by gene transfer. Antifreeze protein genes cloned from the wolffish *Anarhichas lupus* and the winter flounder *Pleuronectes americanus* have been expressed in transgenic *Drosophila* (Rancourt et al. 1990, 1992, Peters et al. 1993) using the *hsp70* promoter and yolk polypeptide promoters of *Drosophila*. While additional work is required to obtain flies that are able to tolerate cold temperatures, the results suggest that subtropical or tropical species of arthropod natural enemies could become useful or adapted in a much broader range of climates.

Altering the longevity of beneficial arthropods might result in more effective biological control of pests in some environments. Research on mechanisms of aging may provide genes useful for modifying the longevity of arthropods. A cloned *catalase* gene inserted into *D. melanogaster* by P-element-mediated transformation provided resistance to hydrogen peroxide, which is implicated in cell damage, although the *catalase* gene alone did not prolong the life span of flies (Orr and Sohal 1992). Transgenic flies carrying three copies each of both the *copper–zinc superoxide dismutase* and the *catalase* genes overexpressed both proteins. The transgenic flies exhibited up to a one-third extension of life span, longer mortality rate doubling time, lower protein oxidative damage, and delayed loss of physical performance (Orr and Sohal 1994). In addition, a 77-kDa protein associated with longevity in *D. melanogaster* was characterized and associated with an autosomal longevity locus; efforts to clone this gene are under way (Okano et al. 1992, Yonemura et al. 1992).

As basic research progresses, other traits that might be important or useful to introduce into beneficial insects will become obvious. Shortening the development rate, enhancing progeny production, altering sex ratio, extending temperature and relative humidity tolerances, and altering host or habitat preferences could enhance biological control (Hoy 1976). It also could be possible to alter gene expression in transgenic insects by inserting DNA that codes for antisense RNA directed against a specific insect gene (Heinrich et al. 1993). However, it is not simple to document that changes in one or more of these attributes would actually improve the performance of a biological control agent.

Likewise, for pest insects, it might be desirable to insert genes that slow development, reduce their ability to overwinter or survive adverse weather, skew the sex ratio to produce a preponderance of nonvectoring males, or reduce their ability to vector pathogens or their propensity to take blood meals. Because these traits are probably determined by multiple genes, they may be more diffi- cult to manipulate than traits such as resistance to pesticides, which are often influenced by single genes. At present, inserting multiple genes into organisms by recombinant DNA techniques is much more difficult than working with a single gene.

Most of the methods discussed above involve inserting exogenous DNA into random sites in the arthropod chromosomes. The recent success in achiev- ing targeted gene conversion in *Drosophila* suggests that it may be possible to use transposable elements as targets for gene conversion in insects and mites (Sentry and Kaiser 1992) (Chapter 10). Targeted gene conversion could improve the efficiency and specificity of inserting genes in both pest and beneficial arthro- pods.

## The Importance of Appropriate Regulatory Signals

Genes consist of coding segments that determine the amino acid sequences in the enzyme or structural proteins produced. However, whether a coding region is transcribed and translated in a specific tissue is determined by a number of regulatory sequences in the DNA, including promoters and enhancers. Some of these regulatory structures are in close proximity to the coding region, while others may be located farther away (Abraham et al. 1993). The stability of messenger RNA is influenced by a variety of signals in the RNA, including the polyadenylation (poly(A)) signals at the $3'$ end of the RNA, which can influ- ence the amount of protein produced. It is crucial to obtain expression of the inserted gene at appropriate times, levels, and tissues. Another factor that may be important in maintaining the inserted DNA in the transgenic line over time is the presence of origins of replication that regulate DNA replication of the chromosomes. If exogenous DNA is inserted into a region of the chromosome far from a site where an origin of replication occurs naturally, the exogenous DNA could be lost over time because it is not replicated.

Regulatory sequences from *Drosophila* can be combined with a protein- coding sequence from a prokaryote such as *E. coli* to form a DNA construct that will function in a eukaryote. However, regulatory sequences from prokaryotes do not function in eukaryotic organisms. Because regulatory sequences may vary from species to species, the source of regulatory sequences chosen for cloning may be as important, or even more important, than the source of the protein- coding sequences (Figure 15.1). Furthermore, some regulatory sequences allow genes to be expressed only in particular tissues or in response to particular stimuli (such as heat shock), while other genes are expressed in most tissues most of the time. If it is important that the inserted gene function in a tissue- or stimulus- specific manner, it is essential to identify tissue- or stimulus-specific promoters.

Currently, the number of suitable regulatory sequences available for genetic manipulation of arthropods is limited. The heat shock 70 (*hsp70*) promoter from *Drosophila* is commonly used as an inducible promoter. It is the strongest promoter known in *Drosophila* and appears to function in all cells. Heat shock proteins are present in all organisms subjected to high temperatures and, while the number of these proteins varies from organism to organism, all produce a 70-kDa protein encoded by an *hsp70* gene family member. It is likely that the *Drosophila hsp70* promoter can be used whenever an inducible promoter is required that will function in all cells.

While the *hsp70* promoter is highly conserved, it may perform differently in different arthropod species. For example, the mosquito *Anopheles gambiae* was transformed with a plasmid containing the *hsp70* promoter of *Drosophila* attached to a microbial *neomycin* resistance gene, which also confers resistance to the antibiotic G418 (Miller et al. 1987). Transgenic mosquitoes expressed the *neo* gene at a low level in adults at 26°C, and a heat shock for 15 minutes at 37°C enhanced the level of expression. Recently, Sakai and Miller (1992) found that survival of transgenic larvae exposed to G418 was increased after heat shock at 41°C, which is higher than the temperature (37°C) typically used to induce genes in *Drosophila*. McInnis et al. (1990) found that three heat shocks produced higher survival rates in Mediterranean fruit flies, *Ceratitis capitata*, transiently transformed with *neo* and treated with geneticin.

Other commonly used regulatory sequences from *Drosophila* are the *actin 5C* promoter, the α1-*tubulin* promoter, and the *metallothionein (Mtn)* promoter (Kovach et al. 1992). Angelichio et al. (1991) compared the function of these promoters and the *fibroin* promoter from *Bombyx mori* in cultured *D. melanogaster* cells and found that the *actin 5C* and the *metallothionein* promoters generated comparable levels of RNA and protein. The α1-*tubulin* promoter generated about fourfold lower levels of RNA and protein. The *fibroin* promoter, cloned from *B. mori*, had no detectable activity in these cells. Another potentially useful promoter is the *B. mori* cytoplasmic *actin A3* gene, which was expressed transiently in embryos (Coulon-Bublex 1993).

The effects of three poly(A) signals isolated from mammals and arthropods were evaluated to determine their impact on stability of transcribed mRNA. Angelichio et al. (1991) compared the poly(A) signals of the SV40 early region, SV40 late region, and the *Drosophila metallothionein* gene. The SV40 late poly(A) constructs yielded protein levels that were three- to fivefold higher than the SV40 early construct. The metallothionein poly(A) and SV40 early constructs produced nearly equivalent levels.

It is often important that genes be expressed in tissue- or cell-specific patterns. Learning how to achieve this type of targeted gene expression in transgenic arthropods might employ a method similar to that used by Brand and Perrimon (1993) to evaluate the impact of a gene on cell fate during development of *Drosophila*. They inserted a gene that encodes the yeast transcriptional activator GAL4 randomly into the *Drosophila* genome to drive GAL4 expression from one of a diverse array of genomic enhancers. This system allows rapid development of strains in which expression of the target gene can be directed to different tissues or cell types.

Chromosome replication in higher eukaryotes is not well understood, but it is known that origins of replication are located at intervals along each chromosome. Origins of replication involved in amplification of chorion genes in *D. melanogaster* were identified by Carminati et al. (1992). During follicle cell differentiation, chorion genes (which code for the eggshell proteins) are amplified by multiple rounds of DNA replication, which results in high levels of protein expression during a very short time. The ACE3 chorion element has been cloned and shown to be sufficient to regulate amplification of the chorion gene cluster on chromosome III of *D. melanogaster* (Carminati et al. 1992). During genetic manipulation of pest or beneficial species, it may be useful to insert ACE3 or similar elements along with the exogenous genes to ensure that replication of this region of the chromosome occurs in order to increase the stability of the exogenous DNA in the transgenic strain.

Identification, cloning, or genetic modification of promoters and other regulatory sequences may increase the precision with which desired proteins are transcribed and expressed in transgenic arthropods. Research to understand the structure and function of regulatory sequences for use in transgenic arthropods should have high priority. Project goals will dictate what type of regulatory sequences are most useful. In some cases, a low-level, constitutive production of transgenic proteins will be useful, while in other cases high levels of protein production will be required after inducement by a specific cue. Researchers will have to evaluate the tradeoffs between high levels of protein production and the subsequent impact on relative fitness of the transgenic arthropod strain based on the specific goals of each program.

## Identifying Transformed Arthropods

After inserting the desired genes, the next issue is how to detect whether the exogenous gene has in fact been incorporated into the germ line. Because transformation methods are relatively inefficient, a screening method is needed to identify transformed individuals. This process is relatively simple in *Drosophila*, where there is a wealth of genetic information and visible markers can be used to identify transgenic individuals. Most pest or beneficial arthropods lack such extensive genetic information or markers.

Transformed individuals could be identified by using a pesticide resistance gene, such as the *opd* gene, as the selectable marker. However, the release of pesticide-resistant pest arthropods into the environment may create concerns about risk. Another option is to use the *neomycin* (*neo*) antibiotic resistance gene. This prokaryotic gene has been shown to function in both *Drosophila* and mosquitoes, and is less likely to provoke concern about the risks of releasing transgenic arthropods into the environment. Another marker is the β-*galactosidase* gene (*lacZ*) isolated from *E. coli* and regulated by the *Drosophila hsp70* promoter, which has been expressed in both *Drosophila* and the phytoseiid predator *Metaseiulus occidentalis* (Presnail and Hoy 1992). The β-*galactosidase* gene can be detected by an assay that produces a blue color in the transformed immature insects and mites. A green fluorescent protein from the bioluminescent jellyfish *Aequorea victoria* produces a product without requiring exogenous substrates or cofactors and could be a useful marker for gene expression (Chalfie et al. 1993). The mosquito *dihydrofolate reductase* gene (*dhfr*) also functions as a dominant selectable marker in transfected cells (Shotkoski and Fallon 1993). If an appropriate marker is not available, transformed lines can be identified with the polymerase chain reaction and subsequent analysis by Southern blot hybridization, or an immunological procedure, as described in Chapters 6 and 9.

## Risks Associated with Releases of Transgenic Arthropods

Risk assessments will be somewhat different for pest and beneficial arthropods. Until recently, most practitioners of biological control asserted that biological control of arthropod pests or weeds by arthropod natural enemies was environmentally safe and risk free if carried out by trained scientists. However, questions about the safety of classic biological control have been raised, particularly where environmentalists are concerned about the preservation of native flora and fauna (Howarth 1991), and the era of accepting classic biological control as environmentally risk free appears to have passed (Ehler 1990, Harris 1985, Hoy 1992a).

An evaluation of the risks associated with releasing parasitoids and predators that have been manipulated with recombinant DNA techniques will likely include, at a minimum, the questions or principles outlined in Table 15.3 (Hoy 1990a, 1992a,b, Tiedje et al. 1989, USDA 1991). Concerns can be summarized as questions about: (1) whether the transgenic population is stable, (2) whether its host or prey range has been altered, and (3) whether its potential to persist in the environment (geographic distribution and climatic tolerances) has been altered. For the foreseeable future, releases will be evaluated by regulatory agencies on a case-by-case basis. Initial permits for releases will be for short-term releases in controlled situations so that unexpected outcomes might be mitigated more readily.

Another concern involves questions about how far and how quickly the transgenic arthropod can disperse from the experimental release site. Less is known about dispersal behavior of many arthropod species than might be needed

**Table 15.3**

Some Issues Relating to Risks of Releasing Transgenic Arthropods into Experimental Field Plots[a]

---

Attributes of the unmodified organism
    What is the origin of the transgenic organism (indigenous or nonindigenous) in the accessible environment?
    What is the arthropod's trophic level and host range?
    What other ecological relationships does it have?
    How easy is it to monitor and control?
    How does it survive during periods of environmental stress?
    What is the potential for gene exchange with other populations?
    Is the arthropod involved in basic ecosystem processes?

Attributes of the genetic alteration
    What is the intent of the genetic alteration?
    What is the nature and function of the genetic alteration?
    How well characterized is the genetic modification?
    How stable is the genetic alteration?

Phenotype of modified organism compared with unmodified organism
    What is the host or prey range?
    How fit and effective is the transgenic strain?
    What is the expression level of the trait?
    Has the alteration changed the organism's susceptibility to control by natural or artificial means?
    What are the environmental limits to growth or reproduction (habitat, microhabitat)?
    How similar is the transgenic strain being tested to phenotypes previously evaluated in field tests?

Attributes of the accessible environment
    Describe the accessible environment, whether there are alternate hosts or prey, wild relatives within dispersal capability of the organisms, and the relationship of the site to the potential geographic range of the transgenic arthropod strain.
    Are there endangered or threatened species present that could be affected?
    Are there vectors or agents of indirect dissemination present in the environment?
    Do the test conditions provide a realistic simulation of nature?
    How effective are the monitoring and mitigation plans?

---

[a]Modified from Tiedje et al., 1989; USDA 1991; and from a discussion held at a conference on "Risks of Releasing Transgenic Arthropod Natural Enemies," held November 13–16, 1993 in Gainesville, Florida.

when releasing a transgenic arthropod. For example, Raymond et al. (1991) suggest that there has been a worldwide migration of *Culex pipiens* mosquitoes carrying amplified organophosphorus resistance genes. If migration, rather than independent selection on a conserved gene, is the basis for the widespread amplification of esterase genes in *Culex* mosquitoes, then dispersal of some transgenic arthropods could be more rapid and extensive than anticipated.

Another risk issue involves the possibility that horizontal transfer of genes may occur between one arthropod species and another (Houck et al. 1991, Plasterk 1993). The P element appears to have invaded *D. melanogaster* populations within the past 50 years, perhaps from a species in the *D. willistoni* group. The evidence for this hypothesis comes from the overlap in geographic ranges of *D. melanogaster* and *D. willistoni*, the strong similarity in DNA sequences of P elements from *D. melanogaster* and *D. willistoni*, the DNA sequence similarity among *D. melanogaster* P elements from diverse geographic locations, the absence of P elements from species closely related to *D. melanogaster*, the highly infectious nature of active P elements when they are experimentally introduced into susceptible *D. melanogaster* populations, and geographic distribution patterns within *D. melanogaster* (see also Chapter 10).

Controversy exists as to whether P elements may have been transferred between *Drosophila* species by the semiparasitic mite *Proctolaelaps regalis* (Houck et al. 1991). Horizontal transfer of P elements from *D. willistoni* to *D. melanogaster* must be a very rare event, requiring that two *Drosophila* females of different species lay their eggs in proximity so that a mite can feed sequentially on one and then on the other (in the correct order). The mite must carry the P element to the recipient egg, which must be in a very early stage of embryonic development, the recipient embryo must incorporate a complete copy of the P element into a chromosome before it is degraded by enzymes in the cytoplasm, the recipient embryo must survive the feeding by the mite, and the adult that develops from the embryo must transmit the P element to its progeny. If each event is rare, and the combined probability is multiplicative, then the probability that a horizontal gene transfer between different arthropod species will occur must be exceedingly rare.

Interspecific transfer of another transposable element (*mariner*) has been suggested as an explanation of its presence in the drosophilid genera *Drosophila* and *Zaprionus* (Maruyama and Hartl 1991). The *mariner* element occurs only in five of eight species in the *D. melanogaster* species group, but is found in the genus *Zaprionus* outside the *Drosophila* group even though *Zaprionus* is not closely related to *Drosophila*. DNA sequences indicate that the *mariner* elements in the two groups are 97% identical, although, by comparison, the nuclear gene *Adh* is not this close phylogenetically, suggesting that there has been horizontal transfer of the *mariner* element. A *mariner* sequence has been discovered in the genome of the lepidopteran *Hyalophora cecropia* (Lidholm et al. 1991) and Robertson

(1993a,b) found that the *mariner* element is present in many insects and several species of mites. Robertson (1993a) found sufficient diversity in *mariner* DNA sequences to classify them into several different subfamilies. The diversity of species containing *mariner* elements suggests that: (1) *mariner* elements have been present in arthropods for a long time, although some lineages have lost these elements, and (2) horizontal transfer of *mariner* elements has occurred. However, these horizontal transfers have occurred relatively infrequently on an evolutionary time scale. Many or most of the *mariners* discovered have become degenerated and inactive over time.

Horizontal transfer of genes may occur when bacterial endosymbionts move from species to species. DNA sequence data suggest that bacterial endosymbionts of mosquitoes, Coleoptera, and *Drosophila* may have been horizontally transferred. An analysis of the 16S rRNA genes specific to prokaryotes from *Culex pipiens*, *Tribolium confusum*, *Hypera postica*, *Aedes albopictus*, two populations of *Drosophila simulans*, and *Ephestia cautella* indicated that their symbionts are all closely related (O'Neill et al. 1992). Horizontal transfer of symbionts may be more widespread than indicated because they are involved in many examples of cytoplasmic incompatibility, in which certain crosses between symbiont-infected individuals lead to death of embryos or distortion of the progeny sex ratio (Chapters 5 and 11). O'Neill et al. (1992) speculated that cytoplasmic incompatibility is due to infection with a specialized bacterium that infects a wide range of different arthropod hosts, including *Corcyra cephalonica*, *Sitotroga cerealella*, *Diabrotica virgifera*, *Attagenus unicolor*, *Rhagoletis pomonella*, *Rhagoletis mendax* and *Anastrepha suspensa*. While these species all carry the symbiont, cytoplasmic incompatibility has not been demonstrated in all of them.

Other potential vectors for horizontal transfer of DNA include the insect viruses. If horizontal transmission of DNA by transposable elements or microorganisms occurs, there is no absolute guarantee that genes inserted into any species will be completely stable. Such naturally occurring horizontal transmission of DNA between species may have provided some of the variability upon which evolution has acted. The extent and nature of this naturally occurring gene transfer are just being determined (Plasterk 1993). It is unlikely, however, that the presence of a transgene in an organism will increase the very small probability that the transgene will be transferred to another species by horizontal transfer, unless it was inserted using an active transposable element. Even then, the probability of horizontal transfer should be very small.

There are no clear guidelines for evaluating the risks of releasing transgenic arthropods for long-term establishment in the environment. Experience indicates that the probability that a 'new' organism will become established in a new environment is small (Williamson 1992). Historical examples of biological invasions of pests or of classic biological control agents demonstrate the lack of predictability; the low level of successful establishment; and the importance of scale, specificity, and the speed of evolution (Ehler 1990). Transgenic arthropods could pose somewhat increased risks because they would be released in large numbers. Williamson (1992) also speculated that the greater the genetic novelty, the greater the possibility of surprising results, and recommended using molecular markers to begin to understand dispersal and the interactions among species in natural communities.

Discussions of risk probably will include questions about survival, reproduction, and dispersal of transgenic populations and their effects on other species in the community. Questions also will be asked about the inserted DNA, its stability, and its possible effect on other species should the genetic material be transferred (Table 15.3). In the United States, both state and federal regulatory agencies, including state departments of agriculture and United States Department of Agriculture—Animal and Plant Health Inspection Service, will have to be consulted for permission to release transgenic arthropods. Questions about the impact of the transgenic arthropod on threatened and endangered species will be asked by state and federal agencies, including the U.S. Department of Interior Fish and Wildlife Service.

Hadrys et al. (1992) point out that several molecular genetic techniques for analyzing behavioral ecology and population biology are now available. Thus, DNA fingerprinting and PCR-RFLP techniques can be used to determine taxonomic identity, assess kinship, analyze mixed genome samples, and create specific probes. The RAPD method of PCR is useful in situations in which limited amounts of DNA are available, for species with minimal genetic information, and because it is relatively efficient and inexpensive (Chapter 9). The use of molecular techniques in ecological studies promises to provide powerful tools to help assess the risks of releasing transgenic arthropods. These techniques and others, such as population genetic models that incorporate information on dispersal rates and gene frequencies (Caprio et al. 1991), will provide methods for improving our knowledge of the ecology and behavior of both pest and beneficial arthropods in pest management programs, whether or not they have been genetically manipulated.

## How to Deploy Transgenic Pest and Beneficial Arthropods

Figure 15.1 illustrates the steps involved in developing transgenic arthropods up to the stage where they are ready to be released into the environment. That illustration is incomplete in several ways. A similar flow diagram, developed for beneficial arthropods that were manipulated by traditional genetic techniques, illustrated additional steps in the genetic manipulation project (Hoy 1990a). In that more complete flow diagram, the manipulated strains are reared in an insectary, evaluated in the laboratory, then in field cages and small field plots, and finally in large-scale plots. Mass rearing requires careful attention to quality control and efforts to maintain adequate genetic variation (Bartlett 1985). After evaluation in large plots, implementation on a commercial scale can occur if the manipulated strains perform as expected. Once the genetically manipulated strain has been employed in a pest management program, a cost–benefit analysis should be conducted (Headley and Hoy 1987).

A critically important step not shown in Figure 15.1 is consideration of *how* to employ the genetically manipulated strain in pest management programs. Ideally, the following questions should be considered when *initiating* the project, because genetic manipulation projects of beneficial or pest arthropods are neither rapid, inexpensive, nor simple.

1. Do you understand the biology, ecology, and behavior of the target species in its natural environment? Do you understand how it disperses, reproduces, and behaves under field conditions? What are its relationships with other organisms in the accessible environment?

2. Can the transgenic strain be mass reared easily and inexpensively?

3. Will the transgenic strain eventually be released into the environment for permanent establishment or will it be released periodically?

4. Will the transgenic strain be released into a geographic region where conspecific populations exist with which it can interbreed?

5. Will the transgenic strain be expected to replace the 'native' population in order to achieve the desired effect?

6. Do you know how to ensure that the transgenic strain will replace the established native population, if that is crucial to the success of the project? Is it possible that the native population can readily develop 'resistance' to the released strain?

7. Is the improved strain likely to be stable?

8. Is the released population likely to become a pest or cause other significant harm in the environment? What could be done to mitigate any possible harm?

Answers to these questions were important to the success of pest management programs involving the SIRM programs with screwworms and fruit flies, or with natural enemies improved by artificial selection. For example, the most successful genetic improvement programs with beneficial arthropods involved predatory mites (Phytoseiidae) selected for resistance to pesticides (Hoy 1990a,b). The programs were successful because a great deal is known about how to rear, release, and monitor phytoseiids. Because the manipulated species are effective biological control agents, a key factor, resistance to pesticides, could be identified that would eliminate one of the critical constraints to their effectiveness in specific pest management programs.

Genetic manipulation of pest arthropods has employed a variety of approaches, but the release of insects that have been sterilized by radiation has been more successful than releases of insects that have had more complex genetic manipulations, such as chromosomal inversions (Pal and Whitten 1974). Sterile male releases have been successful because (1) the species can be reared in large numbers, (2) highly effective methods of sterilization are available, and (3) releases of sterile males can be targeted after the 'native' population has been suppressed by pesticides, cold weather, or some other biotic or abiotic factor.

The efficacy of a 'drive' mechanism to control a pest population has not yet been demonstrated in large-scale trials. Although some small-scale experimental releases have occurred, this type of pest population manipulation remains controversial. Genetic control programs usually require repeated releases of large numbers of pest insects, and in order to carry out a prolonged series of releases of pest insects that are not sterile, it will be necessary to have the cooperation of the human inhabitants of the release area, who may be concerned about the addition of pests to their environment (Pal 1974).

Will it be possible to use cytoplasmic incompatibility to control pest arthropods? This prospect has been discussed since Laven (1951) observed the impact of cytoplasmic incompatibility on *Culex* mosquito populations and suggested that it could be used as a means of controlling mosquitoes. Yen and Barr (1974) identified the cause of the incompatibility in *Culex pipiens* as *Wolbachia pipientis* and some experiments were conducted, but the incompatibility produced was incomplete (Pal 1974). O'Neill and Karr (1990) reported that *Wolbachia* caused reduced egg hatch in *Drosophila simulans* when infected males mated with uninfected females, and Turelli and Hoffman (1991) reported that this infection is spreading rapidly in field populations of *D. simulans* in California, based on data from mtDNA RFLP analyses. Turelli et al. (1992) concluded that cytoplasmic incompatibility induced by *Wolbachia* "therefore provides a mechanism for introducing cytoplasmic factors into natural populations. This may eventually be useful for introducing deleterious factors into pest insect populations."

Richards (1993) suggested that a transgene could be spread through a pest population by transforming a symbiont of the pest. This approach also could be employed to transfer a desirable gene to beneficial species. The 'drive mechanism' could involve cytoplasmic incompatibility, caused by *Wolbachia pipientis*, which infects both male and female insects and is transovarially transmitted. An infected female insect with *Wolbachia* in her ovaries can mate and produce progeny with both infected and uninfected males, but an uninfected female can produce progeny *only* if she mates with uninfected males. Matings between an uninfected female and an infected male lead to few, if any, progeny. The overall result is that infected females will have a reproductive advantage because they can produce progeny with both infected and uninfected males, but uninfected females are successful only if they mate with uninfected males. In theory, this type of cytoplasmic incompatibility could be transferred to a population of insects lacking the *Wolbachia* by microinjection of the transgenic *Wolbachia*, mass rearing of the infected individuals, and release. As a step toward that goal, Braig et al. (1994) transferred *Wolbachia* isolated from the mosquito *Aedes albopictus* into *Drosophila simulans* by microinjection of embryos. The trans-infected flies became bidirectionally incompatible with all other natural strains of *D. simulans*. A number of questions remain to be resolved, including whether resistance to the cytoplasmically transmitted organisms could develop in the pest insects, and whether such incompatibility will be stable.

Possible other cytoplasmic factors that would decrease fitness under specific conditions include genetically altered mitochondria or viruses that would increase susceptibility to chemicals, or cytoplasmic factors (including genetically engineered *Wolbachia*) that decrease resistance to temperature extremes.

Deployment of genetically manipulated arthropods is complicated if some form of reproductive isolation or drive mechanism cannot be provided. One of the reasons genetically modified predatory mites have been successfully employed in pest management programs may be because phytoseiids disperse relatively slowly. Releases of pesticide-resistant strains into pesticide-treated greenhouses, orchards, or vineyards has provided sufficient isolation that the genetically manipulated strains have been able to establish without extensive competition from, or interbreeding with, susceptible 'native' populations (Hoy 1991). Likewise, releases of a pesticide-resistant strain of the parasitoid *Aphytis melinus* into Israeli citrus groves did not involve competition or interbreeding with susceptible populations because this species was not present in Israel.

The population genetic issues were different when a pesticide-resistant strain of walnut aphid parasite, *Trioxys pallidus*, was released into pesticide-treated California walnut orchards for control of the walnut aphid. The outcome of these releases was more complicated. Because the resistant strain is expensive to rear in very large numbers, inoculative releases were made and the released parasites were expected to establish (Hoy et al. 1990, Caprio et al. 1991). Because susceptible populations of parasites were abundant in nearby orchards, the released population could interbreed with them and resistance would be lost unless strong selection was maintained.

Predicting whether, and how, genetically modified pest or beneficial arthropods will establish is difficult. There are at least two models that could be employed in the establishment of a genetically modified strain in situations in which a 'native' population exists: (1) The released strain displaces the 'native' population and replaces it (replacement model). This model assumes that relatively little interbreeding occurs between the released and 'native' populations. (2) Alternatively, the released strain interbreeds with the 'native' population and a hybrid population is produced. By appropriate strong selection, often with pesticide applications, the desired trait is selected for and the resultant population contains the desired gene (introgression model).

Until recently, it has been difficult to tell what was happening when genetically modified arthropod natural enemies became established. The only method for determining whether the released population was present was to conduct bioassays for resistance, which are unable to resolve whether replacement or introgression occurred. However, RAPD-PCR of DNA markers allowed monitoring of establishment and dispersal of pesticide-resistant strains of *Trioxys pallidus* in several California walnut orchards. The results suggest that introgression has occurred in at least two release sites. Thus, molecular genetic techniques may help resolve the population genetics of released populations, as well as providing new tools for genetic manipulation of pest and beneficial arthropods.

## Research Needs

Genetic manipulation projects of pest and beneficial arthropods share many problems and issues. One factor hindering progress is the lack of a 'universal' transformation system. Also, we lack an example that demonstrates that recombinant DNA technology can yield an effective beneficial arthropod or that it can result in the control of a pest arthropod population.

Because the potential risks of releasing transgenic arthropods into the environment have been discussed in only a preliminary manner, it may be appropriate to release a relatively risk-free example first. This might involve the release of a transgenic beneficial arthropod that is carrying either a noncoding segment of exogenous DNA or a marker gene such as β-*galactosidase*. The transgenic strain should not contain an active transposable element vector (Hoy 1992b).

One early candidate for release might be a transgenic strain of the phytoseiid predator *Metaseiulus occidentalis* with a *lacZ* construct. *M. occidentalis* is an obligatory predator, has a low dispersal rate, and is unlikely to become a pest (Hoy 1992b). Ideally, the transgenic strain of *M. occidentalis* could be released into a site where it is unlikely to become permanently established or be able to interbreed with native populations. Such a release would allow a relatively risk-free evaluation of release, monitoring, and mitigation procedures. Later releases could evaluate strains with more useful characters.

Once risk assessment issues and safety have been demonstrated with such a beneficial species, releases of transgenic pest arthropods might be more readily assessed. Releases in the U.S.A. will be evaluated as a two-step process. Initial releases will be experimental and on a small scale. No guidelines are available for evaluating the risks associated with permanent releases of transgenic arthropods into the environment. Permission for long-term and large scale releases may require five to ten years of evaluating small-scale releases. Thus, risk assessment of transgenic arthropods, as it has with transgenic crops and microorganisms, will add a significant cost in both time and resources to pest management projects. It has taken about ten years for the first transgenic crop to become commercially available and may take as long for transgenic arthropods to be released permanently into the environment.

Significant, exciting, and unpredictable advances are being achieved in molecular biology and genetics. As a result of rapid advances in molecular genetic techniques and knowledge of basic developmental mechanisms, it is very difficult to anticipate the opportunities that might arise over the next few years for genetically manipulating pest and beneficial arthropods. However, additional research is required if we are to gain an understanding of the attributes other than resistance to pesticides that we might manipulate. Furthermore, getting a transgenic arthropod into a pest management program will be an awesome challenge, requiring risk assessments, detailed knowledge of the population genetics, biology, and behavior of the target species, and coordinated efforts between molecular and population geneticists, ecologists, regulatory agencies, and pest management specialists.

## References Cited

Abraham, E. G., N. Mounier and G. Bosquet. 1993. Expression of a *Bombyx* cytoplasmic actin gene in cultured *Drosophila* cells: Influence of 20-hydroxyecdysone and interference with expression of endogenous cytoplasmic actin genes. Insect Biochem. Biol. 23: 905–912.

Angelichio, M. L., J. A. Beck, H. Johansen and M. Ivey-Hoyle. 1991. Comparison of several promoters and polyadenylation signals for use in heterologous gene expression in cultured *Drosophila* cells. Nucleic Acids Res. 19: 5037–5043.

Atkinson, P. W., E. R. Hines, S. Beaton, K. I. Matthaei, K. C. Reed and M. P. Bradley. 1991. Association of exogenous DNA with cattle and insect spermatozoa in vitro. Mol. Reprod. Dev. 23: 1–5.

Atkinson, P. W., W. D. Warren and D. B. O'Brochta. 1993. The *hobo* transposable element of *Drosophila* can be cross-mobilized in houseflies and excises like the *Ac* element of maize. Proc. Natl. Acad. Sci. USA 90: 9693–9697.

Baldarelli, R. M. and J. A. Lengyel. 1990. Transient expression of DNA after ballistic introduction into *Drosophila* embryos. Nucleic Acids Res. 18: 5903–5904.

Bartlett, A. C. 1985. Guidelines for genetic diversity in laboratory colony establishment and maintenance. Pp. 7–17. In: *Handbook of Insect Rearing*, Vol. 1. P. Singh and R. F. Moore, Eds. Elsevier, Amsterdam.

Beall, C., C. Fyrberg, S. Song and E. Fyrberg. 1992. Isolation of a *Drosophila* gene encoding glutathione *S*-transferase. Biochem. Genet. 30: 515–527.

Beard, C. B., P. W. Mason, S. Aksoy, R. B. Tesh and F. F. Richards. 1992. Transformation of an insect symbiont and expression of a foreign gene in the Chagas' disease vector *Rhodnius prolixus*. Am. J. Trop. Med. Hyg. 46: 195–200.

Beard, C. B., S. L. O'Neill, R. B. Tesh, F. F. Richards and F. Aksoy. 1993. Modification of arthropod vector competence via symbiotic bacteria. Parasitol. Today 9: 179–183.

Beck, E., G. Ludwig, E. A. Auerswald, B. Reiss and H. Schaller. 1982. Nucleotide sequence and exact localization of the neomycin phosphotransferase gene from transposon Tn5. Gene 19: 327–336.

Beckendorf, S. K. and M. A. Hoy. 1985. Genetic improvement of arthropod natural enemies through selection, hybridization or genetic engineering techniques. Pp. 167–187. In: *Biological Control in Agricultural IPM Systems*. M. A. Hoy and D. C. Herzog, Eds. Academic Press, Orlando.

Berg, D. E. and M. N. Howe, Eds. 1989. *Mobile DNA*. Amer. Soc. Microbiol., Washington, DC.

Besansky, N. J. 1990. A retrotransposable element from the mosquito *Anopheles gambiae*. Mol. Cell. Biol. 10: 863–871.

Besansky, N. J. and F. H. Collins. 1992. The mosquito genome: Organization, evolution and manipulation. Parasitol. Today 8: 186–192.

Braig, H. R., H. Guzman, R. B. Tesh and S. L. O'Neill. 1994. Replacement of the natural *Wolbachia* symbiont of *Drosophila simulans* with a mosquito counterpart. Nature 367: 453–455.

Brand, A. H. and N. Perrimon. 1993. Targeted gene expression as a means of altering cell fates and generating dominant phenotypes. Development 118: 401–415.

Bush, G. L. 1979. Ecological genetics and quality control. Pp. 145–152. In: *Genetics in Relation to Insect Management*. M. A. Hoy and J. J. McKelvey, Jr., Eds. Rockefeller Found. Press, New York.

Caprio, M. A., M. A. Hoy and B. E. Tabashnik. 1991. A model for implementing a genetically-improved strain of the parasitoid *Trioxys pallidus* Haliday (Hymenoptera: Aphidiidae). Amer. Entomol. 34: 232–239.

Carminati, J., C. G. Johnston, and T. L. Orr-Weaver. 1992. The *Drosophila* ACE3 chorion element autonomously induces amplification. Mol. Cell. Biol. 12: 2444–2453.

Chalfie, M., Y. Tu, G. Euskirchen, W. W. Ward and D. C. Prasher. 1994. Green fluorescent protein as a marker for gene expression. Science 263: 802–805.

Cockburn, A. F. and J. A. Seawright. 1990. Application of molecular genetics to insect control. Pp. 227–239. In: *CRC Handbook of Natural Pesticides: Insect Attractants and Repellents*. Vol. VI. D. Morgan and N. B. Mandava, Eds. CRC Press, Boca Raton, FL.

Cockburn, A. F., A. J. Howells and M. J. Whitten. 1984. Recombinant DNA technology and genetic control of pests. Biotechn. Genet. Eng. Rev. 2: 69–99.

Coulon-Bublex, M., N. Mounier, P. Couble and J. C. Prudhomme. 1993. Cytoplasmic actin A3 gene promoter injected as supercoiled plasmid is transiently active in *Bombyx mori* embryonic vitellophages. Roux's Arch. Dev. Biol. 202: 123–127.

Cooley, R. N., R. F. M. van Gorcom, C. A. M. J. J. van den Hondel and C. E. Caten. 1991. Isolation of a benomyl-resistant allele of the β-tubulin gene from *Septoria nodorum* and its use as a dominant selectable marker. J. Gen. Microbiol. 137: 2085–2091.

Crampton, J., A. Morris, G. Lycett, A. Warren and P. Eggleston. 1990. Transgenic mosquitoes: a future vector control strategy? Parasitol. Today 6: 31–36.

Curtis, C. F. 1979. Translocations, hybrid sterility, and the introduction into pest populations of genes favorable to man. Pp. 19–30. In: *Genetics in Relation to Insect Management*. M. A. Hoy and J. J. McKelvey, Jr., Eds. Rockefeller Found. Press, New York.

Dave, K. I., C. E. Miller and J. R. Wild. 1993. Characterization of organophosphorus hydrolases and the genetic manipulation of the phosphotriesterase from *Pseudomonas diminuta*. Chem.-Biol. Interact. 87: 55–68.

Dumas, D. P., J. R. Wild and F. M. Rauschel. 1990. Expression of *Pseudomonas* phosphotriesterase activity in the fall armyworm confers resistance to insecticides. Experientia 46: 729–34.

Eggleston, P. 1991. The control of insect-borne disease through recombinant DNA technology. Heredity 66: 161–172.

Ehler, L. E. 1990. Environmental impact of introduced biological-control agents: Implications for agricultural biotechnology, Pp. 85–96. In: *Risk Assessment in Agricultural Biotechnology*. Proc. Intern. Conf., J. J. Marois and G. Bruyening, Eds. Univ. Calif., Div. Agric. Natur. Res. Publ. No. 1928.

Fallon, A. 1991. DNA-mediated gene transfer: applications to mosquitoes. Nature 352: 828–829.

Feyereisen, R., J. F. Koener, D. E. Farnsworth and D. W. Nebert. 1989. Isolation and sequence of cDNA encoding a cytochrome P450 cDNA from insecticide-resistant strain of the house fly, *Musca domestica*. Proc. Natl. Acad. Sci. USA 86: 1465–1469.

ffrench-Constant, R. H. and T. A. Rocheleau. 1993. *Drosophila* γ-aminobutyric acid receptor gene *Rdl* shows extensive alternative splicing. J. Neurochem. 60: 2323–2326.

ffrench-Constant, R. H., D. P. Mortlock, C. D. Shaffer, R. J. Macintyre and R. T. Roush. 1991. Molecular cloning and transformation of cyclodiene resistance in *Drosophila*: An invertebrate γ-aminobutyric acid subtype A receptor locus. Proc. Natl. Acad. Sci. USA 88: 7209–7213.

ffrench-constant, R. H., T. A. Rocheleau, J. C. Steichen and A. E. Chalmers. 1993a. A point mutation in a *Drosophila* GABA receptor confers insecticide resistance. Nature 363: 449–451.

ffrench-Constant, R. H., J. C. Steichen, T. A. Rocheleau, K. Araonstein and R. T. Roush. 1993b. A single-amino acid substitution in a γ-aminobutyric acid subtype A receptor locus is associated with cyclodiene insecticide resistance in *Drosophila* populations. Proc. Natl. Acad. Sci. USA 90: 1957–1961.

Field, L. M., M. S. Williamson, G. D. Moores and A. L. Devonshire. 1993. Cloning and analysis of the esterase genes conferring insecticide resistance in the peach-potato aphid, *Myzus persicae* (Sulzer). Biochem. J. 194: 569–574.

Fournier, D., F. Karch, J. Bride, L. M. C. Hall, J. B. Berge and P. Spierer. 1989. *Drosophila melanogaster* acetylcholinesterase gene structure, evolution and mutations. J. Mol. Biol. 210: 15–22.

Fournier, D., J. M. Bride, F. Hoffmann and F. Karch. 1992a. Acetylcholinesterase: Two types of modifications confer resistance to insecticide. J. Biol. Chem. 267: 14270–14274.

Fournier, D., J. M. Bride, M. Poirie, J. Berge and F. W. Plapp, Jr. 1992b. Insect glutathione S-trans-ferases, biochemical characteristics of the major forms from houseflies susceptible and resistant to insecticides. J. Biol. Chem. 267: 1840–1845.

Gerrard, B., C. Stewart and M. Deran. 1993. A *Drosophila* P-glycoprotein/multidrug resistance gene homolog. Genomics 17: 83–88.

Gopinathan, K. P. 1992. Biotechnology in sericulture. Curr. Sci. 62: 283–287.

Hadrys, H., M. Balick and B. Schierwater. 1992. Applications of random amplified polymorphic DNA (RAPD) in molecular ecology. Mol. Ecol. 1: 55–63.

Hall, L. M. C. and C. A. Malcolm. 1991. The acetycholinesterase gene of *Anopheles stephensi*. Cell. Mol. Neurobiol. 11: 131–141.

Hall, L. M. C. and P. Spierer. 1986. The *Ace* locus of *Drosophila melanogaster*. Structural gene for acetylcholinesterase with an unusual 5' leader. EMBO J. 5: 2949–2954.

Handler, A. M. and D. A. O'Brochta. 1991. Prospects for gene transformation in insects. Annu. Rev. Entomol. 36: 159–183.

Handler, A. M., S. P. Gomez and D. A. O'Brochta. 1993. A functional analysis of the P-element gene-transfer vector in insects. Arch. Insect Bioch. Physiol. 22: 373–384.

Hanson, S. M., J. P. Mutebi, G. B. Craig, Jr. and R. J. Novak. 1993. Reducing the overwintering ability of *Aedes albopictus* by male release. J. Am. Mosq. Contr. Assoc. 9: 78–83.

Harris, P., 1985. Biocontrol and the law. Bull. Entomol. Soc. Can. 17: 1.

Headley, J. C. and M. A. Hoy. 1987. Benefit/cost analysis of an integrated mite management program for almonds. J. Econ. Entomol. 80: 555–559.

Heinrich, J. C., M. Tabler and C. Louis. 1993. Attenuation of *white* gene expression in transgenic *Drosophila melanogaster*: Possible role of a catalytic antisense RNA. Dev. Genet. 14: 258–265.

Hoffmann, F., D. Fournier and P. Spierer. 1992. Minigene rescues acetylcholinesterase lethal muta-tions in *Drosophila melanogaster*. J. Mol. Biol. 223: 17–22.

Houck, M. A., J. B. Clark, K. R. Peterson and M. G. Kidwell. 1991. Possible horizontal transfer of *Drosophila* genes by the mite *Proctolaelaps regalis*. Science 253: 1125–1129.

Howarth, F. G. 1991. Environmental impacts of classical biological control. Annu. Rev. Entomol. 36: 485–509.

Hoy, M. A. 1976. Genetic improvement of insects: Fact or fantasy. Environ. Entomol. 5: 833–839.

Hoy, M. A. 1990a. Genetic improvement of arthropod natural enemies: becoming a conventional tactic? Pp. 405–417. In: *New Directions in Biological Control*. R. Baker and P. Dunn, Eds. UCLA Symp. Mol. Cell. Biol., New Series, Vol. 112, A. R. Liss, New York.

Hoy, M. A. 1990b. Pesticide resistance in arthropod natural enemies: Variability and selection responses. Pp. 203–236. In: *Pesticide Resistance in Arthropods*. R. T. Roush and B. E. Tabashnik, Eds. Chapman and Hall, New York.

Hoy, M. A. 1991. Genetic improvement of phytoseiids: in theory and practice. Pp. 175–184. In: *Modern Acarology*. Vol. 1. F. Dusbabek and V. Bukva, Eds. Academia, Prague and SPB Academic Publ., The Hague.

Hoy, M. A. 1992a. Commentary: Biological control of arthropods: Genetic engineering and environmental risks. Biol. Control 2: 166–170.

Hoy, M. A. 1992b. Criteria for release of genetically-improved phytoseiids: An examination of the risks associated with release of biological control agents. Exp. Appl. Acarol. 14: 393–416.

Hoy, M. A. 1993. Transgenic beneficial arthropods for pest management programs: An assessment of their practicality and risks. Pp. 357–369. In: *Pest Management: Biologically Based Technologies*. R. D. Lumsden and J. L. Vaughn, Eds. Amer. Chem. Soc. Conf. Proc. Series., Washington, DC.

Hoy, M. A., F. E. Cave, R. H. Beede, J. Grant, W. H. Krueger, W. H. Olson, K. M. Spollen, W. W. Barnett and L. C. Hendricks. 1990. Release, dispersal, and recovery of a laboratory-selected strain of the walnut aphid parasite *Trioxys pallidus* (Hymenoptera: Aphidiidae) resistant to azinphosmethyl. J. Econ. Entomol. 83: 89–96.

Iatrou, K. and R. G. Meidinger. 1990. Tissue-specific expression of silkmoth chorion genes in vivo using *Bombyx mori* nuclear polyhedrosis virus as a transducing vector. Proc. Natl. Acad. Sci. USA 87: 3650–3654.

International Atomic Energy Agency. 1993. *Management of Insect Pests: Nuclear and Related Molecular and Genetic Techniques*. Proc. Symp., Vienna 19–23. Oct. 1992. U. N. Food Agric. Organ. and Intern. Atomic Energy Agency, IAEA Press, Vienna.

Kamdar, P., G. von Allmen and V. Finnerty. 1992. Transient expression of DNA in *Drosophila* via electroporation. Nucleic Acids Res. 20: 3526.

Kidwell, M. G. and J. M. C. Ribeiro. 1992. Can transposable elements be used to drive disease refractoriness genes into vector populations. Parasitol. Today 8: 325–329.

Kilby, N. J., M. R. Snaith and J. A. H. Murray. 1993. Site-specific recombinases: Tools for genome engineering. Trends Genet. 9: 413–421.

Konsolaki, M., M. Sanicola, T. Kozlova, V. Liu, B. Arca, C. Savakis, W. M. Gelbart and F. C. Kafatos. 1992. FLP-mediated intermolecular recombination in the cytoplasm of *Drosophila* embryos. New Biol. 4: 551–557.

Kovach, M. J., J. O. Carlson and B. J. Beaty. 1992. A *Drosophila* metallothionein promoter is inducible in mosquito cells. Insect Mol. Biol. 1: 37–43.

LaChance, L. E. 1979. Genetic strategies affecting the success and economy of the sterile insect release method. pp. 8–18. In: *Genetics in Relation to Insect Management*. M. A. Hoy and J. J. McKelvey, Jr., Eds. Rockefeller Found. Press, New York.

Laski, F. A., D. C. Rio and G. M. Rubin. 1986. Tissue specificity of *Drosophila* P element transposition is regulated at the level of mRNA splicing. Cell 44: 7–19.

Laven, H. 1951. Crossing experiments with *Culex* strains. Evolution 5: 370–375.

Lawrence, P. A. 1992. *The Making of a Fly: The Genetics of Animal Design*. Blackwell Scientific Publ., London.

Lawrence, P. A., M. Ashburner and P. Johnston. 1993. An attempt to hybridize *Drosophila* species using pole cell transplantation. Genetics 134: 1145–1148.

Lidholm, D. A., G. H. Gudmundsson and H. G. Boman. 1991. A highly repetitive, *mariner*-like element in the genome of *Hyalophora cecropia*. J. Biol. Chem. 266: 11518–11521.

Mangan, R. L. 1991. Analysis of genetic control of mating behavior in screwworm (Diptera: Calliphoridae) males through diallel crosses and artificial selection. Theor. Appl. Genet. 81: 429–436.

Maruyama, K. and D. L. Hartl. 1991. Evidence for interspecific transfer of the transposable element *mariner* between *Drosophila* and *Zaprionus*. J. Mol. Evol. 33: 514–524.

McGrane, V., J. O. Carlson, B. R. Miller and B. J. Beaty. 1988. Microinjection of DNA into *Aedes triseriatus* ova and detection of integration. Am. J. Trop. Med. Hyg. 39: 502–510.

McInnis, D. O., D. S. Haymer, S. Y. T. Tam, and S. Thanaphum. 1990. *Ceratitis capitata* (Diptera: Tephritidae): Transient expression of a heterologous gene for resistance to the antibiotic geneticin. Ann. Entomol. Soc. Amer. 83: 982–986.

Meredith, S. E. O. and A. A. James. 1990. Biotechnology as applied to vectors and vector control. Ann. Parasitol. Hum. Comp. 65 (Suppl.I): 113–118.

Mialhe, E., A. Laughinghouse and L. H. Miller. 1993. Use of the biolistic technique for gene transfer to mosquito embryos. Pp. 145–147. In: *Management of Insect Pests: Nuclear and Related Molecular and Genetic Techniques.* Proc. Symp., Vienna 19–23. Oct. 1992. U. N. Food Agric. Organ. and Intern. Atomic Energy Agency, IAEA Press, Vienna.

Michaille, J.J., S. Mathavan, J. Gaillard and A. Garel. 1990. The complete sequence of *mag*, a new retrotransposon in *Bombyx mori*. Nucleic Acids Res. 18: 674.

Miller, L. H., R. K. Sakai, P. Romans, W. Gwadz, P. Kantoff and H. G. Coon. 1987. Stable integration and expression of a bacterial gene in the mosquito *Anopheles gambiae*. Science 237: 779–781.

Miller, L. K. 1988. Baculoviruses as gene expression vectors. Annu. Rev. Microbiol. 42: 177–199.

Milne, C. P., Jr., J. P. Phillips, and P. J. Krell. 1988. Microinjection of early honeybee embryos. J. Apicult. Res. 27: 84–89.

Milne, C. P., F. A. Eishen, J. E. Collis and T. L. Jensen 1989. Preliminary evidence for honey bee sperm-mediated DNA transfer. Int. Symp. Mol. Insect Sci., Tucson, AZ, p. 71 (Abstract).

Monroe, T. J., M. C. Muhlmann-Diaz, M. J. Kovach, J. O. Carlson, J. S. Bedford and B. J. Beaty. 1992. Stable transformation of a mosquito cell line results in extraordinarily high copy numbers of the plasmid. Proc. Natl. Acad. Sci. USA 89: 5725–5729.

Morris, A. C., P. Eggleston and J. M. Crampton. 1989. Genetic transformation of the mosquito *Aedes aegypti* by micro-injection of DNA. Med. Vet. Entomol. 3: 1–7.

Morris, A. C., T. L. Schaub and A. A. James. 1991. FLP-mediated recombination in the vector mosquito, *Aedes aegypti*. Nucleic Acids Res. 19: 5895–5900.

Morton, R. A. 1993. Evolution of Drosophila insecticide resistance. Genome 36: 1–7.

Mouches, C. 1989. Genie génétique et transgénése des insectes pour améliorer les techniques de lutte biologique. C. R. Acad. Agric. France 75: 27–32.

Mouches, C., N. Pasteur, J. B. Berge, O. Hyrien, M. Raymond, B. R. De Saint Vincent, M. De Silvestri and G. P. Georghiou. 1986. Amplification of an esterase gene is responsible for insecticide resistance in a California *Culex* mosquito. Science 233: 778–780

Mouches, C., Y. Pauplin, M. Agarwal, L. Lemieux, M. Herzog, M. Abadon, V. Beyssat-Arnaouty, O. Hyrien, B. R. Desaint Vincent, G. P. Georghiou and N. Pasteur. 1990. Characterization of amplification core and esterase B1 gene responsible for insecticide resistance in *Culex*. Proc. Natl. Acad. Sci. USA 98: 2574–2578.

Mulbry, W. W. and J. S. Karns. 1989. Parathion hydrolase specified by the *Flavobacterium opd* gene: Relationship between the gene and protein. J. Bacteriol. 171: 6740–6746.

Okano, A., I. Yonemura, Y. Shimizu, Y. Yanagidaira, H. Hasekura and B. Boettcher. 1992. Purification and characterization of a protein associated with genetically-determined longevity difference in *Drosophila melanogaster*. Hereditas 117: 251–258.

Okano, K., N. Miyajima, N. Takada, M. Kobayashi and H. Maekawa. 1992. Basic conditions for the drug selection and transient gene expression in the cultured cell line of *Bombyx mori*. In Vitro Cell. Dev. Biol. A 28: 779–781.

O'Neill, S. L. and T. L. Karr. 1990. Bidirectional incompatibility between conspecific populations of *Drosophila simulans*. Nature 348: 178–180.

O'Neill, S. L., R. Giordano, A. M. E. Colbert, T. L. Karr and H. M. Robertson. 1992. 16S rRNA phylogenetic analysis of the bacterial endosymbionts associated with cytoplasmic incompatibility in insects. Proc. Natl. Acad. Sci. USA 89: 2699–2702.

Orbach, M. J., E. B. Porro and C. Yanofsky. 1986. Cloning and characterization of the gene for β-tubulin from a benomyl-resistant mutant of *Neurospora crassa* and its use as a dominant selectable marker. Mol. Cell. Biol. 6: 2452–2461.

Orr, W. C. and R. S. Sohal. 1992. The effects of catalase gene overexpression on life span and resistance to oxidative stress in transgenic *Drosophila melanogaster*. Arch. Biochem. Biophys. 297: 35–41.

Orr, W. C. and R. S. Sohal. 1994. Extension of life-span by overexpression of superoxide dismutase and catalase in *Drosophila melanogaster*. Science 263: 1128–1130.

Pal, R. 1974. WHO/ICMR programme of genetic control of mosquitos in India. Pp. 73–95. In: *The Use of Genetics In Insect Control*. R. Pal and M. J. Whitten, Eds. Elsevier, Amsterdam.

Pal, R. and M. J. Whitten, Eds. 1974. *The Use of Genetics In Insect Control*. Elsevier, Amsterdam.

Peters, I. D., D. E. Rancourt, P. L. Davies and V. K. Walker. 1993. Isolation and characterization of an antifreeze protein precursor from transgenic *Drosophila*: Evidence for partial processing. Biochim. Biophys. Acta 1171: 247–254.

Peterson, K. R., C. H. Clegg, C. Huxley, B. M. Josephson, H. S. Haugen, T. Furukawa and G. Stamatoyannopoulos. 1993. Transgenic mice containing a 248-kb yeast artificial chromosome carrying the human β-globin locus display proper developmental control of the human globin genes. Proc. Natl. Acad. Sci. USA 90: 7593–7597.

Phillips, J. P., J. H. Xin, K. Kirby, C. P. Milne, Jr., P. Krell and J. R. Wild. 1990. Transfer and expression of an organophosphate insecticide-degrading gene from *Pseudomonas* in *Drosophila melanogaster*. Proc. Natl. Acad. Sci. USA 87: 8155–8159.

Plasterk, R. H. A. 1993. Molecular mechanisms of transposition and its control. Cell 74: 781–786.

Presnail, J. K. and M. A. Hoy. 1992. Stable genetic transformation of a beneficial arthropod by microinjection. Proc. Natl. Acad. Sci. USA 89:7732–7726.

Rancourt D. E., I. D. Peters, V. K. Walker and P. L. Davies. 1990. Wolffish antifreeze protein from transgenic *Drosophila*. Bio/Technology 8: 453–457.

Rancourt, D. E., P. L. Davies and V. K. Walker. 1992. Differential translatability of antifreeze protein mRNAs in a transgenic host. Biochim. Biophys. Acta 1129: 188–194.

Raymond, M., A. Callaghan, P. Fort and N. Pasteur. 1991. Worldwide migration of amplified insecticide resistance genes in mosquitoes. Nature 350: 151–153.

Richards, F. F. 1993. An approach to reducing arthropod vector competence: Dispersion of insect-borne diseases could be modified by genetically altered symbionts. Am. Soc. Microbiol. News 59: 509–514.

Robertson, H. M. 1993a. The *mariner* transposable element is widespread in insects. Nature 362: 241–245.

Robertson, H. M. 1993b. Reply to: Infiltration of *mariner* elements. Nature 364: 109–110.

Rothenbuhler, W. C. 1979. Semidomesticated insects: Honeybee breeding. Pp. 84–92. In: *Genetics in Relation to Insect Management*. M. A. Hoy and J. J. McKelvey, Jr., Eds. Rockefeller Found. Press, New York.

Rubin, G. M. and A. C. Spradling. 1982. Genetic transformation of *Drosophila* with transposable element vectors. Science 218: 348–353.

Sailer, R. I. 1961. Possibilities for genetic improvement of beneficial insects. In: *Germ Plasm Resources*. American Assoc. Adv. Sci., Washington, DC.

Sakai, R. K. and L. H. Miller. 1992. Effects of heat shock on the survival of transgenic *Anopheles gambiae* (Diptera: Culicidae) under antibiotic selection. J. Med. Entomol. 29:374–375.

Santamaria, P. 1986. Injecting eggs. Pp. 159–173. In: Drosophila: A *Practical Approach*. D. B. Roberts, Ed. IRL Press, Oxford.

Schedl, A., F. Beermann, E. Thies, L. Montoliu, G. Kelsey and G. Schutz. 1992. Transgenic mice generated by pronuclear injection of a yeast artificial chromosome. Nucleic Acids Res. 20: 3073–3077.

Sentry, J. W. and K. Kaiser. 1992. P element transposition and targeted manipulation of the *Drosophila* genome. Trends Genet. 8: 329–331.

Serdar, C. M., D. C. Murdock and M. F. Rohde. 1989. Parathion hydrolase gene from *Pseudomonas diminuta* MG: Subcloning, complete nucleotide sequence, and expression of the mature portion of the enzyme in *Escherichia coli*. Bio/Technology 7: 1151–1155.

Shotkoski, F. A. and A. M. Fallon. 1993. The mosquito dihydrofolate reductase gene functions as a dominant selectable marker in transfected cells. Insect Biochem. Mol. Biol. 23: 883–893.

Simpson, P. 1993. Flipping fruit-flies: A powerful new technique for generating *Drosophila* mosaics. Trends Genet. 9: 227–228.

Slee, R. and M. Bownes. 1990. Sex determination in *Drosophila melanogaster*. Q. Rev. Biol. 65: 175–204.

Spradling, A. C. and G. M. Rubin. 1982. Transposition of cloned P elements into *Drosophila* germline chromosomes. Science 218: 341–47.

Steller, H. and V. Pirrota. 1985. Fate of DNA injected into early *Drosophila* embryos. Dev. Biol. 109: 54–62.

Tamura, T., T. Kanda, S. Takiya, K. Okano and H. Maekawa. 1990. Transient expression of CAT genes injected into early embryos of the domesticated silkworm *Bombyx mori*. Jap. J. Genet. 65: 401–410.

Theodore, L., A. Ho and G. Maroni. 1991. Recent evolutionary history of the metallothionein gene *Mtn* in *Drosophila*. Genet. Res., Camb. 58: 203–210.

Thompson, M., F. Shotkoski and R. ffrench-Constant. 1993. Cloning and sequencing of the cyclodiene insecticide resistance gene from the yellow fever mosquito *Aedes aegypti*: Conservation of the gene and resistance associated mutation with *Drosophila*. FEBS 325 (3): 187–190.

Tiedje, J. M., R. K. Colwell, Y. L. Grossman, R. E. Hodson, R. E. Lenski, R. M. Mack and P. J. Regal. 1989. The planned introduction of genetically engineered organisms: Ecological considerations and recommendations. Ecology 70: 298–315.

Toung, Y. P. S., T. S. Hsieh and C. P. D. Tu. 1990. *Drosophila* glutathione S-transferase 1-1 shares a region of sequence homology with the maize glutathione S-transferase III. Proc. Natl. Acad. Sci. USA 87:31–35.

Toung, Y. P. S., T. Hsieh and C. P. D. Tu. 1993. The glutathione S-transferase *D* genes: A divergently organized, intronless gene family in *Drosophila melanogaster*. J. Biol. Chem. 268: 9737–9746.

Turelli, M. and A. A. Hoffmann. 1991. Rapid spread of an inherited incompatibility factor in California *Drosophila*. Nature 353: 440–442.

Turelli, M., A. A. Hoffmann and S. W. McKechnie. 1992. Dynamics of cytoplasmic incompatibility and mtDNA variation in natural *Drosophila simulans* populations. Genetics 132: 713–723.

U.S. Department of Agriculture. 1991. Part III. Proposed guidelines for research involving the planned introduction into the environment of organisms with deliberately modified hereditary traits: Notice. Fed. Reg. Vol. 56 (22), Friday, February 1, 1991, pp. 4134–4151.

U.S. Department of Agriculture. 1993. Flightless wasp a step ahead in biocontrol. Agric. Res., June 1993, p. 17.

Walker, V. K. 1989. Gene transfer in insects. Adv. Cell Culture 7: 87–124.

Wang, J. Y., S. McCommas and M. Syvanen. 1991. Molecular cloning of a glutathione S-transferase overproduced in an insecticide-resistant strain of the housefly (*Musca domestica*). Mol. Gen. Genet. 227: 260–266.

Waters, L. C., A. C. Zelhof, B. J. Shaw and L. Y. Chang. 1992. Possible involvement of the long terminal repeat of transposable element 17.6 in regulating expression of an insecticide resistance-associated P450 gene in *Drosophila*. Proc. Natl. Acad. Sci. USA 89: 4855–4859.

Whitten, M. J. 1979. The use of genetically selected strains for pest replacement or suppression. Pp. 31–40. In: *Genetics in Relation to Insect Management*. M. A. Hoy and J. J. McKelvey, Jr., Eds. Rockefeller Found. Press, New York.

Whitten, M. J. 1985. The conceptual basis for genetic control. Chapter 14, pp. 465–527. In: *Comprehensive Insect Physiology Biochemistry and Pharmacology*. G. A. Kerkut and L. I. Gilbert, Eds. Pergamon Press, Oxford.

Williams, K. D. and M. B. Sokolowski. 1993. Diapause in *Drosophila melanogaster* females: A genetic analysis. Heredity 71: 312–317.

Williamson, M. 1992. Environmental risks from the release of genetically modified organisms (GMOs)—The need for molecular ecology. Mol. Ecol. 1: 3–8.

Williamson, M. S., I. Denholm, C. A. Bell and A. L. Devonshire. 1993. Knockdown resistance (*kdr*) to DDT and pyrethroid insecticides maps to a sodium channel gene locus in the housefly (*Musca domestica*). Mol. Gen. Genet. 140: 17–22.

Wood, H. A. and R. R. Granados. 1991. Genetically engineered baculoviruses as agents for pest control. Annu. Rev. Microbiol. 45: 69–87.

Wright, J. W. and R. Pal, Eds. 1967. *Genetics of Insect Vectors of Disease*. Elsevier, Amsterdam

Wu, C. T., M. Budding, M. S. Griffin and J. M. Croop. 1991. Isolation and characterization of *Drosophila* multidrug resistance gene homologs. Mol. Cell. Biol. 11:3940–3948.

Yen, J. H. and A. R. Barr. 1974. Incompatibility in *Culex pipiens*. Pp. 97–118. In: *The Use of Genetics In Insect Control*. R. Pal and M. J. Whitten, Eds. Elsevier, Amsterdam.

Yokoyama, T. 1979. Silkworm selection and hybridization. Pp. 71–83. In: *Genetics in Relation to Insect Management*. M. A. Hoy and J. J. McKelvey, Jr., Eds. Rockefeller Found. Press, New York.

Yonemura, I., A. Okano, Y. Shimizu, H. Hasekura and B. Boettcher. 1992. A difference in the proteins found in young adults of inbred strains of *Drosophila melanogaster* which correlates with genetically-determined, long or short life span. Hereditas 117: 241–250.

Yu, Z., J. D. Podgwaite, and H. A. Wood. 1992. Genetic engineering of a *Lymantria dispar* nuclear polyhedrosis virus for expression of foreign genes. J. Gen. Virol. 73: 1509–1514.

Zalokar, M. 1981. A method for injection and transplantation of nuclei and cells in *Drosophila* eggs. Experientia 37: 1354–1356.

# Glossary

**acentric** A chromosome, or chromosome fragment, that lacks a centromere.

**action potential** A rapid change in the polarity of the membrane of a neuron that facilitates the interaction and transmission of impulses.

**adenine** A purine and one of the nitrogenous bases found in DNA and RNA.

**adenosine triphosphate** See ATP.

**Africanized bees** Honeybees in the western hemisphere that are derived from hybridization of African and European subspecies of *Apis mellifera*. The degree of hybridization is unresolved.

**agarose** A polysaccharide gum obtained from seaweed used as a gel medium in electrophoresis; used to separate DNA molecules on the basis of their molecular weight.

**allele** One of two or more alternative forms of a gene at a particular locus.

**allozyme** Allozymes are a subset of isozymes. Allozymes are variants of enzymes representing different allelic alternatives of the same locus.

**alternative splicing** Gene regulation by means of alternative splicing of exons to produce different amounts of protein or even different proteins.

**amino acid** One of the monomeric units that polymerize to make a protein molecule.

**aminoacyl tRNA synthetase** Enzymes that catalyze the attachment of each amino acid to the appropriate transfer RNA molecule. A tRNA molecule carrying its amino acid is called a charged tRNA.

**anchored PCR** A modification of PCR that allows amplification in situations in which only one sequence is known that is suitable for a primer (rather than two). The procedure involves synthesis of cDNA with the known primer from mRNA. A poly(G) tail is added to the cDNA. The second primer is developed by synthesizing a primer with a poly(C) sequence, which allows amplification of a second DNA strand that is complementary to the cDNA. Subsequent cycles yield amplified DNA from both strands.

**aneuploid**  A condition in which the chromosome number of an organism is not an exact multiple of the typical haploid set for the species.

**angstrom**  Abbreviated as Å; one hundred-millionth of a centimeter, or 0.1 nm.

**anticodon**  The triplet of nucleotides in a transfer RNA molecule that is complementary to and base-pairs with a codon in a messenger RNA.

**apomorphic**  A character that is derived and not ancestral.

**arbitrarily primed PCR (AP-PCR)**  Does not require a particular set of primers, rather it uses a single primer chosen without regard to the sequence to be fingerprinted. By using a single primer and two cycles of low-stringency PCR followed by many cycles of high-stringency PCR, discrete and reproducible products characteristic of specific genomes are produced. As originally described, the primers used are 20-bp sequencing primers.

**arrhenogenic**  A sex-determining system in which females produce male progeny only. Found in the blowfly *Chrysomya rufifacies* (Calliphoridae).

**arrhenotoky**  A form of parthenogenesis in which an unfertilized egg develops into a male by parthenogenesis and a fertilized egg develops into the female. Arrhenotoky is found in many Hymenoptera.

**associative learning**  The process of learning through the formation of associations between stimuli and responses.

**asymmetric PCR**  Single-stranded DNA produced by providing an excess of primer for one of the two DNA strands. Asymmetric primer ratios are typically 50:1 to 100:1. Single-stranded DNA produced can be sequenced directly without cloning.

**ATP**  Adenosine triphosphate is the primary molecule for storing chemical energy in a cell.

**autecology**  The ecology of an individual organism or species.

**autoradiography**  A method for detecting radioactively labeled molecules through exposure of an X-ray-sensitive photographic film.

**autoregulatory control**  Regulation of the synthesis of a gene product by the product itself. In some systems, excess gene product behaves as a repressor and binds to the operator of its own structural gene.

**B chromosomes**  Nonvital supernumerary chromosomes found in many plant and animal species. They are thought to be derived from one of the normal chromosomes, and are often transmitted at higher rates than expected, thus exhibiting 'drive.' Paternal sex ratio (PSR) of the parasitic wasp *Nasonia vitripennis* is an example of a B chromosome.

**back mutation**  Mutations that occur to reverse a point mutation to the original condition.

**bacterial conjugation**  A temporary union between two bacteria, in which genetic material is exchanged; DNA from the 'male' cell transfers all or part of its chromosomes into the recipient 'female'.

**bacteriophage**  A virus whose host is a bacterium. See lambda for description of λ phage.

**baculovirus** An insect-pathogenic virus with a circular double-stranded DNA genome and rod-shaped enveloped virion, found primarily in lepidopterans. These viruses have been engineered for two purposes: (1) as expression vectors to produce large quantities of proteins or (2) as biological pesticides.

**balancer chromosomes** These chromosomes initially were developed by H. J. Muller as a method for maintaining lethal mutations in laboratory stocks without continuous selection. Modern balancer chromosomes have been engineered and are maintained. *Drosophila* balancer stocks contain several recessive visible mutations, one or more inversions, and transpositions on a specific chromosome. These mutations have been induced to suppress crossing over. Balancer chromosomes may also have a clearly visible dominant mutation so that heterozygous flies can be identified easily.

**blastoderm** The layer of cells in an insect embryo that surrounds an internal yolk mass. The cellular blastoderm develops from a syncytium by surrounding the cleavage nuclei with membranes derived from the infolding of the surrounding membrane.

**blunt end** An end of a DNA molecule, at which both strands terminate at the same nucleotide position, with no extension of one of the strands.

**bootstrapping** A statistical method based on repeated random sampling with replacement from an original sample to provide a collection of new estimates of a parameter from which confidence limits can be calculated.

**C-banding** Dark bands on chromosomes produced by strong alkaline treatment at high temperature followed by incubation in sodium citrate and Giemsa staining. C-bands correspond to regions of constitutive heterochromatin.

**C value paradox** C stands for 'constant' or 'characteristic' and denotes the fact that the DNA content (size) of the haploid genome is fairly constant within a species. C values vary widely among species. Size is usually measured in picograms of DNA.

**capping** The modification of the 5' end of the pre-mRNA in eukaryotes when a GTP is added to the molecule via an unusual 5'-5' triphosphate bond. Capping is necessary for the ribosome to bind with the mRNA to begin protein synthesis.

**cDNA** The double-stranded DNA copy of a eukaryotic messenger RNA molecule, produced *in vitro* by enzymatic synthesis and used for producing cDNA libraries or probes for isolating genes in genomic libraries.

**cDNA library** A collection of clones containing dsDNA that is complementary to the mRNA. Such clones will lack introns and regulatory regions of eukaryotic genes. Once cDNA molecules are transcribed, they are inserted into a vector and amplified in *E. coli*. Genes that are inactive will not be represented in a cDNA library, nor will noncoding regions of the genome.

**cell autonomous determination** The establishment of a developmental pathway within a particular cell. Determination is not influenced by substances diffusing from elsewhere.

**cell cycle** The sequence of events between one cell mitotic division and another in a eukaryotic cell. Mitosis (M phase) is followed by a growth ($G_1$) phase, then by DNA synthesis (S phase), then by another growth ($G_2$) phase, and then by another mitosis.

**Central Dogma** Proposed by F. Crick in 1958. It states that the genetic information is contained in DNA, which is transcribed into RNA, which is translated into polypeptides. The transfer of information was proposed to be unidirectional from DNA to polypeptides: polypeptides are unable to direct synthesis of RNA, and RNA is unable to direct synthesis of DNA. The Central Dogma was modified in 1970 when RNA viruses were found to transfer information from RNA to DNA.

**centromere** A region of a chromosome to which spindle fibers attach during mitosis. The position of the centromere determines whether the chromosome will appear as a rod, a J, or a V during migration to the poles in anaphase. In some insects, the spindle fibers attach throughout the length of the chromosome and such chromosomes are called holocentric. Centromeres are usually bordered by heterochromatin containing repetitive DNA.

**chain-terminating method of DNA sequencing** See dideoxy method.

**channels** See ion channels.

**character** A taxonomic character is any attribute of a member of a taxon by which it differs from a member of a different taxon.

**chemotaxis** The movement of a cell or organisms toward or away from a chemical substance.

**chiasmata** Occur during prophase I of meiosis and represent points where crossing over, or exchange of genetic information, between nonsister chromatids occurred. When the synapsed chromosomes begin to separate in late prophase I, they are held together by these connections between the chromatids of homologous chromosomes.

**chorion** A complex structure covering the insect egg.

**chromatids** Chromosome components that have duplicated during interphase and become visible during the prophase stage of mitosis. Chromatids are held together at the centromere.

**chromomere** A region of densely packed chromatid fibers on a chromosome that produce a dark band. Chromomeres are readily visualized on polytene chromosomes."

**chromosome imprinting** The mechanisms involved in chromosomal imprinting are not understood in insects, but imprinting, or labeling of DNA, is associated with methylation of DNA in other organisms. Imprinting may make chromosomes susceptible to endonucleases, which are known to be present in eukaryotic cells.

**chromosome puffs** A localized swelling of a region of a polytene chromosome due to synthesis of DNA or RNA. Puffing is readily seen in polytene salivary gland chromosomes of dipteran insects.

**chromosome walking** A molecular genetic technique that allows a series of overlapping fragments of DNA to be ordered. The technique is used to isolate a gene of interest for which no probe is available but that is known to be linked to a gene which has been identified and cloned. The marker gene is used to screen a genomic library. All fragments containing the known cloned gene are selected and sequenced; the fragments are then aligned and those cloned segments farthest from the marker gene in both directions are subcloned for the next step, and so on. The subclones are used as probes to screen the genomic library to identify new clones containing DNA with overlapping sequences. As the process is repeated, the nucleotide sequences of areas farther and farther away from the marker gene are identified, and eventually the gene of interest will be found.

**circadian clock** Changes in biological or metabolic functions that show periodic peaks or lows of activity based on or approximating a 24-hour cycle.

**clade** An evolutionary lineage derived from a single stem species. A branch of a cladogram.

**cladistic systematics** Systematics that use only shared and derived characters as a basis of constructing classifications. The rate or amount of change subsequent to splitting of phyletic lines is not considered. All taxa must arise from a common ancestral species.

**cladogenic speciation** Branching evolution of new species.

**cladogram** A term used two ways by different authors. Either a dendrogram (tree) produced using the principle of parsimony, or a tree that depicts inferred historical relationships among organisms. Generally, branch lengths in a cladogram are arbitrary, with only the branching order considered significant.

**clone** A population of identical cells often containing identical recombinant DNA molecules.

**cloning vector** A DNA molecule capable of replicating in a host organism; a gene is often inserted into it to construct a recombinant DNA molecule and the vector is then used to amplify (clone) the recombinant DNA.

**cluster analysis** A method of hierarchically grouping taxa or sequences on the basis of similarity or minimum distance. UPGMA is an unweighted pair group method using the arithmetic average. WPGMA is the weighted pair group method using the arithmetic average.

**coding strand** The strand of the DNA molecule that carries the biological information of a gene and which is transcribed by RNA polymerase into mRNA.

**codominant** Alleles whose gene products are both manifested in the heterozygote.

**codon** A triplet of nucleotides that code for a single amino acid.

**coefficient of gene differentiation** Interpopulation diversity using allozyme data is usually measured using the coefficient of gene differentiation ($G_{ST}$). $G_{ST}$ is derived by estimating the average similarity within and between populations. It is an extension of Wright's correlation ($F_{ST}$) between two gametes drawn at random from each subpopulation. The coefficient of differentiation is $G_{ST} = (H_T - H_S)/H_T$ where $H_S$ is the average gene diversity within populations, and $H_T$ is the interpopulation gene diversity.

**competent cells** Bacterial cells in a state in which exogenous DNA molecules can bind and be internalized, thereby allowing transformation.

**complementary base pairing** Nucleotide sequences are able to base pair; A and T are complementary; 5'-ATGC-3' is complementary to 5'-GCAT-3'.

**complementary DNA** See cDNA.

**concatemers** The linking of multiple subunits into a tandem series or chain results in structures called concatemers.

**concerted evolution** Maintenance of homogeneity of nucleotide sequences among members of a gene family in a species even though the sequences change over time. Members of a gene family evolving in a nonindependent fashion.

**conditional lethal** A mutation that may be lethal only under certain environmental conditions.

**conditioned stimulus** A stimulus that evokes a response that was previously elicited by an unconditioned stimulus.

**constitutive heterochromatin** Regions of the chromosome containing mostly highly repeated, noncoding DNA; usually near the telomeres and centromeres.

**contig** Segments of DNA that partially overlap in their sequence.

**convergent evolution** The evolution of unrelated species resulting in structures with a superficial resemblance.

**copy number** The number of plasmids in a cell; the number of genes, transposons, or repetitive elements in a genome.

**core DNA** The DNA in the core nucleosome that is wrapped around the histone octamer. The core nucleosome is connected to others by linker DNA.

**cos sites** The *cos* sites are cohesive end sites or nucleotide sequences that are recognized when a phage DNA molecule is being packaged into its protein coat.

**cosmid** Engineered vectors used to clone large segments of exogenous DNA, derived by inserting *cos* sites from phage λ into a plasmid. The resulting hybrid molecule can be packaged in the protein coat of a phage.

**crossing over** The exchange of polynucleotides between homologous chromosomes during meiosis.

**cytoplasmic incompatibility** Reproductive incompatibility between two populations caused by factors that are present in the cytoplasm. Often associated with microorganisms.

**cytoplasmic inheritance** See also maternally inherited.

**cytoplasmic sex ratio distorters** Cytoplasmic genes that manipulate the sex ratio of their host to promote their own spread. Microbes (*Wolbachia*, spiroplasmas, viruses) often are transovarially and transtadially transmitted that can alter the sex ratio of insects and mites.

**cytosine** A pyrimidine, one of the bases in DNA and RNA.

**degeneracy** Refers to the genetic code and the fact that most amino acids are coded for by more than one triplet codon.

**degenerate primers** Degenerate primers can be used for the PCR when a limited portion of a protein sequence is known for a gene, but the DNA sequence is not known.

**deletion** The loss of a portion of the genetic material from a chromosome. The size can vary from one nucleotide to sections containing many genes.

**denaturation** Breakdown of secondary and higher levels of structure of proteins or nucleic acids by chemical or physical means.

**dendrogram** A branched diagram that represents the evolutionary history of a group of organisms.

**density-gradient centrifugation** Separation of molecules and particles on the basis of buoyant density, often by centrifugation in a concentrated sucrose or cesium chloride solution.

**deoxyribonucleic acid (DNA)** The genetic information is contained within the DNA. DNA consists of chemically linked sequences of nucleotides, each containing a heterocyclic ring of carbon and nitrogen atoms (nitrogenous base), a five-carbon sugar in ring form (a pentose) and a phosphate group. The nitrogenous bases are of two types, pyrimidines (cytosine and thymine) and purines (adenine and guanine). Pyrimidines have a six-member ring, purines have fused five- and six-member rings. Nucleotides are linked together into a polynucleotide chain by a backbone consisting of an alternating series of sugar and phosphate residues. The nitrogenous bases pair by hydrogen bonding, with A pairing with T and G with C, to hold complementary polynucleotide chains in a double helix.

**deoxyribonuclease** An enzyme that breaks a DNA polynucleotide by cleaving phosphodiester bonds.

**deuterotoky** A form of parthenogenesis in which unfertilized eggs can develop into either males or females.

**dicentric** A chromosome or chromatid with two centromeres.

**dideoxy sequencing** Developed by F. Sanger and A. R. Coulson in 1975, and known as the 'plus and minus' or 'primed synthesis' method of DNA sequencing. DNA is synthesized *in vitro* so that it is radioactively labeled and the reaction terminates specifically at the position corresponding to a specific base. After denaturation, fragments of different lengths are separated by electrophoresis and identified by autoradiography. In the 'plus' protocol, only one kind of dNTP is available for elongation of the $^{32}$P-labeled primer. In the 'minus' protocol, one of the four dNTPs is missing, or specific terminator base analogs are used.

**diploid** Having two copies of each chromosome.

**direct repeats** When a transposable element is inserted into a host genome, a small segment, typically 4 to 12 bp, of the host DNA is duplicated at the insertion site. The duplicated repeats are in the same orientation and are called **direct repeats**.

**discontinuous gene** A gene in which the genetic information is separated into two or more different exons by an intervening sequence (intron) which typically is noncoding. Most eukaryotic genes are discontinuous.

**discrete character** A character that is countable.

**distance** A measure of the difference between two objects.

**distance estimates** A phrase used to emphasize the fact that evolutionary history is inferred from experimental or sequence data, and distance is thus an estimate.

**distance matrix methods** Phylogeny construction method that uses estimates of genetic distances such as UPGMA.

**DNA** See deoxyribonucleic acid.

**DNA binding protein** Proteins such as histones or RNA polymerase that attach to DNA as part of their function.

**DNA–DNA hybridization** A method for determining the degree of sequence similarity between DNA strands from two different organisms by the formation of heteroduplex molecules.

**DNA fingerprinting** See fingerprinting.

**DNA ligase** An enzyme that repairs single-stranded discontinuities in double-stranded DNA. DNA ligases also are used in constructing recombinant DNA molecules.

**DNA polymerase** An enzyme that catalyzes the formation of DNA from dNTPs, using single-stranded DNA as a template. Three different DNA polymerases (I, II, and III) have been isolated from E. coli. Eukaryotes contain different DNA polymerases, found in the nucleus, cytoplasm, or mitochondria, which are involved in DNA replication, repair, and recombination.

**DNA polymerase I** The enzyme in E. coli that completes synthesis of individual Okazaki fragments during DNA replication.

**DNA polymerase III** The enzyme that primarily functions in DNA replication of E. coli.

**DNA sequencing** The process of determining the order of nucleotides in a DNA molecule.

**DNA topoisomerase** An enzyme that introduces or removes turns from the double helix by transiently breaking one or both of the strands.

**dosage compensation** A mechanism that compensates for the dosage of genes carried on the X chromosome in XX and XY organisms. In mammals, one or more of the X chromosomes is inactivated. In *Drosophila* males the Y chromosome is hypertranscribed.

**double helix** The base-paired structure consisting of two polynucleotides in the natural form of DNA.

**downstream** Toward the 3′ end of a DNA molecule.

**driver** Unlabeled DNA used in DNA–DNA hybridization.

**dsDNA** Double-stranded DNA.

**ssDNA** Single-stranded DNA.

**ecdysone** A steroid hormone found in insects that initiates and coordinates the molting process and the sequential expression of stage-specific genes.

**EDTA** Ethylene dinitrilotetraacetic acid, a chelating agent, is able to react with metallic ions, even in minute amounts, and form a stable, inert, water-soluble complex.

**electrophoresis** The separation of molecules in an electric field.

**electroporation** Disruption of cell membranes by electric current to insert exogenous DNA.

**enantiomers** Compounds showing mirror-image isomerism.

**endocytobiosis** Microorganisms, including bacteria, rickettsia, mycoplasmas, viruses, and yeasts, live within the cells of many eukaryotic organisms including insects. Symbiosis often is used to mean mutualism, but originally included parasitism and mutualism. Intracellular symbionts have been called endocytobionts, with no assumptions being made about whether the relationship is mutualistic or parasitic.

**endonuclease** An enzyme that degrades nucleic acid molecules by cleaving phosphodiester bonds internally.

**endopolyploidy** The occurrence in a diploid individual of cells containing 4-, 8-, 16-, 32-fold, etc., amounts of DNA in their nuclei. Nurse cells of ovaries are often endopolyploid.

**enhancer** Sequences of DNA that can increase transcription of neighboring genes over long distances up or downstream of the gene and in either possible orientation.

**enhancer trap** A method to identify genes based on their pattern of expression. A reporter gene under the control of a weak constitutive promoter, when brought in proximity to a tissue-specific 'enhancer' element would be regulated by that enhancer, resulting in the expression of the reporter gene in a tissue- and stage-specific pattern similar to that of the native gene normally controlled by the enhancer.

**environmental sex determination** A method of sex determination in which the environment, such as temperature, has a significant effect on the developmental processes leading to one or the other sex.

**epistatic** The nonreciprocal interaction of nonallelic genes. A gene epistatic to another masks the expression of the second gene.

**ethidium bromide** A dye that binds to double-stranded DNA by intercalating between the strands. DNA stained with EtBr fluoresces under UV illumination.

**euchromatin** Regions of a eukaryotic chromosome that appear less condensed and stain less well with DNA-specific dyes than other segments of the chromosome.

**eukaryote** An organism with cells containing a membrane-bound nucleus that reproduces by meiosis. Cells divide by mitosis. Oxidative enzymes are packaged within mitochondria.

**evolutionary systematics** Methods that focus on the order of origin of lineages and amount and nature of change.

**exogenous DNA** DNA from an outside source. In genetic engineering, DNA from one organism is often inserted into another by a variety of methods.

**exon** One of the coding regions of a discontinuous gene.

**exonuclease** A nuclease that degrades a nucleic acid molecule by progressive cleavage along its length, beginning at the 3' or 5' end.

**expression vector** Vectors that are designed to promote the expression of gene inserts. Usually an expression vector has the regulatory sequence of a gene ligated into a plasmid that contains the gene of interest. This gene lacks its own regulatory sequence. The plasmid with this new combination (regulatory sequence + gene) is placed into a host cell such as *E. coli* or yeast, where the protein product is produced.

**extrachromosomal gene** A gene not carried by the cell's chromosomes, such as mitochondrial or plasmid-borne genes.

**F pili** The presence of an F (fertility) factor determines the sex of a bacterium. Cells with F factors (circular DNA molecules that are about 2.5% of the length of the bacterial chromosome) are able to function as males by producing an F pilus. The F pilus is a hollow tube through which chromosomal DNA is transferred during bacterial conjugation.

**F statistics** A set of coefficients that describe how genetic variation is partitioned within and among populations and individuals, such as the $F_{ST}$ and inbreeding coefficients.

**facultatively heterochromatic** Chromosomal material that, unlike euchromatin, shows maximal condensation in nuclei during interphase. Constitutive heterochromatin is composed of repetitive DNA, is late to replicate, and is transcriptionally inactive. Portions of the chromosome that are normally euchromatic may become heterochromatic at a particular developmental stage (= facultative heterochromatin). An example of facultative heterochromatin is the inactivated X chromosomes in the diploid somatic cells of mammalian females.

**fate maps** A technique used to analyze behavior in *Drosophila*. Using a ring X chromosome which is usually lost, individuals can be produced which are partly male and partly female. The pattern of genetic markers can be used to construct a fate map, which correlates precise anatomical sites on the embryonic blastoderm with abnormalities affecting behavior.

**FB transposons** A family of transposons in *Drosophila* that are associated with chromosomal abnormalities.

**fibroin** See silk.

**fingerprinting** Uses the presence of simple tandem-repetitive sequences that are present throughout the genome. The regions show length polymorphisms, but share common sequences. DNA from different individuals is cut and separated by size on a gel. A probe containing the core sequence is used to label those fragments that contain the complementary DNA sequences. The pattern on each gel is specific for a given individual, and can be used to establish parentage.

**FLP recombinase** Yeast FLP recombinase is able to catalyze recombination in which a DNA segment that is flanked by direct repeats of FLP target sites (FRTs) can be excised from the chromosome. If two homologous chromosomes each bear an FRT site, mitotic recombination can occur in *Drosophila*, leading to the introduction of DNA into known and specific sites. FRT sites can be introduced into *Drosophila* chromosomes by P-element-mediated transformation.

**foldback DNA** DNA that contains palindromic sequences that can form hairpin double-stranded structures when denatured DNA is allowed to renature.

**frameshift mutation** A mutation resulting from inserting or deleting a group of nucleotides that is not a multiple of three so that the polypeptide produced will probably have a new set of amino acids specified for downstream of the frameshift.

$F_{ST}$ Coancestry coefficient; a measure of the relatedness of individuals.

**fusion protein** A hybrid protein molecule produced when a gene of interest is inserted into a vector and that displaces the stop codon for a gene already present in the vector. The fusion protein begins at the amino end with a portion of the vector protein sequence and ends with the protein of interest.

**G-banding** Dark bands on chromosomes produced by Giemsa staining; G-bands occur in AT-rich regions of the chromosome.

**gap genes** Gap gene mutants lack large areas of the normal cuticular pattern. Three wild-type gap genes, *Krüppel*, *hunchback*, and *knirps*, regionalize the embryo by delimiting domains of homeotic gene expression and effect position-specific regulation of the pair-rule genes.

**gating** The process of shutting off a function when the value of a specific parameter attains a critical level.

**gel electrophoresis** Separation of molecules on the basis of their net electrical charge and size.

**gene** A segment of DNA that codes for an RNA and/or a polypeptide molecule.

**gene amplification** The production of multiple copies of a DNA segment in order to increase the rate of expression of a gene carried by the segment. The chorion genes of *Drosophila* are amplified in the ovary.

**gene cloning** Insertion of a fragment of DNA containing a gene into a cloning vector and subsequent propagation of the recombinant DNA molecule in a host organism. Recently, cloning of a DNA fragment by the polymerase chain reaction has simplified the technology.

**gene conversion** A genetic process by which one sequence replaces another at an orthologous or paralogous locus, resulting in concerted evolution. May result from mismatch repair.

**gene duplication** The duplication of a DNA segment coding for a gene; gene duplication produces two identical copies which may retain their original function, allowing the organism to produce larger amounts of a specific protein. Alternatively, one of the gene copies may be lost by mutation and become a pseudogene, or a duplicated gene can evolve to perform a different task.

**gene expression** The process by which the information carried by a gene is made available to the organism through transcription and translation.

**gene gun** A method for propelling microscopic particles coated with DNA into cells, tissues, and organelles to produce stable or transient transformation of plants.

**gene library** A collection of recombinant clones derived from genomic DNA or from the cDNA transcript of an mRNA preparation. A complete genetic library is sufficiently large to have a high probability of containing every gene in the genome.

**gene regulation** The mechanism that determines the level and timing of gene expression.

**gene targeting** A technique for inserting changes into a genetic locus in a desired manner. The desired locus is transferred into an embryo by microinjection and is allowed to undergo homologous recombination into the chromosomes, replacing the original allele.

**gene transfer** The movement of a gene or group of genes from a donor to a recipient organism.

**genetic code** The rules that determine which triplet of nucleotides codes for which amino acid during translation.

**genetic distance** A measure of the evolutionary divergence of different populations of a species, as indicated by the number of allelic substitutions that have occurred per locus in the two populations. The most widely used measure of genetic distance is that of Nei (1972), $D = -\ln(I)$.

**genetic diversity ($G_{ST}$)** Variation in populations averaged over different loci.

**genetic marker** An allele whose phenotype is recognized and which can be used to monitor the inheritance of its gene during genetic crosses between organisms with different alleles.

**genetic sex determination system** The mechanism in a species by which sex is determined. In most organisms, sex is genetically, rather than environmentally, determined.

**genic balance model** Sex in Hymenoptera is determined by a balance between nonadditive male-determining genes and additive female-determining genes scattered throughout the genome. Maleness genes (m) have noncumulative effects but femaleness genes (f) are cumulative. In haploid individuals m>f, which results in a male, while in diploids ff>mm, which results in a female.

**genome** The total complement of DNA in an organism.

**genomic footprinting** A technique for identifying a segment of a DNA molecule in a living cell that is bound to some protein of interest. The phosphodiester bonds in the region covered by the protein are protected from attack by endonucleases. A control sample of pure DNA and one of protein-bound DNA are subjected to endonuclease attack. In DNA footprinting, the resulting fragments are electrophoresed on a gel to separate them according to their size. For every bond that is susceptible to restriction, a band is found on the control gel. The gel prepared from the protein-bound DNA will lack bands and these missing bands identify where the protein is protecting the DNA from being cut. The goal of genomic footprinting is to determine the contacts between DNA bases and specific proteins in a living cell. DNA footprinting determines these interactions *in vitro*.

**genomic imprinting** The process by which some genes are found to function differently when they are transmitted by the mother rather than the father, or vice versa. Mechanisms of imprinting may include methylation of the DNA. The more a gene is methylated, the less likely it is to be expressed.

**genomic library** A random collection of DNA fragments from a given species inserted into a vector (plasmids, phages, cosmids). The collection must be large enough to include all the unique nucleotide sequences of the genome.

**genotype** The genetic constitution of an organism.

**geotaxis** The movement of an animal in response to gravity.

**glycosylation** A process in which a sugar or starch is linked to a protein molecule.

**guanine** A purine, one of the nucleotides in DNA and RNA.

**haploid** Cells or organisms that contain a single copy of each chromosome.

**Hardy–Weinberg equilibrium** An equilibrium of genotypes achieved in populations of infinite size in which there is no migration, selection, or mutation after at least one generation of panmictic mating. With two alleles, A and a, of frequency $p$ and $q$, the Hardy–Weinberg equilibrium frequencies of the genotypes AA, Aa, and aa are $p^2$, $2pq$, and $q^2$, respectively.

**helicase** The enzyme responsible for breaking the hydrogen bonds that hold the double helix together so that replication of DNA can occur.

**helper plasmid** A plasmid that is able to supply something to a defective plasmid, thus enabling the plasmid to function.

**heritability** In the *broad sense* ($h_B = V_G/V_P$), the fraction of the total phenotypic variance that remains after exclusion of variances caused by environmental effects. In the *narrow sense*, the ratio of the additive genetic variance to the total phenotypic variance ($V_A/V_P$).

**heterochromatin** The regions of the chromosome that appear relatively condensed and stain deeply with DNA-specific stains.

**heteroduplex DNA** A hybrid DNA–DNA molecule formed from tracer and driver from different individuals or species.

**heterogametic sex** The sex that produces gametes containing unlike sex chromosomes. Many males are XY and thus heterogametic. Lepidopteran females are the heterogametic sex. Crossing over is often suppressed in the heterogametic sex.

**heterologous DNA** DNA from a species other than that being examined.

**heterologous recombination** Recombination between two DNA molecules that apparently lack regions of homology.

**heteroplasmy** The coexistence of more than one type of mitochondrial DNA within a cell or individual.

**heterozygous** A diploid cell or organism that contains two different alleles of a particular gene.

**heterozygosity** Having a pair of dissimilar alleles at a locus; a measure of genetic variation in a population estimated by a single locus or an average over several loci.

**highly-repetitive DNA** DNA made up of short sequences, from a few to hundreds of nucleotides long, which are repeated on an average of 500,000 times.

**histones** Basic proteins that make up nucleosomes and have a fundamental role in chromosome structure.

**histone gene family** See histones.

**Hogness box** A DNA sequence 19–27 bp upstream from the start of a eukaryotic structural gene to which RNA polymerase II binds. The sequence is usually 7 bp long (TATAAAA); named in honor of D. Hogness. Often called TATA box and pronounced 'tah-tah'.

**holocentric** Chromosomes that have diffuse centromeres.

**homeodomain** See homeobox.

**homeobox** A conserved DNA sequence about 180 bp in size found in a number of homeotic genes involved in eukaryotic development.

**homeotic** The replacement of one serial body part by a serially homologous body part.

**homeotic gene** Genes that determine the identification and sequence of segments during embryonic development. Although most genes with a homeodomain are in the homeotic class, a few are found among the segmentation genes. Homeotic genes have been described in a variety of insects other than *Drosophila*, including *Musca, Aedes, Anopheles, Blatella,* and *Tribolium*.

**homeotic mutations** Mutations in which one developmental pattern is replaced by a different, but homologous, one. Homeotic mutations of *Drosophila* and other insects cause an organ to differentiated abnormally and form a homologous organ that is characteristic of an adjacent segment. Examples in *Drosophila* include *Antennapedia* in which the antenna becomes leglike, and *bithorax* in which halteres are changed into winglike appendages.

**homoduplex DNA molecules** A double-stranded DNA molecule in which the two strands come from different sources in DNA–DNA hybridization. Heteroduplex DNA will denature or melt into single strands at lower temperatures than homoduplex DNA from a single source.

**homogametic sex** The sex that produces gametes with only one kind of sex chromosome. The females of many insects are XX and thus homogametic.

**homology** Common ancestry of two or more genes or gene products.

**homologous chromosomes** Two or more identical chromosomes.

**homologous genes** Two genes from different organisms and therefore of different sequence that code for the same gene product.

**homology** Homology has been defined as 'having a common evolutionary origin', but also is often used to mean 'possessing similarity or being matched'.

**homoplasy** Phenomena that lead to similarities in character states for reasons other than inheritance from a common ancestor, including convergence, parallelism, and reversal.

**homozygous** Diploid cells or organisms that contain two identical alleles of a particular gene.

**horizontal transmission** The transfer of genetic information from one species to another.

**housekeeping genes** Genes whose products are required by the cell for normal maintenance.

**hybrid dysgenesis** A syndrome of genetic abnormalities that occurs when hybrids are formed between strains of *Drosophila melanogaster*, one carrying (P) and the other lacking (M) the transposable P element. The abnormalities include chromosomal damage, lethal and visible mutations, and sometimes sterility. Dysgenesis is caused by crossing P males with M females, but the reciprocal cross is not dysgenic.

**hybridization probe** A labeled nucleic acid molecule used to identify complementary or homologous molecules through the formation of stable base pairs.

**hydrogen bonding** A hydrogen bond is a weak electrostatic attraction between an electronegative atom (such as oxygen or nitrogen) and a hydrogen atom attached to a second electronegative atom. In effect, the hydrogen atom is shared between the two electronegative atoms.

**hypertranscription** Transcription of DNA at a rate higher than normal. For many species with an XY sex-determination system, the male compensates for his single X chromosome by hypertranscribing the X chromosome. He produces a nearly equal amount of gene product compared with that produced by females with two X chromosomes.

**hypervariable sequences** Tandem repeats of short minisatellite DNA sequences vary in individuals and are useful for fingerprinting.

**imaginal discs** Cells set off during embryonic development that will give rise, during the pupal stage, to adult organs.

**inbreeding coefficient** The correlation of genes within individuals ($F_{IT}$), or the correlation of genes within individuals within populations ($F_{IS}$). Both $F_{IS}$ and $F_{IT}$ are measures for deviation from expected Hardy–Weinberg proportions.

**independent assortment** See Law of Independent Assortment.

**initiation codon** AUG serves as an initiation codon when it occurs at the start of a gene; it marks the site where translations should begin. AUG also codes for methionine so most newly synthesized polypeptides will have this amino acid at the amino terminus, although it may later be removed by post-translational processing of the protein. AUG is the only codon for methionine, so AUGs that are not initiation codons are also found in the middle of a gene.

**insertion mutation** Alteration of a DNA sequence by inserting one or more nucleotides.

**insertion sequences** Insertion sequences are the simplest transposable elements, carrying no genetic information except what is needed to transpose (i.e., transposase). Usually 700–2500 bp long, denoted by the prefix IS and followed by the type number.

**insertion vectors** Vectors that have a single target site at which foreign DNA is inserted.

*in situ* **hybridization** The pairing of complementary DNA and RNA strands, or the pairing of complementary DNA single strands to produce a DNA–DNA hybrid in intact chromosomes. Pairing is detected by some form of label, often specific radiolabeled nucleic acid molecules. For chromosomal squash preparations on glass slides, the DNAs are denatured and adhering RNAs and proteins are removed. Then the DNA is incubated with tritium-labeled nucleic probes. The radioactive molecules hybridize with the DNA segments of specific chromosomal areas and these are visualized in autoradiographs. It can also be used to identify DNA sequences in DNAs released from lysed bacterial colonies onto nitrocellulose filters.

**introgression** The incorporation of genes of one species into the gene pool of another. If the ranges of two species overlap and fertile hybrids are produced, they will tend to backcross with the more abundant species.

**intercalating agent** A chemical compound which is able to invade the space between adjacent base pairs of a double-stranded DNA molecule; includes ethidium bromide.

**intergenic region** The noncoding region between segments of DNA that code for genes.

**interphase** The stage of the cell cycle when chromosomes are not visible by light microscopy. During interphase, DNA synthesis occurs.

**intron** A region of eukaryotic DNA coding for RNA that is later removed during splicing; it does not contribute to the final RNA product.

**inverse PCR** Allows amplification of an unknown DNA sequence that flanks a 'core' region with a known sequence. The basic method for inverse PCR involves digesting template DNA, circularizing the digested DNA, and amplifying the flanking DNA outside the 'core' region with the primers oriented in the opposite direction of the usual orientation. Primers for inverse PCR are synthesized in the opposite orientation and are homologous to the ends of the core region so that DNA synthesis proceeds across the *uncharacterized* region of the circle rather than across the characterized core region.

**inversion** Alteration of the sequence of a DNA molecule by removal of a segment followed by its reinsertion in the opposite orientation.

**inverted repeat** Two identical nucleotide sequences repeated in opposite orientation in a DNA molecule, either adjacent to one another or some distance apart.

*in vitro* **packaging** The production of infectious particles by enclosing naked DNA in lambda (λ) phage packaging proteins and preheads.

**ion channels** The membrane passages that allow certain ions to cross the membrane.

**ionic selectivity** The ability of ion channels to permit certain ions but not others to cross the membrane.

**isozyme** An isomer of an enzyme. Various structurally related forms of the same enzyme having the same mechanism but differing from each other in chemical or immunological characteristics.

**junk DNA** The proportion of DNA in a genome that *apparently* has no function. Also called parasitic or selfish DNA.

**kilobase** A kilobase (kb) of DNA is equal to 1000 nucleotides.

**kin selection** A theory put forth by W. D. Hamilton (1964) that states that an altruistic act by is favored because it increases the inclusive fitness of the individual performing the social act. Inclusive fitness is the fitness of the individual as well as his effects on the fitness of any genetically related neighbors. The theory is that alleles change in frequency in a population due to a trait's effects on the reproduction of relatives of the individual in which it is expressed rather than on the reproductive success of the individual. A mutation that affects the behavior of a sterile worker bee, even though detrimental to her, could increase the fitness of the worker if her behavior increased the likelihood that a close relative would reproduce. Kin selection could explain the evolution of sociality, which appears to have developed as many as 11 times in the order Hymenoptera.

**Klenow fragment** A portion of bacterial DNA polymerase I derived by proteolytic cleavage. It lacks the 5' to 3' exonuclease activity of the intact enzyme.

**lagging strand** The DNA strand in the double helix which is copied in a discontinuous manner during DNA replication; short segments of DNA produced during the replication are called Okazaki fragments.

**lambda** or λ A double-stranded DNA virus (bacteriophage) that can invade *E. coli*. Once inside the cell, λ can enter a lysogenic cycle or a lytic cycle of replication, which results in death of the host cell. λ has been genetically engineered as a vector for cloning. It is also a microliter unit of measurement, the volume contained in a cube 1 mm on a side.

**leader sequence** An untranslated segment of mRNA from its 5' end to the start codon.

**Law of Independent Assortment** One of Mendel's laws. The members of different pairs of alleles assort independently. Different pairs of alleles assort independently into gametes during gametogenesis, if they are on different chromosomes. The subsequent pairing of male and female gametes is at random, which results in new combinations of alleles.

**Law of Segregation** One of Mendel's laws. The factors of a pair of characters segregate into different gametes, and thus into different progeny, of the two members of each pair of alleles possessed by the diploid parent.

**leading strand** The DNA strand in the double helix that is copied in a continuous fashion during DNA replication.

**leucine zipper** DNA binding proteins that contain four to five leucine residues separated from each other by six amino acids. The leucines on two protein molecules interdigitate and dimerize in a specific interaction with a DNA recognition sequence. Leucine zippers are involved in regulating gene expression.

**ligase chain reaction (LCR)** Also called the ligation amplification reaction (LAR). LCR is an *in vitro* method for amplifying nucleic acids that involves a cyclical accumulation of the ligation products of two pairs of complementary oligonucleotides. The LCR uses oligos that completely cover the target sequence. One set of oligos is designed to be perfectly completely complementary to the left half of a sequence being sought, and a second set matches the right half. Both sets are added to the test sample, along with ligase cloned from thermophilic bacteria. If the target sequence is present, the oligos blanket their respective halves of that stretch, with their ends barely abutting at the center. The ligase interprets the break between the ends as a nick in need of repair and produces a covalent bond. This creates a full-length segment of DNA complementary to the target sequence. The solution is then heated to separate the new full-length strand from the original strand, and both then serve as targets for additional oligos. After sufficient cycles, the full-length ligated oligos are identified using radio-isotopes, fluorescence, or immunological methods. The LCR is particularly useful because the ligase won't join the adjoining oligos if the oligos and the target are not perfectly matched; thus this procedure is highly specific. However, the PCR is faster and requires fewer copies of target DNA than the LCR.

**ligases** Enzymes that form bonds. DNA ligases are enzymes that catalyze the formation of a phosphodiester bond between adjacent 3'-OH and 5'-P termini in DNA. DNA ligases function in DNA repair to seal single-stranded nicks between adjacent nucleotides in a double-stranded DNA molecule.

**ligation** Enzymatic joining together of nucleic acid molecules through their ends.

**likelihood methods** Methods of analyzing DNA sequence data that rely on genetic models and provide a basis for statistical inference. Maximum likeli-hood methods of tree construction assume the form of the tree and then choose the branch length to maximize the likelihood of the data given that tree. These likelihoods are then compared over different possible trees and the tree with the greatest likelihood is considered to be the best estimate.

**linkage group** A group of genes located on a single chromosome.

**linker DNA** The DNA that links nucleosomes; its function is unresolved.

**long germ band development** A pattern of development in insects such as *D. melanogaster* in which the pattern of segmentation is established by the end of blastoderm.

**long-period interspersion genome organization** This organization of the DNA in the genome involves long (> 5600 bp) repeats alternating with very long (> 12 kb) uninterrupted stretches of unique DNA sequences. Long-period interspersion is characteristic of species with small genomes. Short-period interspersion involves a pattern of single-copy DNA, 1000–2000 bp long, alternating with short (200–600 bp) and moderately long (1000–4000 bp) repetitive sequences, which is characteristic of the DNA in most animal species.

**lysogenic** During the lysogenic phase of a bacteriophage, the DNA of a virus is integrated into the chromosome of its bacterial host.

**lytic** A virus in a lytic phase undergoes intracellular multiplication and lysis of the bacterial host cell results.

**major groove** The larger of the two grooves that spiral around the surface of the double helix of the DNA molecule.

**map unit** In linkage maps, a 1% recombination frequency is defined as a map unit or 1 centimorgan.

**maternal effect gene** Genes in the mother that have an effect on the phenotype of her progeny. Usually they are the result of depositing products or maternally derived mRNAs in the egg that are used or transcribed by the embryo.

**maternal gene flow** In honey bees, the gene flow that occurs when a queen and a contingent of workers fly to a new location to establish a new colony (swarm).

**maternally inherited** Characters that are transmitted primarily by cytoplasmic genetic factors, including mitochondria, viruses, and some mRNAs, derived solely from the maternal parent. Also known as cytoplasmic inheritance or extranuclear heredity.

**Maxam and Gilbert sequencing method** A 'chemical' method to sequence DNA developed in 1977 by A. M. Maxam and W. Gilbert. Single-stranded DNA derived from double-stranded DNA and labeled at the 5' end with $^{32}P$ is subjected to several chemical cleavage protocols to selectively make breaks on one side of a particular base. The fragments are separated by size by electrophoresis on acrylamide gels and identified by autoradiography.

**maximum parsimony methods** Taxonomic methods that focus on the character values observed and minimize the number of changes in character state between species over the tree, making the assumption that there have been approximately constant rates of change. The changes at each node in the tree are inferred to be those that require the least number of changes to give each of the two character states of the immediate descendants.

**median melting temperature** The temperature at which 50% of the double helices have denatured; the midpoint of the temperature range over which DNA is denatured.

**meiosis** The sequence of events occurring during two cell divisions to convert diploid cells into haploid cells.

**meiotic drive** Any mechanism that results in the unequal recovery of the two types of gametes produced by a heterozygote.

**messenger RNA (mRNA)** RNA molecules that code for proteins and that are translated on the ribosomes.

**methylation** In bacteria, enzymes (modification methylases) that bind to the DNA attach methyl groups to specific bases. This methylation pattern is unique to and protects the species from its own restriction endonucleases. Methylation also occurs in eukaryotes and may be involved in genomic imprinting. Genes that are methylated are less likely to be active.

**M13 bacteriophage** A single-stranded bacteriophage cloning vehicle, with a closed circular DNA genome of approximately 6.5 kb. M13 produces particles that contain ssDNA that is homologous to only one of the two complementary strands of the cloned DNA and therefore is particularly useful as a template for DNA sequencing.

**M13 universal primer** A primer derived from the M13 bacteriophage is used for sequencing reactions, and has been used to identify satellite DNA sequences in many organisms.

**microgram** ($\mu$g) One-millionth of a gram.

**micron** One-thousandth of a millimeter ($\mu$m).

**middle-repetitive DNA** See moderately repetitive DNA.

**minor groove** The smaller of the two grooves that spiral around the surface of the DNA double helix.

**missense mutation** A codon is changed to one directing the incorporation of a different amino acid, which can result in an inactive or less stable protein.

**mitochondrion** An organelle that occurs in the cytoplasm of all eukaryotes, and is capable of self-replicating. Each mitochondrion is surrounded by a double membrane. The inner membrane is highly invaginated, with projections called cristae that are tubular or lamellar. Mitochrondria are the sites of oxidative phosphorylation which result in the formation of ATP. Mitochondria contain distinctive ribosomes, transfer RNAs, and aminoacyl-tRNA synthetases. Mitochondria depend upon genes within the nucleus of the cells which they inhabit for many essential mRNAs. Proteins translated from mRNAs in the cytoplasm are imported into the mitochondrion. Mitochondria are thought to be endosymbionts derived from aerobic bacteria that associated with primitive eukaryotes and have their own circular DNA molecules. The genetic code of mitochondria differs slightly from the universal genetic code. Mitochondria are transferred primarily through the egg, and thus are maternally inherited.

**mitosis** The sequence of events that occur during the division of a single cell into two daughter cells.

**mobile genetic element** See transposable element.

**moderately repetitive DNA** Nucleotide sequences that occur repeatedly in chromosomal DNA. Repetitive DNA is moderately repetitive or highly repetitive. Highly repetitive DNA contains sequences of several nucleotides repeated millions of times. It is a component of constitutive heterochromatin. Middle-repetitive DNA consists of segments 100–500 bp long repeated 100 to 10,000 times each. This class also includes the genes transcribed into tRNAs and rRNAs.

**molecular biology** A term broadly used to describe biology devoted to the molecular nature of the gene and its biochemical reactions such as transcription and translation.

**molecular clock** The hypothesis that molecules evolve in direct proportion to time so that differences between molecules in two different species can be used to estimate the time elapsed since the two species last shared a common ancestor.

**molecular evolution** That subdivision of the study of evolution that studies the structure and functioning of DNA at the molecular level.

**molecular genetics** Genetic studies that focus on the molecular nature of genes and gene expression.

**molecular phylogeny** An analysis of the relationships of groups of organisms as reflected by the evolutionary history detected in molecules (proteins, DNA).

**molecular systematics** The detection, description, and explanation of molecular diversity within and among species.

**morphogen** Molecules whose local concentration directly determines the local pattern of differentiation.

**mRNA** Messenger RNA.

**mtDNA** Mitochondrial DNA.

**monoclonal antibody** A single antibody produced in quantity by cultured hybridoma cell lines.

**multigene family** A group of genes that are related either in nucleotide sequence or in terms of function; they are often clustered together.

**multiple cloning site** A region of a plasmid that has been engineered with unique restriction sites so that exogenous DNA can be inserted and excised after digestion by one or more restriction endonucleases.

**multiple-locus, multiple-allele model** A model for sex determination in Hymenoptera.

**mushroom body** The corpora pedunculata or mushroom bodies in the insect brain are associated with complex behavior. They are small in Collembola, Heteroptera, Diptera and Odonata, of medium size in Coleoptera, large in Orthoptera, Lepidoptera and solitary Hymenoptera, and are most highly developed in social Hymenoptera. Mushroom bodies are involved with associative learning of odor cues in *Drosophila* and, perhaps, with visual learning in other insects.

**mutagen** A chemical or physical agent able to induce a mutation in a DNA molecule.

**mutant** An organism expressing the effects of a mutated gene in its phenotype.

**mutation** A change in the nucleotide sequence of a DNA molecule. Mutations can involve duplications, deletions, inversions, translocations, and substitutions.

**nanogram (ng)** One-billionth of a gram.

**nanometer (nm)** One-billionth of a meter.

**negative heterosis** The inferiority of a heterozygote over that of the homozygotes with respect to one or more traits such as growth, survival, or fertility.

**Nei's standard genetic distance** $D = -ln(I)$, where genetic identity $I$ is the ratio of the proportions of genes alike between and within populations.

**neutral theory of molecular evolution** A theory that the majority of the nucleotide substitutions in the course of evolution are the result of the random fixation of neutral or nearly neutral mutations, rather than the result of positive Darwinian selection. Many protein mutations are selectively neutral and are maintained in the population by the balance between new mutations and their random extinction. Neutral mutations have a function, but they are equally effective in comparison with the ancestral alleles in the survival and reproduction of the organisms carrying them. Neutral mutations spread within populations by chance because only a relatively small number of gametes are sampled each generation and thus are transmitted to the next generation.

**nick** A break in a single strand of a double-stranded DNA molecule.

**nick translation** A commonly used method of labeling DNA molecules with radioactive isotopes. DNA polymerase I is used to incorporate radiolabeled nucleotides in an *in vitro* reaction.

**nitrogenous base** A purine or pyrimidine compound that forms part of the structure of a nucleotide.

**noncoding strand** The polynucleotide of the DNA double helix that does not carry the genetic information, but which is the complement of the coding strand.

**nonsense mutation** A mutation in a nucleotide sequence that changes a triplet coding for an amino acid into a termination codon so that a truncated polypeptide is produced which can alter the protein's activity.

**Northern blotting** A technique for transferring mRNAs from an agarose gel to a nitrocellulose filter paper sheet via capillary action. The RNA segment of interest is probed with a radiolabeled DNA fragment or gene.

**nuclear genome** The portion of the genome contained in the nucleus of eukaryotes on chromosomes.

**nucleic acid** Either of the polymeric molecules DNA or RNA.

**nucleic acid hybridization** The bonding of two complementary DNA strands, or one DNA and one RNA strand to identify nucleic sequences of interest. Southern blot, Northern blot, and plaque or colony hybridization techniques are all based on nucleic acid hybridization. All employ labeled probes to identify DNA or RNA of interest.

**nucleolus** An RNA-rich, spherical body associated with a specific chromosomal segment, the nucleolus organizer. The nucleolus organizer contains the ribosomal RNA genes, and the nucleolus is composed of the primary products of these genes, their associated proteins, and a variety of enzymes.

**nucleoside** A chemical with a purine or pyrimidine base that is attached to a five-carbon sugar.

**nucleosome** A basic structure in which eukaryotic chromosomes are organized and compacted. Nucleosomes are composed of an octamer of histone proteins with DNA coiled around them and are connected to other nucleosomes by linker DNA.

**nucleotide** A chemical consisting a purine or pyrimidine base that is attached to a five-carbon sugar to which a mono-, di-, or triphosphate is attached. A monomeric unit of DNA or RNA.

**nucleoside** A compound consisting of a purine or pyrimidine base attached to a five-carbon sugar.

**nucleus** The membrane-bound structure of a eukaryotic cell containing DNA organized into chromosomes.

**null allele** An allele that produces no functional product and therefore usually behaves as a recessive.

**Okazaki fragments** Short fragments of DNA that are synthesized during replication of the lagging strand of the DNA molecule.

**open reading frame** A series of codons with an initiation codon at the 5′ end. Often considered synonymous with 'gene' but used to describe a DNA sequence that looks like a gene but to which no function has been assigned.

**origin of replication** A base sequence in DNA that is recognized as the position at which the replication of DNA should begin. In eukaryotes, multiple origins of replication occur on each chromosome.

**P element** P elements are transposable DNA elements first found in *Drosophila melanogaster*, where they can cause hybrid dysgenesis if P-containing strains are crossed with M strains lacking P elements. P elements have been engineered to serve as vectors to insert DNA into the germ line of *Drosophila* embryos.

**pair-rule genes** Mutated pair-rule genes result in repetitive aberrations throughout the germ band, with the removal of integral, alternate segment-width areas. The pair-rule genes (including *runt*, *hairy*, *fushi tarazu*, *even-skipped*, *paired*, *odd-paired*, *odd-skipped*, *sloppy-paired*) are transiently expressed in seven or eight stripes during cellularization of the blastoderm.

**palindrome** A DNA sequence which reads the same in both directions, taking account of the two strands, i.e., 5′-AAAAATTTTTT-3′
                                                     3′-TTTTTTAAAAAA-5′

**paralogy** Homology that arises via gene duplication.

**parasegment** The visible cuticular patterns of sclerites and sutures in an insect do not represent the embryonically determined true segments. Rather, the visible 'segments' are parasegments.

**parasitic** See junk DNA and selfish DNA.

**parsimony** Parsimony dictates that the minimal number of assumptions are made in an analysis.

**paternal gene flow** In honey bees, the gene flow that occurs when drones mate with queens from distant colonies.

**paternal sex ratio (PSR)** A condition carried only by males of the parasitic wasp *Nasonia vitripennis* and transmitted via sperm to fertilized eggs. After an egg is fertilized by a PSR-bearing sperm, the paternally derived chromosomes condense into a chromatin mass, and subsequently are lost. The PSR chromosome itself survives, disrupting normal sex determination by changing fertilized diploid (female) eggs into haploid PSR males. PSR is the first known B chromosome of its kind, and is unusual in its ability to destroy the complete genome of its carrier each generation

**PCR** See the polymerase chain reaction.

**PCR-RFLP** A technique that combines the PCR and RFLP analysis. Genomic DNA is amplified by traditional PCR. Once the DNA is amplified, it is cut with restriction enzymes, electrophoresed, and visualized by ethidium bromide staining. Because the DNA was amplified by the PCR, the DNA fragments can be visualized without having to blot and probe them with a labeled probe, thus making PCR-RFLP more efficient and inexpensive than traditional RFLP analysis.

**peptide bond** The chemical bond that links adjacent amino acids into a polypeptide.

**phagemid** A hybrid vector molecule engineered from plasmid and M13 vectors. Phagemids provide a method for obtaining single-stranded DNA because they contain two replication origins, one a standard plasmid origin that allows production of dsDNA, and the other from M13, which allows the synthesis of ssDNA if the host cell is superinfected with a helper phage.

**phenogram** A branching diagram that links different taxa by estimating overall similarity based on data from characters. Characters are not evaluated as to whether they are primitive or derived.

**phenotype** The observable characteristics of an organism.

**phenetic systematics** Classification based on overall similarities among living organisms. All possible characters are examined and average similarities are calculated, with all characters assumed to be of equal importance.

**phenotype** The observable characteristics of an organism that are determined by both genotype and environment.

**pheromone-binding protein** Two soluble proteins are found in the lymph, a pheromone-degrading esterase and a pheromone-binding protein. The pheromone-binding proteins bind species-specific pheromones and are present in very high concentrations. Volatile hydrophobic odorant molecules have to enter an aqueous compartment and traverse a hydrophilic barrier before reaching olfactory neurons. The function of the pheromone binding proteins is not fully resolved, although they are thought to be involved in carrying the hydrophobic odorant through the sensillum lymph toward the receptor proteins located in the dendrite membranes.

**phosphodiester bond** The chemical bond that links adjacent nucleotides in a polynucleotide.

**phosphorylation** The combination of phosphoric acid with a compound. Many proteins in eukaryotes are phosphorylated.

**phototaxis** The movement of a cell or organism toward or away from light.

**phyletic speciation** The gradual transformation of one species into another without an increase in species number at any time within the lineage. Also called vertical evolution or speciation.

**phylogenetic tree** A graphic representation of the evolutionary history of a group of taxa or genes.

**phylogenetics** The reconstruction of the evolutionary history of a group of organisms or genes.

**phylogeny** The evolutionary history of a group of taxa or genes, and their ancestors.

**physical map** A map of the order of genes on a chromosome. The locations are determined by DNA sequencing, producing overlapping deletions in polytene chromosomes, or electron micrographs of heteroduplex DNAs.

**picogram** $10^{-12}$ gram. A picogram of DNA is approximately $0.98 \times 10^9$ base pairs.

**plaque** A clear spot on an opaque bacterial lawn in a petri dish. A plaque results after a single phage adsorbs to a bacterial cell, infects it, and lyses, releasing progeny phage. The progeny phage infect nearby bacteria and produce more phage until a clear area becomes visible to the naked eye. Each clear area contains many copies of a single phage and, if the phage is a vector containing exogenous DNA, it contains many copies of the foreign DNA.

**plaque hybridization** See plaque screening.

**plaque screening** Screening is employed to identify, by nucleic acid hybridization with radiolabeled probes, those plaques containing specific DNA sequences.

**plasmid** Circular, dsDNA molecules found in bacteria that are often used in cloning.

**plesiomorphic** A character used to reconstruct a phylogeny that is ancestral or primitive.

**point mutation** A mutation that results from changes in a single base pair in a DNA molecule.

**pole cells** The precursors of the germ cells become separated early in embryonic development in *D. melanogaster* into distinctive cells in the posterior of the egg.

**poly(A) tail** The processing of the 3′ end of the pre-mRNA molecule by the addition of as many as 200 adenine nucleotides, which may determine mRNA stability.

**polyacrylamide gel** A gel that results from the polymerization of acrylamide monomers into linear chains and the linking of these chains with n, N′-methylenebisacrylamide (bis). The concentration of acrylamide and the ratio of acrylamide to bis determine the pore size of the three-dimensional network and its sieving effect on nucleic acids of different size.

**polydnaviruses** A newly recognized group of viruses, with double-stranded, circular DNA genomes, found only within certain groups of parasitic Hymenoptera. Virus particles replicate only in the wasp ovary and are secreted into the oviducts from where, during oviposition, the virus is injected into host larvae. It is believed that one or more gene in the virus contributes to the immunosuppressive state of the host, thus allowing the parasitoid eggs and larvae to survive. The polydnaviruses appear to integrate into parasitoid chromosomal DNA, but are also present in extrachromosomal molecules.

**polylinker** A genetically engineered segment in a vector that allows exogenous DNA to be cloned in at that region by one of two or more unique restriction sites.

**polymer** A chemical compound constructed from a long chain of identical or similar units.

**polymerase chain reaction (PCR)** A method for amplifying DNA by means of DNA polymerases such as *Taq* polymerase. PCR fundamentally involves denaturing double-stranded DNA, adding dNTPs, DNA polymerase, and primers. DNA synthesis occurs, resulting in a doubling of the number of DNA molecules defined by the primers. Additional rounds of denaturation and synthesis occur, resulting in a geometric increase in DNA molecules because each newly synthesized molecule can serve as the template for subsequent DNA amplification. Modifications of the PCR reaction have been developed for special purposes. PCR is used to clone genes, produce probes, produce ssDNA for sequencing, and carry out site-directed mutagenesis. DNA sequence differences are used to identify individuals, populations, and species.

**polymorphism** Two or more genetically different classes in the same interbreeding population.

**polynucleotide** A polymer consisting of nucleotide units.

**polyploidy** An increase in the number of copies of the haploid genome. Most individuals are 2n, but species are known that are polyploid (3n, 4n, 5n, 6n) and such species are parthenogenetic because of the difficulty of maintaining normal meiosis. Many insect species have tissues that are polyploid, including the salivary glands and nurse cells of the ovary and fat body, but the germ line tissues remain 2n.

**polysome** An mRNA molecule in the process of being translated by multiple ribosomes.

**polytene chromosomes** Chromosomes in which the chromatid has duplicated up to 1000-fold without separating. Salivary gland chromosomes in *Drosophila* and other Diptera are polytene. The discrete bands of polytene chromosomes allow a physical map of genes to be constructed using light microscopy.

**polyteny** See polytene chromosomes.

**position effect variegation** The change in the expression of a gene when it is moved to a different region of the genome. The change in expression can be stable or variegated. Variegated position effects usually involve the suppression of activity of a wild-type gene when it is placed in contact with heterochromatin because of a chromosomal mutation. Under some conditions, the gene may escape suppression and the final phenotype of the organism may thus be variegated, with patches of normal and mutant tissues.

**post-translational processing** Changes to polypeptide chains after they have been synthesized, such as cleavage of specific regions to convert proenzymes to enzymes, phosphorylation, etc.

**pre-mRNA** The unprocessed transcript of a protein-coding gene.

**primary transcript** The immediate product of transcription of a gene or group of genes which will be processed to give the mature transcript(s).

**primase** The RNA polymerase that synthesizes the primer needed to initiate replication of a DNA polynucleotide during DNA replication.

**primer** A short oligonucleotide that is attached to a ssDNA molecule in order to provide a site at which DNA replication can begin.

**primer-dimer artifacts** Low-molecular-weight DNA products produced during PCR as artifacts when the reaction is carried out with high primer concentrations, too much DNA polymerase in early cycles, and small amounts of template DNA. The primer-dimer is made when the DNA polymerase makes a product by reading from the 3′ end of one primer across to the 5′ end of the other primer. This results in a sequence being produced that is complementary to each primer and can serve as a template for additional primer binding and extension.

**probe** A molecule labeled with radioactive isotopes or another tag that is used to identify or isolate a gene, gene product, or protein.

**prokaryote** An organism whose cells lack a distinct nucleus.

**promoter** A region of DNA crucial to the accuracy and rate of transcription initiation. Usually immediately upstream of the gene itself.

**protease** An enzyme that degrades proteins.

**protein** The polymeric compounds made up of amino acids.

**proteoglycan** A protein that is glycosylated to a variety of polysaccharide chains.

**pseudogene** A nucleotide sequence that is similar to a functional gene, but without accurate information so that it is not functional.

**PSR** See paternal sex ratio.

**puffing** A swelling in the giant polytene chromosomes of salivary glands of many dipterans, indicating that region is being transcribed.

**pulsed field gel electrophoresis** A technique for separating DNA molecules by subjecting them to alternately pulsed, perpendicularly oriented electrical fields. The technique allows separation of the yeast genome into a series of intact chromosomes on a gel. Chromosomes larger than yeast chromosomes are digested with a restriction enzyme before electrophoresis.

**purine** One of the two types of nitrogenous bases that are components of nucleotides.

**pyrimidine** One of the two types of nitrogenous bases that are components of nucleotides.

**Q-banding** Bands on chromosomes produced by quinacrine staining. The staining can be seen only under UV light and is brightest in AT-rich regions.

**quantitative genetics** Analysis of the genetic influence of many genes and substantial environmental variation. It is assumed that Mendel's laws of discrete inheritance apply to complex characteristics, so that many genes, each with small effect, combine to produce observable differences among individuals in a population. Quantitative genetics determines the sum of heritable genetic influence on behavior, regardless of the complexity of genetic modes of action or the number of genes involved. It does not tell us which genes are responsible for the trait.

**radiolabeling** The attachment of a radioactive atom to a molecule, incorporation of $^{32}P$-dNTPs into DNA.

**RAPD-PCR** Derived from the term Random Amplified Polymorphic DNA. This PCR technique uses single primers of arbitrary nucleotide sequence consisting of nine or ten nucleotides with a 50 to 80% GC content, and no palindromic sequences. These 10-mers can act as a primer in the PCR and yield reproducible polymorphisms from random segments of genomic DNA.

**reading frame** A nucleotide sequence from which translation occurs.

**recessive** A trait or gene is recessive if it is expressed in homozygous, but not heterozygous, cells.

**recombination** Exchange of segments between two DNA molecules which can result in progeny of two parents having different combinations of alleles than are found in either parent.

**recombinant DNA molecule** A DNA molecule created by combining DNA fragments that are not normally contiguous.

**recombinant DNA technology** All the techniques involved in the construction, study, and use of recombinant DNA molecules. Often abbreviated rDNA, which can be confused with ribosomal DNA (rDNA).

**recombination** A physical process that can lead to the exchange of segments of two DNA molecules and which can result in progeny from a cross between two different parents with combinations of alleles not displayed by either parent.

**regulatory gene** A gene that codes for a protein that is involved in the regulation of the expression of other genes.

**regulatory mutation** Mutations that affect the ability to control gene expression.

**reinforcement** An event (reward or punishment) that follows a response and increases or decreases the likelihood that it will recur.

**repetitive DNA** DNA sequences that are repeated a number of times in a DNA molecule or in a genome. Some repetitive DNA is associated with heterochromatin, centromeres, and telomeres. Middle repetitive DNA may code for ribosomal RNAs and transfer RNAs.

**replacement vectors** Vectors that have a pair of insertion sites that span a DNA segment that can be exchanged with a foreign DNA fragment.

**replica plating** A technique to produce identical patterns of bacterial colonies on a series of petri plates. A plate containing colonies is inverted and its surface is pressed against a block covered with velveteen. The block can then be used to inoculate up to about eight additional petri plates. By marking the patterns of the colonies on the different plates with different selective properties, it is possible to identify which colonies differ in their responses to these agents.

**replication fork** The region of a dsDNA molecule that is being unwound so that DNA replication can occur.

**replication origin** The site(s) on a DNA molecule where unwinding of the double helix occurs so that replication can occur. There are multiple replication origins on eukaryotic chromosomes.

**reporter gene** A gene used to identify or locate another gene.

**response to selection (R)** The difference in mean phenotypic value between the offspring of the selected parents and the mean phenotypic value of the entire parental generation before selection.

**restriction endonuclease** An enzyme that cuts DNA at only a limited number of specific nucleotide sequences. Also called restriction enzyme.

**restriction fragment length polymorphism (RFLP)** A polymorphism in an individual, population, or species defined by restriction fragments of a distinctive length. Usually caused by gain or loss of a restriction site but could result from an insertion or deletion of DNA between two conserved restriction sites. Differences in DNA RFLPs are visualized by gel electrophoresis.

**restriction site** A specific sequence of nucleotides in a piece of dsDNA which is recognized by a restriction enzyme and which signals its cleavage.

**restriction site mapping** DNA is digested with a series of different restriction endonucleases, the DNA fragments are electrophoresed, and the DNA fragments are ordered to produce a linear physical map of the locations of specific DNA sequences.

**retrotransposon** A type of transposable element that transposes by means of an RNA intermediate. At least ten families of retrotransposons are known in *Drosophila*.

**retroposition** The transfer of genetic information through an RNA intermediate. The genetic information carried by the DNA is transcribed into RNA, which is then reverse-transcribed into cDNA. The result is that the element is duplicated and the copy of the element is transposed.

**retroelement** DNA or RNA sequences that contain a gene for reverse transcriptase. There are different classes of retroelements, including retroviruses and retrons.

**retrosequences** Retrosequences and retrotranscripts are sequences derived through the reverse transcription of RNA and subsequent integration into the genome. They lack the ability to produce reverse transcriptase.

**retrovirus** RNA viruses that use reverse transcriptase during their life cycle. This enzyme allows the viral genome to be transcribed into DNA. The transcribed viral DNA is integrated into the genome of the host cell where it replicates in unison with the genes of the host. The cell suffers no damage from this relationship unless the virus carries an oncogene. If so, it could be transformed into a cancer cell. Retroviruses violate the Central Dogma during their replication. The HIV virus responsible for the AIDS epidemic is a retrovirus.

**reverse transcription** DNA synthesis from an RNA template, mediated by reverse transcriptase.

**reverse transcriptase** An enzyme that synthesizes a DNA copy from an RNA template.

**reversions** Reverse mutation.

**RFLP** See restriction fragment length polymorphism.

**ribonuclease** An enzyme that degrades RNA.

**ribosomal RNA (rRNA)** The RNA that acts as a structural component of ribosomes.

**ribosome** A cellular organelle composed of proteins and RNA in which translation of mRNA occurs. Ribosomes consist of two subunits, each composed of RNA and proteins. In eukaryotes, ribosome subunits sediment as 40S and 60S particles.

**ribosomal RNA** These genes (rRNA genes) are found as tandem repeating units in the nucleolus organizer regions of eukaryotic chromosomes. Each unit is separated from the next by a nontranscribed spacer. Each unit contains three regions coding for the 28S, 18S, and 5.8S rRNAs.

**ring chromosome** An aberrant chromosome with no ends.

**RNA** Ribonucleic acid, one of the two forms of nucleic acids.

**RNA polymerase** An enzyme capable of synthesizing an RNA copy of a DNA template.

**RNA transcript** An RNA copy of a gene.

**rooted tree** Rooted phylogenetic trees have a common ancestor of all taxa under study and define the evolutionary path.

**S phase** The portion of interphase in the cell cycle in which DNA replication occurs. The S phase occurs between the $G_1$ and $G_2$ phases of the interphase. Mitosis occurs after the $G_2$ phase.

**satellite DNA** Highly repeated DNA sequences with such a uniform nucleotide composition that, upon fractionalization of the genomic DNA and separation by density-gradient centrifugation, they form one or more bands that are clearly different from the main band of DNA and from the smear created by other fragments of a more heterogeneous composition. The base composition of satellite DNA differs from that of the majority of DNA in a eukaryotic species, i.e., is either AT-rich or GC-rich. Usually highly repetitive in sequence.

**secondary transposition** Movement of an element after its initial insertion into the chromosome. Secondary transposition can be induced with P elements in *Drosophila*.

**segment polarity genes** These genes appear to determine a linear sequence of repeated positional values within each segment. Segment polarity mutants have repetitive deletions of pattern, but the deletions occur within each segment and are followed by a partial mirror-image duplication of the part that remains. Segment polarity genes (including *engrailed, naked, patched, wingless, gooseberry, patched hedgehog, porcupine, armadillo, fused*) are required either continuously or over extensive periods to maintain the segmental pattern. Most or all are required to maintain patterns in the imaginal tissues.

**segmentation genes** Genes, including the gap, pair-rule, and segment polarity classes of genes, that determine the number and polarity of the body segments during embryonic development in insects.

**selectable marker** A gene that allows identification of specific cells with a desirable new genotype. Many vectors used for genetic engineering carry antibiotic resistance genes or other genes that allow identification of cells containing exogenous DNA.

**selection differential** In artificial selection, the difference in mean phenotypic value between individuals selected as parents of the following generation and the whole population.

**selfish DNA** DNA that may not provide any advantage to its carrier or host but ensures its own survival. Transposable elements are considered to be selfish DNA.

**semiconservative replication** DNA replication in which each daughter double helix consists of one strand from the parent and one newly synthesized strand.

**sensory transduction** Sensory cells transform and amplify the energy provided by a stimulus into an electrical signal. Sensory transduction is probably due to a change in the ionic permeability of the sensory cell membrane, which causes a depolarization of the membrane.

**sericin** See silk.

**sex chromosome** A chromosome that is involved in sex determination.

**short germ band development** A pattern of development found in some insects in which all or most of the metameric pattern is completed after the blastoderm stage by the sequential addition of segments during elongation of the caudal region of the embryo.

**short-period interspersion pattern of genome organization** This form of genome organization has single-copy DNA, 1000–2000-bp long, alternating with short (200–600 bp) and moderately long (1000–4000 bp) repetitive sequences. This pattern is found in the house fly *Musca domestica*, the Australian sheep blowfly *Lucilia cuprina*, and the wild silk moth *Antheraea pernyi*.

**shotgun cloning** Genomic libraries constructed from random fragments of DNA from an organism.

**shotgun libraries** Genomic libraries in which a random collection of a sufficiently large sample of cloned fragments of the DNA is present so that all the genes are represented.

**silent mutation** Changes in DNA that do not influence the expression or function of a gene or gene product.

**silk** The cocoon filament spun by the fifth instar larva of *Bombyx mori* and other silk moths. Each cocoon filament contains two cylinders of fibroin, each surrounded by three layers of sericin. Fibroin is secreted by the cells of the posterior portion of the silk gland. The fibroin gene is present in only one copy per haploid genome, but these silk gland cells undergo 18 to 19 cycles of endomitotic DNA replication before they begin transcribing fibroin mRNAs. The sericin proteins are named because they contain abundant serines (over 30% of the total amino acids). Sericins are secreted by the cells from the middle region of the silk gland.

**similarity** A measure of the resemblance between two objects, usually on a scale of zero to one.

**single-locus, multiple-allele model** A model for sex determination in Hymenoptera.

**single-strand binding proteins** One of the proteins that attaches to ssDNA in the replication fork to prevent reannealing of the DNA during replication.

**site-directed mutagenesis** Mutagenesis to produce a predetermined change at a specific site in a DNA molecule.

**slot blot** A hybridization technique that allows multiple samples of DNA to be applied to nitrocellulose filters in specific sites (slots) by using a vacuum.

**S1 nuclease** A specific nuclease that degrades single-stranded nucleic acids or splits short single-stranded segments in DNA but does not attack any double-stranded molecules. Used to convert sticky ends of duplex DNA to form blunt ends or to trim off single-stranded ends after conversion of single-stranded cDNA to the double-stranded form.

**Southern blotting** A technique developed by E. M. Southern for transferring DNA fragments isolated electrophoretically in an agarose gel to a nitrocellulose filter paper sheet by capillary action. The DNA fragment of interest is then screened with a radioactive nucleic acid probe that is complementary to the fragment of interest. The position on the filter is determined by autoradiography. The related techniques for RNA and proteins have been dubbed 'Northern' and 'Western' blots, respectively.

**specific activity** The ratio of radioactive to nonradioactive molecules of the same kind. Probes with a high specific activity can produce a more intense signal than probes with a low specific activity.

**spliceosome** The RNA and protein particles in the nucleus that remove introns from pre-messenger RNA molecules.

**stable transformation** Transformation that alters the germ plasm of an organism so that the progeny transmit the trait of interest through subsequent generations.

**start codon** The mRNA codon, usually AUG, at which synthesis of a polypeptide begins.

**sterile insect release method (SIRM)** A technique used to control pest insects. Large numbers of mass-produced males are given nonlethal but sterilizing doses of radiation or chemical mutagens and then released in nature. The natural populations are so overwhelmed by these males that females are almost always fertilized by them. The resultant matings produce inviable progeny and a new generation is not produced. Used to eradicate the screwworm from the U.S.A.

**sterile male technique** See sterile insect release method (SIRM).

**sticky end** Single-stranded ends of DNA fragments produced by restriction enzymes; sticky ends are able to reanneal.

**stop codon** One of the three mRNA codons (UAG, UAA, and UGA) that prevents further polypeptide synthesis.

**stringency** As used in hybridization reactions, refers to the conditions that can be altered to influence the ease with which a probe hybridizes to template nucleic acids.

**structural gene** A gene that codes for an RNA molecule or protein other than a regulatory gene.

**subclones** A DNA fragment that has been cloned into one vector may be moved, or subcloned, into a second type of vector in order to perform a different procedure.

**supercoiled** The coiling of a covalently closed circular duplex DNA molecule upon itself so that is crosses its own axis. A supercoil is also called a superhelix. The B form of DNA is a right-handed double helix. If the DNA duplex is wound in the same direction as that of the turns of the double helix, it is positively supercoiled. Twisting of the DNA molecule in a direction opposite to the turns of the strands in the double helix is called negative supercoiling.

**symbiont** An organism living with another organism of a different species.

**sympatry** Living in the same geographic location. Sympatric species have overlapping or coinciding distributions.

**synapsis** The pairing of homologous chromosomes during the zygotene stage of meiosis.

**synecology** The study of relationships among communities of organisms and their environment.

**syncytium** A mass of protoplasm containing many nuclei not separated by cell membranes.

**systematics** The study of classification, based on evolutionary change.

**tandem repeat** Direct repeats in DNA codons adjacent to each other.

*Taq* **DNA polymerase** A DNA polymerase that isolated from the bacterium *Thermus aquaticus* that is tolerant of high temperatures. Used in the polymerase chain reaction.

**targeted gene replacement** The ability to replace or modify genes in their normal chromosomal locations has not been possible with *Drosophila* until recently. The cut-and-paste model of P-element transposition provided a model for inserting a gene into the double-stranded gap left behind by transposition of a P element. The gap can be repaired by using a template provided either by an extrachromosomal element introduced by the investigator or by a sister chromatid, or an homologous chromosome.

**targeted gene transfer** See targeted gene replacement.

**targeted mutagenesis** The ability to replace or modify DNA sequences in their normal chromosomal location.

**TATA box** See Hogness box.

**taxa** The general term for taxonomic groups whatever their rank. Taxon is singular.

**taxonomy** The principles and procedures according to which species are named and assigned to taxonomic groups.

**tDNA-PCR** Universal primers for transfer RNA can be used to generate tDNA by the PCR. The resulting fragments are visualized by gel electrophoresis and produce characteristic fingerprints for different species.

**telomerase** An enzyme that adds specific nucleotides to the tips of chromosomes to form telomeres.

**telomeres** The physical ends of eukaryotic chromosomes. They protect the ends of chromosomes and confer stability. Telomeres consist of simple DNA repeats and the nonhistone proteins that bind specifically to those sequences.

**telomere terminal transferase** See telomerase.

**template** A macromolecular mold for synthesis of another macromolecule. Duplication of the template takes two steps; a single strand of DNA serves as the template for a complementary strand of DNA or mRNA.

**termination codon** One of the three codons in the standard genetic code that indicates where translation of a mRNA should stop, i.e., 5'-UAA-3', 5'-UAG-3', or 5'-UGA-3'.

**thelygenic** Occurs when females produce only female progeny, as in the blowfly *Chrysomya rufifacies*.

**thelytoky** Parthenogenesis in which no functional males are known; unmated females produce female progeny only, or rarely, a few males.

**30-nm fiber** Condensation of DNA in eukaryotic chromosomes involves formation of 30-nm fibers from supercoils of six nucleosomes per turn. The 30-nm fiber somehow is condensed further.

$T_m$ The interpolated temperature along a DNA melting curve at which 50% of the duplex DNA formed in a DNA–DNA hybridization is double stranded. The difference in $T_m$ between homoduplex and heteroduplex curves is called $\Delta T_m$.

**tracer DNA** In DNA–DNA hybridization, single-stranded, single-copy DNA from one species is radioactively labeled (tracer DNA) and hybridized with unlabeled DNA (driver DNA) from the same species or from different species. DNA-DNA hybridization is used to determine the degree of sequence identity between DNAs.

**trailer segment** A nontranslated sequence at the 3′ end of mRNA following the termination signal, exclusive of the poly(A) tail.

**transcript** An RNA copy of a gene.

**transcription** The process of producing an RNA copy of a gene.

**transcriptional activator proteins** Elements that stimulate transcription by binding with particular sites in the DNA.

**transfection** Infection of bacteria with viral nucleic acid that lacks a protein coat.

**transfer RNA (tRNA)** A family of small RNA molecules (usually more than 50 types per cell) that serve as adapters for bringing amino acids to the site of protein synthesis on the ribosome.

**transformant** An individual organism produced by introducing exogenous DNA.

**transformation** The process of changing the genetic makeup of an organism by introducing foreign DNA. Transformation may be transient or stable (transferred to succeeding generations.)

**transient transformation** Transient transformation involves changing the genetic makeup by introducing foreign DNA. If the genetic information is not incorporated into the germ line, transformation is unstable or temporary.

**translation** The process by which the amino acid sequence in a polypeptide is determined by the nucleotide sequence of a messenger RNA molecule on the ribosome.

**translational regulation** Gene regulation by controlling translation. Translation of an mRNA can be tied to the presence of a specific molecular signal; the longevity of an mRNA molecule can be regulated or overall protein synthesis can be regulated.

**transovarial transmission** Transmitted to the next generation through the egg.

**transposable element** An element that can move from one site to another in the genome. Transposable elements (TEs) have been divided into two classes, those that transpose with an RNA intermediate, and those that transpose as DNA.

**transposase** An enzyme that catalyzes transposition of a transposable element from one site to another in a DNA molecule.

**transposon** A transposable element carrying several genes, including at least one coding for a transposase enzyme. Many elements are flanked by inverted repeats. *Drosophila melanogaster* contains multiple copies of 50–100 different kinds of transposons.

**transposon jumping** Moving stably-inserted P-element vectors from one site to another within the genome to explore position effects.

**transposon tagging** A method of cloning genes from *Drosophila* after they have been 'tagged' by having the P element inserted into them.

**transposition** The movement of genetic material from one chromosomal location to another.

**triplex DNA** In triplex DNA, the usual AT and CG base pairs of duplex DNA are present, but in addition a pyrimidine strand is bound in the major groove of the helix. DNA sequences that potentially can form triplex DNA structures appear to be common, are dispersed at multiple sites throughout the genome, and comprise up to 1% of the total genome.

**unique genes** Genes present in only one copy per haploid genome, which includes most of the structural (protein-encoding) genes of eukaryotes.

**unrooted tree** A phylogenetic tree in which the location of the most recent common ancestor of the taxon is unknown.

**UPGMA** The use of distance measurements to group taxonomic units into phenetic clusters by the Unweighted Pair-Group Method using an Arithmetic average.

**upstream** Toward the 5′ end of a DNA molecule.

**uracil** A pyrimidine that is one of the nitrogenous bases found in RNA.

**vector** A DNA molecule capable of autonomous replication in a cell and which contains restriction enzyme cleavage sites for the insertion of foreign DNA.

**vitellogenin** The major yolk proteins are derived from vitellogenins, which are produced by the fat body and secreted for uptake by maturing oocytes.

**Western blots** A process in which proteins are separated electrophoretically and a specific protein is identified with a radioactively labeled antibody raised against the protein in question.

**wobble hypothesis** A hypothesis to explain how one tRNA may recognize two different codons on the mRNA. Anticodons are triplets with the first two positions pairing according to base-pairing rules. The third position 'wobbles' and can recognize any of a variety of bases in different codons so that it can bind to either of two or more codons.

**X-gal** A lactose analog (5-bromo-4-chloro-3-indolyl-β-D-galactopyranoside). X-gal is cleaved by β-galactosidase into a product that is bright blue. If exogenous DNA has inserted into and disrupted the β-*galactosidase* gene, λ plaques will appear white or colorless. Plaques without recombinant vectors will be blue.

**Z chromosome** One of the sex chromosomes found in heterogametic ZW female insects.

**Z-DNA** A structural form of DNA in which the two strands are wound into a left-handed helix rather than a right-handed form.

**zinc finger protein** Proteins with tandemly repeating segments that bind zinc atoms. Each segment contains two closely spaced cysteine molecules followed by two histidines. Each segment folds upon itself to form a fingerlike projection. The zinc atom is linked to the cysteines and histidines at the base of each loop. The zinc fingers serve in some way to enable the proteins to bind to DNA molecules, where they regulate transcription.

# Appendix

Significant Events In Genetics,
Molecular Biology,
and Insect Molecular Genetics

Any scientific discipline relies on a wide array of facts, appropriate technology, and effective theory which provokes testing of hypotheses. Particular scientific publications and events are deemed significant according to an individual's personal interests and biases. Thus, this list represents only one viewpoint. However, it should become obvious that much of the work on insect molecular genetics is of very recent origin and relies on a solid foundation of fundamental research.

**1865** Mendel reports his work

**1869** DNA found in a pus cell extract

**1873** Mitosis first described (Schneider)

**1881** Polytene chromosomes described in a chironomid (Balbiani)

**1887** Weismann postulates that a reduction in chromosome number must occur in all sexual organisms

**1894** Three bases in DNA identified

**1986** The cell theory is described (Schwann)

**1900** Mendel's work is rediscovered by DeVries, Correns and Tschermak DNA found to contain a pentose sugar

**1901** DeVries adopts the term mutation to describe sudden, spontaneous changes in hereditary material

**1902** McClung notes that two different types of sperm are produced with different numbers of chromosomes and speculates the chromosomes may be involved in sex determination; also suggests that sex is determined at fertilization in insects.

**1903** Genes are located on chromosomes (Sutton)

**1905** Bateson introduces term 'genetics' to describe study of heredity

**1909** Johannsen proposes Mendel's factors be called 'genes'

**1910** Morgan begins work with *Drosophila melanogaster* and discovers sex linkage

**1911** Five genes mapped on a *Drosophila* chromosome

**1915** 85 different genes of *D. melanogaster* mapped to four chromosomes
Goldschmidt describes intersexes in the gypsy moth, *Lymantria dispar*

**1927** Muller discovers X rays are mutagenic

**1928** The term 'heterochromatin' is introduced (Heitz)

**1930** R. A. Fisher publishes *The Genetical Theory of Natural Selection*

**1931** Cytological proof of crossing over obtained (Stern, Creighton, and McClintock)

S. Wright publishes *Evolution in Mendelian Populations*

**1933** *Drosophila* salivary gland chromosomes are discovered (Painter)

Sex determination of *Bombyx mori* is determined

Electrophoresis is invented (Tiselius)

T. H. Morgan receives the Nobel Prize for developing the theory of the gene

**1935** A gene is proposed to be a physical unit

Biochemical genetics of eye color in *Drosophila* and *Ephestia* determined (Beadle, Ephrussi, Kuhn, Butenandt)

Bridges publishes salivary gland chromosome maps for *D. melanogaster*

**1939** Basic studies on bacteriophage biology published (Ellis and Delbruck)

**1941** Each gene controls the activity of a single enzyme (Beadle and Tatum)

**1943** Bacterial mutations demonstrated (Luria and Delbruck)

**1944** Avery, MacLeod and McCarty discover the 'transforming principle' is DNA, which is the first evidence that DNA is the genetic material

**1950** Base ratios in DNA are shown to be consistently different in different species (Chargaff)

**1952** Hersey and Chase show phage genes are made of DNA

X-ray diffraction pictures indicate that DNA is a helix (Franklin)

Beerman suggests that stage and tissue specificities of chromosomal puffing in polytene chromosomes represent differential gene activities

**1953** Watson and Crick deduce the double helix structure of DNA

**1956** Kettlewell studies industrial melanism in the peppered moth in England, showing that bird predation is determined by the moths' conspicuousness

**1957** DNA replication demonstrated to be semiconservative (Meselson and Stahl)

DNA polymerase I isolated (Kornberg)

**1958** Crick proposes the 'Central Dogma' of information transfer: DNA → RNA → protein

Semiconservative replication of DNA demonstrated (Meselson and Stahl)

Crick predicts the discovery of transfer RNA

**1959** Mitochondrial DNA discovered

'Transitions' and 'transversions' as types of mutations are defined (Freese)

Ochoa and Kornberg receive Nobel Prizes for *in vitro* synthesis of nucleic acids

**1960** Specific puffing patterns induced in polytene chromosomes by injecting *Chironomus* larvae with ecdysone (Beerman)

**1961** Genetic code consists of three-letter codons (Crick, Barnett, Brenner and Watts-Tobin)

The *engrailed* gene of *D. melanogaster* shown to cause a shift from one developmental prepattern to a different, related, prepattern (Tokunaga)

**1962** Nobel Prizes awarded to Watson, Crick and Wilkins for studies on DNA structure

**1964** Holliday proposes a model for crossing over that involves breakage and reunion between DNA molecules in homologous chromosomes

W. D. Hamilton proposes genetical theory of social behavior

**1965** UAG and UAA shown to be termination codons for growing polypeptides

Complete sequence for alanine transfer RNA determined from yeast (Holley)

Jacob, Monod, and Lwoff receive Nobel Prizes for microbial genetics work

**1966** DNA ligase isolated (Weiss and Richardson)

Wobble hypothesis proposed to explain degeneracy in genetic code (Crick)

Repetitive DNA demonstrated in vertebrates

Lewontin and Hubby use electrophoretic methods to survey natural populations of *D. pseudoobscura* for protein variations, concluding that 8–15% of all loci are heterozygous

**1967** Successful *in vitro* synthesis of biologically active DNA

Khorana et al. solve the genetic code

**1968** First restriction endonuclease isolated (*Hind* II)

Kimura proposes neutral theory of evolution

Okazaki reports that newly synthesized DNA contains many fragments that are replicated discontinuously and spliced together

**1969** *In situ* hybridization technique developed to localize specific nucleotide sequences (Gall and Pardue)

Delbruck, Luria, and Hershey receive Nobel Prizes for work on viral genetics

**1970** RNA-dependent DNA polymerase found (Baltimore and Temin)

**1972** mRNA for silk fibroin from *Bombyx mori* isolated and the gene characterized (Suzuki, Brown and Gage)

**1973** Cohen, Chang, Boyer, and Helling construct first functional hybrid bacterial plasmid by *in vitro* joining of restriction fragments from different plasmids

**1974** Chromatin composed of DNA and two each of histones H2A, H2B, H3 and H4 in nucleosomes (Kornberg)

λ bacteriophage first engineered as a cloning vehicle (Murray and Murray)

**1975** Asilomar conference issues guidelines on minimizing potential risks of recombinant DNA research

Hybridization analysis of specific DNAs separated by electrophoresis (Southern)

Plus and minus method for sequencing DNA (Sanger and Coulson)

mRNAs for heat shock proteins hybridize to specific puffs on *Drosophila* polytene chromosomes (McKenzie, Henikoff and Meselson)

**1976** First genetic engineering company formed (Genentech)

Eukaryotic gene segments synthesized *in vitro* (Efstratiadis, Kafatos, Maxam and Maniatis)

**1977** The chemical method for sequencing DNA described ( Maxam and Gilbert)

Discovery of intervening sequences (introns) in 28S rRNA genes of *D. melanogaster* (Hogness, Glover and White)

Discovery of Northern blot method of identifying specific RNA sequences

1978 Dispersed repetitive DNAs in *Drosophila* analyzed, initiating new insights into mutability, transposition, transformation, hybrid dysgenesis, and retroviruses in eukaryotes (Finnegan, Rubin, Young, and Hogness)

Maniatis, Hardison, Laay, Lauer, O'Connell, Quon, Sim and Efstratiadis develop procedure for eukaryotic gene isolation by producing cloned libraries and screening them by hybridization with specific nucleic acid probes

1979 Avise, Lansman and Shade use restriction endonucleases to measure mitochondrial DNA sequence relatedness in natural populations

Genetic code in mitochondria has unique, nonuniversal features

Complete sequence for plasmid cloning vector pBR322 obtained (Sutcliffe)

1980 U.S. Supreme Court rules that genetically modified microorganisms can be patented

Berg, Gilbert, and Sanger receive Nobel prizes for experimental manipulation of DNA

Chakrabarty files patent for genetically engineered microorganisms able to consume oil

1981 L. Margulis publishes *Symbiosis in Cell Evolution*, summarizing evidence that mitochondria, chloroplasts, and kinetosomes evolved from prokaryotic endosymbionts

1982 First drug made by recombinant DNA marketed—human insulin (Eli Lilly)

P strains of *Drosophila* shown to contain transposable P elements that cause hybrid dysgenesis (Bingham, Kidwell, and Rubin)

Genetic transformation of *Drosophila* with P-element vectors (Rubin and Spradling)

1983 Barbara McClintock receives the Nobel Prize for her discovery, in the 1940s, of transposable elements in corn

1985 Polymerase chain reaction (PCR) developed (K. Mullis)

A general model for sex determination in insects proposed (Nothiger and Steinmann-Zwicky)

*Drosophila yakuba* mitochondrial genome sequenced (Clary and Wolstenholme)

1987 Transformation of the mosquito *Anopheles gambiae* by microinjection (Miller, Sakai, Romans, Gwadz, Kantoff and Coon)

**1988** *Taq* DNA polymerase used in polymerase chain reaction (Sakai, Gelfand, Stoffel, Scharf, Higuchi, Horn, Mullis and Erlich)

Transformation of the mosquito *Aedes triseriatus* by microinjection (McGrane, Carlson, Miller and Beatty)

**1989** Transformation of the mosquito *Aedes aegypti* by microinjection (Morris, Eggleston and Crampton)

**1990** Resistance to organophosphorus insecticides in *Culex* mosquitoes shown to be due to esterase gene amplification (Mouches, Pasteur, Berge, Hyrien, Raymond, De Saint Vincent, DeSilvestri and Georghiou)

Site-directed mutagenesis in *Drosophila* achieved (Engels, Johnson-Schlitz, Eggleston and Sved)

Two PCR techniques developed (RAPD-PCR and arbitrarily primed PCR) that allow genome mapping and ecological studies of arthropod populations for which limited genetic information is available (Welsh and McClelland; Williams, Kubelik, Livak, Rafalski and Tingey)

**1991** Amplified insecticide resistance genes (esterases) in the mosquito *Culex pipiens* shown to be present in mosquito populations around the world by migration (Raymond, Callaghan, Fort and Pasteur)

'Africanized' honey bees shown to spread from South America to North America as unbroken African maternal lineages based on mtDNA analyses (Hall and Smith)

FLP-mediated transformation of *Aedes aegypti* (Morris, Schaub and James)

**1992** Transformation of a beneficial arthropod by maternal microinjection (Presnail and Hoy)

Transformation of a symbiont of *Rhodnius*, a vector of Chagas' disease (Beard, O'Neill, Tesh, Richards and Aksoy)

DNA from 20–40 million year old bee in amber amplified by the PCR (Cano, Poinar, Roubik and Poinar)

**1993** Mitochondrial genome of *Apis mellifera* sequenced (Crozier and Crozier)

The *mariner* transposable element is widespread in insects (Robertson)

Kary Mullis receives Nobel Prize for the PCR

# Index

## E